MEMS AND MICROSYSTEMS

MEMS AND MICROSYSTEMS
Design, Manufacture, and Nanoscale Engineering

Second Edition

TAI-RAN HSU
Microsystems Design and Packaging Laboratory
Department of Mechanical and Aerospace Engineering
San Jose State University

WILEY

JOHN WILEY & SONS, INC.

Copyright © 2008 by John Wiley & Sons, Inc. All rights reserved.

Published by John Wiley & Sons, Inc., Hoboken, New Jersey.
Published simultaneously in Canada.

Library of Congress Cataloging-in-Publication Data:

Hsu, Tai-Ran.
  Mems and microsystems : design, manufacture, and nanoscale engineering / Tai-Ran Hsu.—
2nd ed.
        p. cm.
  Includes bibliographical references and index.
  ISBN: 978-0-470-08301-7 (cloth)
  1. Microelectronics. 2. Microelectromechanical systems. 3. Microelectronic
packaging.   I. Title.
  TK7874.H794 2008
  621.381—dc22

                                 2007017041

Printed in the United States of America.

10 9 8 7 6 5 4 3 2 1

My wife, Grace Su-Yong,
who has backed me up for decades
with infinite encouragement, support, and love.

My three loving children, Jean, May, and Leigh,
for their support and stimulating discussions on cyberspace
communication technology and molecular biology.

My students who motivated me to write this book.

老子天下篇：（公元前 450 年）

一尺之錘

日取其半

萬世不絕

An excerpt from the chapter of "Under Heaven" by Lao-Tze, 450 B.C.:

"A meter-long stick,
cut in half every day.
The last cutting to eternity."

An ancient Chinese wisdom that inspired my belief in the fantasy land of the nanoworld.

CONTENTS

PREFACE

Five years have elapsed since the publication of the first edition of this book. During this time, we have witnessed significant changes in the science and technology of miniaturization that include microsystems technology and nanotechnology. New nanoproducts have been developed and produced by research institutions and industries around the world. Many of these products have already made their way to the consumer marketplace, showing impressive performance.

In response to the increasing needs of engineers to acquire basic knowledge and experience in nanotechnology, an introductory chapter on nanoscale engineering is included in this new edition. This chapter begins with a brief introduction of the history and evolution of nanotechnology, followed by brief descriptions of the fabrication techniques for producing nanoproducts such as quantum dots, nanowires, and tubes. These products have their own unique characteristics and applications.

Engineering design principles in molecular dynamics, fluid flows, and heat transmission in nanoscale substances are presented. These analytical tools provide engineers with overviews of fundamentals in engineering design of nanostructures, which are significantly different from those for micro- and mesoscale analyses as presented in Chapters 4 and 5. Issues such as self-assembly of nanostructures—an essential element in the eventual mass production of nanostructures—and the potential socioeconomic impact of nanotechnology on humanity are also included in this chapter.

Chapter 11 includes a major expansion on the assembly, packaging, and testing (APT) of microelectromechanical systems (MEMSs) and microsystems. A quarter of a century of mixed success in commercializing MEMS and microsystems has shown more clearly than ever that APT is a major stumbling block for mass production and cost-effective packaging of microsystems products. Much research-and-development (R&D) effort is needed to overcome this road block. The chapter presents a number of critical issues involved in microassembly technology. It also presents the scientific bases of enabling techniques used in the bonding and sealing of microscale products. In situ testing during fabrication and postproduction of MEMS and microsystems products for structural and performance reliabilities appear to be other major development areas for engineers and scientists.

Further additions in the new edition include the introduction of microgyroscopes and micro–heat pipes in Chapter 2, design methodologies for thermally actuated multilayered strips in Chapter 4, and the use of popular SU-8 polymer material in Chapter 7.

I would like to express my gratitude for the plentiful constructive feedback that I have received from colleagues in this country and abroad who adopted the first edition of the book for their classes. The translation of the first edition of the book into both traditional and contemporary versions of Chinese has prompted valuable feedback from instructors and readers from the Greater Chinese community. I also appreciate my colleague Professor Ji Wang's efforts in thoroughly proofreading the entire text of the first edition and his valuable suggestions for further improvements. They have helped me make this edition more of a useful contribution to the two emerging technologies of miniaturization.

In closing, I am grateful to McGraw-Hill Companies for reverting the copyright of the first edition of the book to me due to its policy change for upper division textbook publications and the Institution of Engineering and Technology in the United Kingdom for its permission for my extensive use of materials in the book on "MEMS Packaging," which I had the pleasure of editing in 2004. Lastly, I am pleased by the offer to publish the second edition of the book by John Wiley & Sons.

Tai-Ran Hsu
San Jose, California
April 2007

PREFACE TO THE FIRST EDITION

The technological advances of microsystems engineering in the past decade have been truly impressive in both paces of development and new applications. Microsystems engineering involves the design, manufacture, and packaging of microelectromechanical systems (MEMS) and the peripherals. Applications of microsystems in aerospace, automotive, biotechnology, consumer products, defense, environmental protection and safety, healthcare, pharmaceutical and telecommunications industries prompted many experts to account for a staggering $82 billion in revenue for the microsystems and related industries in the Year 2000.

The strong demand for MEMS and microsystems by a rapidly growing market has generated strong interest, as well as a need for engineering educators to offer courses on this subject in their respective institutions. Likewise, many practicing engineers have shown similar interest and a desire to acquire the necessary knowledge and experience in the design and manufacture of microsystems. These individuals have faced great difficulty in realizing their goals and objectives because of a lack of comprehensive books that synergistically integrate the wide spectrum of disciplines in science and engineering that microsystem engineering draws on. A book that provides the reader with a broad perception of microsystem technology and the methodologies for the design, manufacture, and packaging of microsystem products is thus very much in demand. Such demand motivated me to develop this textbook.

The objective of this book is to provide upper division undergraduate and entry-level graduate students in mechanical, electrical, manufacturing, and related engineering disciplines with necessary fundamental knowledge and experience in the design, manufacture, and packaging of microsystems. Emphasis has been placed on the application of students' knowledge and experience acquired in previous years to the design and manufacture of microsystems. The layout and the organization of the related topics, and the many design examples presented in this book, will usher practicing engineers too into the field of microsystems engineering.

The book is intended as a textbook for classes covered by one 15-week long semester at both the undergraduate and graduate levels. Students are expected to possess prerequisite knowledge in college mathematics, physics, and chemistry, as well as in engineering subjects such as fundamental material science, electronics and machine design.

The book consists of eleven chapters:

Chapter 1 begins with an overview of microsystems and the evolution of microfabrication that led to the production of microsystems. It concludes with a preview of the current and potential markets for various types of microsystems.

Chapter 2 presents working principles for currently available microsensors, actuators and motors, valves, pumps, and fluidics used in microsystems.

Chapter 3 offers an overview of engineering science topics applicable to microsystems design and fabrication.

Chapter 4 includes engineering mechanics that are relevant to microsystems design and packaging. Topics in this chapter include mechanics of deformable solids and mechanical vibration theories. Also presented are the basic formulations of thermomechanics and fracture mechanics of interfaces of thin films that are common in microstructures. It concludes with an outline of the finite element method for stress analysis.

Chapter 5 deals with the application of thermofluids engineering principles in microsystems design. It begins with an overview of fluid mechanics theories and the capillary effect on micro fluid flows. Also presented is the rarefaction effect of gases in submicrometer and nanoscales. Modified formulations for heat conduction in submicrometer scaled systems are presented.

Chapter 6 relates to the scaling laws that are used extensively in the conceptual design of microdevices and systems. Students will learn both the positive and negative consequences of scaling down certain physical quantities that are pertinent to microsystems.

Chapter 7 deals with materials used for common microcomponents and devices. Active and passive substrates as well as packaging materials are presented in the chapter. Other materials—such as piezoresistives, piezoelectric and polymers—for microsystems are described. Chapter 8 presents an overview of microfabrication processes for micromanufacturing. Chapter 9 covers the three common micromanufacturing techniques of bulk micromanufacturing, surface micromachining, and the LIGA process.

Chapters 10 and 11 offer essential elements involved in the design and packaging of microsystems. The use of CAD and the finite element method in these endeavors is presented. Selected case studies and examples in the design and packaging of micro pressure sensors and fluidics demonstrate the applications of these methodologies in product design and packaging.

To cover these diversified subjects in the textbook in a 15-week long semester with 3-hour-per-week lectures presents a major challenge to the instructor. For this reason, "Suggestions to Instructors" based on my personal experience in teaching this course to both levels of students, is offered in the subsequent section. Selection of subjects for teaching a MEMS or microsystems course in either a 15-week long semester or a 10-week long quarter are proposed in that section.

The task of preparing a book of this breadth cannot possibly be accomplished by a single person's effort. I enjoyed communicating with able professionals like Professor Ali Beskok on nano fluid dynamics and Dr. Krishnamoorthy on the design and modeling of capillary electrophoresis in microfluidics. It was also a great pleasure to have dedicated and capable students like Ta-jen Tai, Valerie Barker, Mathew Smith, Jeanette Wood, Jacob Griego, and Yen-chang Hu, who contributed by either filling in detailed computations or checking the numerical accuracy of a number of examples that I developed for the book. I also appreciate the help that I received from Mindy Kwan in digitizing many photographs and diagrams used in the book. The contributions of valuable photographs and drawings by various sectors of the MEMS industry in this country and abroad are appreciated as acknowledged in the book. Development of this book was a part of a project on the development of an undergraduate curriculum stem on mechatronics during 1995–1998, in which I played a principal role. Support of this project by the Division of Undergraduate Education of the National Science Foundation (NSF) is gratefully acknowledged.

I am also grateful for the indefatigable efforts of many reviewers, who went above and beyond the call of duty in making this book as comprehensive and lucid as possible.

Norman Tien, Cornell University
Robert S. Keynton, University of Louisville
Imin Kao, State University of New York at Stony Brook
Dennis Polla, University of Minnesota
Zeynep Celik-Butler, Southern Methodist University
Ashok Srivastava, Louisiana State University
Ryszard Pryputniewicz, Worcester Polytechnic Institute
Michael Y. Wang, University of Maryland
Michael Histand, Colorado State University
Liwei Lin, University of California at Berkeley
Masood Tabib-Azar, Case Western Reserve University
George F. Watson, copy-editor

I am especially indebted to Cheah Choo Lek of Ngee Ann Polytechnic of Singapore, for his meticulous and thoughtful reading of the manuscript and many suggestions for improvement. Last, but not the least, I wish to acknowledge the excellent services provided by the editorial and production staff at McGraw-Hill. The support and able assistance by Jonathan Plant are greatly appreciated.

Tai-Ran Hsu
San Jose, California

SUGGESTIONS TO INSTRUCTORS

This book is written as a textbook for upper division undergraduate and entry-level graduate students. It is also intended to be a reference book that will usher practicing engineers to the growing field of microsystems and nanoscale engineering. Its content is designed for 15-week-long semester classes at both undergraduate and graduate levels. With significant omission of materials, the book can also be used for classes covering a 10-week-long quarter.

The textbook will be more effective for students in the classes with the following academic background and experience.

1. Undergraduate students in good upper division academic standings with sound knowledge of college mathematics, physics, and chemistry.
2. Students who have satisfied the prerequisite courses, or their equivalent, in fundamentals of electrical and electronic engineering, material sciences, engineering mechanics and machine design.
3. Students with working experience in mathematics software such as MathCAD or MatLAB.

Teaching a course on MEMS and elementary nanotechnology with extreme diversity of subjects such as in this course in a 15-week-long semester or 10-week-long quarter with 3-hour-per-week lectures presents a major challenge to instructors. The problem may be further compounded by the likelihood of having students with diversified academic backgrounds and experiences in the same class. The following Schedule A and B offer suggested topics that the instructor may cover in either one semester or in one quarter. Instructor would use his or her discretion to assign the unlisted sections in these tables either as the omitted materials or as a reading materials to the students.

Schedule A For 15-Week-Long Semesters

Week No.	Undergraduate Classes	Graduate Classes
1	Chapter 2	Chapter 2
2	Sections 3.3, 3.5, 3.6, 3.7, 3.8	Sections 3.3, 3.5, 3.6, 3.7, 3.8
3	Sections 4.1, 4.2 and 4.3	Section 4.1 and 4.2, 4.3 and 4.4.
4	Sections 4.4, 4.6, 4.7, 5.1	Remaining of Chapter 4 and Sections 5.1 and 5.2
5	Sections 5.6, 5.7, 5.8	Remaining of Chapter 5
6	Chapter 6	Chapter 6 and Sections 7.1 to 7.3
7	Sections 7.1 to 7.5	Sections 7.4 to 7.11
8	Sections 7.6 to 7.11	Sections 8.1 to 8.5
9	Sections 8.1 to 8.5	Sections 8.6 to 8.10
10	Sections 8.6 to 8.10	Chapter 9, Sections 10.1 and 10.2
11	Chapter 9	Remaining of Chapter 10
12	Sections 10.2, 10.3, 10.4, 10.7, 10.8	Sections 11.1 to 11.8
13	Sections 11.1 to 11.8	Sections 11.9 to 11.19
14	Sections 11.9 to 11.19	Sections 12.1 to 12.8
15	Sections 12.1 to 12.6, 12.9 to 12.12	Remaining of Chapter 12

Schedule B For 10-Week-Long Quarters

Week No.	Undergraduate Classes	Graduate Classes
1	Sections 2.2 to 2.6	
2	Sections 3.5, 3.6 and 3.7 Sections 4.2.3, 4.4.3	Sections 3.5, 3.6, 3.7 Sections 4.2.3, 4.4.3, Sections 5.1, 5.4 to 5.9
3	Sections 5.4, 5.6 and 5.7	Chapter 6
4	Sections 7.1 to 7.9	Chapter 7
5	Sections 7.10 and 7.11, 8.1 to 8.3	Sections 8.1 to 8.5
6	Sections 8.4 to 8.10	Sections 8.6 to 8.10, 9.1 and 9.2
7	Chapter 9	Sections 9.3, 9.4, 10.1 to 10.4
8	Sections 10.2, 10.4, 10.5.2, 10.7 and 10.8	Sections 10.2, 10.3, 10.4, 10.5.2, 10.7 and 10.8
9	Sections 11.2 to 11.9, 11.11 to 11.14, 11.16, 11.21	Sections 11.2 to 11.9, 11.11 to 11.14, 11.16, 11.21
10	Sections 12.2 to 12.5, 12.11 and 12.12	Chapter 12

Instructors, of course, may use their own discretion in selecting the topics other than those suggested in the above tables to suit their specific preferences and schedules.

The author would like to encourage the use of the blackboard for examples offered by the textbook, rather than have these examples swiftly disseminated to the students by view graphs or slides. Design projects on carefully chosen topics will be a valuable experience for students in learning topics such as specific microfabrication processes in depth. These projects may be assigned to groups of two or three students. Due credits, e.g., 20 percent of the total mark, need be assigned to the projects. Extra time slots, or a "Project Day," may be reserved for the presentation of these projects for peer review among all students in the class. These presentations will benefit all students in the class as they learn from each other through the presentations followed by active discussion and questioning. Instructors may incorporate ideas and results of these projects as additional materials for their lectures in future years.

CHAPTER 1

OVERVIEW OF MEMS AND MICROSYSTEMS

1.1 MEMS AND MICROSYSTEMS

The term MEMS is an abbreviation of microelectromechanical system. A MEMS contains components of sizes ranging from one micrometer (μm) to one millimeter (mm) (1 mm = 1000 μm). A MEMS is constructed to achieve a certain engineering function or functions by electromechanical or electrochemical means.

The core element in MEMSs generally consists of two principal components: a sensing and/or an actuating element and a signal transduction unit. Figure 1.1 illustrates the functional relationship between these components in a microsensor.

Microsensors are built to sense the existence and the intensity of certain physical, chemical, or biological quantities such as temperature, pressure, force, sound, light, nuclear radiation, magnetic flux, and chemical and biological compositions. Microsensors have the advantages of being sensitive and accurate with minimal amount of required sample substance. They can also be mass produced in batches with large volumes.

There are many different types of microsensors developed for a variety of applications, and they are widely used in industry. Common sensors include biosensors, chemical sensors, optical sensors, and thermal and pressure sensors. Working principles of these sensors will be presented in Chapter 2. In a pressure sensor, an input signal such as pressure from a source is sensed by a microsensing element, which may include simply a thin silicon diaphragm only a few micrometers as illustrated in Figures 1.22c and 2.8. The deflection of the diaphragm induced by the applied pressure is converted into a change of electrical resistance by micropiezoresistors that are implanted in the diaphragm. These piezoresistors constitute a part of the transduction unit. The change of electrical resistance in the piezoresistors induced by the change of the induced maximum stresses can be further converted into corresponding voltage changes using a micro–Wheatstone

1

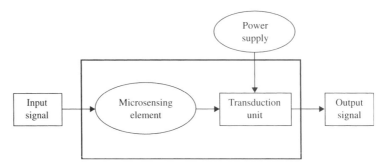

Figure 1.1. MEMS as a microsensor.

bridge circuit also attached to the sensing element as another part of the transduction unit (see Figure 2.9). The output signal of this type of microsensor is thus in the voltage change corresponding to the input pressure. Typical micro–pressure sensors are shown as packaged products in Figure 1.2.

There are many other types of microsensors that are either available in the marketplace or are being developed. They include chemical sensors for detecting chemicals or toxic gases such as CO, CO_2, NO, O_3, and NH_3, either from exhaust from a combustion or a fabrication process or from the environment for air quality control.

Biomedical sensors and biosensors have enjoyed significant share in the microsensor market in recent years. Micro–biomedical sensors are mainly used for diagnosis analyses. Because of their miniature size, these sensors typically require minute amount of sample and can produce results significantly faster than the traditional biomedical instruments. Moreover, these sensors can be produced in batches, resulting in low unit cost. Another cost saving is that most of these sensors are disposable; manual labor involving cleaning and proper treatments for reuse can be avoided. Biosensors are extensively used in analytical chemistry and biomedicine as well as genetic engineering.

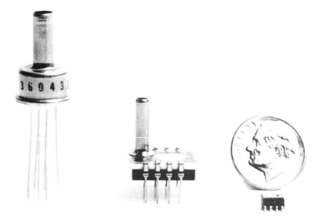

Figure 1.2. Micro–pressure sensor packages.

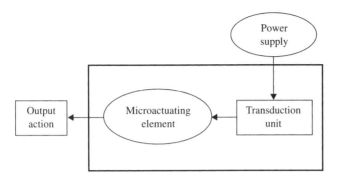

Figure 1.3. MEMS as a microactuator.

Working principles for many of these microsensors will be presented in Section 2.2. Chapters 4 and 5 will present the design methodologies for these sensors.

Figure 1.3 illustrates the functional relationship between the sensing element and the transduction unit in a microactuator. The transduction unit converts the input power supply into the form such as voltage for a transducer, which functions as the actuating element.

There are several ways by which microdevice components can be actuated, as will be described in Section 2.3. One popular actuation method involves using electrostatic forces generated by charged parallel conducting plates or electrodes separated by dielectric material such as air. The arrangement is similar to that of capacitors. The application of input voltage to the plates (i.e., the electrodes in a capacitor) from a direct-current (DC) power source can result in electrostatic forces that prompt relative motion of these plates in the normal direction of aligned plates or parallel movement for misaligned plates. These motions are set to accomplish the required actions. Electrostatic actuation is used in many microactuators. One such application is in a microgripper as illustrated in Figure 1.4. Chapter 2 will include several other actuation methods and for other types of devices.

A *microsystem* is a miniature engineering system that usually contains MEMS components designed to perform specific engineering functions. Despite the fact that many MEMS components can be produced in the size of micrometers, microsystems are typically in "mesoscales." (There is no clear definition of what constitutes the mesoscale. It implies a scale that is between micro- and macroscales. Conventional wisdom, however, suggests that a mesoscale is in the size range of millimeters to a centimeter.) In a somewhat restricted view, Madou (1997) defines a microsystem to include three major components: microsensors, actuators, and a processing unit. The functional relationship of these three components is illustrated in Figure 1.5.

Figure 1.5 shows that signals received by a sensor or sensors in a microsystem are converted into forms compatible with the actuator through the signal transduction and processing unit. One example of this operation is the airbag deployment system in an automobile, in which the impact of the car in a serious collision is "felt" by a micro–inertia sensor built on the principle of a microaccelerometer, as described in Chapter 2. The sensor generates an appropriate signal to an actuator that deploys the airbag to protect the driver and the passengers from serious injuries.

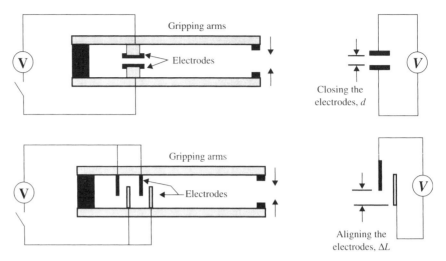

Figure 1.4. MEMS using electrostatic actuation.

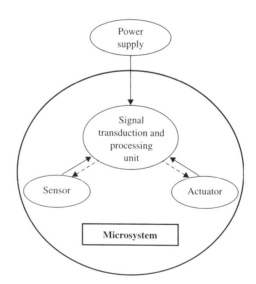

Figure 1.5. Essential components of a microsystem.

Figure 1.6 shows a micro–inertia sensor produced for this purpose. The sensor, which contains two microaccelerometers, is mounted to the chassis of the car. The accelerometer on the left measures the deceleration in the horizontal (x) direction, whereas the unit on the right measures the deceleration in the y direction. Both these accelerometers were mounted on the same integrated circuit (IC) chip with signal transduction and processing units. The entire chip has an approximate size of 3×2 mm, with the microaccelerometers taking about 10% of the overall chip area. The working principle of this type of microaccelerometer will be explained in Section 2.5.

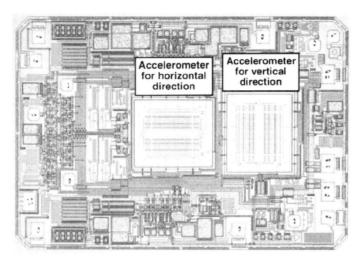

Figure 1.6. Analog Devices ADXL276/ADXL 250 microaccelerometer. (Courtesy of Analog Devices, Norwood, MA.)

Most microsystems are designed and constructed to perform single functions such as presented in this section. There is a clear trend in the industry to incorporate signal processing and close-loop feedback control systems in a microsystem to make the integrated system "intelligent." The arrangement in Figure 1.7 illustrates such a possibility.

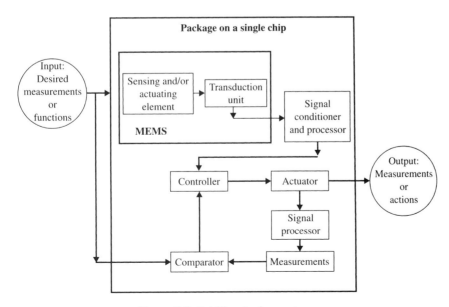

Figure 1.7. Intelligent microsystems.

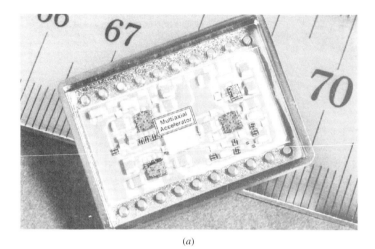

(a)

(b)

Figure 1.8. Intelligent inertia sensors used in automobile airbag deployment systems. (*a*) Inertia sensor on chip. (Courtesy of Karlsruhe Nuclear Research Center, Karlsruhe Germany.) (*b*) Packaged sensor on chip. (Courtesy of Analog Devices, Norwwod, MA.)

Many microsystems have been built on the "lab-on-the-chip" concept. The entire unit can be contained in a silicon chip of size less than 0.5×0.5 mm. Figure 1.8 shows such an example for two different designs of microaccelerometers, or inertia sensors used in airbag deployment systems of automobiles.

1.2 TYPICAL MEMS AND MICROSYSTEMS PRODUCTS

Research institutions and industry have made relentless effort in the past two decades in developing and producing smaller and better MEMS devices and components. Many of these miniaturized devices such as silicon gear trains and tongs were reported by Mehregany et al. (1988). Following is a list, with brief descriptions, of the MEMS devices and components that were produced in recent years. The list, of course, is constanlty becoming longer as the micromanufacturing technology advances.

1.2.1 Microgears

Figure 1.9*a* shows a gear made to a size that is significantly smaller than an ant's head, whereas Figure 1.9*b* shows a two-level gear made from ceramics. The pitch of the gears is on the order of 100 μm. Both these gears were produced using the LIGA (a German acronym of Lithographie Galvanoformung Abformung), process as will be described in detail in Chapter 9.

1.2.2 Micromotors

Figure 1.10 shows an electrostatic-driven micromotor produced by the LIGA process (Bley, 1993). All three components—the rotor (the center gear), the stator, and the torque transmission gear—are made of nickel. The toothed rotor, which has a pitch diameter of 700 μm, is engaged to a gear wheel with 250 μm diameter. The latter wheel transmits the torque produced by the motor. The gap between the rotor and the axle and between the rotor and the stator is 4 μm. The height of the unit is 120 μm.

1.2.3 Microturbines

A microturbine was produced to generate power (Bley, 1993). As shown in Figure 1.11, the turbine is made of nickel. The rotor has a diameter of 130 μm. A 5-μm gap is provided between the axle and the rotor. The turbine has a height of 150 μm. The entire unit was made of nickel. The maximum rotational speed reached 150,000 revolutions per minute (rpm) with a lifetime up to 100 million rotations.

1.2.4 Micro-Optical Components

These components are extensively used in high-speed signal transmissions in the telecommunication industry. Here, we will show two such micro-optical components in Figures 1.12 and 1.13. In Figure 1.12 is a micro-optical switch using a silicon-based manufacturing process. These switches are used to regulate incident light from optical fibers (shown as cylinders in the figure) to appropriate receiving optical fibers.

Figure 1.13 shows microlenses made of transparent polymer, poly (methyl methacrylate) (PMMA). Each lens shown on the left has a diameter of 150 μm. These arrays of lenses are combined into micro-objectives for endoscopy with an optical resolution down to 3 μm. These lenses can also be used for copiers, laser scanners, and printers. At the right is a combination of one such lens with a micro-objective for neurosurgery (Bley, 1999).

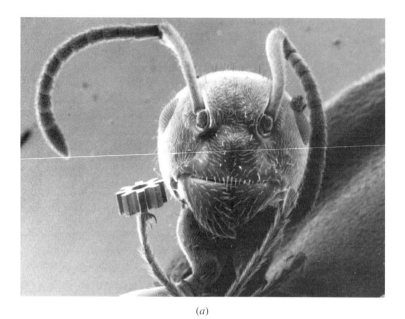

(*a*)

(*b*)

Figure 1.9. Microgears produced by the LIGA process: (*a*) microgear at tip of ant's leg; (*b*) two-level ceramic gear. (Courtesy of Karlsruhe Nuclear Research Center, Karlsruhe, Germany.)

Figure 1.10. Micromotor produced by LIGA process. [Reproduced with permission from Bley (1993).]

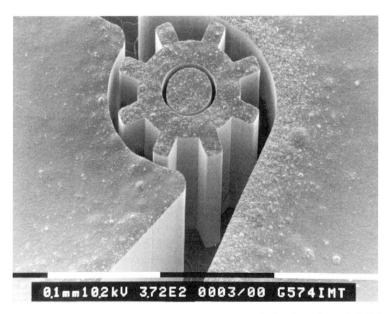

Figure 1.11. Microturbine. [Reproduced with permission from Bley (1993).]

Figure 1.12. Lucent MEMS 1×2 optical switch. (Courtesy of Bell Laboratories of Lucent Technologies.)

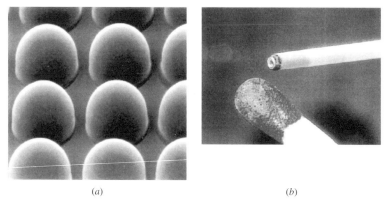

(a)   (b)

Figure 1.13. Arrays of high-aperture microlenses. [Reproduced with permission from Bley (1999).]

1.3 EVOLUTION OF MICROFABRICATION

The size of components of microsystems has been decreasing continuously, with some of the MEMS components made in submicrometer sizes. Fabrication of device components of these minute sizes is clearly beyond traditional mechanical means such as machining, drilling, milling, forging, casting, and welding. The technologies used to produce these minute components are called *microfabrication technologies*, or *micromachining*. For example, the three-dimensional microstructures can be produced by removing part of the base material by a physical or chemical etching process, whereas thin-film deposition techniques are used to build layers of materials on the base materials. We will learn these microfabrication processes and micromanufacturing technologies in detail in, respectively, Chapters 8 and 9.

Current micromanufacturing processes involve a significant number of microfabrication processes that were developed for the microelectronics industry in the last 50 years. Microfabrication processing is thus an important part of microsystems design, manufacture, packaging, and assembly.

Many would attribute the origin of modern-day microfabrication to the invention of transistors by W. Schockley, J. Bardeen, and W. H. Brattain in 1947. Indeed, microfabrication technology is inseparable from the IC fabrication technology that is used to produce microelectronics. The IC concept was first evolved from the production of the monolithic circuit at RCA in 1955 after the invention of transistors. The first IC was produced three years later by Jack Kilby of Texas Instruments. It was in the same year that Robert Noyce of Fairchild Semiconductors produced the first planar silicon-based IC. Today, the next generation of IC, that is, the ULSI (ultra-large-system integration), can contain 10 million transistors and capacitors on a typical chip of size that is smaller than a fingernail.

The term *micromachining* first appeared in the public literature in 1982, although some of its key processes were in existence long before. For example, a key fabrication process known as *isotropic etching* of silicon was first used in 1960, and the improved *anisotropic etching* was invented in 1967.

1.4 MICROSYSTEMS AND MICROELECTRONICS

It is a well-recognized fact that microelectronics is one of the most influential technologies of the twentieth century. The boom of the MEMS industry in recent years would not have been possible without the maturity of microelectronic technology. Indeed, many engineers and scientists in today's MEMS industry are veterans of the microelectronics industry, as the two technologies do share many common fabrication technologies. However, overemphasis on the similarity of the two technologies is not only inaccurate, but it can also seriously hinder further advances of microsystems development. We will notice that there are significant differences in the design, packaging, and assembly of microsystems from that of ICs and microelectronics. It is essential that engineers recognize these differences and develop the necessary methodologies and technologies accordingly. Table 1.1 summarizes the similarities and differences between the two technologies.

We may observe from Table 1.1 that there are indeed sufficient differences between the two technologies. Some of the more significant differences between these two technologies are as follows:

1. Microsystems involve more different materials than microelectronics. Other than the common material of silicon, other materials such as quartz and GaAs are used as substrates in microsystems. Polymers and metallic materials are common in microsystems produced by LIGA processes. Packaging materials of microsystems include glasses, plastics, and metals, which are excluded in microelectronics.

2. Microsystems are designed to perform a greater variety of functions than ICs. The latter are limited to specific electrical functions.

TABLE 1.1. Comparison of Microsystems and Microelectronics

Microsystems and MEMS (Silicon Based)	Integrated Circuits
Complex three-dimensional structures	Primarily 2-dimensional structures
Packaged products may involve moving components, access to light beams, and sealing fluids and chemicals	Stationary structure once they are packaged
Perform a variety of specific biological, chemical, electromechanical, and optical functions	Transmit electricity for specific electrical functions
Delicate components required to interface with working media	Delicate dies are protected from contacting media
Use single-crystal silicon dies, silicon compounds, ceramics, and other materials, e.g., quartz, polymers, and metals	Use single-crystal silicon dies, silicon compounds, ceramics, and plastics
Many more components to be packaged and assembled	Fewer components to be packaged and assembled
Lack of engineering design methodology and standards	Mature design methodologies
Simpler patterns over the substrates but require integration with complex electrical circuits for signal transduction	Complex patterns with high-density electrical circuitry over substrates
No industrial standard to follow in design, material selections, fabrication processes, and packaging and assembly	Industrial standards available
Most are in low-volume batch production on customer need bases	Mass production
Available manufacturing techniques of bulk micromanufacturing, surface micromachining, and the LIGA process	Manufacturing techniques are proven and well documented
Packaging and assembly techniques are at infant stage of development	Packaging and assembly techniques are well established
Lack of testing and measurement techniques and tools	Well-documented techniques available
Involve all disciplines of science and engineering	Primarily involve electrical and chemical engineering

3. Many microsystems involve moving parts such as microvalves, pumps, and gears. Many require fluid flow through the systems such as in biosensors and analytic systems. Micro-optical systems need to provide input/output (I/O) ports for light beams. Integrated circuits and microelectronics, on the other hand, do not have any moving component or access for lights or fluids.

4. Integrated circuits are primarily two-dimensional in structure and are confined to the silicon die surface, whereas most microsystems involve complicated geometry in three-dimensions. Mechanical engineering design is thus an essential part in the product development of microsystems.

5. The core elements in ICs are shielded by passivation materials from the surroundings once they are packaged. The sensing elements and many core elements in microsystems, however, are required to be in contact with working media, which creates many technical problems in design and packaging.

6. Manufacturing and packaging of ICs and microelectronics are mature technologies with well-documented industry standards. The production of microsystems, on the other hand, is far from being at that level of maturity. There are generally three distinct manufacturing techniques, as will be described in Chapter 9. Because of the great variety of structural and functional aspects in microsystems, the packaging and assembly of these products remain in the infant stage at the present time.

The slow advance in the development of microsystems technology is mainly attributed to the complex nature of these systems. As we will learn from Section 1.5, there are many science and engineering disciplines involved in the design, manufacture, and packaging of microsystems.

1.5 MULTIDISCIPLINARY NATURE OF MICROSYSTEMS DESIGN AND MANUFACTURE

Despite the fact that micromanufacturing has evolved from IC fabrication technologies, there are several other science and engineering disciplines involved in today's microsystems design and manufacturing. Figure 1.14 illustrates the applications of principles of natural science and several engineering disciplines in this process.

With reference to Figure 1.14, natural science is deeply involved in the following areas:

1. Electrochemistry is widely used in electrolysis by ionizing substances in some micromanufacturing processes. Electrochemical processes are also used in the design of chemical sensors. More detailed description will be given in Chapters 2 and 3.

2. Electrohydrodynamics principles are used as the driving mechanisms in fluid flows in microchannels and conduits such as in capillary fluid flows, as will be described in Chapters 2, 5, and 10.

3. Molecular biology is intimately involved in the design and manufacture of biosensors and biomedical equipment, as will be presented in Chapter 2. Most of the basic molecular biology principles are used in the development of nanotechnology in areas such as self-replication and assemblies, as will be elaborated in Chapter 12.

4. Plasma physics involves the production and supply of ionized gases with high energy. It is required for etching and depositions in many microfabrication processes. The generation of plasma will be covered in Chapter 3, whereas the application of plasma in microfabrication will be described in Chapter 8.

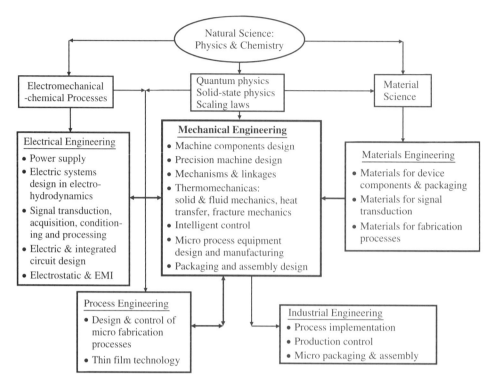

Figure 1.14. Principal science and engineering disciplines involved in microsystems design and manufacture.

5. Scaling laws provide engineers with a sense for the scaling down of physical quantities involved in the design of microdevices. We will realize from Chapter 6 that not all physical quantities can be scaled down favorably.

6. Quantum physics is used as the basis for modeling certain physical behaviors of materials and substances in microscales, as will be described for microfluid flows and heat transportation in solids in Chapter 12.

7. Molecular physics provides many useful models in the description of materials at micro- and nanoscales as well as the alteration of material properties and characteristics used in microsystems, as will be described in Chapters 3, 7, and 12. Molecular dynamics theories are the principal modeling tool for describing mechanical behavior of materials at the nanoscale.

Five engineering disciplines are involved in microsystems design, manufacture, and packaging, as described below:

1. Mechanical engineering principles are used primarily in the design of microsystems structures and the packaging of the components. These tasks involve many aspects of design analyses as indicated in the central box in Figure 1.14. Intelligent

control of microsystems has not been well developed, but it is an essential part of *micromechatronics systems,* which are defined as intelligent microelectromechanical systems.

2. Electrical engineering involves electrical power supply and the functional control and signal processing circuits design. For the integrated microsystems, for example, "laboratory-on-the-chip," the IC and microelectronics circuitry that integrates microelectronics and microsystems makes electrical engineering a major factor in the design and manufacturing processes.

3. Chemical engineering is an essential component in microfabrication and micromanufacturing, as will be described in Chapters 8 and 9. Almost all such processes involve chemical reactions. Some of the microdevice packaging techniques also rely on special chemical reactions, as will be described in Chapter 11.

4. Materials engineering offers design engineers with the selection and characterization of available materials that are amenable to microfabrication and micromanufacturing as well as packaging. Theories of molecular physics are often used in the design of the material's characteristics, such as doping the semiconducting materials to change the electrical resistivity of the material. Materials engineering plays a key role in the development of chemical, biological, and optical sensors, as will be described in Chapter 2.

5. Industrial engineering relates to the assembly and production of microsystems. Optimum design of microfabrication processes and control with required equipment and facility is essential in microsystems production. Much MEMS production falls into the category of medium volume with medium variety, and flexible micromanufacturing systems appear to be both beneficial and desirable.

1.6 MICROSYSTEMS AND MINIATURIZATION

It is fair to say that microsystems, as we have defined in Section 1.1, are a major step toward the ultimate miniaturization of machines and devices, such as dust-size computers and needle-tip-size robots that have fascinated many futurists in their seemingly science fiction articles and books. The current microsystems technology, though at a relative early stage, has already set the tone for the development of device systems at the nanoscale for this new century. [A nanometer (nm) is 10^{-9} meter.] The maturity of nanotechnology will certainly result in the realization of much of the superminiaturized machinery that engineers and scientists have fantasized at the present time.

According to *Webster's Dictionary*, the word *miniature* means a copy on a much-reduced scale. In essence, miniaturization is an art that substantially reduces the size of the original object yet retains the characteristics of the original (and more) on the reduced copy. The need for miniaturization has become more prominent than ever in recent decades, as engineering systems and devices have become more and more complex and sophisticated. The benefits of having smaller components and hence a device or machine with enhanced capabilities and functionalities are obvious from engineering perspectives:

- Smaller systems tend to move more quickly than larger systems because of smaller mass and thus low mechanical inertia.

- Likewise, smaller systems react to change of thermal conditions much rapidly than larger systems because of low thermal inertia. It is therefore more effective in actuation by thermal forces.

- Smaller systems do not produce significant overall thermal distortion because of their small sizes.

- The minute sizes of small devices encounter fewer problems in vibration because resonant vibration of a system is inversely proportional to the mass. Smaller systems with lower masses have much higher natural frequencies than those expected from most machines and devices in operation. Silicon-based miniature condenser microphones as described in Chapter 2 are virtually free from resonant diaphragm vibration for this reason.

- In addition to the more accurate performance of smaller systems, their minute sizes make them particularly suitable for applications in medicine and surgery and microelectronics assemblies in which miniaturized tools are necessary.

- Miniaturization is also desirable in satellites and spacecraft engineering to satisfy the prime concerns about high precision and payload size.

- The high accuracy of miniaturized systems in motion and dimensional stability makes them particularly suitable for telecommunication systems.

The process of miniaturizing engineering systems has been an ongoing effort by engineers. We have witnessed many of the results of miniaturization in consumer products with much better performance but much reduced size. Such products include hair dryers, cameras, radios, and telephone sets. We will see from the following examples and appreciate how these efforts have produced even more staggering results in other fundamentally important engineering products in the last 50 years. There is no doubt that with the rapid development of microsystems technologies, such process will be accelerated at an even more spectacular rate.

Perhaps the foremost important miniaturization began with the development of ICs in the mid-1950s. As described in Section 1.3, a ULSI of the size of a small fingernail can contain 10 million transistors. Microprocessors of similar size can now perform over 300 millions operations per second. A leading microprocessor manufacturer in the United States reported its success in early 2007 in producing transistors that have the size of 45 nm, which is equivalent to having 2000 wires in the space of a human hair.

The advances in the IC and microprocessor technologies have led to a spectacular level of miniaturization of complex digital electronics computers. Two engineers, J. Presper Eckert and John Mauchly, publicly demonstrated the first general-purpose electronic digital computer called ENIAC (Electronic Numerical Integrator and Calculator) in 1946 at the Moore School of Electrical Engineering at the University of Pennsylvania. The U.S. Army during World War II originally funded this project. As shown in Figure 1.15, the U-shaped computer was 80 ft long by 8.5 ft high and several feet wide. Each of

Figure 1.15. The ENIAC digital computer in 1947. (Used by permission of the University of Pennsylvania's School of Engineering and Applied Science, formerly known as the Moore School of Electrical Engineering.)

the twenty 10-digit accumulators was 2 ft long. In total, ENIAC had 18,000 vacuum tubes.

At the time of the ENIAC fiftieth anniversary celebration, Dr. Van der Spiegel worked with a group of students at the same university to reduce the original ENIAC to the size of a tiny chip with equivalent functional power (shown in Figure 1.16) (Van der Spiegel et al., 1998).

In sharp contrast to ENIAC, a typical notebook or laptop computer, shown in Figure 1.17, is only about 14 in. long by 10 in. wide by 1 in. thick. In size, ENIAC, estimated at about 4000 ft^3, was more than five orders of magnitude bigger than laptop computers today, but the latter have a computational speed that is six to eight orders of magnitudes faster than ENIAC. This staggering miniaturization took place in just about 50 years!

The microfabrication technology that evolved from IC fabrications in the last two decades has made machine components in sizes of micrometers and submicrometers possible. Many of these devices and components have been shown in the foregoing sections. While most sensors and actuating components use silicon as primary material, other materials such as nickel and ceramics have been used. High-aspect-ratio microstructures (the *aspect ratio* is the ratio of the dimensions in the height to those of the surface) have been successfully manufactured by the LIGA process, as shown in

Figure 1.16. The "miniaturized" ENIAC. (Used by permission of the University of Pennsylvania's School of Engineering and Applied Science, formerly known as the Moore School of Electrical Engineering.)

Figure 1.17. Laptop computer as of 2000.

Figures 1.9–1.11. This special micromanufacturing process will be described in detail in Chapter 9. The broader use of materials along with the possibility of producing MEMS with high aspect ratio by the LIGA process has prompted the production of miniaturized machines.

Another driving force for miniaturizing industrial products is the strong market demand for smaller, intelligent, robust, and multifunctional consumer products. Figure 1.18 illustrates the evolution of mobile personal telecommunication equipment. On the left is

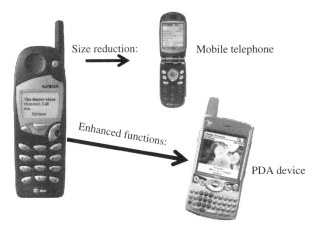

Figure 1.18. Evolution of mobile telecommunication devices.

the mobile telephone set at its debut 15 years ago. It was a breakthrough in telecommunication technology at that time, as for the first time consumers could use a cordless hand-held device to transceive voice messages almost everywhere in the world. The popularity of mobile phones has led to a speedy evolution to their current state-of-the-art of smaller but more versatile telephone sets and powerful personal digital assistants (PDAs), as shown in the figure. Although the reduction in size of these telecommunication devices has not been as spectacular as in digital computers, the functions that they can perform are truly spectacular; a current mobile telephone, which is about one-quarter of the size of the original sets, can be used, in addition to transceive voices, as a clock, a calendar, a video camera, and an instant positioning device with the use of the global positioning system (GPS) through satellites. The PDA devices combine computing, telephone/fax, Internet, and networking features. A typical PDA can function as a cellular phone, fax sender, Web browser, and personal organizer. Unlike laptop computers, most PDAs began as pen based, using a stylus rather than a keyboard for input. This means that they also incorporated handwriting recognition features. Some PDAs can also react to voice input by using voice recognition technologies. Most PDAs of today are available in either a stylus or keyboard version.

We may quickly realize that the only solution to produce these smaller, intelligent, and multifunctional products such as mobile telephones and PDAs is to pack many miniature function components into the same device.

Another spectacular micromachine is the microcars that were produced by the Denso Corporation in Japan. Figure 1.19 shows cars the size of rice grains. The smaller car shown in the figure weighs 33 mg. The overall dimensions are 4.785 mm long by 1.73 mm wide by 1.736 mm high. The body is made of nonelectrolysis nickel plating. The nickel film is 30 μm thick. These two cars were fabricated using microprecision machining with semiconductor processes. The electrical motor that drives the cars uses magnetic induction, with a rotor of 0.6 mm in diameter. Detailed design methodology for these cars can be found in an article by Teshigahara et al. (1995).

(a)

(b)

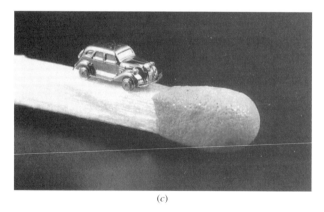

(c)

Figure 1.19. Miniature electric cars: (*a*) comparing with the size of rice grains; (*b*) comparing with the size of an American 5c coin; (*c*) miniature car resting near the of matchstick. (Courtesy of Denso Research Laboratories, Denso Corporation, Aichi, Japan.)

1.7 APPLICATION OF MICROSYSTEMS IN AUTOMOTIVE INDUSTRY

The MEMS and microsystems products have become increasingly dominant in every aspect of the marketplace as technologies for microfabrication and miniaturization continue to be developed. At the present time, two major commercial markets for these products are (1) the information technology industries involving the productions of inkjet printer heads and read/write devices in data storage and (2) the automotive industry. We will focus our attention to the latter due to the high variety of products.

The automotive industry has been the major user of MEMS technology in the last two decades because of the size of its market. A 1991 report indicated that the industry had a production of 8 million vehicles per year with 6 million of these in the United States (Sulouff, 1991). The number of vehicles produced in the United States rose to 17 million in 2005 (National Geographic, 2005). The primary motivation for adopting MEMS and microsystems in automobiles is to make automobiles safer and more comfortable for riders and to meet the high fuel efficiencies and low emissions standards required by governments. In all, the widespread use of these products can indeed make the automobile "smarter" for consumers' needs. The term *smart cars* was first introduced in the cover story of a special issue of a trade magazine (Smart Cars, 1988). Many of the seemingly fictitious predictions of the intelligent functions of a smart car are in place in today's vehicles.

Smart vehicles are built on the extensive use of sensors and actuators. Various kinds of sensors are used to detect the environment or road conditions, and the actuators are used to execute whatever actions are required to deal with these conditions. Microsensors and actuators allow automobile makers to use smaller devices, and thus more of them, to cope with the situation in much more effective ways. Comprehensive summaries on the use of sensors and actuators are available in an early report (Paulsen and Giachino, 1989). Sensors of various kinds are extensively used in automobile engines, as illustrated in Figure 1.20. Figure 1.21 illustrates the application of pressure sensors in an automobile.

The design and production of automobiles satisfying the above expectations present a major challenge to engineers as these vehicles are expected to perform in extreme environmental conditions as outlined in several reports (Giachino and Miree, 1995; Chiou, 1999). Table 1.2 presents some of these stringent requirements for endurance tests for MEMS components used in the automotive industry.

In addition to the stringent endurance testing conditions for sensors and actuators presented in Table 1.2, they are expected to perform in extremely harsh environmental conditions. Design conditions include a temperature range of -40 to $+125°$ C and a dynamic loading up to $50g$ (MacDonald, 1990). Table 1.2 presents some of the endurance testing conditions for pressure sensors (Chiou, 1999). The pressure range for automobiles extends from $100\,kPa$ for a manifold absolute pressure (MAP) to $10\,MPa$ for brake pressure sensors.

Applications of microsystems in automobiles can be categorized into the following four major areas: (1) safety, (2) engine and power train, (3) comfort and convenience, and (4) vehicle diagnostics and health monitoring.

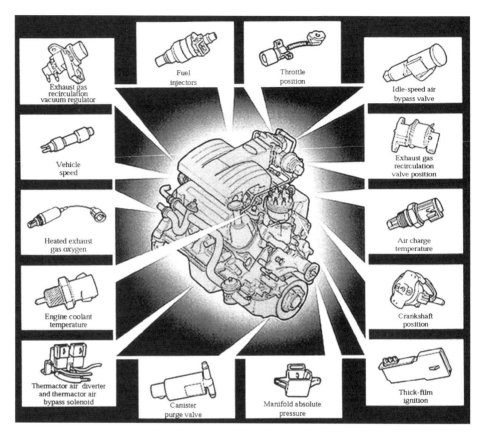

Figure 1.20. Sensors in an automobile engine and powertrain. [Reproduced from Paulsen and Giachino (1989) with permission.]

1.7.1 Safety

- An airbag deployment system is introduced in automobiles to protect the driver and passengers from injury in the event of serious vehicle collision. The system uses microaccelerometers or micro–inertia sensors similar to those illustrated in Figures 1.6 and 1.8.
- An antilock braking system using position sensors allows the driver to steer the vehicle with ceased wheels while braking.
- Suspension systems, using displacement, position and pressure sensors, and microvalves, mitigate violent vibrations of the vehicle that cause structural damages and discomfort to the riders.
- Object avoidance using pressure and displacement sensors allows the driver to detect obstructions along the path of the vehicle.
- An automatic navigation system using microgyroscope and GPS can navigate the moving vehicle in hazardous and rough terrain.

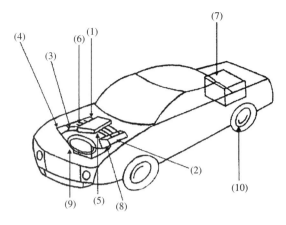

(1) Map or TMAP sensor
(2) EGR (DPFE) sensor
(3) Fuel rail pressure sensor
(4) BAP sensor
(5) Combustion sensor

(6) GDI pressure sensor
(7) EVAP sensor
(8) Engine oil sensor
(9) Transmission sensor
(10) Tire pressure sensor

Figure 1.21. Pressure sensors in automotive applications. [Reproduced from Chiou (1999) with permission.]

TABLE 1.2. Typical Durability Requirements

Tests	Conditions	Duration
High-temperature bias	100°C, 5 V	1000 h
Thermal shock	−40 to 150°C	1000 cycles
Temperature and humidity	85°C and 85% relative humidity (RH) with and without bias	1000 h
Pressure, power, and temperature cycles	20 kPa to atmospheric pressure, 5 V, −40 to 150°C	3000 h
Hot storage	150°C	1000 h
Cold storage	−40°C	1000 h
Pressure cycling	700 kPa to atmospheric pressure	2 million cycles
Overpressure	2 × maximum operating pressure	
Vibration	$2g$ to $40g$, frequency sweep, depending on locations	30 + hours in each axis
Shock	$50g$ with 10-ms pulses	100 times in three planes
Fluid/media compatibility	Air, water, corrosive water, gasoline, methanol, ethanol, diesel fuel, engine oil, nitric and sulfuric acids	Varies with applications

Source: Chiou, 1999.

1.7.2 Engine and Power Trains

Various sensors have been used in engines of modern automobiles, as illustrated in Figure 1.20 (Paulsen and Giachino, 1989). Following are a few of these sensors:

- Manifold control with pressure sensors
- Airflow control
- Exhaust gas analysis and control (see Figure 1.21)
- Crankshaft position
- Fuel pump pressure and fuel injection control
- Transmission force and pressure control
- Engine knock detection for higher power output (Keneyasu et al., 1995)

The MAP was one of the first microsensors adopted by the automotive industry, in the early 1980s. It measures MAP along with the engine speed in rpm to determine the ignition advance. This ignition timing with optimum air–fuel ratio can optimize the power performance of the engine with low emissions. Early designs of MAP sensors were called SCAP (silicon capacitive absolute pressure) sensors. A typical SCAP is shown in Figure 1.22 (Paulsen and Giachino, 1989). The manifold gas enters the sensor at the tubular intake, shown at the top of Figures 1.22*a* and *b*. The intake gas exerts a pressure to a silicon diaphragm, which acts as one of the two electrodes in the chamber, as shown in Figure 1.22*c*. The deflection of the diaphragm from the applied pressure results in a change in the gap between the two electrodes. The applied pressure can be correlated to the change of the capacitance of the electrodes in the capacitor. Appropriate bridge circuits, such as shown in Figure 2.11, are used to convert the change of the capacitance signal output to electric voltages. Figure 1.22*d* shows the relative size of the SCAP unit.

The capacitive pressure sensors have two major shortcomings: bulky size and nonlinear voltage output with respect to the input pressure (Figure 2.13). Consequently, piezoresistive transducer (PRT) pressure sensors that provide DC voltage output that is proportional to the applied pressure have gradually replaced them. The use of piezoresistives as the signal transducer has significantly reduced the size of the sensors.

1.7.3 Comfort and Convenience

- Seat control (displacement sensors and microvalves)
- Rider comfort (sensors for air quality, airflow, temperature, and humidity controls)
- Security (remote status monitoring and access control sensors)
- Sensors for defogging windshields
- Satellite navigation sensors

1.7.4 Vehicle Diagnostics and Health Monitoring

- Engine coolant temperature and quality
- Engine oil pressure, level, and quality
- Tire pressure (Figure 1.21)
- Brake oil pressure

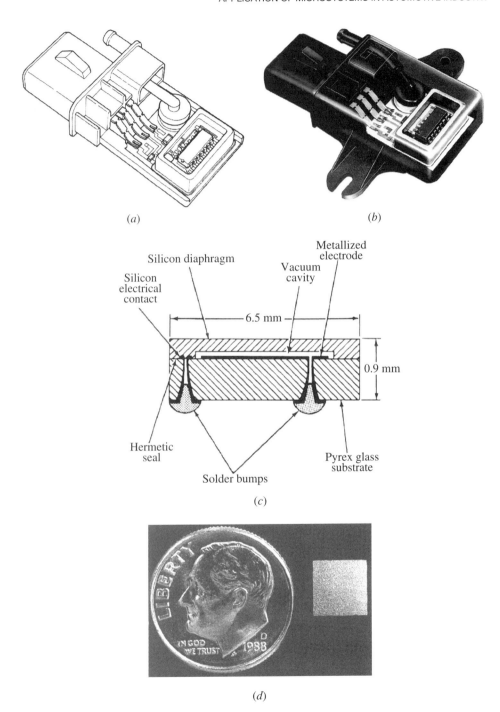

Figure 1.22. Silicon capacitive manifold absolute pressure sensor. (Courtesy of Ford Motor Company.)

- Transmission fluid (Figure 1.21)
- Fuel pressure (Figure 1.21)

1.7.5 Future Automotive Applications

A report (Powers and Nicastri, 1999) indicated that there were 52 million vehicles produced worldwide in 1996, and this number was expected to increase to 65 million by year 2005 because of the growth of the global economy. Consumer expectation on the safety, comfort, and performance of cars will continue to grow. There is every reason to believe that vehicles in the future will contain even more microprocessors with many more microsensors and actuators (Figure 1.23) to be truly smart. Figure 1.24 illustrates the sensors; most of them will be in the microscale range in future vehicles.

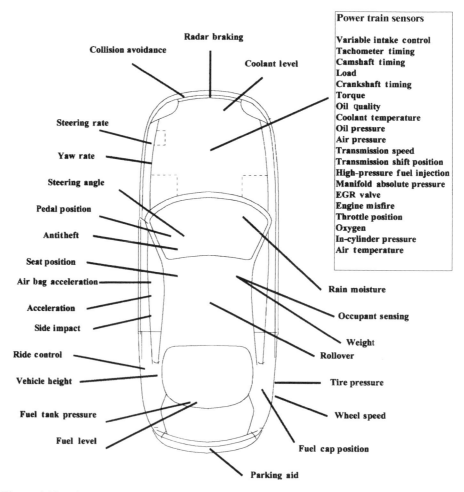

Figure 1.23. Microprocessor control components for future vehicles. (Courtesy of Ford Motor.)

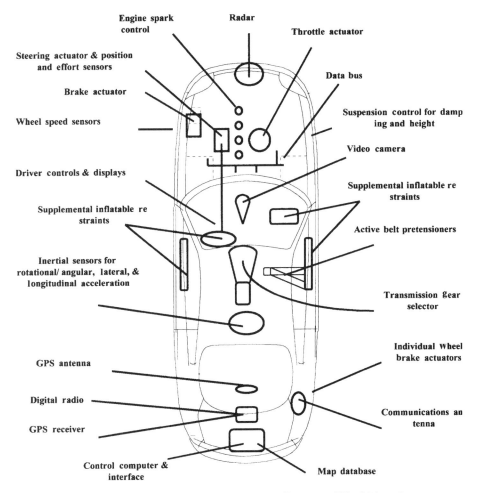

Figure 1.24. Sensors for future vehicles. (Courtesy of Ford Motor.)

1.8 APPLICATION OF MICROSYSTEMS IN OTHER INDUSTRIES

Other commercial applications of MEMS such as biomedical and genetic engineering are emerging at an astounding rate in recent years. We will offer applications of MEMS and microsystems in five industrial sectors.

1.8.1 Application in Health Care industry

- Disposable blood pressure transducer (DPT): lifetime 24–72 hours; approximate annual production 17 million units per year, unit price <$10
- Intrauterine pressure sensors (IUPs): to monitor pressure during child delivery; current market about 1 million units per year

- Angioplasty pressure sensors: to monitor pressure inside a balloon inserted in a blood vessel; current market about 500,000 units per year
- Infusion pump pressure sensors: to control the flow of intravenous fluids and permit several drugs to be mixed into one flow channel; approximate market size about 200,000 units per year
- Other products:

    Diagnostics and analytical systems, such as capillary electrophoresis systems (Section 2.2.2 and Chapter 10)

    Human care support systems

    Catheter tip pressure sensors

    Sphygmomanometers

    Respirators

    Lung capacity meters

    Barometric correction of instrumentation

    Medical process monitoring (e.g., drug production by growth of bacteria)

    Kidney dialysis equipment

1.8.2 Application in Aerospace Industry

- Cockpit instrumentation

    Pressure sensors for oil, fuel, transmission, and hydraulic systems

    Air speed measurement

    Altimeters

- Safety devices (e.g., sensors and actuators for ejection seat controls)
- Wind tunnel instrumentation (e.g., shear stress sensors)
- Sensors for fuel efficiency and safety
- Microgyroscope for navigation and stability control
- Microsatellites

A comprehensive summary of possible uses of MEMS and microsystems in space hardware in the near term (less than 10 years) is presented below (Helvajian and Janson, 1999):

- Command and control systems with MEMtronics for ultraradiation and temperature-insensitive digital logic and on-chip thermal switches for latch-up and reset
- Inertial guidance systems with microgyroscopes, accelerometers, and fiber-optic gyros
- Attitude determination and control systems with micro sun and Earth sensors, magnetometers, and thrusters

- Power systems with MEMtronic blocking diodes, switches for active solar cell array reconfiguration, and electric generators
- Propulsion systems with micro–pressure sensors, chemical sensors for leak detection, arrays of single-shot thrusters, continuous microthrusters, and pulsed microthrusters
- Thermal control systems with micro–heat pipes, radiators, and thermal switches
- Communications and radar systems with very high bandwidth, low-resistance radio frequency switches, micromirrors, and optics for laser communications, micro–variable capacitors, inductors, and oscillators
- Space environment sensors with micromagnetometers and gravity-gradient monitors (nano-g accelerometers)

There has been considerable effort as well as progress in recent years in the development of micropropulsion systems that include micropropellants, nozzles, jet engines, and thrusters (Janson et al., 1999). Principal supporting mechanical engineering subjects include microfluid modeling, microcombustion, and rocket science.

1.8.3   Application in Industrial Products

- Manufacturing process sensors: process pressure transmitters ($>200,000$ units per year)
- Sensors for:

> Hydraulic systems
>
> Paint spray
>
> Agricultural irrigations and sprays
>
> Refrigeration systems
>
> Heating, ventilation, and air conditioning systems
>
> Water-level controls
>
> Telephone cable leak detection

1.8.4   Application in Consumer Products

- Scuba diving watches and computers
- Bicycle computers
- Fitness gears using hydraulics, washers with water-level controls
- Sport shoes with automatic cushioning control
- Digital tire pressure gages
- Vacuum cleaning with automatic adjustment of brush beaters
- Smart toys
- Smart home appliances

1.8.5 Application in Telecommunications

- Optical switching and fiber-optic couplings
- Radio frequency (RF) MEMS in wireless systems (Rebeiz, 2003)
- Tunable resonators

1.9 MARKETS FOR MICROSYSTEMS

As new MEMS and microsystems products have become available, the market for these products has been expanding rapidly. Table 1.3 shows the world market projections to year 1999.

We observe from Table 1.3 that the compound annual growth rate since 1992 is 15.9%. Unit prices continued to drop as volume of production increased.

The MEMS market size takes a dramatic expansion if one follows the definition of MEMS and microsystems products by NEXUS, a European Community that initiated the Network of Excellence in Multifunctional Microsystems. It includes all products that are microstructures in design, including monolithic and hybrid components and silicon-based devices, as well as anything micromachined. The market for established MEMS product between 1996 and 2002 is tabulated in Table 1.4.

NEXUS also offered a market growth of MEMS and microsystems products between 2000 and 2005, as shown in Figure 1.25.

We may observe from Figure 1.25 that information technology (IT) peripherals that include read–write heads for hard disk drives and inkjet printer heads dominate the

TABLE 1.3. Worldwide Silicon-Based Micro Sensor Market

Year	Unit (×1000)	Revenues ($million)	Average Unit Price ($)	Revenue Growth Rate (%)
1989	3026	570.50	188.53	—
1990	5741	744.60	129.70	30.5
1991	6844	851.70	124.44	14.4
1992	7760	925.40	119.25	8.6
1993	8816	977.10	110.83	5.6
1994	10836	1116.20	103.00	14.2
1995	13980	1316.30	94.16	17.9
1996	18127	1564.40	86.30	18.9
1997	23514	1857.60	79.00	18.7
1998	30355	2199.80	72.47	18.4
1999	38792	2593.80	66.86	17.9

Source: Frost & Sullivan Market Intelligence, 1993. (www.frost.com/prod/servlet/frost-home)

TABLE 1.4. Market Growth for Established MEMS Products, 1996–2002

MEMS Products	Year 1996 (10^6 units/revenue in $\$10^6$)	Year 2002 (10^6 units/revenue in $\$10^6$)
HDD heads[a]	530/4500	1500/12000
Inkjet printer heads	100/4400	500/10000
Heart pacemakers	0.5/1000	0.8/3700
In vitro diagnostics	700/450	4000/2800
Hearing aids	4/1150	7/2000
Pressure sensors	115/600	309/1300
Chemical sensors	100/300	400/800
Infrared imagers	0.01/220	04/800
Accelerometers	24/240	90-/430
Gyroscopes	6/150	30/360
Magnetoresistive sensors	15/20	60/60
Microspectrometers	0.006/3	0.15/40
Total	1595/$13,033	6807/$34,290

[a]Read–write heads for hard disk drives.
Source: http://www.wtec.org/loyal/mcc/mems_eu/pages/chapter-6.html.

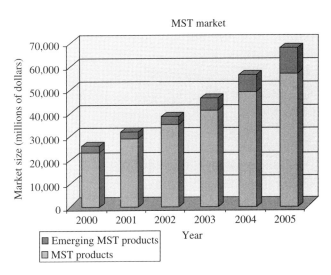

Figure 1.25. Market growth for established and emerging MEMS and microsystems products. (*Source:* Nexus: http://www.smalltimes.com/document_display.cfm?document_id = 3424.)

microsystems (MST) market. Biomedical products such as cardiac pacemakers and hearing aids and other new emerging biomedical products such as lab-on-a-chip analytic systems follow closely the market growth of IT peripherals. Emerging MST products in Figure 1.25 include optical mouse sensors, household appliances, fingerprint sensors, and implantable drug delivery systems. Another potentially significant market for MST products is the flat-panel display for household and industrial applications.

PROBLEMS

Part 1 Multiple Choice

1. The largest market share of MEMS products currently belongs to (a) micropressure sensors, (b) IT peripherals, (c) microaccelerometers.

2. MEMS components range in size (a) from 1 μm to 1 mm, (b) from 1 nm to 1 μm, (c) from 1 mm to 1 cm.

3. One nanometer is (a) 10^{-6} m, (b) 10^{-9} m, (c) 10^{-12} m.

4. When we say a device is mesoscale, we mean the device has the size in the range of of (a) 1 μm to 1 mm, (b) 1 nm to 1 μm, (c) 1 mm to 1 cm.

5. The origin of microsystems can be traced back to the invention of (a) transistors, (b) integrated circuits, (c) silicon piezoresistors.

6. A modern integrated circuit may contain (a) 100,000, (b) 1,000,000, (c) 10,000,000 transistors and capacitors.

7. Miniaturization of computers was possible mainly due to (a) better storage systems, (b) replacing vacuum tubes with transistors, (c) the invention of integrated circuits.

8. In general, a microsystem consists of (a) one, (b) two, (c) three components.

9. The microsensor that is commonly used in airbag deployment systems in automobiles is (a) pressure sensor, (b) inertia sensor, (c) chemical sensor.

10. "Lab-on-a-chip" means (a) performing experiments on a chip, (b) integration of microsensors and actuators on a chip, (c) integration of microsystems and microelectronics in a chip.

11. The origin of modern microfabrication technology is (a) invention of transistors, (b) invention of integrated circuits, (c) invention of micromachining.

12. The first significant miniaturization occurred with (a) integrated circuits, (b) laptop computers, (c) mobile telephones.

13. The term *micromachining* first appeared in (a) the 1970s, (b) the 1980s, (c) the 1990s.

14. The term LIGA is (a) a process for micromanufacturing, (b) a microfabrication process, (c) a material treatment process.

15. A typical single ULSI chip may contain (a) one million, (b) ten million, (c) one hundred million transistors.

16. The development of ICs began in (a) the 1960s, (b) the 1970s, (c) the 1950s.

17. The first digital computer, ENIAC, was developed in (a) the 1960s, (b) the 1950s, (c) the 1940s.

18. The *aspect ratio* of a microsystem component is defined as the ratio of (a) the dimension in the height to those of the surface, (b) the dimensions of the surface to that of the height, (c) the dimension in width to that of the length.

19. The market value of microsystems is intimately related to (a) volume demand, (b) special features, (c) performance of the products.

20. The most challenging issue facing microsystems technology is (a) the small size of the products, (b) the lack of practice application, (c) the multidisciplinary nature.

Part 2 Short Problems

1. Give three examples of objects that you recognize to be approximately 1 mm in size.

2. Explain the difference between MEMS and microsystems.

3. What are the most obvious distinctions between microsystems and microelectronics technologies?

4. Why cannot microelectronics technology be adopted in the design and packaging of MEMS and microsystems products?

5. Why cannot traditional manufacturing technologies such as mechanical milling, drilling, and welding be used to produce microsystems?

6. Give at least four distinct advantages of miniaturization of machines and devices.

7. Give two examples of the miniaturization of consumer products that you have observed and appreciated.

8. Ask your doctor to give at least one example of the application of MEMS in biomedicine that offers significant advantage over traditional devices or methods or conduct your own research.

9. Conduct research to configure an airbag deployment system in an automobile.

10. Describe the proper role that you can play in this multidisciplinary microsystems technology with your academic and professional background.

CHAPTER 2

WORKING PRINCIPLES
OF MICROSYSTEMS

2.1 INTRODUCTION

As we learned in Chapter 1, the rapid advance in microfabrication technologies has enabled many new MEMS products to emerge in the marketplace. Here, we will present the working principles of some typical MEMS devices and microsystems. The reader will soon realize that the design and manufacture of MEMS and microsystems products involve the application of multidisciplinary science and engineering principles, as described in Section 1.5. Learning the working principles of many of these devices thus offers engineers insights into the design and manufacture of new MEMS products. However, we will only present the information necessary to describe these working principles. Detailed descriptions of the processes involved in the fabrication and manufacture of these devices can be found in two principal sources (Madou, 1997, Kovacs, 1998). Description of product geometry as well as the fabrication techniques used in producing many microdevices in early products is available in an excellent survey (Trimmer, 1997). Readers should also be aware that some of the microdevices presented in this chapter were chosen for their clever ideas and many were not successful in the marketplace.

2.2 MICROSENSORS

Microsensors are the most widely used MEMS devices today. According to Madou (1997), a sensor is a device that converts one form of energy into another and provides the user with a usable energy output in response to a specific measurable input. One such example is to convert the energy that is required to deflect the thin diaphragm in a pressure sensor into an electrical energy (signal) output. A sensor system that includes the sensing element and its associated signal processing hardware is illustrated in

Figure 1.1. A smart sensor unit would include automatic calibration, interference signal reduction, compensation for parasitic effects, offset correction, and self-test. All these intelligent functions make this sensor unit an intelligent microsystem, as illustrated in Figure 1.7.

Microsensors are used to measure many physical quantities, as described in Chapter 1. Generally, they cover 10 different measured domains (White, 1987); their principal applications are presented below. There are many microsensors available to perform various functions in a variety of industries. Detailed descriptions of many of these sensors could be found in the literature (Madou, 1997; Kovacs, 1998). Here, we will bring focus on the working principles of some of these sensors.

2.2.1 Acoustic Wave Sensors

Acoustic wave sensors are very sensitive and intrinsically reliable for many emerging industrial products. The telecommunications industry is the major user of this type of sensor, accounting for 3 billion units annually. Primary application of these sensors is for bandpass filters in mobile telephones and base stations (Drafts, 2000). The rapid market growth for acoustic wave sensors also includes automotive applications in torque and tire pressure sensors, medical applications in chemical and biological sensors, and industrial and commercial applications in sensing vapor, humidity, temperature, and mass. Acoustic wave devices are also used to actuate fluid flow in microfluidic systems.

Strictly speaking, the term *acoustic wave* is somewhat misleading because the waves that are produced in the sensor are not related to sound, as the word *acoustic* implies. Rather, the waves generated in the sensors are more related to *stresses* or *deformation* of the media in which the waves propagate. Whatever wave forms are involved in the sensor, we will use the term acoustic wave in subsequent descriptions.

The working principle of acoustic wave sensors involves the detection of the alterations of propagating acoustic waves through a medium that encounter change of properties such as mass and viscoelasticity (*viscoelasticity* is the change of elastic properties of a material with time) induced by the presence of foreign substances, or the measurands, such as humidity or chemicals, or change of geometry and dimensions of the transmission media.

Actuation energy is required to generate acoustic waves in the sensor. It is provided by two principal mechanisms: piezoelectric and magnetostrictive. The former mechanism is a more popular method for generating acoustic waves. Using piezoelectric crystals, such as quartz, is a common means for transducing mechanical to electrical energies and vice versa. We will learn more about piezoelectric crystals in Section 2.3.3 and Chapter 7.

Figure 2.1 illustrates the working principle of a common surface acoustic wave (SAW) sensor using piezoelectric crystal as the transmission medium. Two sets of electrodes made of thin electric conducting films such as aluminum are deposited on the surface of a thick piezoelectric film, which in turn is deposited on a substrate made of a material such as alumina. One of the two electrode sets is to generate the actuating waves and the other set is used to detect wave properties, such as frequency and amplitude. These sets of electrodes consist of conducting arms connected to a number of closely spaced "fingers." For the set that is used to generate the waves, the gaps between these fingers

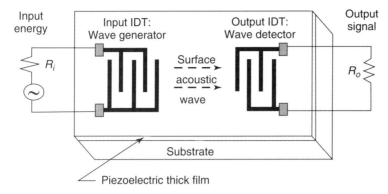

Figure 2.1. Typical SAW sensor.

will close or open as a result of the induced electrostatic forces upon the application of an electric field from an alternating current (AC) power source. Since the electrode set is attached to the surface of the piezoelectric film, any change in the gap between the fingers results in periodic stretching and squeezing of the surface of the piezoelectric thick film to which they are attached. Consequently, a wave of dimensional change is generated and it propagates through the surface of the piezoelectric film. The mechanical energy associated with this wave of dimensional change of the thick piezoelectric film is transmitted to the location where the other set of electrodes is situated, as shown in the figure. This electrode set, acting as the output interdigital transducer (IDT), detects and records the frequency and amplitude of the wave motion that has transmitted through the piezoelectric film. Any change in the characteristics of the transmitted wave motion due to the presence of foreign substance, that is, the measurand, can be measured and related to the nature and quantity of the measurands.

2.2.2 Biomedical and Biosensors

The biomedical industry has become a major user of microsystems after the automotive industry in recent years. The term *BioMEMS* has been used extensively by the MEMS industry and academe. It encompasses (1) biosensors, (2) bioinstruments and surgery tools, and (3) systems for biotesting and analysis for quick, accurate, and low-cost testing of biological substances. BioMEMS presents a great challenge to engineers as the design and manufacture of this type of sensor and instrument require knowledge and experience in molecular biology and physical chemistry as well as engineering.

Major technical issues involved in the application of MEMS in biomedicine are as follows:

1. Functionality for biomedical operations
2. Adaptivity to existing instruments and equipment
3. Compatibility with biological systems of patients

4. Controllability, mobility, and easy navigation for operations such as those required in a laparoscopy surgery

5. Fabrication for MEMS structures with high aspect ratios, which is defined as the ratio of the dimensions in the depth of the structure to the dimensions of the surface

Microsensors constitute the most fundamental element in any bioMEMS product. There are generally two types of sensors used in biomedicine: (a) *biomedical sensors* and (b) *biosensors*. Biomedical sensors are used to detect biological substances, whereas biosensors may be broadly defined as any measuring devices that contain a biological element (Buerk, 1993). These sensors usually involve biological molecules such as antibodies or enzymes, which interact with analytes that are to be detected.

Biomedical Sensors. Biomedical sensors can be classified as biomedical instruments used to measure biological substances as well as for medical diagnosis purposes. These sensors can analyze biological samples in quick and accurate ways. These miniaturized biomedical sensors have many advantages over the traditional instruments. They require typically a minute amount of sample and can perform analyses much faster with virtually no dead volume. There are many different types of biomedical sensors in the marketplace.

Electrochemical sensors work on the principle that certain biological substance, such as glucose in human blood, can release certain elements by chemical reactions. These elements can alter the electricity flow pattern in the sensor, which can be readily detected. An example of such a sensor is illustrated below (Kovacs, 1998).

In Figure 2.2, a small amount of blood sample is introduced to a sensor with a polyvinyl alcohol solution. Two electrodes are present in the sensor: a platinum film electrode and a thin Ag–AgCl film as the reference electrode. The following chemical reaction takes place between the glucose in the blood sample and the oxygen in the polyvinyl alcohol solution:

$$\text{Glucose} + O_2 \rightarrow \text{gluconolactone} + H_2O_2$$

The H_2O_2 produced by the above chemical reaction is electrolyzed by applying a potential to the platinum electrode, with the production of positive hydrogen ions, which will flow toward this electrode. The amount of glucose concentration in the blood sample can thus be measured by measuring the current flow between the electrodes.

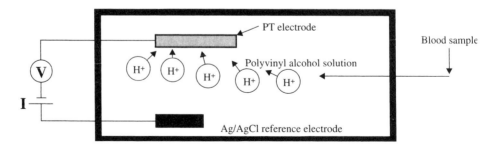

Figure 2.2. Biomedical sensor for measuring glucose concentration.

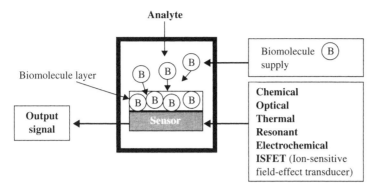

Figure 2.3. Schematics of biosensors.

Biosensors. Biosensors work on the principle of the interaction of the analytes that need to be detected and the biologically derived biomolecules, such as enzyme of certain forms, antibody, and other forms of proteins. These biomolecules, when attached to the sensing elements, can alter the output signals of the sensor when they interact with the analytes. Figure 2.3 illustrates how these sensors function. Proper selection of biomolecules for sensing elements (e.g., chemical, optical, indicated at the right boxes in the figure) can be used for the detection of specific analytes. In-depth description of biosensors is available in a book by Buerk (1993).

Biotesting and Analytical Systems. These systems separate various species in biological samples and are commonly used in DNA sequencing. Analytes include various biological substances and the human genome. Operation of these systems involves the passage of minute samples on the order of nanoliters in capillary tubes or microchannels driven by electrohydrodynamic means such as *electro-osmosis* and *electrophoresis*. Electrohydrodynamics involves the driving of ionized fluid with the application of electric fields. This form of flow will be described in Chapter 3. Isolation and separation of species in capillary tubes or microchannels are achieved by the inherent difference of the electro-osmotic mobility of these species. Optical means such as fluorescence detection are used to identify these species after the separation. Often, these minute electrohydrodynamic systems are built with microelectronics circuits for signal transduction, conditioning, and processing. Hundreds and thousands of these capillary tubes can be constructed on a single chip for parallel testing. Microbioanalytical systems typically require minute amount of samples, yet they offer accurate and nearly instant results.

A simple analyte system used in biotesting and analysis uses a capillary electrophoresis (CE) network, as illustrated in Figure 2.4. The system consists of two capillary tubes of diameters on the order of 30 μm or microchannels of similar sizes. The shorter channel is connected to the sample injection reservoir A and the analyte waste reservoir A′, whereas the longer channel is connected to the buffer solvent reservoirs B and B′. A biological sample consisting of species S_1, S_2, S_3, . . . with distinct electro-osmotic mobilities is injected into reservoir A. Application of an electric field between the terminals at reservoir A and A′ prompts the flow of an injected sample flow from A to A′. A congregation

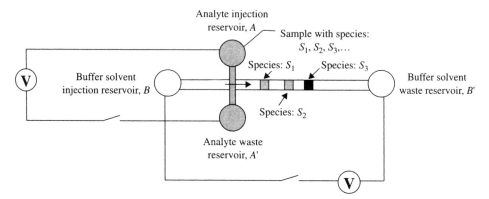

Figure 2.4. Schematic of CE system.

of the sample forms at the intersection of the two channels because of higher resistance to the flow at that location. A high-voltage electric field is then switched to terminals B and B′. This electric field can drive the congregated sample in the buffer solvent to flow from reservoir B to B′. The species in the sample can separate in this portion of the flow because of its inherent difference in electro-osmotic mobility.

The above description illustrates the working principle of CE analytical systems. There are obviously other forms of CE systems, as well as other means of biotesting and analytic devices, as described in Kovacs (1998).

2.2.3 Chemical Sensors

Chemical sensors are used to sense particular chemical compounds, such as various gas species. The working principle of this type of sensor is very simple. As we may observe from our day-to-day lives, many materials are sensitive to chemical attacks. For example, most metals are vulnerable to oxidation when exposed to air for a long time. A significant oxide layer built up over the metal surface can change material properties, such as the electrical resistance of the metal. This natural phenomenon illustrates the principle on which many microchemical sensors are designed and developed. One may assert that oxygen gas can, in principle, be sensed by measuring the change of electrical resistance in a metallic material as a result of the chemical reaction of oxidation. The detection of the presence of oxygen by a chemical sensor, of course, needs to be much more rapid than that of the natural oxidation of a metal, and the physical sizes of the samples are on the microscale.

Material sensitivity to specific chemicals is used as the basic principle for many chemical sensors. Following are four typical cases worth mentioning (Kovacs, 1998):

1. **Chemiresistor Sensors in Figure 2.5:** Organic polymers are used with embedded metal inserts. These polymers can cause changes in the electric conductivity of a metal when it is exposed to certain gases. For example, a special polymer called phthalocyanine is used with copper to sense ammonia (NH_3) and nitrogen dioxide (NO_2) gases.

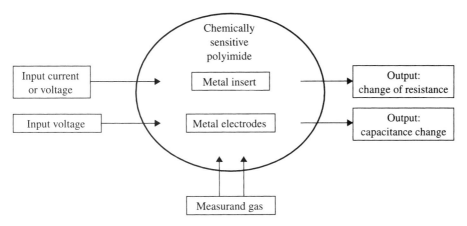

Figure 2.5. Working principle of chemical sensors.

2. **Chemicapacitor Sensors also in Figure 2.5:** Some polymers can be used as the dielectric material in a capacitor. The exposure of these polymers to certain gases can alter the dielectric constant of the material, which in turn changes the capacitance between the metal electrodes. An example is to use polyphenylacetylene (PPA) to sense gas species such as CO, CO_2, N_2, and CH_4.

3. **Chemimechanical Sensors:** Certain materials (e.g., polymers) change shape when they are exposed to chemicals (including moisture). One may detect such chemicals by measuring the change in the dimensions of the material. An example of such a sensor is the moisture sensors using pyraline (PI-2722).

4. **Metal Oxide Gas Sensors:** This type of sensor works on a principle similar to chemiresistor sensors. Several semiconducting metals, such as SnO_2, change their electric resistance after absorbing certain gases. The process is faster when heat is applied to enhance the reactivity of the measurand gases and the transduction semiconducting metals. Figure 2.6 illustrates a microsensor based on the semiconducting material SnO_2 (Kovacs, 1998).

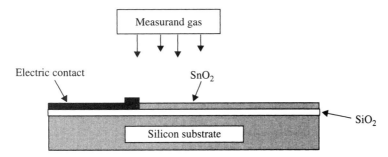

Figure 2.6. Typical metal oxide gas sensor. [After Kovacs (1998).]

TABLE 2.1. Available Metal Oxide Gas Sensors

Semiconducting Metals	Catalyst Additives	Gas to be Detected
$BaTiO_3/CuO$	La_2O_3, $CaCO_3$	CO_2
SnO_2	$Pt + Sb$	CO
SnO_2	Pt	Alcohols
SnO_2	Sb_2O_3	H_2, O_2, H_2S
SnO_2	CuO	H_2S
ZnO	V, Mo	Halogenated hydrocarbons
WO_3	Pt	NH_3
Fe_2O_3	Ti doped + Au	CO
Ga_2O_3	Au	CO
MoO_3	None	NO_2, CO
In_2O_3	None	O_3

Source: Kovacs, 1998.

Better results are obtainable if metallic catalysts are deposited on the surface of the sensor. Such deposition can speed up the reactions and hence increase the sensitivity of the sensor. Table 2.1 provided a list of metal oxide sensors available for sensing various gases (Kovacs, 1998).

2.2.4 Optical Sensors

The principles of interactions between the photons in light and the electronics in solids that receive the light have been well developed in quantum physics. Devices that can convert optical signals into electronic output have been developed and utilized in many consumer products, such as television. Micro-optical sensors have been developed to sense the intensity of lights. Semiconductive materials such as *p*- or *n*-crystalline silicon that provide strong photon–electron interactions are used as the sensing materials. The *p*-crystalline silicon contains a deficit of one electron in its atoms from the natural state, whereas the *n*-crystalline silicon has a surplus of one electron in its atoms from its natural state. Doping processes are used to produce *p*- or *n*-crystalline silicon, as will be described in Section 3.5. Figure 2.7 illustrates the four fundamental optical sensing devices (Kovacs, 1998).

The photovoltaic junction in Figure 2.7*a* can produce an electric potential when the more transparent substrate of semiconductor A is subjected to incident photon energy. The produced voltage can be measured from the change of electrical resistance in the circuit by an electrical bridge circuit. Figure 2.7*b* illustrates a special material that changes its electrical resistance when it is exposed to light. The photodiodes in Figure 2.7*c* are made of *p*- and *n*-doped semiconductor layers. The phototransistors in Figure 2.7*d* are made up of *p*-, *n*-, and *p*-doped layers. As illustrated in these figures, incident photon energy can

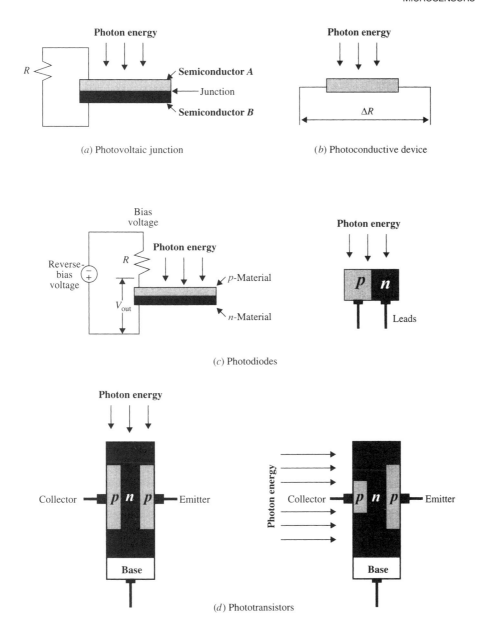

(a) Photovoltaic junction

(b) Photoconductive device

(c) Photodiodes

(d) Phototransistors

Figure 2.7. Optical sensing devices. [After Kovacs (1998).]

be converted into electric current output from these devices. All the devices illustrated in Figure 2.7 can be miniaturized in size and have extremely short response time in generating electrical signals. They are excellent candidates as micro-optical sensors.

Conversion of energies from the incident photons to that of electrons in the materials will be presented in the discussion of quantum physics in Chapter 12.

Selection of materials for optical sensors is principally based on the *quantum efficiency*, which is the material's ability to generate electron–hole pairs (electron output) with input photons. Semiconducting materials such as silicon (Si) and gallium arsenide (GaAs) are common materials used for optical sensors. Gallium arsenide has superior quantum efficiency and thus higher gains in the output but is more costly to produce. Alkali metals such as lithium (Li), sodium (Na), potassium (K), and rubidium (Rb) are also used for this type of sensor. The most commonly used alkali metal, however, is cesium (Cs).

The photovoltaic effect created by layers of *p*- and *n*-crystalline silicon wafers, similar to the arrangement illustrated in Figure 2.7*c*, has been used in the construction of *solar photovoltaic cells*. These cells are typically light in weight with better energy conversion efficiency than other solar collecting methods. The cost for producing solar photovoltaic cells has drastically reduced in recent years to extend their early application in space program powering satellites and other extraterrestrial vehicles to supply electric power for individual homes and business. Solar photovoltaic modules with assemblage of solar voltaic cells are increasingly used to provide supplement power when they are plugged into the bulk electricity grid.

2.2.5 Pressure Sensors

As we learned in Chapter 1, micro–pressure sensors are widely used in the automotive and aerospace industries. Most of these sensors function on the principle of mechanical deformation and stresses of thin diaphragms induced by the measurand pressure. Mechanically induced diaphragm deformation and stresses are then converted into electrical signal output through several means of transduction.

There are generally two types of pressure sensors: absolute and gage pressure sensors. The absolute pressure sensor has an evacuated cavity on one side of the diaphragm. The measured pressure is the "absolute" value with vacuum as the reference pressure. In the gage pressure type, no evacuation is necessary. There are two different ways to apply pressure to the diaphragm. With back-side pressurization as illustrated in Figure 2.8*a*, there is no interference with the signal transducers, such as piezoresistors, that are normally implanted at the top surface of the diaphragm. The other way of pressurization, that is, front-side pressurization in Figure 2.8*b*, is used only under very special circumstances because of the interference of the pressurizing medium with the signal transducers. Signal transducers are rarely placed on the back surface of the diaphragm because of space limitation as well as awkward access for interconnects.

As shown in Figure 2.8, the sensing element is usually made of thin silicon die varying in size from a few micrometers to a few millimeters square. A cavity is created from one side of the die by means of a microfabrication process. The top surface of the cavity forms the thin diaphragm that deforms under the applied pressure from the measurand fluid. The thickness of the silicon diaphragm usually varies from a few micrometers to tens of micrometers. A constraint base made of ceramics or glass (Pyrex glass is a common material) supports the silicon dies. The deformation of the diaphragm by the applied pressure is transduced into electrical signals by various transduction techniques, as will be described later in this section. The assembly of the sensing elements, as shown

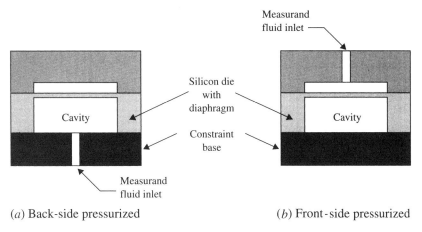

Figure 2.8. Cross sections of typical micro–pressure sensor.

in Figure 2.8, together with signal transduction element is then packaged into a robusting casing made of metal, ceramic, or plastic with proper passivation of the die.

Figure 2.9 schematically illustrates a packaged pressure sensor. The top view of the silicon die shows four piezoresistors (R_1, R_2, R_3, and R_4) implanted beneath the surface of the silicon die. These piezoresistors convert the stresses induced in the silicon diaphragm by the applied pressure into the change of electrical resistance, which is then converted into voltage output using a Wheatstone bridge circuit, as shown in the figure. The piezoresistors are essentially miniaturized semiconductor strain gages, which can produce the change of electrical resistance induced by mechanical stresses. In the case illustrated in Figure 2.9, the resistors R_1 and R_3 are subjected to a stress field by the applied pressure. Such a stress field causes an increase of electrical resistance in these resistors, whereas the resistors R_2 and R_4 experience the opposite of resistance changes because of their orientations. These changes of resistance as induced by the applied measurand pressure are measured from the Wheatstone bridge in the dynamic deflection operation mode as

$$V_o = V_{\text{in}} \left(\frac{R_1}{R_1 + R_4} - \frac{R_3}{R_2 + R_3} \right) \tag{2.1}$$

where V_o and V_{in} are respectively measured voltage and supplied voltage to the Wheatstone bridge.

We will learn the characteristics of piezoresistive materials in Chapter 7. Thin wire bonds are used to transmit the voltage change from the piezoresistor bridge through the two metal pads and the interconnect pins, as shown in the cross-sectional view of the sketch. A packaged pressure sensor of this type is shown on the left in Figure 1.2. We will deal with the packaging of these sensors in detail in Chapter 11. Micro–pressure sensors with piezoresistors have high gains and exhibit a good linear relationship between the in-plane stresses and the resistance change output. However, they have a major drawback: temperature sensitivity.

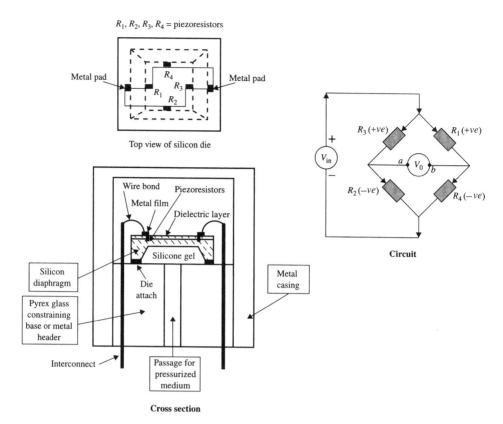

Figure 2.9. Assembly of a typical micro–pressure sensor.

Signal transduction in micro–pressure sensors can vary, depending on the sensitivity and accuracy required from the sensor. There are several other signal transduction methods developed for these sensors. We will illustrate two of them here.

Figure 2.10 illustrates a micro–pressure sensing unit utilizing capacitance changes for pressure measurements. Two electrodes made of thin metal films are attached to the bottom of the top cover and the top of the diaphragm. Any deformation of the diaphragm due to the applied pressure will narrow the gap between the two electrodes, which leads to a change of capacitance across the electrodes. This method has the advantage of being relatively independent of the operating temperature. The capacitance C in a parallel-plate capacitor can be related to the gap d between the plates by the expression

$$C = \varepsilon_r \varepsilon_0 \frac{A}{d} \tag{2.2}$$

where ε_r is the relative permittivity of the dielectric medium and ε_0 is the permittivity of free space (vacuum); $\varepsilon_0 = 8.85$ pF/m (pF = picofarad = 10^{-12} farad)

The relative permittivity for several substances is given in Table 2.2.

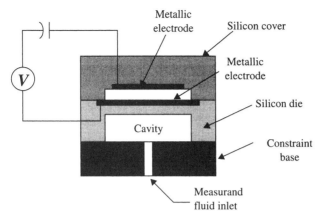

Figure 2.10. Micro–pressure sensors using capacitance signal transduction.

TABLE 2.2. Relative Permittivity ε_r of Selected Dielectric Materials

Material	Relative Permittivity
Air	1.0
Paper	2–3.5
Porcelain	6–7
Mica	3–7
Transformer oil	4.5
Water	80
Silicon	12
Pyrex	4.7

Example 2.1. Determine the capacitance of a parallel-plate capacitor. The two plates have identical dimensions of $L = W = 1000$ μm with a gap $d = 2$ μm. Air is the dielectric medium between the two plates.

Solution: From Table 2.2, we have the relative permittivity $\varepsilon_r = 1.0$. From the value $\varepsilon_0 = 8.85$ pF/m and Equation (2.2), the capacitance can be calculated to be 4.43 pF.

Capacitors are common transducers as well as actuators in microsystems. The capacitance variation in a capacitor can be measured by simple circuits, such as that illustrated in Figure 2.11 (Bradley et al., 1991).

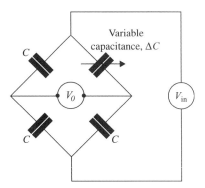

Figure 2.11. Typical bridge for capacitance measurements.

The electrical bridge in Figure 2.11 is similar to the Wheatstone bridge in Figure 2.9 for resistance measurements. The variable capacitance can be measured by measuring the output voltage V_o. The following relationship can be used to determine the variable capacitance in the circuit:

$$V_o = \frac{\Delta C}{2(2C + \Delta C)} V_{in} \tag{2.3}$$

where ΔC is the capacitance change in the capacitor in the micro–pressure sensor and C is the capacitance of the other capacitors in the bridge. The bridge is subjected to a constant supply voltage V_{in}.

Figure 2.12 illustrates a MAP sensor used in an automobile (Chiou, 1999). This sensor uses a capacitor as signal transducer. The detailed arrangement of the silicon die in the pressure chamber is similar to that shown in Figure 1.22c.

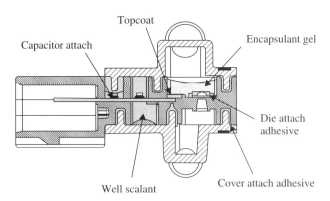

Figure 2.12. Manifold absolute pressure sensor with capacitor transducer. (Courtesy of Motorola Corporation.)

Micro–pressure sensors with capacitance signal transduction are not nearly as sensitive to the applied pressure as those with piezoresistors. However, capacitance transducers are not nearly as sensitive to operating temperatures as with piezoresistors. On the other hand, one should be aware of a major shortcoming of capacitance transducers: The voltage output from a capacitance bridge circuit is nonlinear to the change of capacitance due to pressure variation. This nonlinear I/O relationship is indicated in Equation (2.3). The following example will illustrate such a nonlinear relationship.

Example 2.2. Determine the voltage output of a capacitance bridge circuit as illustrated in Figure 2.11 with variation of the gap between two flat-plate electrodes as described in Example 2.1.

Solution: We may use Equations (2.2) and (2.3) to establish the following table for voltage output:

Gap d in electrodes (μm)	Capacitance C (pF)	Change of Capacitance, ΔC (pF)	Voltage ratio V_o/V_{in}
2.00	4.425	0	0
1.75	5.057	0.632	0.029
1.50	5.900	1.475	0.056
1.00	8.850	4.425	0.100
0.75	11.800	7.375	0.119
0.50	17.700	13.275	0.136

Figure 2.13 shows the nonlinear relationship between the change of gap d and the corresponding change of capacitance.

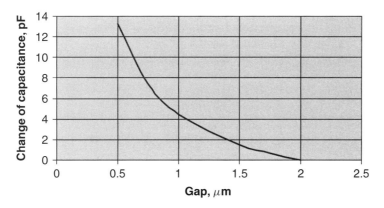

Figure 2.13. Nonlinear output of a pressure sensor using capacitance signal transduction.

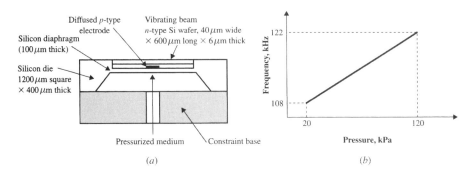

Figure 2.14. Pressure sensors using vibrating beam signal transducer: (*a*) pressure-sensing element; (*b*) reported sensitivity.

Figure 2.14*a* illustrates the construction of a micro–pressure sensor using a vibrating beam for signal transduction (Petersen et al., 1991). A thin *n*-type silicon beam is installed across a shallow cavity at the top surface of the silicon die. A *p*-type electrode is diffused at the surface of that cavity under the beam. The *p*- and *n*-type silicon layers are doped with boron and phosphorus, respectively, as will be described in detail in Chapter 3. Both *p*- and *n*-type silicon are electrically conductive. The beam is made to vibrate at its resonant frequency by applying an AC signal to the diffused electrode in the beam before application of the pressure to the diaphragm. The stress induced in the diaphragm (and the die) will be transmitted to the vibrating beam. The induced stress along the beam causes the shift of the resonant frequency of the beam. The shift of the resonant frequency of the beam can be correlated to the induced stress and thus to the pressure applied to the silicon diaphragm. Formulas for calculating the corresponding resonant frequency shift in a vibrating beam will be presented in Chapter 4.

This type of signal transduction is insensitive to temperature and provides excellent linear output signals, as shown in Figure 2.14*b*. The disadvantage of such sensors is the cost to fabricate them.

2.2.6 Thermal Sensors

Thermocouples are the most common transducer used to sense heat. They operate on the principle of the *electromotive force* (emf) produced at the open ends of two dissimilar metallic wires when the junction of the wires (called the bead) is heated (see Figure 2.15*a*). The temperature rise at the junction due to heating can be correlated to the magnitude of the produced emf, or voltage. These wires and the junction can be made very small in size. By introducing an additional junction in the thermocouple circuit, as shown in Figure 2.15*b*, and exposing that junction to a different temperature than the other, one would induce a temperature gradient in the circuit itself. This arrangement of thermocouple with both hot and cold junctions can produce the *Seebeck effect*, discovered by T. J. Seebeck in 1821. The voltage generated by the thermocouple can be evaluated as $V = \beta \Delta T$, where β is the Seebeck coefficient and ΔT is the temperature difference between the hot and cold junctions. In practice, the cold-junction temperature is maintained constant (e.g., at $0°C$) by dipping that junction in ice water. The coefficient β

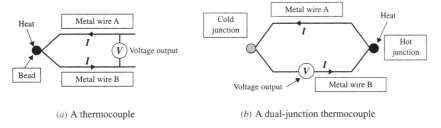

(*a*) A thermocouple   (*b*) A dual-junction thermocouple

Figure 2.15. Schematics of thermocouples.

depends on the thermocouple wire material and the range of temperature measurements. Seebeck coefficients for common thermocouples are given in Table 2.3.

One serious drawback of thermocouples for microthermal transducers is that the output of thermocouples decreases as the sizes of the wires and the beads are reduced. Thermocouples alone are thus not ideal for microthermal sensors.

A microthermopile is a more realistic solution for miniaturized heat sensing. Thermopiles operate with both hot and cold junctions, but they are arranged with thermocouples in parallel and voltage output in series. This arrangement is illustrated in Figure 2.16. Materials for thermopile wires are the same as those used in thermocouples, for instance, copper/constantan (type T) and chromel/alumel (type K), as shown in Table 2.3.

The voltage output from a thermopile can be obtained by the expression

$$\Delta V = N\beta \Delta T \tag{2.4}$$

where

N = number thermocouple pairs in thermopile

β = thermoelectric power (or Seebeck coefficient) of two thermocouple materials, V/K (from Table 2.3)

ΔT = temperature difference across thermocouples, K

TABLE 2.3. Seebeck Coefficients for Common Thermocouples

Type	Wire Material	Seebeck Coefficient (μV/°C)	Temperature Range (°C)	Voltage Range (mV)
E	Chromel/constantan	58.70 at 0°C	−270–1000	−9.84–76.36
J	Iron/constantan	50.37 at 0°C	−210–1200	−8.10–69.54
K	Chromel/alumel	39.48 at 0°C	−270–1372	−6.55–54.87
R	Platinum (10%)–Rh/Pt	10.19 at 600°C	−50–1768	−0.24–18.70
T	Copper/constantan	38.74 at 0°C	−270–400	−6.26–20.87
S	Pt (13%)–Rh/Pt	11.35 at 600°C	−50–1768	−0.23–21.11

Source: Kreith, 1998.

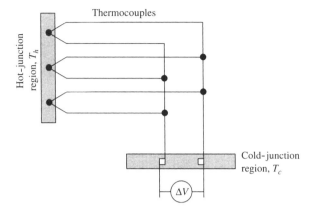

Figure 2.16. Schematic of a thermopile.

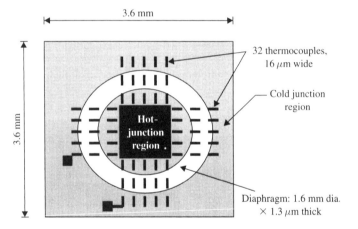

Top view

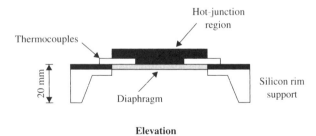

Elevation

Figure 2.17. Schematic of a microthermopile.

Choi and Wise (1986) produced a microthermopile, as graphically represented in Figure 2.17. A total of 32 polysilicon–gold thermocouples were used in the thermopile. The overall dimension of the silicon chip on which the thermopile was built is 3.6 mm × 3.6 mm × 20 μm thick. Typical output signal of 100 mV was obtained from a 500 K blackbody radiation source $Q_{in} = 0.29$ mW/cm^2 with a response time at about 50 ms.

A pressure sensor similar to that shown in Figure 2.10 could be used to detect the temperature of a working medium. In such cases, the deformation of the silicon diaphragm is induced by thermal sources instead of pressures as in pressure sensors. The thermal forces that cause the deflection of the diaphragms can be related to the capacitance of the deflected electrodes. A more sensitive means, however, is to use a thermal pile, or a thermocouple, as the sensing element.

2.3 MICROACTUATION

Webster's Dictionary (Mish, 1995) defines *actuator* as "a mechanical device for moving or controlling something." The actuator is a very important part of a microsystem that involves motion. Four principal means are commonly used for actuating motions of microdevices: (1) thermal forces, (2) shape memory alloys, (3) piezoelectric crystals, and (4) electrostatic forces. Electromagnetic actuation is widely used in devices and machines of macroscales. However, it is rarely used in microdevices because of unfavorable miniaturization scaling laws, as will be described in Chapter 6. Here, we will briefly describe the working principles of these four actuation methods.

An actuator is designed to deliver a desired motion when driven by a power source. Actuators can be as simple as an electrical relay switch or as complex as an inkjet printer head. The driving power for actuators varies depending on the specific applications. An on–off switch in an electric circuit can be activated by the deflection of a bimetallic strip as a result of resistance heating the strip with passing electric current. On the other hand, most electrical actuators, such as motors and solenoid devices, are driven by electromagnetic induction governed by Faraday's law.

For devices at micro- or mesoscales, there is little room for the electrical conducting coils required for electromagnetic induction. Consequently, other kinds of driving power sources have to be developed. We will introduce three commonly used sources for microactuation.

2.3.1 Actuation Using Thermal Forces

Bimetallic strips are actuators based on thermal forces. These strips are made by bonding two materials with distinct thermal expansion coefficients. The strip will bend when heated or cooled from the initial reference temperature due to incompatible coefficients of thermal expansions of the materials that are bonded together. It will return to its initial reference shape once the applied thermal force is removed. The same principle has been used to produce several microactuators, such as microclamps or valves. In these cases, one of the strips is used as a resistance heater. The other strip could be made from common microstructural material such as silicon or polysilicon (Riethmuller et al., 1987).

Figure 2.18. Thermal actuation of a dissimilar material.

The behavior of thermally actuated bimetallic strips is illustrated in Figure 2.18. The two constituent materials have coefficients of thermal expansion α_1 and α_2, respectively, with $\alpha_1 > \alpha_2$. The beam made of the bimetallic strips will deform from its original straight shape to a bent shape, shown on the right in the figure, when it is heated by external sources. The beam is expected to return to its original shape after the removal of the heat.

2.3.2   Actuation Using Shape Memory Alloys

Microactuation can be produced more accurately and effectively by using shape memory alloys (SMAs) such as Nitinolor, or TiNi alloys. These alloys tend to return to their original shape at a preset temperature. The working principle of a microactuator using SMAs, is illustrated in Figure 2.19. An SMA strip originally in a bent shape at a designed preset temperature T is attached to a silicon cantilever beam. The beam is set straight at room temperature. However, heating the beam with the attached SMA strip to the temperature T would prompt the strip's "memory" to return to its original bent shape. The deformation of the SMA strip causes the attached silicon beam to deform with the strip, and microactuation of the beam is thus achieved. This type of actuation has been used extensively in producing microrotary actuators, microjoints and robots, and microsprings (Gabriel et al., 1988).

2.3.3   Actuation Using Piezoelectric Effect

Certain crystals, such as quartz, that exist in nature deform with the application of an electrical voltage. The reverse is also valid; that is, an electric voltage can be generated

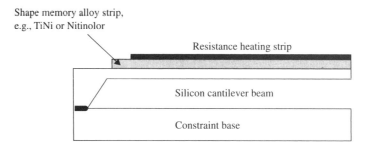

Figure 2.19. Microactuation using SMAs.

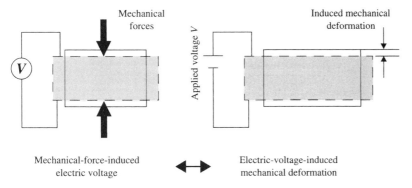

Figure 2.20. Piezoelectric effect.

across the crystal when an applied force deforms the crystal. This phenomenon, called the piezoelectric effect, is illustrated in Figure 2.20.

Piezoelectricity was discovered by brothers Pierre and Paul-Jacque Curie in 1880. Pierre Curie was the famous physicist, Marie Curie's husband. The term *piezoelectricity* was coined in 1881 by Wilhelm Hankel. Piezoelectricity is a principal means of microactuation in many MEMS and microsystems.

Let us refer to Figure 2.21 to illustrate the working principle of a microactuator by piezoelectricity. In the arrangement is a flexible silicon beam attached to a thin film of piezoelectric strip sandwiched between two electrically conductive electrodes. An applied voltage across the piezoelectric crystal prompts a deformation of the crystal, which can in turn bend the attached silicon cantilever beam. Piezoelectric actuation is used in a micropositioning mechanism and microclamp reported in Higuchi et al. (1990)

Piezoelectric crystals are essential materials for microactuators. More information on the crystal structures and mathematical formulation for the determination of electromechanical characteristics of common piezoelectric crystals will be presented in Chapter 7.

2.3.4 Actuation Using Electrostatic Forces

Electrostatic forces are used as the driving forces for many actuators. Accurate assessment of electrostatic forces is an essential part of the design of many micromotors and actuators.

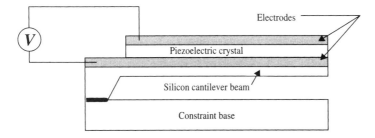

Figure 2.21. Actuator using piezoelectric crystal.

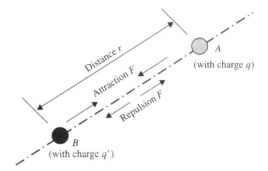

Figure 2.22. Two particles in an electric field.

A review of the fundamentals of electrostatics is thus necessary to use this important power source effectively in the design of microactuators.

Coulomb's Law. Electrostatic force F is defined as the electrical force of repulsion or attraction induced by an electric field E. As we have learned from physics, an electric field E exists in a field carrying positive and negative charges. Charles Augustin Coulomb (1736–1806) discovered this phenomenon and postulated the mathematical formulation for determining the strength of the force F between two charged particles.

With reference to Figure 2.22, where two charged particles A and B are in an electric field, the induced electrostatic force, according to Coulomb, can be expressed as

$$F = \frac{1}{4\pi \varepsilon_0} \frac{qq'}{r^2} \tag{2.5}$$

where $\varepsilon_0 = 8.85 \times 10^{-12}\, C^2/N\text{-}m^2$ in free space in vacuum (this is equivalent to 8.85 pF/m used in a capacitor). The symbol r in Equation (2.5) is the distance vector between the two charged particles in the field.

The electrostatic force F is repulsive if both charges q and q' carry positive or negative charges. It becomes attractive if the two charges have opposite signs.

Electrostatic Forces in Parallel Plates. Figure 2.23 represents two charged plates separated by a dielectric material (i.e., an electric insulating material) with a gap d. The plates become electrically charged when an emf, or voltage, is applied to the plates. This action will induce capacitance in these charged plates, which can be expressed as

$$C = \varepsilon_r \varepsilon_0 \frac{A}{d} = \varepsilon_r \varepsilon_0 \frac{WL}{d} \tag{2.6}$$

where A is the area of the plates and ε_r is the relative permittivity. The relative permittivity ε_r in Equation (2.6) for common dielectric materials is presented in Table 2.2.

One may imagine that the difference of electric charges between the top and bottom plates in Figure 2.23 can be maintained as long as a voltage is applied to the system. However, the charges that are stored in either plate can be discharged instantly by short

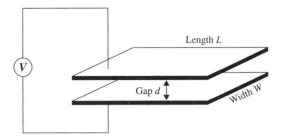

Figure 2.23. Electric potential in two parallel plates.

circuiting the plates with a conductor. One may thus realize that an electric potential does exist in the situation, as illustrated in Figure 2.23. The energy associated with this electric potential can be expressed as

$$U = -\frac{1}{2}CV^2 = -\frac{\varepsilon_r \varepsilon_0 WLV^2}{2d} \tag{2.7}$$

A negative sign is attached in Equation (2.7) because there is a loss of the potential energy with increasing applied voltage.

The associated electrostatic force that is normal to the plates (in the d direction) can thus be derived from the potential energy expressed in Equation (2.7) as

$$F_d = -\frac{\partial U}{\partial d} = -\frac{1}{2}\frac{\varepsilon_r \varepsilon_0 WLV^2}{d^2} \tag{2.8}$$

Example 2.3. A parallel capacitor is made of two square plates with dimension $L = W = 1000$ μm (or 1 mm), as shown in Figure 2.24. Determine the normal electrostatic force if the gap d between these two plates is 2 μm. The plate is separated by static air, which is a dielectric material, but allows the variation of the gap d with negligible resistance.

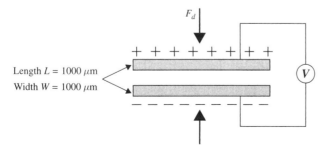

Figure 2.24. Normal electrostatic force in a parallel-plate capacitor.

Solution: The normal electrostatic force exerted on the plates can be calculated by using Equation (2.8) with $\varepsilon_r = 1.0$ for air as the dielectric material and $\varepsilon_0 = 8.85$ pF/m or 8.85×10^{-12} C^2/N-m^2:

$$F_d = -\frac{1.0 \times (8.85 \times 10^{-12})(1000 \times 10^{-6})(1000 \times 10^{-6})V^2}{2 \times (2 \times 10^{-6})^2}$$

$$= -1.106 \times 10^{-6}V^2 \qquad \text{Newtons (N)}$$

Thus, an 11-mN (millinewton) force is generated with 100V applied to the plates.

Equation (2.8) also can be used to derive expressions for electrostatic forces in the width W and length L directions. These forces are induced with partial alignment of the plates in the respective directions (Trimmer and Gabriel, 1987). The following general form of Equation (2.8) can be used to derive these forces:

$$F_i = -\frac{\partial U}{\partial x_i} \tag{2.9}$$

where the index i refers to the direction in which misalignment occurs, for example, the width direction W or the length direction L. The negative sign in Equation (2.9) indicates that any increase of the relative movement of the plate electrodes, that is, in the x direction, results in reduction in the overlap area, which leads to the reduction of the potential function U.

Thus, by referring to the designation of forces indicated in Figure 2.25, we can write expressions for the two forces in the two directions:

$$F_W = \frac{1}{2}\frac{\varepsilon_r \varepsilon_0 L V^2}{d} \quad \text{in the width direction,} \tag{2.10}$$

and

$$F_L = \frac{1}{2}\frac{\varepsilon_r \varepsilon_0 W V^2}{d}\text{in the length direction} \tag{2.11}$$

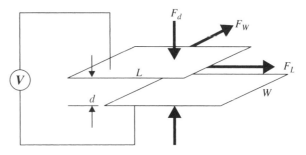

Figure 2.25. Electrostatic forces on parallel plates.

The normal force F_d in Equation (2.8) can be derived by the same partial derivative given in Equation (2.9). The in-plane electrostatic forces F_W and F_L given in Equations (2.10) and (2.11) exist only in the direction of misalignment. Consequently, F_W exists if the plate electrodes are misaligned in the width direction. One may also notice from these expressions that F_W is independent of the width dimension W. Likewise, F_L is independent of the length L.

These electrostatic forces are the prime driving forces of micromotors, as will be demonstrated for microdevices in Sections 2.4–2.6. They are also used to drive what are called comb drivers in microgrippers using DC power sources or in resonators using AC power sources. One drawback of electrostatic actuation is that the force that is generated by this method usually is low in magnitude. Its application is thus primarily limited to actuators for optical switches, such as shown in Figure 1.12, and below for microgrippers in Figure 2.27 and resonators in Figure 2.39.

2.4 MEMS WITH MICROACTUATORS

We will present a few microdevices that function on the principles of microactuation as described in the foregoing section.

2.4.1 Microgrippers

The electrostatic forces generated in parallel charged plates can be used as the driving forces for gripping objects, as illustrated in Figure 2.26. As the figure shows, the required gripping forces in a gripper can be provided either by normal forces (Figure 2.26a) or by the in-plane forces from pairs of misaligned plate electrodes (Figure 2.26b).

The arrangement that uses normal gripping forces from parallel plates, Figure 2.26a, appears to be simple in practice. A major disadvantage of this arrangement, however, is the excessive space that the electrodes occupy in a microgripper. Consequently, it is rarely used. The other arrangement, with multiple pairs of misaligned plates, is commonly

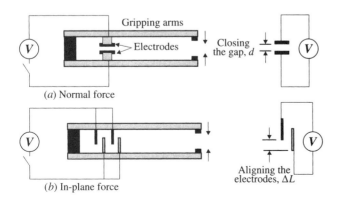

Figure 2.26. Gripping forces in a microgripper.

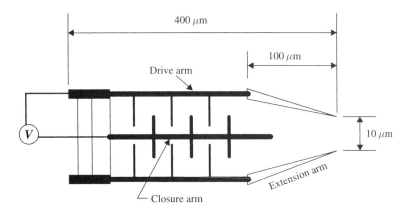

Figure 2.27. Schematic of a microgripper.

used in microdevices. This arrangement is frequently referred to as the *comb drive*. A comb drive is used to generate acoustic waves, as illustrated in Figure 2.1, and in the construction of the microgripper, as illustrated in Figure 2.27 (Kim et al., 1991). The gripping action at the tip of the gripper is initiated by applying a voltage across the plates attached to the drive arms and the closure arm. The electrostatic forces generated by these pairs of misaligned plates tend to align themselves, causing the drive arms to bend, which in turn closes the extension arm for gripping. These microgrippers can be adapted to micromanipulators or robots in micromanufacturing processes or microsurgery. The length of the gripper produced by Kim et al. was 400 μm. It had a tip opening of 10 μm.

> **Example 2.4.** For the comb-driven actuator in Figure 2.28, determine the voltage supply required to pull the moving electrode 10 μm from its outstretched position of the spring. The spring constant k is 0.05 N/m. The comb drive is operated in air. The gap d between the electrodes and the width W of the electrodes are 2 and 5 μm, respectively.
>
> *Solution:* The required traveling distance of the moving electrodes is $\delta = 10 \times 10^{-6}$ m, which is equivalent to a spring force $F = k\delta = 0.05$ N/m $\times 10 \times 10^{-6}$ m $= 0.5 \times 10^{-6}$ N.
>
> There are two sets of electrodes with four electrodes (a nickname of "fingers") in the system; each set (i.e., pair of fingers), needs to generate 0.5 F, or 0.25×10^{-6} N. By using Equation (2.11) with $\varepsilon_r = 1.0$, $\varepsilon_0 = 8.85 \times 10^{-12}$ C/N-m^2, $W = 5 \times 10^{-6}$ m, and $d = 2 \times 10^{-6}$ m, we obtain the following expression for the required voltage supply:
>
> $$0.25 \times 10^{-6} = \frac{1}{2} \frac{1 \times 8.85 \times 10^{-12} \times 5 \times 10^{-6}}{2 \times 10^{-6}} V^2$$
>
> which gives the required voltage $V = 150.33$ V.

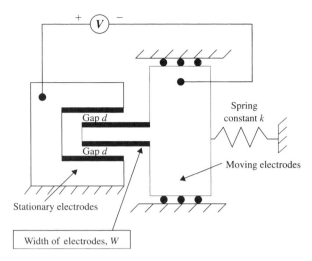

Figure 2.28. Schematic of a comb drive actuator.

The required voltage supply of 150.33 V is unusually high for microcomb drives. One effective way to reduce the supply voltage is to use multiple sets of electrodes. By following similar procedures presented in the solution of Example 2.4, we can show the relationship of the supply voltage reduces dramatically as the number of sets of electrodes increases, as demonstrated in Figure 2.29. It takes 113 sets of electrodes, or 114 fingers, to reduce the required voltage supplies from 150.33 to 20 V.

2.4.2 Miniature Microphones

A microphone is a sensor that senses sound. It appears in our daily lives through the sound we hear on radio, television, and telephone communications. Human effort in receiving sounds from distances can be traced back many hundreds of years. The first engineered microphone can be attributed to Alexander Graham Bell in his patented device in 1876 (Eargle, 2004). The technology of designing and building better microphones has never ceased advancing since then.

Miniaturization of microphones has become a major effort by many engineers and scientists due to the rapid growth of applications in health care, in particular, involving hearing aids, and information technology in miniature portable computers and mobile telecommunications. Miniature microphones at the micrometer scale are the essential components of these products. Often, the size and performance of microphones in products such as hearing aids and mobile telephones become the critical success factors in commercialization and marketing.

The demand to further miniaturize microphones in mobile telephones remains strong. Miniature microphones have the following principal applications:

- Mobile telephones (current global annual market at 200 million units)
- Personal digital assistants

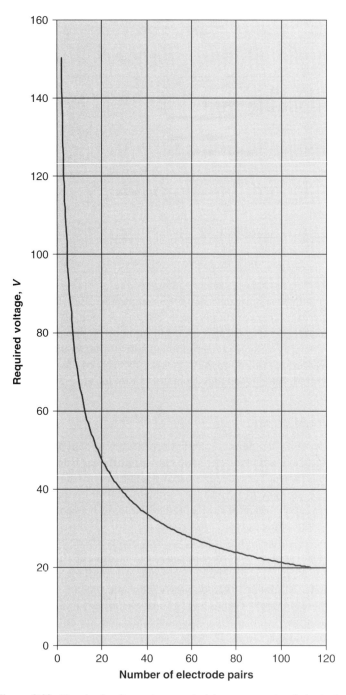

Figure 2.29. Required voltages to a comb drive versus sets of electrodes.

- Web-enabled phones (WAPs)
- Tablets and audio devices
- Advanced hearing aids (7 million units in 2002, as indicated in Table 1.4)
- Notebook computers
- Acoustic sensors
- Consumer electronics, toys
- Smart home appliances

The market for miniature microphones was 50 million units in 2004 and is projected to increase to 350 million units in 2008.

The typical packaged size of these microphones is about 1 mm × 1 mm × 0.5 mm thick. They are made using either polymers (Hsieh et al., 1997) or single-crystal silicon (Hsu et al., 1998).

Figure 2.30 illustrates the structure of the core elements of a typical silicon-based condenser-type microphone. Principal components involve a thin silicon diaphragm and a thin perforated back plate. The diaphragm and the back plate are normally made of electric conducting polycrystalline silicon so they may act as electrodes in a capacitor. Both these components are fixed at the edge of a housing unit, which provides a cavity beneath the back plate and a vent for pressure equalization.

The incoming sound from surrounding sources induces an air pressure on the thin diaphragm, resulting in a deflection of the diaphragm. This instantaneous deflection of the diaphragm in turn causes variation of the gap between the diaphragm and the back plate. Such variation of the gap between the diaphragm and the back plate can cause a change of the capacitance as an output signal, as indicated in Figure 2.30. Since the input air pressure is in the wave form with an inherent frequency, the mechanical response of the diaphragm is in the form of vibration with oscillatory amplitudes and frequencies.

Most microphones are designed for input sound levels in the range of 20–80 decibels (dB) at a frequency range of 150–1000 Hz. The noise level in an average office surrounding is 50 dB, whereas 80 dB is for a noisy office environment. Sound level in

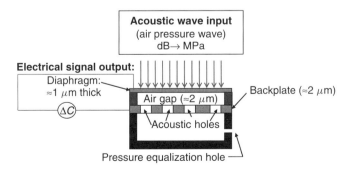

Figure 2.30. Core element of a miniature microphone.

decibels can be converted to air pressure using the expression (Eargle, 2004)

$$dB = 20 \ \log_{10} \left(\frac{P}{P_0} \right) \tag{2.12}$$

where P is the associated air pressure in pascals, P_0 is the reference pressure associated with a sound wave at a threshold of audibility (i.e., 0 dB) and at a frequency of 1000 Hz.

Despite the similarity in arrangement of core components between miniature microphones and pressure sensors (Figure 2.8), signal processing for miniature microphone is much more complicated than that in pressure sensors because of the required high signal–noise ratio for clarity in voice transmission (Pedersen et al., 1998). Consequently, many miniature microphones use piezoresistive materials for signal generators instead of capacitance (Arnold et al., 2001; Papila et al., 2003) because piezoresistive materials offer higher sensitivity and linear relationship between the electronic output signals and the incoming signals, as described in Section 2.2.5.

Miniature microphone arrays, rather than single microphones, are used in aerodynamic testing with wind tunnels (Arnold et al., 2003) and smart hearing aids (Chowdhury et al., 2002). These applications require even more sophisticated signal processing techniques, as offered in a reference book by Johnson and Dudgeon (1993).

2.4.3 Micromotors

Two types of micromotors are used in microsystems: linear motors and rotary motors.

The actuation forces for micromotors are primarily electrostatic forces. The sliding force generated in pairs of electrically energized misaligned plates, such as illustrated in Figure 2.25, produces the required relative motion in a linear motor. Figure 2.31 illustrates the working principle of the linear motion between two sets of parallel base plates which are separated by a noncompressible, frictionless dielectric material such as thin film made of quartz. Each of the two sets of base plates contains a number of electrodes made of electric conducting plates. All these electrodes have a length W. The bottom base plate has an electrode pitch w, whereas the top base plate has a slightly different pitch, say, $w + w/3$. The two sets of base plates are initially misaligned by $w/3$,

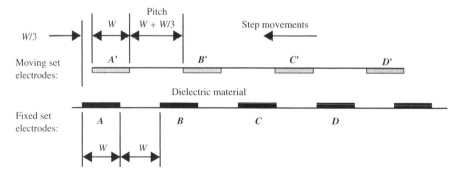

Figure 2.31. Working principles for electrostatic micromotors.

as shown in Figure 2.31. We set the bottom plates as stationary so the top plate can slide over the bottom plate in the horizontal plane. Thus, on energizing, the pair electrodes A and A′ can cause the motion of the top plate moving to the left until A and A′ are fully aligned. At that moment, the electrodes B and B′ are misaligned by the same amount $w/3$. One can energize the misaligned pair B–B′ and prompt the top plate to move by another $w/3$ distance toward the left. We may envisage that by then the C–C′ pair is misaligned by $w/3$ and subsequently energizing that pair would produce a similar motion of the top plate to the left by another distance $w/3$. The motion will be completed by yet another sequence of energizing the last pair D–D′. We may thus conceive that with carefully arranged electrodes in the top and bottom base plates with proper pitches, one can create the necessary electrostatic forces that are required to provide the relative motion between the two sets of base plates. It is readily seen that the smaller the preset misalignment of the electrode plates, the smoother the motion becomes. Rotary micromotors can be made to work following a similar principle.

Detailed design of these motors has been presented in an article by Trimmer (1997). A major problem in micromotor design and construction is the bearings for the rotors. Electric levitation principles have been used for this purpose (Kumar and Cho, 1991).

Micromotors built on the principles of electrostatics forces are described in detail by Fan et al. (1988), Mehregany et al. (1990), and Gabriel et al. (1988). Fabrication techniques used to produce these motors are discussed by Lober and Howe (1988).

Rotary motors driven by electrostatic forces can be constructed similar to linear motors, as shown in Figure 1.10. Figure 2.32 shows a top view of an electrostatically driven

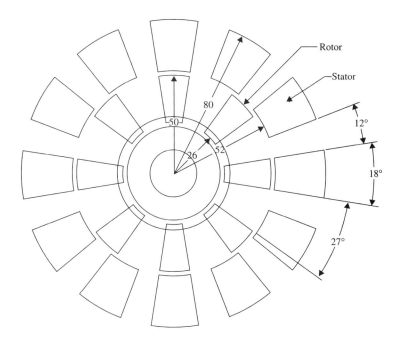

Figure 2.32. Schematic of a microrotary motor.

micromotor. As can be seen from the figure, electrodes are installed in the outer surface of the rotor poles and the inner surface of the stator poles. As in linear motors, pitches of electrodes in rotor poles and stator poles are mismatched in such a way that they will generate an electrostatic driving force by misalignment of the energized pairs of the electrodes between those on the rotor and the stator. The reader will notice that the ratio of poles in the stator to those in the rotor is 3 : 2. The air gap between rotor poles and stator poles can be as small as 2 μm. The outside diameter of the stator poles is in the neighborhood of 100 μm, whereas the length of the rotor poles is about 20–25 μm.

One serious problem that is encountered by engineers in the design and manufacture of microrotary motors is the wear and lubrication of the bearings. Typically these motors rotate at over 10,000 revolutions per minute (rpm). With such high rotational speed, the bearing quickly wears off, which results in wobbling of the rotors. Much effort is needed to solve this problem. Consequently, microtribology, which deals with friction, wear, and lubrication, has become a critical research area in microtechnology. Zum Gahr (1993) provided a comprehensive introduction of this critical subject of microtribology.

2.5 MICROACTUATORS WITH MECHANICAL INERTIA

Mechanical inertia force that equals the product of the acceleration (or deceleration) of a moving solid and its mass, as expressed by Newton's second law, is often used as the driving force in sensing the forces and torques induced by a motion or the positioning of moving components in a device. Here, we will describe the working principles of two important microsystems that are built on such principles: microaccelerometers that involve linear acceleration or deceleration and microgyroscopes that involve angular acceleration or deceleration.

2.5.1 Microaccelerometers

An accelerometer is an instrument that measures the acceleration (or deceleration) of a moving solid. Microaccelerometers are used to detect the associated dynamic forces to a mechanical system in motion. These accelerometers are widely used in automotive industry, as described in Chapter 1. For example, acceleration sensors in the $\pm 2g$ range are used in the suspension system and the antilock braking system (ABS), whereas $\pm 50g$ range acceleration sensors are used to actuate airbags for driver and passenger safety in the event of collision with other vehicles or obstacles. The notation g represents gravitational acceleration, with a numerical value of $32\,\text{ft/s}^2$, or $9.81\,\text{m/s}^2$. We present microaccelerometers in a separate section, because this type of device is often classified as an *inertia sensor*, yet it contains actuation elements.

Most accelerometers are built on the principles of mechanical vibration, as will be described in detail in Chapter 4. The principal components of an accelerometer are a *mass* supported by *springs*. The mass is often attached to a *dashpot* that provides the necessary damping effect. The spring and the dashpot are in turn attached to a casing, as illustrated in Figure 2.33.

In microaccelerometers, significantly different arrangements are necessary because of the very limited space available in microdevices. A minute silicon beam with an

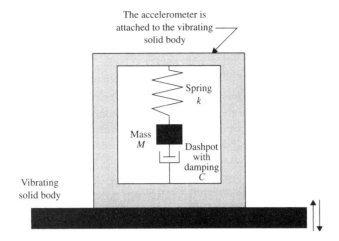

Figure 2.33. Typical arrangement of an accelerometer.

attached mass (often called *proof mass*) constitutes a spring–mass system, and the air in the surrounding space is used to produce the damping effect. The structure that supports the mass acts as the *spring*. A typical microaccelerometer is illustrated in Figure 2.34; in it, the mass is attached to a cantilever beam or plate, which is used as a "spring." A piezoresistor is implanted on the beam or plate to measure the deformation of the attached mass, from which the amplitudes and thus the acceleration of the vibrating mass can be correlated. A mathematical formula that relates the vibrating mass to the acceleration of the casing is available in Chapter 4. Since acceleration (or deceleration) is related to the driving dynamic force that causes the vibration of the solid body to which the casing is attached, accurate measurement of acceleration can thus enable engineers to measure the applied dynamic force. It is not surprising to find that microaccelerometers are widely used as a trigger to activate airbags in automobiles in an event of collision and to sense the excessive vibration of the chassis of a vehicle from its suspension systems.

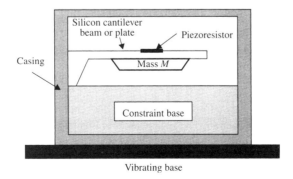

Figure 2.34. Schematic of a microaccelerometer.

There are many different types of accelerometers available commercially. Signal transducers used in microaccelerometers include piezoelectric, piezoresistive, capacitive, and resonant members (Madou, 1997). Principles of signal conversion of these schemes will be presented in Chapter 11.

A widely used microaccelerometer (or inertia sensor) in the marketplace is the integrated microaccelerometer in the airbag deployment system in automobiles. As shown in Figures 1.6 and 1.8, these devices are integrated with signal transduction and the associated electronic circuits. The sensing element, that is, the accelerometer, has a special configuration, as illustrated in Figure 2.35. The working principle of this type of microaccelerometer is presented below.

With reference to Figure 2.35a, a thin beam is attached to two tethers at both ends. The tethers are made of elastic materials and are anchored at one side, as shown in the figure. The thin beam acts as the proof mass with an electrode plate attached. The

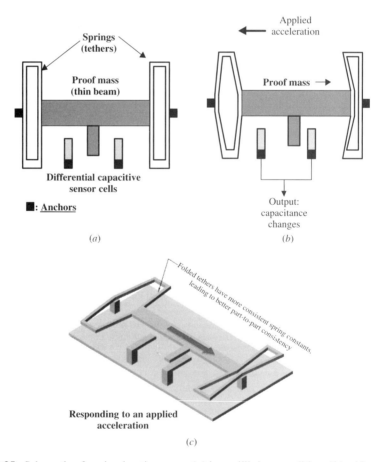

Figure 2.35. Schematic of a microinertia sensor: (*a*) in equilibrium condition; (*b*) with acceleration toward the left; (*c*) movement of a proof with acceleration of base toward the left (Doscher, 1999).

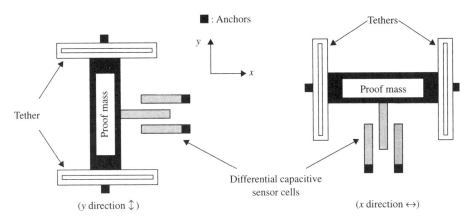

Figure 2.36. Schematic of a dual-axial motion sensor (Doscher, 1999).

electrode plate that is attached to the proof mass is placed between two fixed electrodes. In the event of acceleration of the unit, the proof mass will move in the direction opposite to the acceleration, as shown in Figures 2.35*b* and 2.35*c*. The movement of the proof mass induced by the acceleration (or deceleration) can be correlated with the capacitance change between the two pairs of the electrodes in this arrangement.

We realize that the proof mass moves in the direction opposite to the acceleration or deceleration of the unit. The arrangement in Figure 2.35 will only measure the acceleration in the direction along the length of the proof mass. The arrangement in Figure 2.36 is designed for the measurements in both the *x* and *y* directions when both units are attached to the same base.

A more compact arrangement is illustrated in Figure 2.37 in which the proof mass is replaced by a square plate that can displace in both the *x* and *y* directions. Electromechanical design principles are presented in the references (Doscher, 1999; Chau et al., 1995).

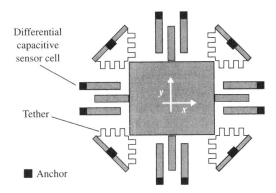

Figure 2.37. Schematic of a compact dual-axial motion sensor. (Courtesy of Analog Devices, Norwood, MA.)

2.5.2 Microgyroscopes

Gyroscopes are devices used to measure the angular rotation rate. They are widely used in airplanes, spacecrafts, missiles, automobiles, and consumer electronics. Microgyroscopes are the miniaturization of gyroscopes that have been packaged with electrical transducers. These devices range in size from a few micrometers to 1 cm. The minute size and ultralight weight of microgyroscopes make them particularly suited for microsatellites and micro–air vehicles for long flights with low power consumption. An astronaut can wear several microgyroscopes for positioning or monitoring his or her activities during space travel. Silicon sensing systems using microgyroscopes are the key controlling element in Segway Human Transporters (http://www.baesystems.com/newsroom/2002/jul/220702 news8.htm).

There are three types of gyroscopes: rotational, optical, and vibrating. Rotational microgyroscopes are similar in design as traditional gyroscopes, typically with a rotating mass with its axle of rotation supported by a gimbal, which in turn is supported by a gyro frame. Because of the high-speed rotational component involved in this design, this type of microdevice requires a near-frictionless bearing. This becomes too complex in design and is costly in manufacturing. Optical microgyroscopes are the most accurate, but they are not commonly used by industry because of the size and cost involved in producing these devices. Vibrating or tuning-fork-type microgyroscopes are the most frequently used by industry today because of their compliance to batch micromanufacturing processes and minimal structure complexity.

Vibratory microgyroscopes work on the principle of the Coriolis effect, or Coriolis acceleration. For a solid mass preceding along one reference axis (x) at a velocity V, a rotation of this mass ($\mathbf{\Omega}$) in the plane x–y can cause a Coriolis acceleration a_c in the direction along the perpendicular axis y. The associated Coriolis force with a_c is, by Newton's law, $F_c = ma_c$, where m is the mass of the moving solid. The Coriolis acceleration in this case is $\mathbf{a}_c = 2\mathbf{V} \times \mathbf{\Omega}$, which leads to the Coriolis force

$$\mathbf{F}_c = 2m\mathbf{V} \times \mathbf{\Omega} \tag{2.13}$$

The vectorial expression in Equation (2.13) is depicted in Figure 2.38.

In Figure 2.38, the x axis is the direction of the preceding mass, the z axis is the axis of solid rotation, and the y axis is the direction of the Coriolis force.

We realize that the rotation of the moving solid, $\mathbf{\Omega}$, can be computed by the preceding velocity V and the measured induced Coriolis force $\mathbf{F}_c$ from Equation (2.13).

Special design features are necessary for microgyroscopes because of the limitation of available space in order to accommodate the requirements of high linear velocity of the solid mass and the accurate measurements of the induced Coriolis forces induced by slight rotation of the gyro frame. Figure 2.39 illustrates a typical structure of the core of a vibrating microgyroscope.

The structure in Figure 2.39 illustrates the schematic of the top view of a vibrating (or tuning fork) type microgyroscope. The two axes x and y correspond to those illustrated in Figure 2.38. The core element consists of a proof mass attached to a floating frame through two tether springs in the y direction. The floating frame is in turn attached to

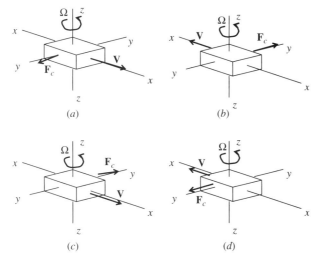

Figure 2.38. Induced Coriolis forces in a linearly moving solid undergoing a rotation.

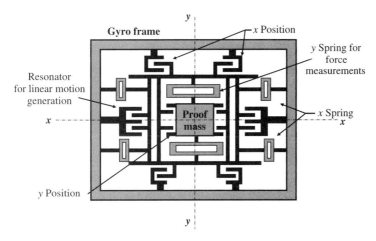

Figure 2.39. Schematic of a "tuning fork" type microgyroscope.

the gyro frame by means of a comb-driven resonator and four x springs, as shown in the figure.

The comb-driven resonator provides linear motion of the proof mass along both x directions, as shown in all four cases in Figure 2.38. Any slight rotation of the gyro frame, and thus the vibrating mass, will generate a Coriolis effect and thus a Coriolis force along either the positive or negative y direction depending on the direction of the rotation of the proof mass, as illustrated in Figure 2.38. The magnitude of the induced Coriolis forces is determined by measuring the deflection of the tether springs in the

y-position capacitors and the equivalent spring constant of the tethers. Through both the x- and y-position capacitors, one could also measure the magnitudes of the rotation and thus the rate of rotation.

2.6 MICROFLUIDICS

Microfluidic systems are widely used in biomedical, precision manufacturing processes, and pharmaceutical industries. Principal applications of microfluidic systems are for chemical analysis, biological and chemical sensing, drug delivery, molecular separation in DNA analysis, amplification, sequencing of DNA, synthesis of nucleic acids, and environmental monitoring (Kovacs, 1998). Microfluidics is also an essential part of precision control systems for automotive, aerospace, and machine tool industries. The principal advantages of microfluidic systems are as follows:

1. 'The ability to work with small samples, which leads to significantly smaller and less expensive biological and chemical analyses.
2. Most microfluidic systems offer better performance with reduced power consumption.
3. Most microfluidic systems for biotechnical analyses can be combined with traditional electronics systems on a single piece of silicon, called "lab-on-a-chip," or LOC (Lipman, 1999).
4. Since many of these systems are produced in batches, they are disposable after use, which ensures safety in application and savings in cleaning and maintenance costs.

A microfluidic system consists of nozzles, pumps, channels, reservoirs, mixers, oscillators, and valves in micro- or mesoscales. Henning (1998) defines the scope of microfluidic systems in a slightly different way. By his definition, a fluidic system comprises the following major components:

1. *Microsensors* used to measure fluid properties (pressure, temperature, and flow). Many of these sensors are built on the working principles described in Section 2.2.

2. *Actuators* used to alter the state of fluids. Microvalves are built on the principles described in Section 2.6.1 and micropumps and compressors operating on a principle described in Section 2.6.2.

3. *Distribution channels* regulating flows in various branches in the systems. Capillary networks such as illustrated in Figure 2.4 are common in microfluidics. Microchannels of noncircular cross sections such as those illustrated in Figure 2.40 are used in many microfluidic systems. They typically have open cross-sectional areas in square micrometers. The channels direct fluid flow of a few hundred nanoliters to a few microliters. As will be described in Chapter 10, microchannels of noncircular cross sections are usually produced by chemical etching in open channels. The two open channels are adhered to provide the close conduits, as shown in Figures 2.40a, b, and c, whereas other conduits

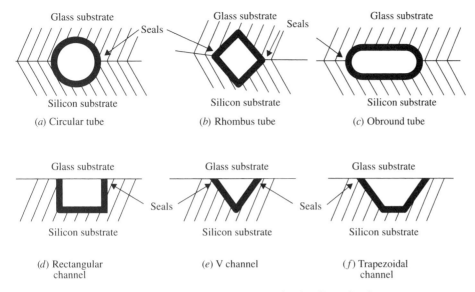

Figure 2.40. Microchannels with Noncircular Cross Sections.

in Figure 2.40*d*, *e* and *f* are produced by adhering the etched open channels to flat glass plates. These channels can be produced in length of less than a millimeter. The vee channel in Figure 2.40*e* is commonly used in micro–heat pipes to dissipate heat from heat-generating electronic components in minute space.

Electrohydrodynamic forces provided by electro-osmosis or electrophoresis in biotesting and analytical systems (see Section 2.2.2) are used extensively to drive the minute fluid samples through these microchannels. The working principles of electrohydrodynamics will be presented in Chapter 3.

4. *Systems integration* includes integrating the microsensors, valves, and pumps through the microchannel links. This integration also involves the required electrical systems that provide electrohydrodynamic forces, the circuits for transducing and processing the electronic signals, and control of the microfluid flow in the system.

Microfluidic systems can be built with a variety of materials such as quartz and glasses, plastics and polymers, ceramics, semiconductors, and metals. Working principles of microvalves and pumps are presented in Sections 2.6.1 and 2.6.2, respectively. The design of these systems requires special considerations, as we will learn from the scaling laws in Chapter 6. The surface-to-mass ratio changes drastically when the systems scale down. For example, due to the well-known capillary effect for liquid flows through minute tubes or channels, surface tension and viscosity become two major governing factors in the design of the flow system. Electrohydrodynamic pumping is an effective way of moving fluids in microchannels, as will be described in Section 3.8.2. The well-known Navier–Stokes equation presented in Section 5.5 for fluid dynamics can no longer be used in predicting the dynamics of fluid flow in microsystems (Pfahler et al., 1990).

Modified theoretical formulation of microchannel flows will be presented in microfluid dynamics in Chapter 5 and a design case will be presented in Chapter 10.

2.6.1 Microvalves

Microvalves are primarily used in industrial systems that require precision control of gas flow for manufacturing processes or in biomedical applications such as controlling blood flow in an artery. A growing market for microvalves is in the pharmaceutical industry, where these valves are used as a principal component in microfluidic systems for precision analysis and separation of constituents. Microvalves operate on the principles of microactuation. Jerman (1991) reported one of the earlier microvalve designs. As illustrated in Figure 2.41, the heating of two electrical resistor rings attached to the top diaphragm can cause a downward movement to close the passage of flow. Removal of heat from the diaphragm opens the valve again to allow the fluid to flow. In Jerman's design, the diaphragm is 2.5 mm in diameter and 10 μm thick. The heating rings are made of aluminum 5 μm thick. The valve has a capacity of 300 cm^3/min at a fluid pressure up to 100 psi, and 1.5 W of power is required to close the valve at 25 psig pressure. Detailed design of this type of valve and its performance can be found in a paper by Jerman (1990). Another type of valve with pumping actuation by electromagnetic solenoid is described in a paper by Pourahmadi et al., (1990).

A rather simple microvalve design uses a thermal actuation principle (Henning et al., 1997; Henning, 1998). The cross section of this type of valve is schematically shown in Figure 2.42. This design is used to control the flow rate from a normally opened valve (as shown) to a fully closed state. The downward bending of the silicon diaphragm regulates the amount of valve opening. Bending of the diaphragm is activated by heat supplied to a special liquid in the sealed compartment above the diaphragm. The heat source in this case is electric resistance foils attached at the top of the device.

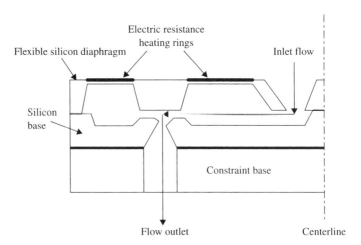

Figure 2.41. Schematic of a microvalve.

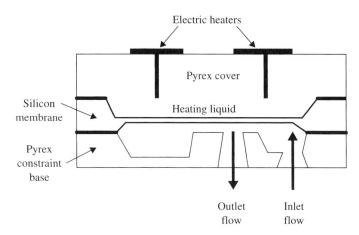

Figure 2.42. Thermally actuated microvalve.

2.6.2 Micropumps

A simple micropump can be constructed by using the electrostatic actuation of a diaphragm, as illustrated in Figure 2.43. The deformable silicon diaphragm forms one
electrode in a capacitor. It can be actuated and deformed toward the top electrode by
applying a voltage across the electrodes. The upward motion of the diaphragm increases
the volume of the pumping chamber and hence reduces the pressure in the chamber. This
reduction of pressure causes the inlet check valve to open to allow inflow of fluid. The
subsequent cutoff of the applied voltage to the electrode prompts the diaphragm to return
to its initial position, which causes a reduction of the volume in the pumping chamber.
This reduction of volume increases the pressure of the entrapped fluid in the chamber.
The outlet check valve opens when the entrapped fluid pressure reaches a designed value
and fluid is released. A pumping action can thus be accomplished. Zengerle et al. (1992)
reported the design of this device. The pump in Figure 2.43 has a square shape with a
diaphragm 4 mm × 4 mm × 25 μm thick. The gap between the diaphragm and the electrode is 4 μm. The actuation frequency is 1–100 Hz. At 25 Hz, a pumping rate of 70
μL/min is achieved.

Another type of micropump, called a *piezopump* (Madou, 1997), is built on the principle of producing wave motion in the flexible wall of minute tubes in which the fluid
flows. Piezoelectric materials coated outside the tube wall generate the wave motion.
The wave motion of the tube wall exerts forces to the contained fluid for the required
motion. We will present a more detailed description of this type of pumping actuation in
Chapter 5.

2.6.3 Micro–Heat Pipes

The heat pipe is a device that can transfer large quantity of heat through small surface
areas with small temperature differences. A typical heat pipe consists of a circular pipe
with an annular layer of wicking material covering the inside. Typical wicking materials

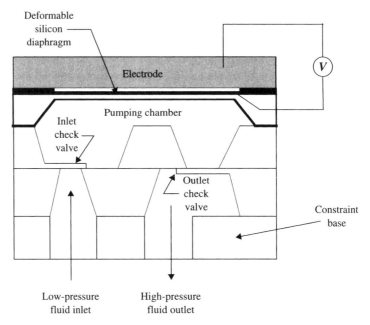

Figure 2.43. Schematic of a micropump.

are porous, containing minute pores that provide capillary flow for liquid flow. Capillary flow of fluids will be described in Chapter 5. One end of the pipe is in contact with heat source and the other end is in contact with a heat sink. The section between the heat source and sink is thermally insulated. The heat supplied by the source causes the adjacent liquid in the central cross section of the pipe to evaporate and flows toward the heat sink. The vapor temperature drops with dissipating heat to the sink, resulting in condensation to the liquid phase. The condensed liquid flows back through the wicking material toward the heat source to start another cycle of heat flow. Design analysis of heat pipe performance is available in a book by Kreith and Bohn (1997).

Like traditional heat pipes, a micro–heat pipe is also a sealed vessel in which the contained fluid flows in two phases, vapor and liquid, along its length. The vessel has typical cross sections in triangular and trapezoidal geometry, as depicted in Figures 2.40e and f, with hydraulic diameter on the order of 100 μm (hydraulic diameter d_p is defined in Chapter 5). Unlike conventional heat pipes, micro–heat pipes in general do not require wicking materials to assist the return of condensate to the evaporator section. Rather, the returned liquid is driven by capillary pressure generated in the sharp corners of the microscaled pipe cross section, as illustrated in Figure 2.44. Ethanol and methanol are common liquids used in micro–heat pipes. This novel idea was first reported by Cotter (1984).

The continuing trend of miniaturizing electronic transistors and devices as described in Chapter 1 has resulted in a serious technical problem of dissipating excessive heat generated by some components at intensities as high as 100 W/cm^2, as in notebook

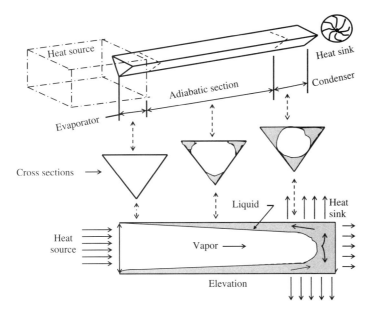

Figure 2.44. Axial and transverse sections of a micro–heat pipe.

computers and PDAs, and the trend is continuing. Micro–heat pipes are considered to be an effective way of resolving this problem (Le Berre et al., 2003). These heat pipes can be made as small as a few micrometers with less than 1 cm in length. They are thus particularly suited for the small space available for heat dissipation in miniature electronic devices.

PROBLEMS

Part 1 Multiple Choice

1. The fundamental working principle of sensors is (a) to convert one form of energy to another form, (b) to convert signals, (c) to convert signs.

2. Acoustic wave sensors are used to detect (a) sound, (b) light, (c) chemical compositions.

3. Acoustic wave sensors work on the principle of measuring the change of resonant frequencies of a solid induced by (a) material property change, (b) sound transmission, (c) change of acoustic characterization of material.

4. The driving force involved in the acoustic wave sensors is (a) sound waves, (b) stress waves, (c) acoustic pressure waves.

5. Medical diagnosis uses (a) biosensors, (b) biomedical sensors, (c) both these sensors.

6. Biomedical and biosensors are (a) the same thing, (b) different in names, (c) different in applications.

7. Working of biosensors require, (a) biomolecules, (b) electrochemical compounds, (c) chemical compounds.

8. Chemical sensors work on the principle of (a) interaction of chemical and electrical properties of materials, (b) chemical and mechanical interaction, (c) mechanical and electrical interaction.

9. Any material that has a change of electrical properties after being exposed to particular gases can be used as (a) chemical sensor, (b) biosensor, (c) thermal sensor.

10. Optical sensors work on the principle of (a) input heat generated by light, (b) input photon energy by light, (c) impact of electrons on solid surface.

11. A photovoltaic device is one that (a) detects heats, (b) detects voltage, (c) generates electricity.

12. Pressure sensors work on the principle of (a) deflecting a thin diaphragm, (b) heating a thin diaphragm, (c) magnetizing a thin diaphragm by the pressurized medium.

13. The deflection of the thin diaphragm in micropressure sensors is measured by (a) mechanical means, (b) optical means, (c) electrical means.

14. Thermal sensors work on the principle of (a) thermal mechanics, (b) thermometers, (c) thermal electricity.

15. Thermopiles have (a) one, (b) two, (c) three junctions.

16. It takes (a) one, (b) two, (c) three different materials to make thermal actuation work.

17. A shape memory alloy is a material that has (a) memory of its shape at the temperature of fabrication, (b) programmed memory of its original shape, (c) memory of its original properties.

18. Piezoelectric actuation works on the principle of (a) electric heating, (b) mechanical–electrical conversion, (c) electrical–mechanical conversion.

19. As the gap between the electrodes grows smaller, the electrostatic forces for actuation (a) grow stronger, (b) grow weaker, (c) do not change.

20. The decibel (dB) is the unit of (a) force, (b) temperature, (c) sound.

21. The input load to a miniature microphone is (a) sound wave, (b) pressure wave, (c) acoustic force.

22. Miniature condenser microphones work on the principle of measuring the change of (a) pressure of vapor phase to liquid, (b) capacitance between the diaphragm and the back plate, (c) electrical resistance of the diaphragm with incident pressure waves.

23. Electrostatic motors work on the principle of (a) closing the gaps, (b) alignment of opposing electrodes, (c) both closing and alignment of opposing electrodes.

24. Microaccelerometers are used to measure (a) the velocity, (b) the position, (c) the dynamic forces associated with a rigid body moving at variable speed.

25. Tethers are used as (a) electric conductors, (b) reinforcement, (c) springs in microaccelerometers.

26. Microgyroscopes (a) sense rotation, (b) sense linear translation, (c) combine linear translation and rotation of a solid in motion.

27. The microgyroscope works on the principle of (a) Coriolis effect, (b) aerodynamic effect, (c) torsional effect of a moving solid with rotation.

28. The resonator in a tuning fork microgyroscope is to provide (a) rotation, (b) Coriolis force, (c) linear velocity of the proof mass.

29. Microfluidics is used extensively in (a) thermomechanical, (b) biomedical, (c) electromechanical analysis.

30. A major problem in microchannel flow is (a) capillary effect, (b) friction effect, (c) pressure distribution.

31. A heat pipe is a device to (a) contain heat, (b) move heat, (c) generate heat.

32. Wicking materials in a heat pipe is to provide (a) capillary force, (b) cooling effect, (c) required pressure drop for the flow of condensate in heat pipes.

33. Micro–heat pipes in general (a) do, (b) do not, (c) occasionally require wicking materials.

34. Micro–heat pipes are effective means of (a) generating heat, (b) dissipating, (c) absorbing excessive heat generated in miniature electronic devices.

35. Micro–heat pipes use (a) sharp corners, (b) small cross sections, (c) short length to generate necessary capillary forces to return the condensate.

Part 2 Descriptive Problems

1. Give examples of at least two sensors that match the definition given in Section 2.2.

2. What are the advantages and disadvantages of using (a) piezoresistors and (b) capacitors as signal transducers?

3. Describe the three principal signal transduction methods for micropressure sensors. Provide at least one major advantage and one disadvantage of each method.

4. What are principal applications of microsensors, actuators, and fluidics?

5. Why are electrostatic forces used to run micromotors rather than conventional electromagnetic forces? Why is this actuation technique not used in macrodevices and machines?

6. Explain why the change of the state of stresses in a silicon diaphragm in a micropressure sensor results in the change of its natural frequency.

7. Explain why the back plate in the miniature microphone in Figure 2.30 is perforated.

8. Plot the relationship between the output voltage of a capacitor transducer versus the change in the gap d using the geometry and dimensions of the plate electrodes in

Example 2.2. What will happen when $d \rightarrow 0$? Observe the shape of the plotted curve.

9. Estimate the voltage output for the microthermopile shown in Figure 2.17 if copper wires are used for the thermocouples with the hot-junction temperature at $120°C$ while the-cold junction temperature is maintained at $20°C$.

10. Describe the four popular actuation techniques for microdevices. Provide at least one major advantage and one disadvantage of each technique.

11. Calculate the electrostatic forces on the plate electrodes with an applied DC voltage at $70\,V$. The geometry and dimensions of the plate electrodes are described in Example 2.1. The plates are initially misaligned by 20% in both length and width directions. Pyrex glass is used as the dielectric material, so there is no gap change of the electrode plates with the applied voltage. We assume that the friction between the electrode plates and the Pyrex dielectric film is negligible.

12. Calculate the required voltage in Example 2.4 if an arrangement similar to that shown in Figure 2.27 is used instead of what is illustrated in Figure 2.28. Assume that the stiffness of the driving arms can be made equivalent to a spring constant of $0.05\,N/m$ as given in Example 2.4 and the rotation of the electrodes resulting from the bending of the driving arms is negligible.

13. Estimate the required applied voltage to the microgripper illustrated in Figure 2.27 with the dimensions given in Figure 2.45 as the first attempt by a design engineer. Both the drive arms and the closure arm and the attached electrodes are made of silicon. The electrodes are coated with thin copper films. The dielectric medium between the electrodes is air. Properties of all materials are available in Table 7.3. You will find the required applied voltage with the given features in Figure 2.45 to be unrealistically high. What would the design engineer do to reduce the required applied voltage to less than $40\,V$?

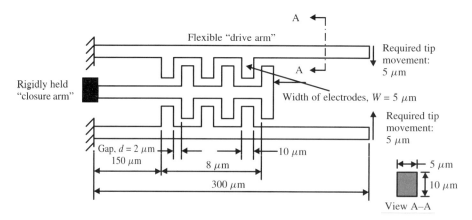

Figure 2.45. Microgripper design.

14. Use Equation (2.13) to estimate the induced Coriolis force in a microgyroscope similar to the one illustrated in Figure 2.39 with an angular rotation of the proof mass about the z axis at 10^{-2} rad in a counterclockwise direction when viewed from the top. Assume the mass of the proof mass is 1 mg and it vibrates at a frequency of 10000 cycles/s with maximum amplitude of ± 100 μm.

15. Estimate the displacement of the proof mass in the y direction in problem 14 if an equivalent spring constant of the tether spring in Figure 2.39 is 100 N/m.

CHAPTER 3

ENGINEERING SCIENCE FOR MICROSYSTEMS DESIGN AND FABRICATION

3.1 INTRODUCTION

The design of most successful microsystem products not only requires the application of theories and principles of mechanical, electrical, materials, and chemical engineering but also involves the theories and principles of physics, chemistry, and biology. While it is not possible for any engineer to develop expertise in all these fields, we will nevertheless need to understand the fundamentals of these science disciplines from which many of the design principles and microfabrication techniques are developed. Consequently, what will be covered in this chapter are science topics that are not often familiar to engineers yet are closely related to the design, manufacture, and packaging of microsystems as presented in subsequent chapters. For brevity, each selected topic will be presented as an overview. Readers are encouraged to look for in-depth descriptions of these topics in the cited references.

3.2 ATOMIC STRUCTURE OF MATTER

Atoms are the bases of all substances that are known to exist in the universe. Everything on Earth is made from 96 stable and 12 unstable elements. Each element has a different atomic structure. The basic structure of an atom involves a *nucleus* and the orbiting *electrons*. The nucleus consists of *protons* and *neutrons*. *The difference in atomic structures results in the different properties of the elements.* Figure 3.1 illustrates the simplest atomic structure of a hydrogen atom. As can be seen from the figure, the nucleus is located at the core of the atom whereas the lone electron revolves in an *orbit* that is away from the nucleus.

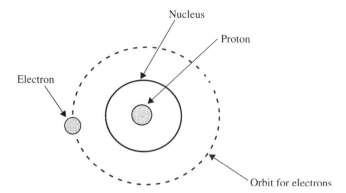

Figure 3.1. Structure of a hydrogen atom.

In general, atoms contain more than one electron. Electrons in atoms can exist in more than one orbit, as illustrated in Figure 3.2. Electrons orbit at different distances from the nucleus, depending on their energy level. An electron with less energy orbits close to the nucleus, whereas one with greater energy orbits farther away. The higher energy electrons farthest from the nucleus are the ones that interact with neighboring atoms to form solid structures. The electrons at the outermost orbit in the atoms that are shared with the neighboring atoms are called *valence electrons*. Electrons can be moved from one orbit to another orbit with a change of energy that keeps the atom in its natural state. Materials with high mobile electrons are called *conductors*, whereas those with immobile electrons are called insulators or *dielectric* materials.

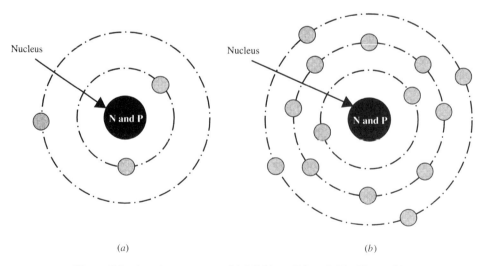

Figure 3.2. Atomic structures of (*a*) lithium (Li) and (*b*) silicon (Si).

The outer orbit of atoms has diameters varying from 2×10^{-8} to 3×10^{-8} cm (or 2–3 Å, 1 Å $= 10^{-4}$ μm), which is about 1000 times greater than that of the nucleus. The core of an atom is the *nucleus*, which contains *neutrons* and *protons* (N and P as shown in Figure 3.2). Neutrons in the atom carry no electric charge. On the other hand, protons in the nucleus are positive electrical charge carriers whereas the electrons carry negative charges. The respective masses for protons and electrons are 1.67×10^{-24} and 9.11×10^{-28} g. In a neutral state, the total numbers of protons and electrons in an atom are equal to each other. (Note that there is no neutron in the nucleus of a hydrogen atom.) Therefore, the total positive charge in protons equals the total negative charge carried by the electrons. For instance, a hydrogen atom, as shown in Figure 3.1, contains only one electron and one proton. The next simplest atomic structure, that of lithium (Figure 3.2*a*), consists of 3 electrons in two orbits and 3 protons in the nucleus, whereas the silicon atoms contain 14 electrons in three orbits with 14 protons in the nucleus (Figure 3.2*b*). The total number of electrons in the atom is designated as the *element or atomic number* of the specific material in a periodic table. The periodic table that consists of 103 elements provides much useful information to engineers. A typical periodic table is shown in Figure 3.3.

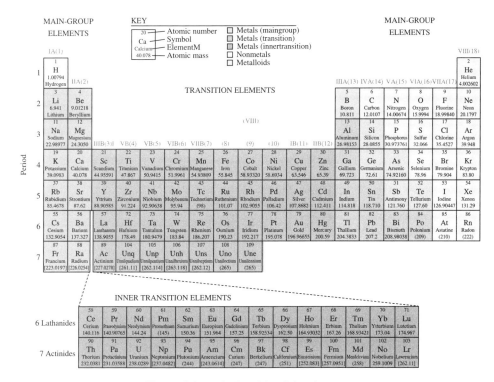

Figure 3.3. Periodic table of the elements.

Following is a set of rules that are applicable to the structure of atoms (Van Zant, 1997). These rules help engineers understand the nature of the elements existing on Earth:

1. In the periodic table, shown in Figure 3.3, each element contains a specific number of protons and no two elements have the same number of protons.
2. Elements with the same number of electrons at the outer orbit have similar properties.
3. Elements are stable with a filled outer orbit of eight electrons.
4. Atoms seek to combine with other atoms to create the stable condition of a full outer orbit, that is, to have *eight* electrons in their outer orbit.

3.3 IONS AND IONIZATION

An *ion* is an electrically charged atom. A negative ion is an atom that contains more electrons than that in its neutral state. A positive ion, on the other hand, is an atom that contains fewer electrons than is necessary to maintain the neutral state.

Ionization is the process of producing ions. Two common methods are available for the production of discrete ions or ion beams: (1) by electrolysis processes and (2) by electron beams. We will describe electrolysis in Section 3.8.1. In either case, external energy is required to initiate and maintain the ionization process. The production of ions from certain substances (gases are common substances to be ionized) by electron beams can be illustrated by the process shown in Figure 3.4.

With reference to Figure 3.4, electron beams are generated by heating the cathode in an electron gun. The electrons released from the cathode by the input thermal energy are guided by a set of electrodes to an accelerator, in which the high-voltage electric field supplies the necessary kinetic energy to accelerate the flow of passing electrons.

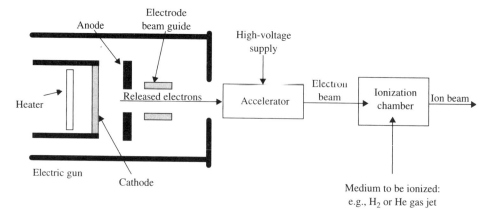

Figure 3.4. Schematic of ionization using electron beams.

The electron beam containing high kinetic energy collides with the molecules of the medium to be ionized in the ionization chamber and thereby ionizes the medium after knocking out electrons from the medium atoms. Hydrogen and helium gases are popular ion sources, as indicated in the figure. Other gases may be used; for example, BF_3 is often used to extract positive boron ions.

Ionization energy is defined as the energy needed to remove the outermost electron from an atom of the ionized medium. The energy required to remove the first electron from the outermost orbit is much less than that required for removing additional electron from the same orbit. An optimal ionization of a gas requires approximately 50–100 electron volts (eV) of input energy ($1\,\text{eV} = 1.6022 \times 10^{-19}$ joules)

3.4 MOLECULAR THEORY OF MATTER AND INTERMOLECULAR FORCES

We may observe the fact that force or energy of various forms can deform a substance, whether it is a solid, a liquid, or a gas. These deformed substances can restore, either completely or partially, to their original shapes once the applied force or energy is removed. This phenomenon leads us to hypothesize that all matter is made up of particles that are interconnected by chemical bonds that can be compressed or stretched by the application of forces or energies. These particles are referred to as the *molecules* of a substance. Molecules are made of atoms, such as the silicon atoms in a silicon wafer, illustrated in Figure 3.5a. Molecules can also involve atoms of chemical compounds, such as the hydrogen and oxygen atoms in the case of water molecules, illustrated in Figure 3.5b.

In either case, as illustrated in Figure 3.5, the atoms are situated in *lattices* that separate them from their neighboring atoms or molecules. The lattice acts as the chemical bonds that hold the atoms together in molecules. The bonding forces are called *atomic cohesive forces*, *intermolecular forces*, or *van der Waals forces*. Van der Waals forces are a major factor in dealing with the serious problem of *stiction* of thin films in

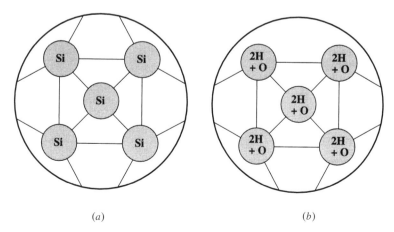

(a)   (b)

Figure 3.5. Molecular structures of matter: (*a*) atoms in a silicon solid; (*b*) molecules in water.

surface micromachining, as will be described in Chapter 9. They are also major factors in formulating molecular dynamics in Chapter 12.

The intermolecular forces are of chemical and electrostatic in nature. Consequently, the distance between molecules, d, plays an important role in determining the magnitude of these forces. Being electrostatic in nature, these forces are based upon Coulomb's law of *attraction* between unlike charges and *repulsion* between like charges, as described in Chapter 2. Let us first consider a situation in which a pair of molecules in a natural state are placed a distance d_0 apart. The variation of the force between the molecules can be qualitatively illustrated as in Figure 3.6.

Figure 3.6 illustrates that pulling the molecules apart will cause the attraction force to increase because of the necessity to overcome the inherent cohesion of the molecules. Further separation, however, results in rapid decrease of the attraction once the atomic cohesion is overcome. The latter state is particularly true for gaseous molecules. The "pushing" of these molecules closer from their natural state, on the other hand, can induce repulsive forces. These forces increase dramatically with the shortened distance between the molecules, as illustrated in Figure 3.6.

In spite of our perception that intermolecular forces are of electrostatic nature, the magnitude of these forces does not always follow Coulomb's law as presented in Chapter 2. For atoms with electric charges (e.g., ions), in the molecules, the magnitude of these forces can vary from $1/d$ to $1/d^6$, where d is the separation distance between such atoms in the molecules. Brown and LeMay (1981) presented a summary of these forces, given in Table 3.1.

Like electrons in atoms, molecules are not stationary even in their natural state. The atoms in the molecules vibrate from the bonds that keep them together. Thus, all molecules are associated with kinetic energies due to this vibration. Table 3.2 outlines the physical behaviors of substances in different states. We will describe these energies at the atomic level in Chapter 12 on quantum physics.

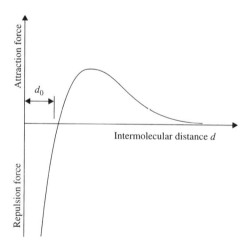

Figure 3.6. Variation of intermolecular forces with separation.

TABLE 3.1. Intermolecular Forces for Molecules with Charge-Carrying Atoms

Type of Interaction	Force or Energy Dependence on Distance	Examples
Ion–ion	$1/d$	Na^+Cl^-, $Mg^{2+}\ O^{2-}$
Ion–dipole[a]	$1/d^2$	Na^+–H_2O
Ion–induced dipole	$1/d^4$	K^+–SF_6
Dipole–dipole	$1/d^6$	HCl–HCl, H_2O–HM_2O

[a]Dipole is a system of two equal and opposite charges placed a very short distance apart.

TABLE 3.2. Physical Behaviors of Molecules

Substances	Atomic Cohesive Forces	Vibration Level	Level of Kinetic Energy
Solids	Strong	Least vigorous	Small
Liquids	Moderate	Very vigorous	Moderate
Gases	Weak	Most vigorous	Large

3.5 DOPING OF SEMICONDUCTORS

The three types of engineering materials that we frequently use for electromechanical systems are (1) electrical conducting materials, (2) electrical insulation or dielectric materials, and (3) semiconducting materials. The classification of these materials is established according to the material's ability to conduct electricity, which is related to the resistance of the material to the movement of electrons. Common materials representing these three classes are presented in Table 3.3 with their respective electrical resistance.

Materials of particular importance to MEMS and microsystems are *semiconductors*. These materials have some natural electrical conductivity but cannot conduct electricity as well as conductors. However, they can be made to be a conductor, with their electrical resistivities reduced to the order of 10^{-3} Ω-cm from those shown in Table 3.3 by implantation of certain foreign impurities. The process of turning semiconducting materials into electrically conducting materials is called *doping*. By virtue of the foreign impurity added to the semiconducting base material and specific doping patterns, one may control both the intensity and path of electric current flow through the semiconductor materials. Microtransistors and microcircuits produced in ICs are formed by using this type of doping process. In MEMS and microsystems, doping of semiconducting materials such as silicon substrates can also alter the material's resistance to chemical or physical etching, which is a common technique in microfabrication. Doping of silicon is thus often used as a barrier to etching as an *etching stop*, as will be described in detail in Chapter 9.

Doping is an essential process to produce p–n junctions in microelectronics. Doping of semiconductors can be achieved by altering the number of electrons in their atoms by implanting foreign atoms with different number of electrons. These foreign materials

TABLE 3.3. Typical Electrical Resistivities of Insulators, Semiconductors, and Conductors

Classification	Materials	Approximate Electrical Resistivity, ρ (Ω-cm)
Conductors	Silver (Ag)	10^{-6}
	Copper (Cu)	$10^{-5.8}$
	Aluminum (Aℓ)	$10^{-5.5}$
	Platinum (Pt)	10^{-5}
Semiconductors	Germanium (Ge)	$10^{1.5}$
	Silicon (Si)	$10^{4.5}$
	Gallium arsenide (GaAs)	$10^{8.0}$
	Gallium phosphide (GaP)	$10^{6.5}$
Insulators	Oxide	10^{9}
	Glass	$10^{10.5}$
	Nickel (pure)	10^{13}
	Diamond	10^{14}
	Quartz (fused)	10^{18}

are called *dopants*. The atom that gives up the electrons is referred to as the *donor*. The atom that receives extra electrons becomes a negatively charged atom.

We will illustrate the principle of how a semiconductor material such as silicon can be doped to become a conductor. Recall that silicon has four valence electrons in its outer orbit (see Figure 3.2*b*). If a material such as boron with three valence electrons in its outer orbit is doped into the silicon, the combined material has one electron in deficit, and a *hole* for the electron is created. A *p*-type semiconductor is thus created. Figure 3.7*a* illustrates such a doping process.

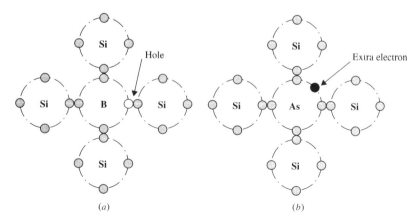

Figure 3.7. Doping of silicon: (*a*) *p*-type doping; (*b*) *n*-type doping.

Negatively charged silicon (i.e., the *n*-type doping) can be created by a doping process that is similar but uses arsenic or phosphorus as the dopant. Both these doping materials have five valence electrons in their outer orbits. The doping of silicon with these materials will result in an extra electron in the combined material, as illustrated in Figure 3.7*b*.

The imbalance of electrons in a doped semiconductor facilitates the flow of electrons and thereby increases the conductivity of the material. The degree of the increased conductivity can be related to the reduction of electrical resistivity in the material. Figure 3.8 presents the change of resistivity of silicon with various doses of boron for *p*-type doping

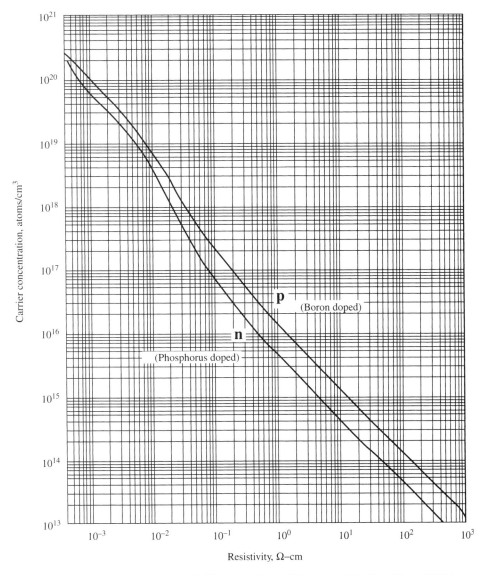

Figure 3.8. Electric resistivity of silicon vs. doses of dopants. (After Van Zant, 1997.)

and phosphorus for *n*-type doping. One may readily see from this figure that the heavier the dose in the doping, the less the resistivity and thus more electrically conductive the silicon becomes.

Doping of semiconductors can be achieved by either the diffusion process at elevated temperature or ion implantation at room temperature, as described in detail in Chapter 8.

3.6 DIFFUSION PROCESS

The diffusion process refers to the introduction of a controlled amount of foreign material into selected regions of another material. It is a phenomenon that is frequently present in our day-to-day life. The spread of a drop of dark ink in a pot of clear water is one example of the diffusion process. In such a case, we can not only readily observe the rapid spread of the dark ink in the pot but also sense the dilution of the dark ink in the mixed liquid with elapsed time. The oxidation of metal in a natural environment is another example of a gas–solid diffusion process. In general, diffusion can take place from liquids to solids, gases to solids, and liquids to liquids, as illustrated by the above examples.

Diffusion is a common process that is used in doping silicon wafers with foreign substances as described in the foregoing section. In microfabrication, diffusion is used to oxidize silicon wafer surfaces, depositing desired thin films of different materials to the base substrates and building up epitaxial layers over single-crystal substrates in IC fabrication. It plays a major role in popular *chemical vapor deposition* (CVD) processes. Many micromixers and microfluidic devices require the use of the diffusion process. Here, we will present the mathematical model that represents this very important physical–chemical process.

Fick's law is used as the basis for the mathematical modeling of diffusion processes. It states that the concentration of a liquid A into a liquid B with distinct concentration (Figure 3.9) is proportional to the difference of the concentrations of the two liquids but inversely proportional to the distance over which the diffusion effect takes place.

Thus, with reference to Figure 3.9 and with an assumption that $C_1 > C_2$, Fick's law can be expressed in mathematical form as

$$C_a \propto \frac{C_{a,x_0} - C_{a,x}}{x_0 - x} \tag{3.1a}$$

or

$$C_a \propto \frac{\Delta C}{\Delta x} \tag{3.1b}$$

where

C_a = concentration of liquid A at a distance x away from the initial contacting surface per unit area and time t

x_0 = position of initial interface of two liquids

$C_{a,x_0}, C_{a,x}$ = respective concentrations of liquid A at x_0 and x

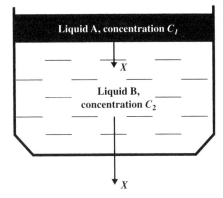

Figure 3.9. Diffusion of liquids of different concentration.

Equation (3.1b) can be expressed in a different form for an incremental variation of the concentration C_a along the x axis as

$$C_a = -D\frac{\Delta C}{\Delta x} \tag{3.2}$$

where the constant D is the *diffusivity* of liquid A. The diffusivity D is often treated as a material property. For most materials, the diffusivity, increases with temperature.

In reality, the duration of diffusion, that is, the time t of the diffusion process, plays an important role in the variation of the concentration of liquid A, as illustrated in Figure 3.10.

In doping semiconductors by diffusion, the semiconductor substrates usually are heated to a carefully selected temperature, and the dopant is made available at the surface

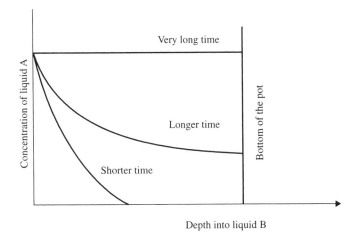

Figure 3.10. Variation of concentration of foreign materials in diffusion process.

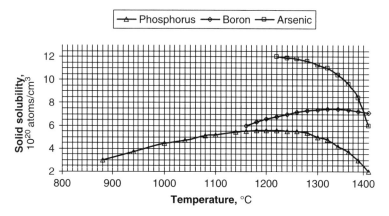

Figure 3.11. Solid solubility of selected materials. (After Van Zant, 1997.)

of the substrate. A mask made of a material that is resistant to the diffusion of the dopant covers the substrate surface during the doping process. The opening made on the mask allows the dopant to be diffused into the substrate surface and thereby controls the region to be doped. The dopant can diffuse into the substrate until a maximum concentration is reached. This maximum concentration of dopant through diffusion is called *solid solubility*. Figure 3.11 shows the solid solubility of common materials that are used as the foreign substances implanted in silicon substrates in microelectronics and microsystem fabrications. One may readily observe that an optimum temperature of the substrate surface does exist for maximum solubility of each material.

A similar expression to Equation (3.2) for one-dimensional solid-to-solid diffusion can be derived as (Kovacs, 1998)

$$J = -D\frac{\partial C}{\partial x} \tag{3.3}$$

where

J = atoms or molecules, or *ion flux*, of foreign materials to be diffused into substrate material, atoms/m^2-s

D = diffusion coefficient, or diffusivity, of foreign material in substrate material, m^2/s

C = concentration of foreign material in substrate, atoms/m^3

The square root of the diffusion coefficient, $\sqrt{D}$, for selected materials can be obtained from Figure 3.12. (The unit for the diffusion coefficient in the figure is μm/h$^{1/2}$). Since the effectiveness of diffusion is proportional to the diffusion coefficient D in Equation (3.3) and the value of D increases with temperature, as shown in Figure 3.12, we realize that a more efficienct diffusion process is possible with higher temperature. However, the solubility of the materials in the diffusion process, as given in Figure 3.11, limits how high the temperature may be in a diffusion process without losing optimal efficiency.

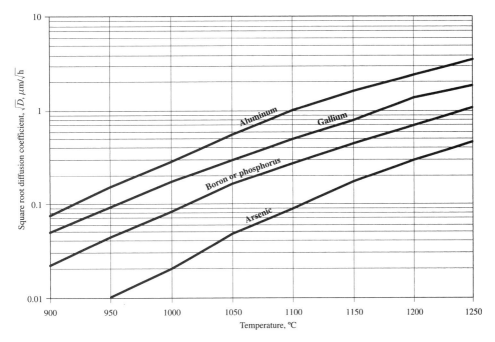

Figure 3.12. Diffusion coefficients of selected materials.

The diffusion of a foreign material into a silicon substrate in microfabrication is illustrated in Figure 3.13.

The concentration C of the foreign material diffused into the substrate material in the situation illustrated in Figure 3.13 varies in three dimensions in the (x,y,z) space frame, in which y and z coordinates are normal to the x coordinate. It requires the solution of a complex partial differential equation involving four variables (x, y, z, t) with t the elapsed time into the diffusion process.

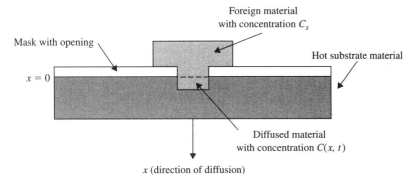

Figure 3.13. Diffusion of a solid material into a substrate.

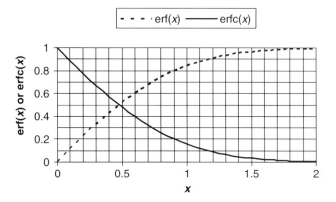

Figure 3.14. Error functions with selected variables.

A quick estimation of the concentration C of the foreign material in the substrate in Figure 3.13 is to assume that the dominant variation of this concentration occurs in the depth, the x direction, in the substrate material with time t. Consequently, the concentration of foreign material in the substrate at a given depth and time, $C(x,t)$, can be obtained by solving the diffusion equation (3.4). This equation is derived from Fick's law in the form

$$\frac{\partial C(x,t)}{\partial t} = D \frac{\partial^2 C(x,t)}{\partial x^2} \tag{3.4}$$

The solution of Equation (3.4) can be shown to take the form

$$C(x,t) = C_s \ \text{erfc}\left(\frac{x}{2\sqrt{Dt}}\right) \tag{3.5}$$

with initial condition $C(x,0)=0$ and boundary condition $C(0,t)=C_s$, the solid solubility at the diffusion temperature (Figure 3.11). The second boundary condition, $C(\infty,t)=0$, reflects a physical situation where the effective zone of diffusion in the substrate material is highly localized at a depth.

The solution in Equation (3.5) is expressed in the complementary error function erfc(X), which is defined as $1 - \text{erf}(X)$, with erf(X) being the *error function*. Values of erf(X) can be found in many mathematical tables, such as in Abramowitz and Stegun (1964). Figure 3.14 provides approximate values of erf(X) and therefore erfc(X).

Example 3.1. Phosphorus is to be doped into a silicon wafer substrate using a diffusion process. The substrate is heated at 1260°C for 30 min in the presence of the dopant. Estimate the variation of the concentration of the dopant at the depth x in the substrate.

Solution: From Figure 3.11, we get the solid solubility of phosphorus as 5.45×10^{20} atoms/cm^3 at 1260°C, which is the maximum achievable concentration for that material at this temperature. Thus, we have $C_s = 5.45 \times 10^{20}$ atoms/cm^3. Also from Figure 3.12, we have $(D)^{1/2} = 0.34 \ \mu$ m/h$^{1/2}$ at that temperature.

From the above, we have

$$2\sqrt{Dt} = 2\sqrt{\frac{(0.34)^2 \times 30}{60}} = 0.481 \mu m$$

Therefore the concentration of phosphorus at a depth x after 30 min into diffusion can be calculated by using Equation (3.5):

$$C(x, 0.5) = C_s \ \text{erfc}\left(\frac{x}{2\sqrt{Dt}}\right) = (5.45 \times 10^{20}) \ \text{erfc}\left(\frac{x}{0.481}\right)$$

By referring to Figure 3.14, we can estimate the concentration of phosphorus at a depth of 0.075 μm to be

$$C(0.075, 0.5) = (5.45 \times 10^{20}) \ \text{erfc}\left(\frac{0.075}{0.481}\right) = (5.45 \times 10^{20}) \ \text{erfc}(0.156)$$

$$= (5.45 \times 10^{20}) \times 0.82 = 4.47 \times 10^{20} \ \text{atoms/cm}^3$$

Example 3.2. Estimate the concentration of phosphorus diffused into the silicon wafer as described in Example 3.1 at a depth $x = 0.075 \ \mu$m after 30, 60, 90, 120, 150, and 180 min into the process.

Solution: The concentration of phosphorus in the silicon wafer at $x = 0.075 \ \mu$m may be approximated by using Equation (3.5) with $C_s = 5.45 \times 10^{20}$ atoms/cm^3 as in Example 3.1 and $2\sqrt{Dt} = 2\sqrt{(0.34)^2 t} = 0.68\sqrt{t}$. We thus have

$$C(0.075, t) = (5.45 \times 10^{20}) \ \text{erfc}\left(\frac{0.075}{0.68\sqrt{t}}\right) = (5.45 \times 10^{20}) \ \text{erfc}\left(\frac{0.1103}{\sqrt{t}}\right)$$

where t has the unit of hours.

The concentration of phosphorus at 0.075 μm beneath the surface of the silicon wafer at the specified time into the diffusion process can thus be determined by this expression. Results are depicted in Figure 3.15.

We may observe from Figure 3.15 that the rate of increase of concentration rises faster at the early hours. The rate increase slows down considerably after 2.5 h into the diffusion.

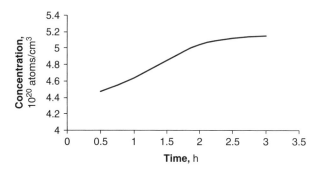

Figure 3.15. Concentration of phosphorus in silicon wafer.

Example 3.3. If boron and arsenic are the dopants in the diffusion process described in Example 3.1, what will be the estimated concentrations of these dopants at depth $x = 0.075$ μm beneath the surface of the silicon wafer at the time into the diffusion specified in Example 3.2?

Solution: We may follow similar procedures stipulated in Examples 3.1 and 3.2 for the solution. However, the solubility and diffusivity of boron and arsenic are different from those for phosphorus used in the two preceding examples and are as follows:

	Solubility at 1260°C (Figure 3.11)	Diffusivity $\sqrt{D}$ at 1260°C (Figure 3.12)
Boron (B)	7.2×10^{20} atoms/cm^3	0.34 μm/h$^{1/2}$
Arsenic (As)	11.8×10^{20} atoms/cm^3	0.12 μm/h$^{1/2}$

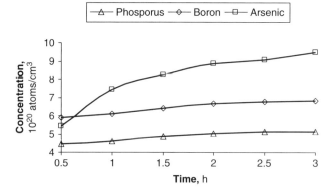

Figure 3.16. Concentrations of boron and phosphorus in silicon wafer.

Computed results, together with that obtained from Example 3.2 for phosphorus, are graphically expressed in Figure 3.16.

It is interesting to observe from Figure 3.16 that boron, with the same diffusivity as phosphorus, shows a similar rate of increase with time. Arsenic, on the other hand, with lower diffusivity but higher solubility, exhibits a significant difference in the rate change of concentration at the same depth in the wafer.

3.7 PLASMA PHYSICS

Plasma is a gas that carries an electrical charge. It contains approximately equal number of electrons and positively charged ions. So, as a whole, plasma is a mixture of neutral ionized gas. Plasma is a key ingredient in microfabrication, as it contains a large number of positive ions with extremely high kinetic energy that can be used to perform the following essential functions in microfabrication:

1. Assist in depositing foreign materials to a base material such as silicon substrates
2. Assist in penetrating desirable foreign substances into a base material to facilitate the implantation of the foreign materials
3. Remove a portion of base material by knocking out the atoms from that material

One advantage of using plasma in microfabrication is the relative ease for one to manipulate its flow by using electrostatic forces or magnetic fields. Plasma-assisted etching and sputtering and plasma-enhanced vapor deposition are popular and effective techniques in microfabrication, as will be described in detail in Chapter 8.

Figure 3.4 showed how ions can be generated for the production of plasma. In most cases, free electrons are first produced by an electron gun, as illustrated in Figure 3.4. These electrons travel at a high speed after passing an accelerator. The fast electrons enter the chamber as illustrated in Figure 3.4, where they knock more electrons from the ionized medium (e.g., a carrier gas of H_2, He, or BF_3). According to Ruska (1987), the following activities can happen in a plasma generator, as illustrated in Figure 3.17: (1) ionization, (2) dissociation, (3) excitation, and (4) recombination.

Ionization. In this process, an electron is knocked loose from the atom of the medium M, resulting in a positive charged molecule, or ion:

$$e^- + M \rightarrow M^+ + 2e^- \tag{3.6}$$

Dissociation. A molecule of the medium, M_2, breaks down into smaller fragments with or without ionization:

$$e^- + M_2 \rightarrow M + M + e^- \rightarrow M^+ + M + 2e^- \tag{3.7}$$

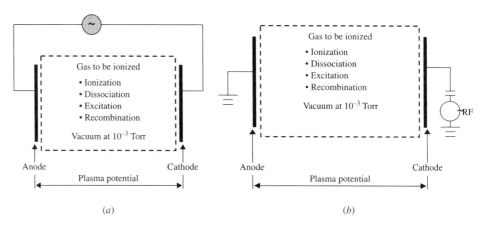

Figure 3.17. Schematic of a plasma generator: (*a*) using electric field; (*b*) Using RF energy source.

Excitation. The molecules hold together but absorb energy from the fast electrons present in the chamber, which results in an excited electronic state (M*):

$$e^- + M \rightarrow M^* + e^- \tag{3.8}$$

Recombination

$$e^- + M^+ \rightarrow M(\text{or } M^*) \tag{3.9}$$

Since electrons are the sources of ionization that sustain the plasma, continuous supply of electrons must be maintained. Consequently, energy sources such as a high-voltage electric field (Figure 3.17*a*) or RF sources (Figure 3.17*b*) must be applied to the generator at all times. Such energy is necessary to allow the cathode in the generator to release electrons on a continuous basis.

3.8 ELECTROCHEMISTRY

Electrochemistry, the study of chemical reactions caused by the passage of an electric current, is an important subject in microsystems design and fabrication. Electrochemical reactions are used in many engineering processes. The principal applications of electrochemistry in microfabrication and microsystems design are in the following areas:

1. In the *electroplating* of polymer molds with thin metal layers by electrolysis in the LIGA process, as will be described in detail in Chapter 9
2. In electrohydrodynamic pumping, which includes both *electrophoresis* and *electro-osmosis*, and for driving capillary flow of fluids in microfluidic systems

The electrochemical reactions in the aforementioned electrolysis process are of particular importance in microfabrication. Electrohydrodynamic pumping is extensively used in the analysis of various species in biomedical solvents, or analytes, in the biotechnology and pharmaceutical industries. The working principle of these analytical systems was described in Chapter 2. We will highlight here both electrolysis and electrohydrodynamic processes.

3.8.1 Electrolysis

An electrolysis process involves the production of chemical changes in a chemical compound or solution by its oppositely charged constituents (or ions) moving in opposite directions under an electric potential difference. As we have learned from the foregoing section, a solution that carries ions is electrically conducting. A solution that conducts electric current is called an *electrolyte* and the vessel that holds the electrolyte is called an *electrolytic cell*. Electrolytes may be liquid solutions or fused salts or ionically conducting solids.

Passing electric current through the fluid in the electrolytic cell can produce electrically charged ions in the electrolyte. Since electric current is the flow of electrons, the free electrons in the current can alter the atomic structures in the fluid molecules. This can lead to atoms with unbalanced electrons, and free ions with electric charges are thus produced in the solution.

In addition to the electrolytic cell, a pair of submerged electrodes is required in order to provide an electric potential in an electrolysis process. A simple example of electrolysis is the decomposition of molten sodium chloride (NaCl) in an electrolytic cell, as illustrated in Figure 3.18.

Electrodes are placed at the two ends of the electrolytic cell. An electric potential is established after connecting the electrodes to a DC source. This source, which usually is a battery, acts as an electron pump that pushes electrons into one electrode, making it a *cathode*, and pulling electrons from the other electrode, to make it an *anode*. Passing the

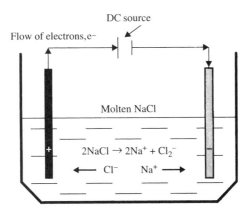

Figure 3.18. Electrolysis process.

electric current decomposes the molten salt into positively charged Na ions and negatively charged Cl ions:

$$2NaCl \rightarrow 2Na^+ + Cl_2^- \qquad (3.10)$$

The Na^+ ions from the decomposed molten salt pick up electrons from the cathode and result in the decrease of electrons at that electrode. As the Na^+ ions in the vicinity of the cathode are depleted, additional Na^+ ions migrate in. In a similar fashion, there is a simultaneous movement of Cl^- ions toward the anode. The salt (Na) and the chlorine (Cl) are thus collected at the respective cathode and anode throughout the electrolysis process.

Electrolysis is a very useful technique in separating and extracting chemical compounds. The value of this process for microsystems design and fabrication, however, lies in the fact that an ionized fluid may be set to motion by an electric potential field. The motion of ions in electrolytes under an electric field leads to a very important technique known as *electrohydrodynamics* of microfluid flow, as will be described in Section 3.8.2

3.8.2 Electrohydrodynamics

Electrohydrodynamics (EHD) deals with the motion of fluids driven by the applied electric field to the fluids. There are two principal applications of EHD in microsystems: (1) electro-osmotic pumping and (2) electrophoretic pumping. These unique pumping techniques are used to move chemical and biological fluids in channels with extremely small crosssections, ranging from square micrometers to square milimeters, at flow rates on the order of cubic micrometers per second. Microfluidics described in Chapter 2 are widely used in the pharmaceutical industry and biochemistry analyses at extremely small sample quantities on the order of a few hundred nanoliters [1 nanoliter (nL) $= 10^{-9}$ liter].

Electrohydrodynamic pumping involves no moving mechanical parts such as rotating impellers. Often, it is the only effective way to move fluids in extremely small channels because of the capillary effect. This effect in fluid flow in small conduits is principally due to the surface tension and the van der Waals forces in fluid molecules, as will be described in detail in Chapter 5. Consequently, conventional volumetric mechanical pumping cannot be used effectively for fluid movement through these extremely small cross sections. One specific application is the capillary electrophoresis (CE) process for rapid, accurate chemical and biological analysis, as will be described in detail in Chapter 10.

Free electric charges in solvents can be produced in several ways, for example, by electrolytes, as described in Section 3.8.1. Dielectric liquids subjected to very high electric voltage can produce a controllable concentration of ions in the fluid. Another way to generate charges is by electrifying layered liquids with spatial gradients in electric conductivity and permittivity, as indicated in the equation (Bart et al., 1990)

$$\frac{\partial \rho}{\partial t} + v\nabla\rho + \frac{\Omega}{\varepsilon}\rho = \mathbf{E} \cdot \nabla\varepsilon - \mathbf{E} \cdot \nabla\Omega \qquad (3.11)$$

where

ρ = free charge density

$\mathbf{E}$ = electric field

ε = permittivity of fluids

Ω = ohmic conductivity of fluids

$\nabla\rho$, $\nabla\varepsilon$, $\nabla\Omega$ = gradient of free charge density, permittivity, and ohmic conductivity, respectively, in fluids

Figure 3.19 illustrates the generation of an ionic solution in an electro-osmotic pumping in a conduit (Bart et al., 1990). The solution consists of two materials with different electric resistivities, Ω_1 and Ω_2, and permittivities, ε_1 and ε_2. A material interface is formed after the application of an electric field across the conduit walls. Free charges are produced by the gradient of electric conductivity, which is represented by the difference of the resistivities and permittivities between the two materials. The signs of the surface charges in the material layers are determined by the permittivities of the materials. For instance, if material layer 1 were close to being an insulator, then the surface charge induced would be opposite to the sign of the electrode array.

Once free charges are generated in the fluid, pumping of the fluid along the longitudinal direction can be accomplished by applying an electric field to the electrodes in the array along the length of the channel, as shown in Figure 3.19. Figure 3.20 illustrates the electro-osmotic pumping of homogeneous fluids and Figure 3.21 illustrates the electrophoretic pumping of heterogeneous fluids.

Electro-Osmotic Pumping. Electro-osmotic pumping is used to move electrically neutral fluids through channels of extremely small cross sections. The condition is that the walls of the conduit or channel must be attached with immobile charges. A glass wall such as shown in Figure 3.20 can produce such immobile charges along the surface if it is coated either with ionized materials (e.g., deprotonated silanol groups), or by strong absorbed charged species that are present in the fluid (Manz et al., 1994). In such cases, the electro-osmotic motion of fluid in microchannels of capillary tubes can be

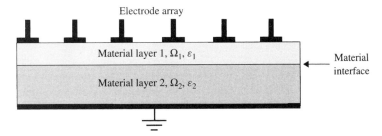

Figure 3.19. EHD Pumping of a solvent (Bart et al., 1990).

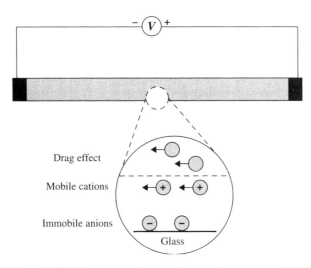

Figure 3.20. Electro-osmotic pumping. (After Kovacs, 1998.)

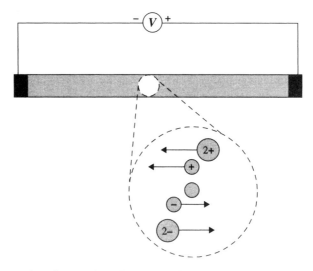

Figure 3.21. Electrophoretic pumping of ions in a fluid. (After Kovacs, 1998, and Manz et al., 1994.)

produced in any electrolyte fluid, as described in Section 3.8.1. Because the gradient of the concentration of electric charges decreases toward the center of the conduit, a dual layer of fluid with varying concentration of charges is formed, as shown in Figure 3.20. The charges in the double layer can be moved with the applied electric charges along the longitudinal direction. The momentum of the moving charges can drive the solvent through the flow channel. A unique feature of electro-osmotic flow is that a uniform velocity profile of the moving fluid across the cross section of the channel or the tube is obtained.

Electrophoretic Pumping. This method of pumping is often used to separate minute foreign particles or species from the bulk fluid. Microfluidic systems with electrophoresis are widely used in biomedical and pharmaceutical industries, as described in Section 2.6. This pumping technique is used to drive heterogeneous fluids through conduits or channels of minute cross sections. The movement of ions of the particles in a heterogeneous medium is prompted by an applied high-voltage electric field, as illustrated in Figure 3.21. The ions with different charges move in opposite directions along the channel in the separation process. When the flow is fully developed, the ions in the stream can automatically separate due to their inherent electro-osmotic mobility under the influence of the applied electric field. It is a highly desirable situation in chemical and biological analyses, in which separation of various minute species is often a difficult task.

The rate of flow in both electro-osmotic and electrophoretic pumping is linearly proportional to the applied electric field. These pumping methods involve no moving mechanical parts. They are especially suitable for miniaturized systems. Only a small sample of analyte is required in most cases in microfluidic systems.

The velocity of the moving free charges in electrophoretic pumping can be determined by the expressions (Kovacs, 1998)

$$v_i = \frac{z_i q}{6\pi r_i \mu} E \tag{3.12}$$

where

v_i = velocity of ion i of charge z_i in homogeneous fluid

E = applied electric field in volts

r_i = actual or hydraulic radius of conduit

μ = viscosity of fluid

q = charge on electron, $= 1.6022 \times 10^{-19}\,\text{C}$

It is obvious from Equation (3.12) that the velocity of the free ions and thus the fluid flow is linearly proportional to the applied electric field in both electrohydrodynamic pumping techniques. We will revisit these pumping systems in Sections 5.6 and 10.7.2.

PROBLEMS

Part 1 Multiple Choice

1. Everything on Earth is made from (a) 86, (b) 96, (c) 106 stable elements, and each element has different atomic structure.

2. The core of an atom is (a) neutron, (b) nucleus, (c) electron.

3. Elements have different properties because they have different (a) atomic structure, (b) chemical composition, (c) physical composition.

4. Elements that have similar properties with the same number of (a) electrons, (b) protons, (c) nucleus in the outer orbit of their respective atomic structures.

5. A nucleus contains (a) neutrons and protons, (b) electrons and protons, (c) neutrons and electrons.

6. Protons carry (a) positive, (b) negative, (c) no charge.

7. Electrons carry (a) positive, (b) negative, (c) no charge.

8. Neutrons carry (a) positive, (b) negative, (c) no charge.

9. The outer orbit of atoms has a diameter that is (a) 100, (b) 1000, (c) 10,000 times that of the nucleus.

10. A periodic table consists of (a) 96, (b) 103, (c) 108 elements.

11. Silicon atoms contain (a) 8, (b) 10, (c) 14 electrons.

12. Silicon atoms have (a) 4, (b) 6, (c) 8 electrons in their outer orbit.

13. Valence electrons are the electrons situated in (a) innermost, (b) outermost, (c) energized orbit in an atom.

14. A valence electron is the electron of an atom that (a) shares, (b) contacts, (c) moves across the neighboring atoms.

15. Silicon atoms have (a) zero, (b) two, (c) four valence electrons.

16. Free electrons may be produced by (a) heating, (b) cooling, (c) pressurizing a cathode.

17. An ion carries (a) electric charge, (b) magnetic charge, (c) electrostatic charge.

18. Ionization energy is the energy required to remove (a) neutrons, (b) protons, (c) electrons from the outermost orbit of an atom.

19. Molecules are made of bounded (a) electrons, (b) atoms, (c) nucleus.

20. The forces that bound the atoms in a molecule are called (a) intermolecular forces, (b) electrostatic forces, (c) electromagnetic forces.

21. Intermolecular forces are (a) van der Waals forces, (b) electrostatic, (c) electromagnetic in nature.

22. Intermolecular forces, in general, are (a) proportional, (b) equal to, (c) inversely proportional to the distances between molecules.

23. The physical behavior of solid molecules are typically (a) strong in kinetic energy and atomic cohesive forces, (b) weak in kinetic energy and atomic cohesive forces, (c) weak in kinetic energy but strong in atomic cohesive forces.

24. Positive silicon can be produced by doping (a) boron atoms, (b) phosphorus atoms, (c) either kind of atom.

25. Negative silicon can be produced by doping (a) boron atoms, (b) phosphorus atoms, (c) either kind of atom.

26. Silicon is a semiconducting material. It can be made more electrically conductive by (a) a doping process, (b) a diffusion process, (c) an electric implantation process.

27. The n-type silicon is (a) less, (b) more, (c) about equally conductive as p-type silicon when it is doped with the same dose of dopant.

28. The "donor" in a doping process is the atom that (a) loses, (b) gains, (c) produces electrons.

29. Diffusion is a good way to (a) coat, (b) implant, (c) remove foreign material in silicon substrates.

30. Diffusion of foreign material into silicon is to create (a) p–n, (b) p–p, (c) n–n junctions for a transistor.

31. Diffusion analysis is based on (a) Fourier's law, (b) Fick's law, (c) Hooke's law.

32. The solubility of a solid offers (a) minimum, (b) optimum, (c) soluble temperature in a diffusion process.

33. Diffusivity of materials is a measure of (a) efficiency, (b) speed, (c) temperature in a diffusion process.

34. Plasma is a gas that (a) does, (b) does not, (c) may carry electric charges.

35. To maintain a plasma, one needs to keep supplying (a) high temperature, (b) high pressure, (c) high electrical field to the plasma chamber.

36. Electrochemistry involves (a) chemical reactions, (b) ionization, (c) decomposition of any substance caused by the passage of an electric current.

37. Electrolysis involves the production of (a) chemicals, (b) chemical changes, (c) ionization in a substance by the application of an electric potential.

38. Electrolysis uses (a) an AC, (b) a DC, (c) either an AC or a DC power supply.

39. Electrolyte is (a) an electrode, (b) the container, (c) the solution that conducts electric current in an electrolysis process.

40. An anode is the (a) positive, (b) negative, (c) neutral electrode.

41. A cathode is the (a) positive, (b) negative, (c) neutral electrode.

42. Electrohydrodynamics deals with (a) dissolution, (b) motion, (c) solidification of a fluid under an applied electric field.

43. The principal use of electrohydrodynamics in microsystems is to (a) conduct electrolysis of minute chemicals, (b) move minute amounts of fluid, (c) detect minute amounts of fluid.

44. Electro-osmotic pumping is used to move minute amounts of (a) homogeneous, (b) heterogeneous, (c) any fluid in capillary passages.

45. Electropheretic pumping is used to move minute amounts of (a) homogeneous, (b) heterogeneous, (c) any fluid in capillary passages.

Part 2 Descriptive Problems

1. Express the sizes and weights of hydrogen and silicon atoms in nanometers (nm), with $1\,nm = 10^{-9}\,m$.

2. What do *atomic numbers* and *group numbers* in a periodic table mean?

3. What is the desirable level of electric resistivity to make a semiconductor electrically conducting?

4. Explain the physical meaning of the negative sign attached to Fick's law in Equations (3.2) and (3.3)?

5. Why is the doping process important to the semiconductor industry and how is this done?

6. What are the two principal techniques used in doping silicon wafers in the semiconductor industry? Offer one advantage and one disadvantage of each technique.

7. Determine the optimum temperatures of silicon substrates for which doping of arsenic, phosphorus, and boron is to be carried out by a diffusion process. Also, what is the solubility of these dopants at these optimal temperatures?

8. Plot the distributions of the concentration of phosphorus atoms at the depth of the silicon substrate in Example 3.1 at the times 30, 60, 90, 120 150, and 180 min into the diffusion process.

9. Determine the time required to dope boron in the silicon substrate so that the resistivity of the doped silicon at the depth of 2 μm is 10^{-3} Ω-cm.

10. Describe how ions are produced in an electrolysis process.

11. Describe the principles of electrophoresis and electro-osmosis. Where are these processes used in MEMS and microsystems?

12. Explain the capillary effect in microfluid flow and why conventional mechanical pumping cannot move fluids in small channels with a capillary effect.

CHAPTER 4

ENGINEERING MECHANICS FOR MICROSYSTEMS DESIGN

4.1 INTRODUCTION

As we learned in Chapter 2, most microsystems are made of three-dimensional structures that often involve heat transmission as well as solid–fluid interactions. Moreover, many of the components in microsystems are essentially machine components in the microscale, such as gears, springs, bearings and linkages, and mechanisms. It is thus necessary to use mechanical engineering and machine design principles in dealing with the design and packaging of microsystems.

In this and subsequent chapters, we will learn how mechanical engineering principles in solid and fluid mechanics and heat transfer derived from theories of continua can be used in the design of various components of micro- and mesoscales in microsystems. We will also discuss the limits of using these phenomenological models derived for continua for components in the small-scale range of single-digit micrometers. In these circumstances, many of the science principles presented in Chapters 3 and 12 will be used as the bases for such modifications for small structures on the order of submicrometers.

Engineering mechanics, which involves both solid and fluid mechanics, is the basis for the mechanical design of microsystems. Proper functioning of microsystems and their structural integrity, as well as reliable systems packaging, require the application of engineering mechanics principles. The field of engineering mechanics is a specialized area of engineering science. It is not possible for us to deal with the entire subject in this section. What we will present just "scratches the surfaces" with a few selected topics that are useful in the design and packaging of microdevices and systems.

Mechanics, by a traditional definition, is a branch of engineering science that studies the relationship between the applied *forces* to a substance and the resulting *motions*. The word motion in microsystems can involve either rigid-body motion, as in a microaccelerometer, or the deformation of solids, as in deformable diaphragms in micro–pressure sensors (Figure 2.8). The principles of dynamics and mechanical vibration are frequently applied in the design and operation of microaccelerometers and pressure sensors using beams vibrating at resonant frequencies for high sensitivities in output (Figure 2.14). Solid mechanics principles are also extensively used in the design of system packaging.

Fluid mechanics, on the other hand, is involved in the design of microvalves and microfluidics. Often, the classical theories and formulations that are derived for idealized loading and geometry are not sufficient for the analyses required in the design of MEMS and microsystems. In such cases, numerical techniques such as computational fluid dynamics (CFD) and the finite element method are used for refined solutions. Commercial CFD and finite element codes are commonly used by industry in the design of these products.

We will also learn how the principles of heat transfer derived for solids at the macroscale can be used in assessing heat transfer and the associated temperature field in MEMS structures. Substantial modification of these principles for components at the single-digit micrometer scale will be outlined in Chapter 12. The application of fluid mechanics and heat transfer in microsystems design will be presented in Chapter 5.

Units for Solid Mechanics Problems. It is prudent for engineers to be consistent in adopting a set of units for stress analyses of microsystems. The following International System (SI) units are recommended for such applications.

The unit *newton* (N) is used for forces. A force of 1 N is defined as a force required to give one kilogram (kg) of mass an acceleration of one meter per square second (m/s^2). Since 1 kg of mass has a weight of 9.81 N, we can readily recognize that 1 kg force equals 9.81 N. The reader will realize that gravitational acceleration $g = 9.81$ m/s^2 is used in the above formulation of the force unit. Recommended units for thermophysical quantities used in microsystems design and conversion between between SI units and the Imperial System can be found in Appendixes 1 and 2.

The unit *meter* (m) is recommended for length. Thus, 1 μm $= 10^{-6}$ m, and 1 mm $= 1000$ μm. The unit *kilogram* (kg) is recommended for mass in all computations.

The *pascal* (Pa) is used for pressures and stresses; 1 Pa is equivalent to 1 N/m^2, from which 1 MPa $= 10^6$ Pa $= 10^6$ N/m^2.

4.2 STATIC BENDING OF THIN PLATES

In Chapter 2, we learned that many micro–pressure sensors work on the principle of converting the stresses in the deformed thin silicon diaphragms, induced by the applied pressure, to the desired form of electronics output. For most cases, these diaphragms, either in circular, square, or rectangular shapes, can be treated approximately as thin plates subjected to lateral bending by uniformly applied pressure.

Several closed-form solutions of plate bending can be found in a useful reference (Timoshenko and Woinowsky-Krieger, 1959). The governing differential equation for the deflection of a rectangular plate subject to lateral bending can be expressed as

$$\left(\frac{\partial^2}{\partial x^2} + \frac{\partial^2}{\partial y^2} \right) \left(\frac{\partial^2 w}{\partial x^2} + \frac{\partial^2 w}{\partial y^2} \right) = \frac{p}{D} \tag{4.1}$$

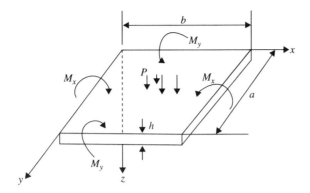

Figure 4.1. Bending of a rectangular plate.

where $w = w(x,y)$ is the lateral deflection of a flat plate due to the uniformly distributed applied pressure p. The x–y plane defines the plate, as shown in Figure 4.1.

The parameter D is the flexural rigidity of the plate, which can be expressed as

$$D = \frac{Eh^3}{12(1 - v^2)} \tag{4.2}$$

where E and v are Young's modulus and Poisson's ratio of the plate material, respectively, and h is the thickness of the plate, as shown in Figure 4.1.

The components of bending moments about the x and y axes (see Figure 4.1) and the bending stresses can be computed from the solution of the deflection, $w(x,y,)$ obtained from the solution of Equation (4.1) as shown below:

The bending moments:

$$M_x = -D\left(\frac{\partial^2 w}{\partial x^2} + v\frac{\partial^2 w}{\partial y^2}\right) \tag{4.3a}$$

$$M_y = -D\left(\frac{\partial^2 w}{\partial y^2} + v\frac{\partial^2 w}{\partial x^2}\right) \tag{4.3b}$$

$$M_{xy} = D(1 - v)\frac{\partial^2 w}{\partial x\, \partial y} \tag{4.3c}$$

Bending stresses:

$$(\sigma_{xx})_{\text{max}} = \frac{6(M_x)_{\text{max}}}{h^2} \tag{4.4a}$$

$$(\sigma_{yy})_{\text{max}} = \frac{6(M_y)_{\text{max}}}{h^2} \tag{4.4b}$$

$$(\sigma_{xy})_{\text{max}} = \frac{6(M_{xy})_{\text{max}}}{h^2} \tag{4.4c}$$

The solution of the plate deflection, $w(x,y)$, obtained from Equation (4.1) with appropriate boundary conditions, is used to obtain the bending moments from Equation (4.3). The maximum bending stresses can thus be computed from Equation (4.4) with respective maximum bending moments. The solution of Equation (4.1) used to be a tedious effort. This task has become manageable with the aid of personal computer and software packages such as MATLAB or MATHCAD.

Simplified formulas for maximum stresses or deflection due to bending in beams can be found in several sources (Avallone and Baumeister, 1996; Roark, 1965) as well as for plates (Roark, 1965). Given below are three cases. The computed magnitude of maximum stresses and the locations in which these stresses occur in the deformed structure enable engineers to place measurement stations at these locations for maximum signal outputs.

4.2.1   Bending of Circular Plates with Edge Fixed

The following solutions are derived for a circular plate with a radius a and thickness h (Roark, 1965). The plate is deflected by the application of a uniform pressure loading p (see Figure 4.2).

The maximum radial stress is

$$(\sigma_{rr})_{\max} = \frac{3W}{4\pi h^2} \tag{4.5a}$$

at the edge, and the maximum hoop (tangential) stress is

$$(\sigma_{\theta\theta})_{\max} = \frac{3\nu W}{4\pi h^2} \tag{4.5b}$$

at the edge.

Both these stresses at the center of the plate become

$$\sigma_{rr} = \sigma_{\theta\theta} = \frac{3\nu W}{8\pi h^2} \tag{4.6}$$

The maximum deflection occurs at the center with the value

$$w_{\max} = -\frac{3W(m^2 - 1)a^2}{16\pi E m^2 h^3} \tag{4.7}$$

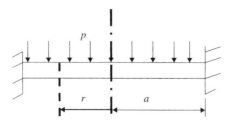

Figure 4.2. Circular plate subjected to uniform pressure loading.

where $W = (\pi a^2)p$ and $m = 1/\nu$. The negative sign attached to the deflection in Equation (4.7) indicates its downward direction.

> **Example 4.1.** Determine the minimum thickness of the circular diaphragm of a micro–pressure sensor made of silicon as illustrated in Figure 4.3. The diaphragm has a diameter of 600 μm and its edge is rigidly fixed to the silicon die. The diaphragm is designed to withstand a pressure of 20 MPa without exceeding the plastic yielding strength of 7000 MPa. The silicon diaphragm has a Young's modulus $E = 190{,}000$ MPa and a Poisson's ratio $\nu = 0.25$.

Solution: From Equation (4.5a), the thickness of a thin plate can be expressed as

$$h = \sqrt{\frac{3W}{4\pi(\sigma_{rr})_{\max}}}$$

or from Equation (4.5b)

$$h = \sqrt{\frac{3\nu W}{4\pi(\sigma_{\theta\theta})_{\max}}}$$

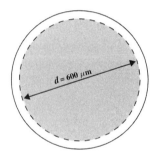

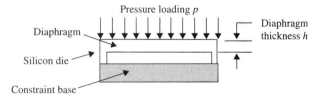

Figure 4.3. Circular diaphragm for a pressure sensor.

Since we will design the diaphragm so that its maximum stress is below the plastic yielding strength of the material, or $(\sigma_{rr})_{\max} \le \sigma_y = 7000$ MPa and $(\sigma_{\theta\theta})_{\max} \le \sigma_y = 7000$ MPa, and also because Poisson's ratio $\nu = 0.25$, that is, $\nu < 1.0$, it is obvious that the first of the two expressions shown above for the thickness h should be used.

The equivalent load W in the expression can be evaluated as

$$W = (\pi a^2)P = 3.14 \times (300 \times 10^{-6})^2 \times (20 \times 10^6) = 5.652 \text{ N}$$

From Equation (4.5a)

$$h = \sqrt{\frac{3 \times 5.652}{4 \times 3.14 \times (7000 \times 10^6)}} = 13.887 \times 10^{-6} \text{ m}$$

or the diaphragm minimum thickness $h = 13.887 \, \mu$m.

The maximum deflection of the diaphragm, w_{max}, can be determined using Equation (4.7):

$$w_{max} = -\frac{3W(m^2 - 1)a^2}{16\pi Em^2h^3}$$

with $m = 1/v = 1/0.25 = 4.0$:

$$w_{max} = -\frac{3 \times 5.652(16 - 1) \times (300 \times 10^{-6})^2}{16 \times 3.14 \times 190{,}000 \times 10^6 \times 16 \times (13.887 \times 10^{-6})^3} = -55.97 \, \mu\text{m}$$

4.2.2   Bending of Rectangular Plates with All Edges Fixed

A closed-form solution for deflection of this case (see Figure 4.4) is available in Timoshenko and Woinowsky-Krieger (1959). A simplified solution can be found in Roark (1965).

The maximum stresses occur at the center of the longer edges:

$$(\sigma_{yy})_{max} = \beta \frac{pb^2}{h^2} \tag{4.8}$$

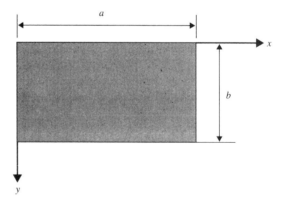

Figure 4.4. Bending of a rectangular plate.

TABLE 4.1. Coefficients for Maximum Stress and Deflection in Rectangular Plate

a/b	1	1.2	1.4	1.6	1.8	2.0	∞
α	0.0138	0.0188	0.0226	0.0251	0.0267	0.0277	0.0284
β	0.3078	0.3834	0.4356	0.4680	0.4872	0.4974	0.5000

and the maximum deflection of the plate occurs at the centroid:

$$w_{\max} = \alpha \frac{pb^4}{Eh^3} \tag{4.9}$$

The coefficients α and β in Equations (4.8) and (4.9) can be determined from Table 4.1.

> ***Example 4.2.*** A rectangular diaphragm with edge dimensions $a = 752\,\mu m$ and $b = 376\,\mu m$ is illustrated in Figure 4.4. These edge dimensions will result in the same plane area as the circular diaphragm in Example 4.1. The thickness of the diaphragm, applied pressure, and material properties remain identical to those in Example 4.1. Determine the maximum stress and deflection in the diaphragm.
>
> *Solution:* We have edge ratio $a/b = 2.0$, which leads to $\alpha = 0.0277$ and $\beta = 0.4974$ from Table 4.1. The diaphragm thickness $h = 13.887\,\mu m$ and the applied pressure $p = 20\,MPa$ are used in determining the maximum stress and deflection.
>   Thus, we may obtain the maximum stress using Equation (4.8) as
>
> $$(\sigma_{yy})_{\max} = \beta \frac{pb^2}{h^2} = 0.4974 \frac{(20 \times 10^6)(376 \times 10^{-6})^2}{(13.887 \times 10^{-6})^2} = 7292.8 \times 10^6 \text{ Pa}$$
>
> or $(\sigma_{yy})_{\max} = 7292.8\,MPa$, which is greater than the plastic yielding strength of silicon at $\sigma_y = 7000\,MPa$—a situation that can be interpreted as "unsafe."
>   The maximum deflection of the diaphragm is determined from Equation (4.9):
>
> $$\begin{aligned} w_{\max} &= \alpha \frac{pb^4}{Eh^3} = \alpha \frac{pb}{E}\left(\frac{b}{h}\right)^3 \\ &= \frac{0.0277 \times (20 \times 10^6) \times 376 \times 10^{-6}}{190{,}000 \times 10^6}\left(\frac{376 \times 10^{-6}}{13.887 \times 10^{-6}}\right)^3 \\ &= 21.76 \times 10^{-6} \text{ m} \end{aligned}$$
>
> The maximum deflection at the center of the diaphragm is $w_{\max} = 21.76\,\mu m$, which is less than which occurred in the case with a circular diaphragm.

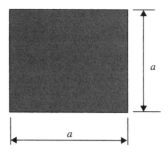

Figure 4.5. Bending of a square plate.

4.2.3   Bending of Square Plates with Edges Fixed

Square diaphragms are common in micro–pressure sensors (see Figure 2.9). This geometry is favored by the industry because it produces maximum stresses, and therefore maximum output, over diaphragms of other geometry with the same plane area.

A typical square plate is illustrated in Figure 4.5

By letting the two edges of the plate to be equal to each other, that is, $a = b$ in the Equations (4.8) and (4.9), we can get the following expressions for the maximum stress and deflection for a square plate.

The maximum stress occurs at the middle of each edge (Roark, 1965):

$$\sigma_{\max} = \frac{0.308\,pa^2}{h^2} \tag{4.10}$$

and the maximum deflection of the plate occurs at the center of the plate:

$$w_{\max} = -\frac{0.0138\,pa^4}{Eh^3} \tag{4.11}$$

The stress at the center of the plate can be derived as

$$\sigma = \frac{6p(m+1)a^2}{47mh^2} \tag{4.12}$$

and the strain at the center is

$$\varepsilon = \frac{1-\nu}{E}\sigma \tag{4.13}$$

Example 4.3. A square silicon diaphragm with $532\,\mu$m edge length is subjected to the same pressure loading $p = 20\,$MPa as in Examples 4.1 and 4.2. The diaphragm has the same thickness, $13.887\,\mu$m. All material properties are identical to those given in Example 4.1. Determine the maximum stress and deflection in the diaphragm under the applied pressure.

Solution: The maximum stress in the diaphragm can be determined using Equation (4.10) with $a = 532 \times 10^{-6}$ m, $h = 13.887 \times 10^{-6}$ m, and $p = 20 \times 10^6$ Pa:

$$\sigma_{max} = \frac{0.308 pa^2}{h^2} = \frac{0.308 \times (20 \times 10^6)(532 \times 10^{-6})^2}{(13.887 \times 10^{-6})^2} = 9040 \times 10^6 \text{ Pa}$$

or $\sigma_{max} = 9040$ MPa, which is much greater than the yield strength of the silicon die at $\sigma_y = 7000$ MPa.

The maximum deflection of the diaphragm can be obtained from Equation (4.11):

$$w_{max} = -\frac{0.0138 pa^4}{Eh^3} = -\frac{0.0138 pa}{E}\left(\frac{a}{h}\right)^3$$

$$= -\frac{0.0138(20 \times 10^6) \times 532 \times 10^{-6}}{190000 \times 10^6}\left(\frac{532 \times 10^{-6}}{13.887 \times 10^{-6}}\right)^3$$

$$= -43 \times 10^{-6} \text{ m}$$

or $w_{max} = 43 \, \mu$m.

We may summarize the results obtained from Examples 4.1–4.3 in the following way based on the same diaphragm area, thickness, and applied pressure:

Geometry of diaphragm	Maximum stress (MPa)	Maximum deflection (μm)
Circular	7000	55.97
Rectangular ($a/b = 2.0$)	7293	21.76
Square	9040	43.00

It is conceivable from the above summary that the circular diaphragm results in the most favorable situation from a traditional mechanical design point of view because of the lowest maximum induced stress in the plate in all three cases. The square diaphragm, on the other hand, results in the highest induced maximum stress in all three cases. It would normally be viewed to be the least favored geometry from both the strength and deflection points of view. However, higher stress and deflection in a bent diaphragm would lead to higher signal outputs, as described in Section 2.2.5. Consequently, square diaphragms are a popular geometry adopted by the micro–pressure sensor industry.

The reader may also notice that the induced maximum stresses in the diaphragms in Examples 4.2 and 4.3 exceed the yield strength of the material. A design engineer may mitigate these stresses by either introducing stiffeners as will be described in Chapter 11 or reducing the size of the diaphragms.

Example 4.4. Determine the deflection and maximum stress in a square diaphragm used in a micro–pressure sensor as illustrated in Figure 4.6*a*. The geometry and dimensions of a square diaphragm are shown in Figure 4.6*b*. The expected maximum applied pressure loading to the micro–pressure sensor is $p = 70$ MPa. The silicon diaphragm has the following material properties: Young's modulus $E = 190,000$ MPa and Poisson's ratio $\nu = 0.25$.

Solution: In reality, the entire top or the front surface of the silicon diaphragm is subjected to the applied pressure. However, we will assume that only the central portion of the diaphragm with an area of $783 \times 783 \ \mu m^2$ is deflectable. The active load-carrying portion of the silicon diaphragm thus has edge length $a = 783 \ \mu m$ or 783×10^{-6} m and thickness $h = 266 \ \mu m$. The maximum stress in the diaphragm can be computed from Equation (4.10) to be

$$\sigma_{max} = \frac{0.308 \times 70 \times 10^6 \times (783 \times 10^{-6})^2}{(266 \times 10^{-6})^2} = 186.81 \text{ MPa}$$

at the midpoint of each edge and the maximum deflection at the center of the diaphragm,

$$w_{max} = -\frac{0.0138 \times 70(783 \times 10^{-6})^4}{190,000(266 \times 10^{-6})^3} = -10153 \times 10^{-11} \text{ m or} - 0.1015 \ \mu m$$

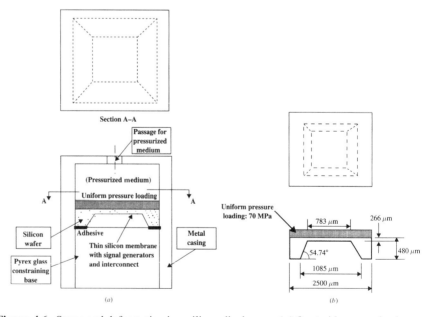

Figure 4.6. Stress and deformation in a silicon diaphragm: (*a*) front-side pressurized sensor; (*b*) dimensions of silicon diaphragm.

One may readily see from the computed results that the maximum stress in the diaphragm is much below the yield strength of the material, which is 7000 MPa. The maximum deflection of 0.1015 μm is small because of the small area that is subjected to the pressure.

4.3 MECHANICAL VIBRATION

The theory of mechanical vibration is the basis for microaccelerometer design. Application of these devices, such as the inertia sensors used in airbag deployment systems in automobiles, is presented in Chapter 1, and the working principles are described in Chapter 2. We will review the fundamental principles of mechanical vibration and its application in microaccelerometer design in this section.

4.3.1 General Formulation

The simplest mechanical vibration system is the mass–spring system illustrated in Figure 4.7a. The mass that is hung from the spring with a spring constant k vibrates from its initial equilibrium position because of the application of a small instantaneous disturbance to the mass. The displacement of mass at a given time t, $X(t)$, can be obtained from the solution of the following equation of motion derived from Newton's second law:

$$m\frac{d^2X(t)}{dt^2} + kX(t) = 0 \tag{4.14}$$

The general solution of Equation (4.14) has the form

$$X(t) = C_1\cos(\omega t) + C_2\sin(\omega t) \tag{4.15}$$

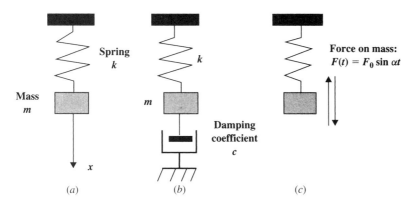

Figure 4.7. Simple mechanical vibration systems: (a) free vibration of mass–spring; (b) free vibration of mass–spring–dashpot; (c) forced vibration of mass–spring.

where the "circular frequency" of the vibrating mass is derived as

$$\omega = \left(\frac{k}{m}\right)^{1/2} \tag{4.16}$$

and C_1 and C_2 are arbitrary constants to be determined by appropriate initial conditions.
The frequency of the vibrating mass is given as

$$f = \frac{\omega}{2\pi}$$

The circular frequency ω in Equation (4.16) is often referred to as the *natural frequency* of the system—a very important quantity in accessing the resonant vibration of solid structures including microdevices. It has a unit of radians per second (rad/s).

Example 4.5. Determine the amplitude and frequency of vibration of a 10-mg mass suspended from a spring with a spring constant $k = 6 \times 10^{-5}$ N/m. The vibration of the mass is initiated by a small "pull" of the mass downward by an amount $\delta_{st} = 5\ \mu$m.

Solution: The description of the physical situation in this example matches the one that Equation (4.14) represents, with initial displacement

$$X(0) = \delta_{st} = 5 \times 10^{-6}\ \text{m}$$

and initial velocity

$$\dot{X}(0) = 0$$

The solution in Equation (4.15) is applicable for the problem with $C_2 = 0$ and $C_1 = \delta_{st} = 5 \times 10^{-6}$ m from the above conditions.
The amplitude of the vibrating mass at any given time t is thus

$$X(t) = (5 \times 10^{-6}) \cos \omega t \tag{4.17}$$

where the circular frequency ω is determined using Equation (4.16) as

$$\omega = \sqrt{\frac{k}{m}} = \sqrt{\frac{6 \times 10^{-5}}{10^{-5}}} = 2.45\ \text{rad/s}$$

and the corresponding frequency is

$$f = \frac{\omega}{2\pi} = \frac{2.45}{2 \times 3.14} = 0.39\ \text{cycle/s (cps)}$$

The natural frequency ω_n of the above spring–mass system is the same as the circular frequency, or

$$\omega_n = 2.45 \text{ rad/s}$$

Example 4.6. Determine the same parameters of Example 4.5 for the balanced mass–spring system illustrated in Figure 4.8 with spring constant $k_1 = k_2$.

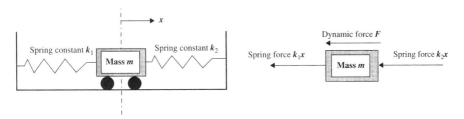

Figure 4.8. Balanced spring–mass system.

Solution: The free-body diagram shown in Figure 4.8 leads to the equation of motion for the mass as

$$m\ddot{X}(t) + (k_1 + k_2)X(t) = 0 \tag{4.18}$$

The solution of Equation (4.18) is similar to that for Equation (4.14), with $k = k_1 + k_2$.

Hence, the amplitude of the vibrating mass remains the same as in Equation (4.17), but the circular frequency, and thus the natural frequency, becomes

$$\omega_n = \sqrt{\frac{k_1 + k_2}{m}} = \sqrt{\frac{(6+6) \times 10^{-5}}{10^{-5}}} = 3.464 \text{ rad/s}$$

One will realize that the solution in Equation (4.15) for the free vibration of a mass–spring system will be oscillatory with constant maximum amplitudes. The oscillation of the mass about its initial equilibrium position will extend indefinitely with time, which obviously is not realistic.

Now, if we introduce a dashpot into the system as shown in Figure 4.7*b*, the dashpot will induce a resistance to the motion of the vibrating mass or a *damping* effect that results in the reduction in amplitudes during the vibration.

Let us assume that the dashpot has a damping coefficient c that will generate a retarding damping force proportional to the velocity of the vibrating mass. The equation of motion in Equation (4.14) is modified to give

$$m\frac{d^2X(t)}{dt^2} + c\frac{dX(t)}{dt} + kX(t) = 0 \tag{4.19}$$

The instantaneous position of the mass, $X(t)$ in Equation (4.19), will take one of the following three forms, depending on the magnitude of the *damping parameter*, defined as $\lambda = c/(2m)$.

Case 1—$\lambda^2 - \omega^2 > 0$, an "overdamping" situation:

$$X(t) = e^{-\lambda t}(C_1 e^{t\sqrt{\lambda^2-\omega^2}} + C_2 e^{-t\sqrt{\lambda^2-\omega^2}}) \tag{4.20a}$$

where C_1 and C_2 are arbitrary constants and ω is given in Equation (4.16).

Figure 4.9 illustrates the solution of Equation (4.20a). We can envisage from these diagrams that there is no oscillatory motion of the mass and the amplitude of vibration of the mass drops rapidly from its initial position in this case. Overdamping is thus desirable in the design of machines and devices, including microsystems that are vulnerable to excessive vibration. The design engineer should select a proper damper in the design of such microdevices.

Case 2—$\lambda^2 - \omega^2 = 0$, a critical damping situation:

$$X(t) = e^{-\lambda t}(C_1 + C_2 t) \tag{4.20b}$$

The above solution is illustrated in Figure 4.10. These representations indicate that the amplitude of vibration decreases initially followed by a slight increase before eventual decay. It is not as desirable a situation as the overdamping case.

Case 3—$\lambda^2 - \omega^2 < 0$, an underdamping situation:

$$X(t) = e^{-\lambda t}(C_1 \cos\sqrt{\omega^2 - \lambda^2}t + C_2 \sin\sqrt{\omega^2 - \lambda^2}t) \tag{4.20c}$$

Figure 4.11 illustrates the underdamping of a system. We realize from Figure 4.11 that, although the amplitude of vibration decays continuously, the mass remains in an oscillatory motion for an indefinite time. This situation obviously is the least desirable as far as machine design is concerned.

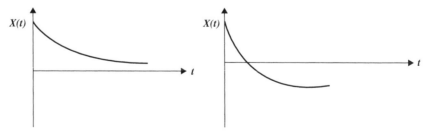

Figure 4.9. Overdamping situation.

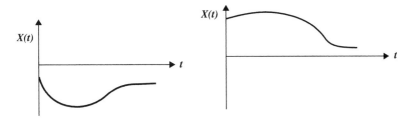

Figure 4.10. Critical damping situation.

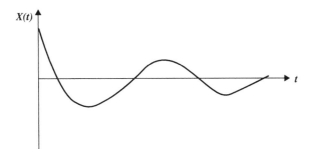

Figure 4.11. Underdamping situation.

4.3.2 Resonant Vibration

Let us consider a case in which the simple mass–spring system in Figure 4.7a is subjected to a force with a harmonic frequency α as shown in Figure 4.7c. The frequency α is often referred to as the *frequency of the excitation force* in a forced vibration.

The equation of motion for the instantaneous position of the mass, $X(t)$, can be expressed as

$$m\frac{d^2X(t)}{dt^2} + kX(t) = F_0 \sin(\alpha t) \tag{4.21}$$

where F_0 is the maximum amplitude of the applied force.

Solving Equation (4.21) for $X(t)$ gives

$$X(t) = \frac{F_0}{\omega(\omega^2 - \alpha^2)}(-\alpha \sin \omega t + \omega \sin \alpha t) \tag{4.22}$$

It is seen from Equation (4.22) that $X(t) \to 0/0$, that is, it is indeterminate when $\alpha = \omega$. However, upon applying the L'Hopital rule, the following solution is obtained for this special case:

$$X(t) = \frac{F_0}{2\omega^2} \sin \omega t - \frac{F_0}{2\omega}t \cos \omega t \tag{4.23}$$

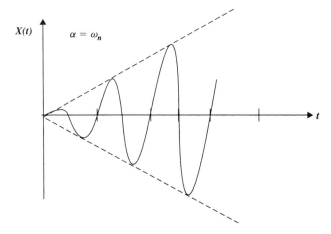

Figure 4.12. Rapid increase of amplitude in a resonant vibration.

One may observe that $X(t) \rightarrow \infty$ (i.e., a very large amplitude of vibration) in a very short time when $\alpha = \omega$, as illustrated in Figure 4.12. This phenomenon is referred to as the *resonant vibration* of a mass–spring system.

Although the equation for resonant vibration is derived for a simple mass–spring system, resonant vibration occurs in other forms of structures whenever the condition of the applied force frequency α equals the natural frequency ω of the structure. For microdevices of complex geometry, there are in theory an infinite number of *modes* in which resonant vibration can take place. Each of these modes is associated with a distinct deformed geometry of the device during the resonant vibration. These multimode resonant vibrations of structures can be attributed to the infinite number of natural frequencies inherent within the structural system.

The method used to determine these natural frequencies associated with various modes of vibration is similar to that used for the simple spring–mass systems. The only difference is that the term ω_n is used to represent the natural frequencies of a structure of complex geometry:

$$\omega_n = \sqrt{\frac{K}{M}} \tag{4.24}$$

The frequency ω_n in Equation (4.24) is called the *natural frequency* of a structure at the nth mode, where n is the mode number, with $n = 1,2, \ldots$. The analysis involving the determination of ω_n ($n = 1,2, \ldots$) is referred to as the *modal analysis* of a structure. In the modal analysis of a structure such as a microdevice, the *stiffness constant K* and mass M in Equation (4.24) are replaced by the respective stiffness matrix [**K**] and mass matrix [**M**] of the structure. These matrices are obtainable from a finite element analysis. Resonant vibration in the structure occurs whenever the frequency of the applied source of vibration, α, equals any of the natural frequencies of the structure, ω_n ($n = 1,2, \ldots$).

The consequence of the resonant vibration of a structure can be catastrophic. Thus, engineering design of structures generally attempts to avoid such occurrence by raising the natural frequencies of the structure so that all conceivable frequencies of vibration induced by the applied excitation forces will not reach even the lowest of all modes of the natural frequencies. This is required in the miniature microphone design described in Chapter 2. However, exceptions are made in the design of many other microdevices, for example, the design of microaccelerometers and acoustic wave sensors as described in Chapter 2, where close-to-resonant vibration of the instrument can result in larger amplitude of vibration of the mass and thus provide more significant and sensitive output signals.

4.3.3 Microaccelerometers

The principle of an accelerometer can be demonstrated by a simple mass attached to a spring that in turn is attached to a casing, as illustrated in Figure 4.13a. The mass used in accelerometers is often called the *seismic* or *proof* mass. In most cases, the system also includes a dashpot to provide the desirable damping effect. A dashpot with damping coefficient c is normally attached to the mass in parallel with the spring (Figure 4.13b).

For miniature, accelerometers the coil spring and dashpot in Figures 4.13a and b occupy too much space. Alternative component arrangements are those shown in

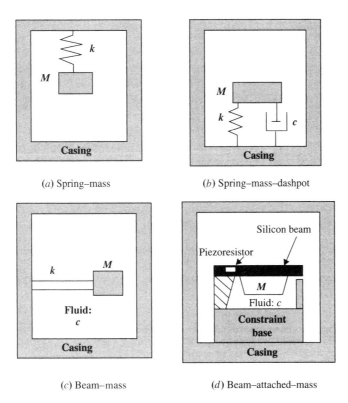

(a) Spring–mass   (b) Spring–mass–dashpot

(c) Beam–mass   (d) Beam–attached–mass

Figure 4.13. Schematic configurations of accelerometers.

Figures 4.13*c* (Petersen, 1982; Putty and Chang, 1989) and 4.13*d* (Barth, 1990). The basic configuration of microaccelerometers includes a casing that contains either a cantilever beam and an attached mass as illustrated in Figure 4.13*c* or a hanging mass from a thin beam or diaphragm as shown in Figure 4.13*d*. In a beam–mass or diaphragm–mass system, the elastic beam or diaphragm replaces the spring and the entrapped air or fluid provides the damping effect.

An alternative arrangement of a more compact microaccelerometer is illustrated in Figure 2.35, where the beam mass is attached to two tether springs made of thin silicon frames. The beam–mass system moves in the longitudinal direction when it is subjected to acceleration or deceleration in the same direction.

If the casing of an accelerometer is attached to a vibrating machine or device, when such a machine or device is subject to a dynamic or impact loads, the mass (called *proof mass*) will vibrate. The acceleration of the vibrating mass in the accelerometer can be measured from its instantaneous position using a potentiometer or comparable means. Theoretical correlation of the amplitude of vibration of the mass and the acceleration of the casing can be found in many textbooks on vibration.

4.3.4 Design Theory of Accelerometers

We will present in this section the theoretical formulation of the amplitude of vibration of the proof mass in an accelerometer. This formulation can also be used as the basis of the design of microaccelerometers.

Figure 4.14*a* illustrates a typical accelerometer that consists of a proof mass supported by a spring and a dashpot. The casing of the vibration system is attached to a vibrating machine with an amplitude of vibration $x(t)$ that can be described by a harmonic motion, expressed mathematically as

$$x(t) = X \sin \omega t \tag{4.25}$$

where X is the maximum amplitude of vibration of the base, t is time, and ω is the circular frequency of vibration of the base.

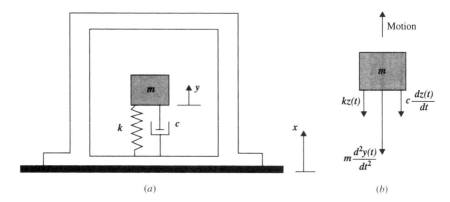

(a)   *(b)*

Figure 4.14. Typical accelerometer: (*a*) schematic of accelerometer; (*b*) forces on seismic mass.

If we designate $y(t)$ to be the amplitude of vibration of the mass m from its initial equilibrium position, then the relative, or net, motion of the mass m with reference to the base can be expressed as

$$z(t) = y(t) - x(t) \tag{4.26}$$

from which we have

$$\dot{z}(t) = \dot{y}(t) - \dot{x}(t) \quad \ddot{z}(t) = \ddot{y}(t) - \ddot{x}(t)$$

where

$$\dot{z}(t) = \frac{dz(t)}{dt} \quad \ddot{z}(t) = \frac{d^2 z(t)}{dt^2}$$

and so on.

The equation of motion of the proof mass can be derived from Newton's law of dynamic force equilibrium, as illustrated in Figure 4.14b, as

$$-kz(t) - c\dot{z}(t) - m\ddot{y}(t) = 0 \tag{4.27}$$

Substituting the relation $\ddot{y}(t) = \ddot{z}(t) + \ddot{x}(t)$ into Equation (4.27) leads to

$$m\ddot{z}(t) + c\dot{z}(t) + kz(t) = -m\ddot{x}(t) \tag{4.28}$$

Since the attached machine vibrates with an amplitude $x(t) = X \sin \omega t$, as in Equation (4.25), Equation (4.28) takes the form

$$m\ddot{z}(t) + c\dot{z}(t) + kz(t) = mX\omega^2 \sin \omega t \tag{4.29}$$

Equation (4.29) is a second-order nonhomogeneous differential equation, and its solution consists of two parts: the complementary solution (CS) and the particular solution (PS).

The CS of Equation (4.29) can be obtained from the homogeneous portion of that equation:

$$m\ddot{z}(t) + c\dot{z}(t) + kz(t) = 0$$

which is similar to Equation (4.19) with solution presented in the form of Equations (4.20a–4.20c). In any of these three cases for the CS, the amplitude of vibration $z(t)$ is illustrated in Figures 4.9–4.11.

The part critical to accelerometer design is in the PS of Equation (4.29). This part of the solution can be obtained by assuming that

$$z(t) = Z \sin(\omega t - \phi) \tag{4.30}$$

where ϕ is the phase angle difference between the input motion $x(t) = X \sin \omega t$ and the relative motion $z(t)$.

The maximum amplitude of the relative motion of the mass, that is, Z, can be determined by substituting Equation (4.30) into Equation (4.29). It can be shown in the two forms

$$Z = \frac{\omega^2 X}{\sqrt{\left(k/m - \omega^2\right)^2 - (\omega c/m)^2}} \tag{4.31a}$$

$$\phi = \tan^{-1} \frac{\omega c/m}{k/m - \omega^2} \tag{4.31b}$$

Alternatively, the above solution can be expressed as

$$Z = \frac{\omega^2 X}{\omega_n^2 \sqrt{\left[1 - (\omega/\omega_n)^2\right]^2 + \left[2h(\omega/\omega_n)\right]^2}} \tag{4.32a}$$

$$\phi = \tan^{-1} \frac{2h(\omega/\omega_n)}{1 - (\omega/\omega_n)^2} \tag{4.32b}$$

where

$$\omega_n = \sqrt{\frac{k}{m}}$$

is the natural frequency of the undamped free vibration of the accelerometer and

$$h = \frac{c}{2m\omega_n} = \frac{c}{c_c}$$

is the ratio of the damping coefficients of the damping medium in the microaccelerometer to its critical damping, with $c_c = 2m\omega_n$ [see the case represented by Equation (4.20b)].

One may readily observe that, when the system approaches resonant vibration, that is, $\omega \to \omega_n$, the amplitude of vibration $Z \to X/(2h)$ from Equation (4.32a). Since h is related to the damping effect, a free vibration with $h = 0$ will lead to "infinitely large" amplitude of vibration of the mass. Hence the selection of the damping parameter h is critical in accelerometer design.

The effect of damping on the amplitude of vibration of the mass is qualitatively illustrated in Figure 4.15. Curves included in the figure were drawn from Equation (4.32a). One may observe from this illustration that when the frequency of vibration of the system ω is much greater than the natural frequency of the accelerometer, that is, $\omega \gg \omega_n$, the maximum relative amplitude Z is approximately equal to the maximum amplitude of vibration of the proof mass, or $Z = X$. Precise graphs of Figure 4.15 with Z/X versus ω/ω_n as well as $\omega_n Z/\ddot{X}$ can be found in several mechanical vibration textbooks. These graphs serve a useful purpose in the design of accelerometers. On the other hand, if the frequency of vibration of the machine, ω, is much smaller than the natural frequency of the microaccelerometer, that is, $\omega \ll \omega_n$, we will observe the following relationship (Dove and Adams, 1964):

$$Z \approx -\frac{a_{\text{base,max}}}{\omega_n^2} \tag{4.33}$$

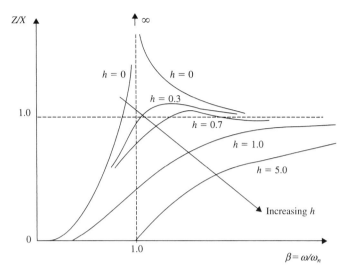

Figure 4.15. Amplitude of vibration of an accelerometer.

where $a_{base,max}$ is the maximum acceleration of the machine on which the accelerometer is attached.

Example 4.7. Derive a formula for estimating the natural frequency of a microaccelerometer involving a beam mass as illustrated in Figure 4.13c with negligible damping effect.

Solution: We may idealize the configuration in Figure 4.13c as a simple cantilever beam subjected to an equivalent concentrated load $W = Mg$ at the free end, as shown in Figure 4.16, where W is the weight of the mass M and g is gravitational acceleration.

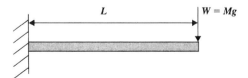

Figure 4.16. Beam spring.

Recall the maximum deflection of the beam, $\delta = WL^3/(3EI)$, where E and I are the respective Young's modulus of the beam material and the area moment of inertia of the beam cross section.

Since the spring constant of a material, k, is defined as the ratio of load to deflection, we will have the equivalent spring constant for the beam:

$$k = \frac{W}{\delta} = \frac{3EI}{L^3} \tag{4.34}$$

The above equivalent spring constant leads to the approximate natural frequency ω_n of the microaccelerometer:

$$\omega_n = \sqrt{\frac{k}{M}} = \sqrt{\frac{3EI}{ML^3}} \tag{4.35}$$

where M is the proof mass attached to the beam and the mass of the beam is neglected.

The reader is reminded that the above example is only good as an approximation, as many microaccelerometers are constructed with suspended mass from a cantilever plate instead of a beam. The equivalent spring constant derived from the simple beam theory obviously is not applicable to these cases. A silicon-based microaccelerometer with a proof mass attached to a cantilever plate was constructed (Bryzek et al., 1992). It had a diaphragm (i.e., the cantilever plate) of $3.4 \times 3.4\,\text{mm} \times 1.25\,\text{mm}$ thick. The displacement of the mass M and hence the acceleration of the vibrating diaphragm are correlated to the associated strains in the diaphragm by the piezoresistors implanted at the root of the cantilever plate by a diffusion process. There were two "overrange" stops installed in the device to prevent possible damage of the sensor from mishandling. Because of the complex geometry of the accelerometer, a finite element method was used to determine the natural frequencies of the device.

Example 4.8. Determine the equivalent spring constant k and the natural frequency ω_n of a cantilever beam element in a microaccelerometer as illustrated in Figure 4.17. The beam is made of silicon with a Young's modulus of 190,000 MPa.

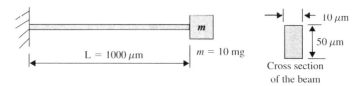

Figure 4.17. Beam spring and proof mass.

Solution: The moment of inertia of the beam crosssection is given as

$$I = \frac{(10 \times 10^{-6})(50 \times 10^{-6})^3}{12} = 0.1042 \times 10^{-18}\ \text{m}^4$$

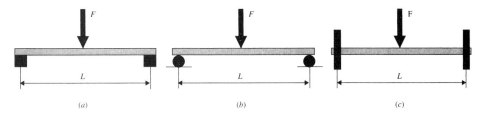

Figure 4.18. Beam springs for microaccelerometers: (*a*) beam spring; (*b*) simply supported ends; (*c*) fixed ends.

The spring constant k of the beam can be determined by using Equation (4.34) as

$$k = \frac{3(190{,}000 \times 10^6)(0.1042 \times 10^{-18})}{(1000 \times 10^{-6})^3} = 59.39 \text{ N/m}$$

The natural frequency ω_n of the beam spring–mass system is calculated using Equation (4.24) as

$$\omega_n = \sqrt{\frac{k}{m}} = \sqrt{\frac{59.39}{10^{-5}}} = 2437 \text{ rad/s}$$

The unit kilogram was used for the mass m in the above calculation.

Other types of beam springs can be used for microaccelerometer design. The reader will readily derive the equivalent spring constants for the centrally loaded beams illustrated in Figure 4.18a to be $k = (48EI)/L^3$ for the simply supported beam spring and $k = (192EI)/L^3$ for the fixed-end beam spring.

Example 4.9. Determine the natural frequency of a "force-balanced" microaccelerometer similar to the one illustrated in Figure 2.35. An idealized form of the structure is illustrated in Figure 4.19. Dimensions of the beam springs are indicated in Figure 4.20.

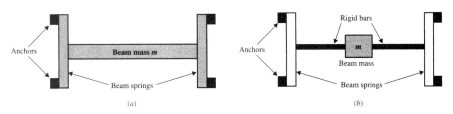

Figure 4.19. Basic structure of a force-balanced microaccelerometer: (*a*) beam mass–spring system; (*b*) idealized situation.

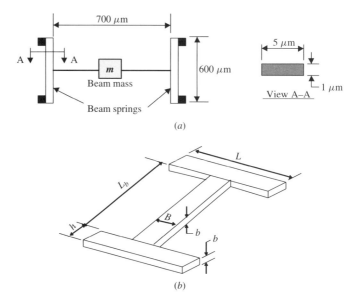

Figure 4.20. Dimensions of a force-balanced microaccelerometer: (*a*) idealized configuration; (*b*) perspective view.

The entire structure is assumed to be made of silicon with Young's modulus $E = 190,000$ MPa. The proof mass m is assumed to be concentrated at the centroid of the moving beam, as illustrated in Figure 4.19*b*.

Solution: With the given dimensions of the accelerometer in Figure 4.20*a*, we have, according to Figure 4.20b, the following dimensions computing the natural frequencies of the accelerometer:

$$b = 1 \times 10^{-6} \text{ m} \quad B = 100 \times 10^{-6} \text{ m}$$

$$L = 600 \times 10^{-6} \text{ m} \quad L_b = 700 \times 10^{-6} \text{ m}$$

The proof mass of the moving beam is given as $m = \rho V$, where ρ, the mass density of silicon, is 2.3 g/cm^3 from Table 7.3 and $V = BbL_b$ is the volume of the beam. We compute the proof mass $m = 16.1 \times 10^{-11}$ kg.

The area moment of inertia of the beam spring, I, is first computed:

$$I = \frac{bh^3}{12} = \frac{(1 \times 10^{-6})(5 \times 10^{-6})^3}{12} = 10.42 \times 10^{-24} \text{ m}^4$$

Case 1 Beams are simply supported at the ends, Figure 4.18b: The equivalent spring constant k is determined as:

$$k = \frac{48EI}{L^3} = \frac{48(190,000 \times 10^6)(10.42 \times 10^{-24})}{(600 \times 10^{-6})^3} = 0.44 \text{ N/m}$$

The natural frequency of the arrangement in Figure 4.20 is similar to a mass vibrating between two springs, as described in Example 4.6 and Figure 4.8. We thus have

$$\omega_n = \sqrt{\frac{2k}{m}} = \sqrt{\frac{2 \times 0.44}{16.1 \times 10^{-11}}} = 23{,}380 \text{ rad/s}$$

Case 2 Beams are rigidly fixed at the ends, Figure 4.18c: The equivalent spring constants k takes a different form:

$$k = \frac{192EI}{L^3} = \frac{192(190{,}000 \times 10^6)(10.42 \times 10^{-24})}{(600 \times 10^{-6})^3} = 1.76 \text{ N/m}$$

The corresponding natural frequency ω_n is given as

$$\omega_n = \sqrt{\frac{2k}{m}} = \sqrt{\frac{2 \times 1.76}{16.1 \times 10^{-11}}} = 147{,}860 \text{ rad/s}$$

We notice that the natural frequencies of this beam accelerometer design in both cases are very high to avoid resonant vibrations.

Example 4.10. Determine the displacement from its neutral equilibrium position of the mass of the microaccelerometer in Example 4.9 one (1) ms after the deceleration from its initial velocity of 50 km/h to a standstill.

Solution: We have derived the equivalent spring constant, $k_{eq} = 1.76 \text{ N/m}$, from case 2 of Example 4.9. Thus, by using this equivalent spring constant and the situation simulated in Figure 4.8, we have the equation of motion given in Equation (4.14) in a new form:

$$m \frac{d^2 X(t)}{dt^2} + 2k_{eq} X(t) = 0$$

or

$$\frac{d^2 X(t)}{dt^2} + \omega^2 X(t) = 0$$

where

$$\omega = \sqrt{\frac{2k_{eq}}{m}} = \sqrt{\frac{2 \times 1.76}{16.1 \times 10^{-11}}} = 147{,}860 \text{ rad/s}$$

The two initial conditions for the above differential equation are

$$X(t)|_{t=0} = 0 \quad \text{for initial displacement}$$

$$\frac{dX(t)}{dt}\bigg|_{t=0} = 50 \text{ km/h} = 13.8888 \text{ m/s} \quad \text{for initial velocity}$$

The general solution of the differential equation is similar to that shown in Equation (4.15); that is,

$$X(t) = C_1 \cos(\omega t) + C_2 \sin(\omega t)$$

The arbitrary constants C_1 and C_2 can be determined by the above two initial conditions to be $C_1 = 0$ and $C_2 = 9.3932 \times 10^{-5}$, which leads to the following expression for the instantaneous position of the centroid of the beam mass:

$$X(t) = 9.3932 \times 10^{-5} \sin(147{,}860t)$$

The position of the beam mass at $t = 1$ ms $= 10^{-3}$ is obtained from the above expression to be $X(10^{-3}) = 9.3932 \times 10^{-5} \sin(147{,}860 \times 10^{-3} \text{ rad}) = 9.3932 \times 10^{-5} \sin(8476°) = -2.597 \times 10^{-5}$ m, or 26 μm from its initial equilibrium position opposite to the direction of deceleration.

4.3.5 Damping Coefficients

The damping coefficient c in Equation (4.19) has a significant effect on the physical behavior of a mechanical vibration system. Figure 4.15 illustrates that the amplitude of the vibrating proof mass is sensitive to the h factor, which is related to the damping coefficient of the surrounding medium.

Damping is a form of resistance induced by the friction between the surface of the vibrating mass and the surrounding fluid, which can be either liquid or gas. In microaccelerometer design, damping can occur in two distinct ways:

(a) *Squeeze-film damping* for those microaccelerometers involving the compression of the surrounding damping fluids by the vibrating mass, as illustrated in Figure 4.13

(b) *Shear resistance* for the force-balanced type of microaccelerometers, depicted in Figures 1.6, 2.35, and 4.19

In either way, the damping coefficient c can be determined from the simple relationship

$$F_D(t) = cV(t) \tag{4.36}$$

where F_D is the resistance force to the moving mass, c is the damping coefficient, and $V(t)$ is the velocity of the moving mass.

We will derive the expressions for estimating the damping coefficients in the design of microaccelerometers.

Damping Coefficient in Squeeze Film. The following derivations are based on the information provided in two sources (Newell, 1968; Starr, 1990). The system illustrated in Figure 4.21 represents a vibrating strip with length $2L$ and width $2W$ that squeezes

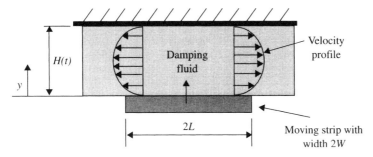

Figure 4.21. Squeeze-film damping.

the damping fluid in a narrow gap $H(t)$. If $y(t)$ is the instantaneous position of the strip, then the velocity of the moving strip is expressed as $\dot{y}(t)$.

For noncompressible damping fluid media, the following expression for the resistance force F_D is obtained:

$$F_D = 16 f \left(\frac{W}{L}\right) W^3 L \left(\frac{dy(t)}{dt}\right) H_0^3 \qquad (4.37)$$

where H_0 is the nominal thickness of the fluid film.

The derivative of $y(t)$ in Equation (4.37) represents the velocity of the moving strip. Thus the squeeze damping coefficient c can be obtained from Equations (4.36) and (4.37):

$$c = 16 f \left(\frac{W}{L}\right) W^3 L H_0^3 \qquad (4.38)$$

Numerical values of the function $f(W/L)$ in Equation (4.38) versus W/L are given in Table 4.2. Clearly, the damping coefficient in noncompressible squeeze films is independent of the fluid properties.

TABLE 4.2. Geometric Function for Squeeze Film Damping

W/L	$f(W/L)$	W/L	$f(W/L)$
0	1.00	0.6	0.60
0.1	0.92	0.7	0.55
0.2	0.85	0.8	0.50
0.3	0.78	0.9	0.45
0.4	0.72	1.0	0.41
0.5	0.60		

Example 4.11. Estimate the damping coefficient of a microaccelerometer using a cantilever beam spring as illustrated in Figure 4.22.

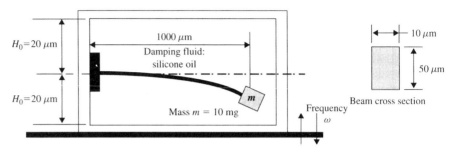

Figure 4.22. Microaccelerometer with liquid damping fluid.

Solution: The damping coefficient for the case in Figure 4.22 can be determined using Equation (4.38) with $L \approx 500 \times 10^{-6}$ m and $W = 5 \times 10^{-6}$ m. The function $f(W/L) = 0.992$ by interpolation from Table 4.2 with $W/L = 0.01$.

The nominal film thickness H_0 can be determined by the average maximum deflection of the beam along the length or simply using the gap $H_0 = 20\,\mu$m, as shown in Figure 4.22. The adoption of the latter value leads to a very small value of $c = 8 \times 10^{-33}$ N-s/m.

For a squeeze film made of compressible fluids such as air, a *squeeze number S* is introduced in the form (Starr, 1990)

$$S = \frac{12\mu\omega L^2}{H_0^2 P_a} \tag{4.39}$$

where

μ = dynamic viscosity of gas film, N-s/m^2

ω = frequency of vibrating mass, rad/s

L = characteristic length, m

P_a = ambient gas pressure, Pa

The damping effect is included in the stiffness enhancement Δk of the equivalent spring constant k as

$$\frac{\Delta k}{k} = \alpha f_k(\varepsilon)\left(\frac{\omega}{\omega_n}\right) hS \tag{4.40}$$

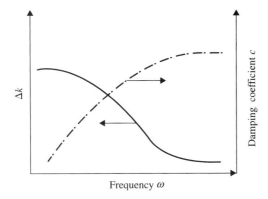

Figure 4.23. Squeeze-film damping coefficient by compressible fluids.

where the constant $\alpha = 0.8$ for long narrow strips such as beam springs, $h = c/(2m\omega_n)$, with ω_n the natural frequency of the accelerometer, and ε is the ratio of the displacement of the moving mass to the film thickness.

The function $f_k(\varepsilon)$ can be determined by the expression

$$f_k(\varepsilon) = \frac{1 + 3\varepsilon^2 + 3\varepsilon^4/8}{(1 - \varepsilon^2)^3} \tag{4.41}$$

Equation (4.40) is qualitatively plotted against the input frequency ω in Figure 4.23.

The reader needs to exercise caution because the above expressions are derived from the theories of continuum fluid dynamics based on the assumption of nonslip fluid flow. The validity of using continuum fluid mechanics theories is limited by the film thickness, or the gap $H(t)$ in Figure 4.21. Depending on the type of gas films, the theory applies only when the gap $H > H_{threshold}$, where $H_{threshold}$ is governed by the *mean free path* of the gas molecules, as will be defined in Chapter 12. For air at 25°C and 1 atm, the mean free path is about 0.09 μm (or 90 nm). Slip flow of gases may take place in film gaps that are less than 100 times the mean free path length. Thus, the above formulation is valid only when the air film thickness $H > 9$ μm.

We notice the appearance of fluid dynamic viscosity μ in Equation (4.39). Indeed, this fluid property is an important factor in assessing the damping of a vibrating system. Table 4.3 presents the dynamic viscosity of seven selected fluids. The same property for many other fluids can be found in *Mark's Standard Handbook for Mechanical Engineers* (Avallone and Baumeister, 1996).

Microdamping in Shear. For microaccelerometers that involve a thin vibrating beam mass (often referred to as the proof mass) moving in the surrounding fluid, as in

TABLE 4.3. Dynamic Viscosity for Selected Fluids (10^{-6} N-s/m^2)

Compressible fluids	0°C	20°C	60°C	100°C	200°C
Air	17.08	18.75	20.00	22.00	25.45
Helium	18.60	19.41	21.18	22.81	26.72
Nitrogen	16.60	17.48	19.22	20.85	24.64
Noncompressible fluids	0°C	20°C	40°C	60°C	80°C
Alcohol	1772.52	1199.87	834.07	591.80	432.26
Kerosene	2959.00	1824.23	1283.18	971.96	780.44
Fresh water	1752.89	1001.65	651.65	463.10	351.00
Silicone oil,[a]	20°C: 740				

[a]From Pfahler et al., 1990.

the force-balanced accelerometers in Figures 2.35 and 4.19, the damping effect is not provided by squeezing the contacting fluid. Rather, it is induced by the resistance in shear between the contacting surfaces of the beam mass and the surrounding fluid.

Let us consider the situation illustrated in Figure 4.24, in which the moving mass m moves in the surrounding fluid at a velocity V. The assumed nonslip fluid flow condition results in linear velocity profiles on both faces of the beam, as shown in the figure.

The shear stress τ_s at either the top or bottom face of the beam mass can be expressed as

$$\tau_s(y) = \mu \frac{du(y)}{dy} \tag{4.42}$$

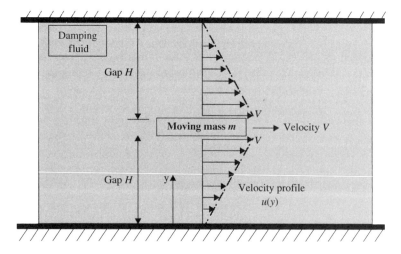

Figure 4.24. Moving mass in nonslip fluid flow.

where μ is the dynamic viscosity of the damping fluid and $u(y)$ the velocity profile in the fluid, with the fluid velocity V at the fluid–solid interface.

The velocity profile in the present case follows a linear relationship, that is, $u(y) = Vy/H$, where H is the width of the gap between the top or bottom surface of the beam and the wall of the enclosure.

We can solve for the shear stress at the contacting surfaces, τ_0, using Equation (4.42) and the above velocity function:

$$\tau_0 = \frac{\mu V}{H}$$

from which we can determine the equivalent shear forces F_D acting on both the top and bottom faces of the beam mass:

$$F_D = \tau_0(2Lb) = \frac{2\mu Lb}{H} V$$

where L, b are the length and width of the beam mass, respectively.

The damping coefficient c can thus be determined using Equation (4.36):

$$c = \frac{F_D}{V} = \frac{2\mu Lb}{H} \quad \text{N-s/m} \tag{4.43}$$

Example 4.12. Estimate the damping coefficient in the force-balanced microaccelerometer described in Example 4.9. The core element of the accelerometer is placed in an enclosure, as illustrated in Figure 4.25, for air and silicone oil as the damping fluid. Assume that the microaccelerometer operates at 20°C.

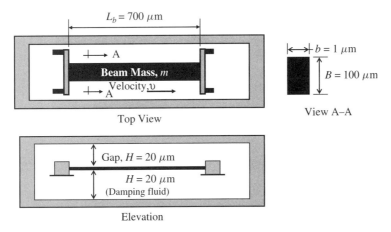

Figure 4.25. Damping of a force-balanced microaccelerometer.

Solution: From Figure 4.25, we have the following physical dimensions: width of the beam mass, $B = 100 \times 10^{-6}$ m; length of the beam, $L_b = 700 \times 10^{-6}$ m; and width of the gap, $H = 20 \times 10^{-6}$ m.

The viscosity for air at $20°C$ is $\mu_{\text{air}} = 18.75 \times 10^{-6}$ N-s/m^2 from Table 4.3. Thus the damping coefficient c is determined from Equation (4.43):

$$c = \frac{2\mu_{\text{air}} L_b B}{H} = \frac{2(18.75 \times 10^{-6})(700 \times 10^{-6})(100 \times 10^{-6})}{20 \times 10^{-6}}$$

$$= 2.625 \times 10^{-12} \text{ N-s/m}$$

For the case involving silicone oil as the damping fluid, we have $\mu_{\text{si}} = 740 \times 10^{-6}$ N-s/m^2 at $20°C$ from Table 4.3, which leads to the damping coefficient

$$c = \frac{2\mu_{\text{si}} L_b B}{H} = \frac{2(740 \times 10^{-6})(700 \times 10^{-6})(100 \times 10^{-6})}{20 \times 10^{-6}}$$

$$= 1.036 \times 10^{-10} \text{ N-s/m}$$

As we have observed from the last two examples, the damping effect in micro-accelerometers is indeed very small. This situation leads to a very small value of h in Equations (4.32a-b). Consequently one may expect large signal output from a microaccelerometer with a low value of h, as can be observed from Figure 4.15. The low damping coefficients actually enhance the signal outputs in microaccelerometers.

Example 4.13. Determine the displacement from its neutral equilibrium position of the beam mass of a microaccelerometer, as described in Example 4.12, 1 ms after the deceleration from its initial velocity of 50 km/h to a standstill. The accelerometer is submerged in silicone oil functioning as a damping medium.

Solution: We obtained the damping coefficient with silicone oil in the arrangement illustrated in Figure 4.25 as $c = 1.036 \times 10^{-10}$ N-s/m.

The equation of motion for the vibrating beam mass is similar to that shown in Example 4.10 with an additional term for damping:

$$m\frac{d^2X(t)}{dt^2} + c\frac{dX(t)}{dt} + 2k_{\text{eq}}X(t) = 0 \tag{a}$$

where m is the mass of the beam mass and k_{eq} is the equivalent spring constant of the beam springs attached to the beam mass.

Equation (a) may be rewritten in the form

$$\frac{d^2X(t)}{dt^2} + \frac{c}{m}\frac{dX(t)}{dt} + \omega^2 X(t) = 0 \tag{b}$$

We have the numerical values of m and ω computed in Example 4.10, which lead to $\omega' = 147,860$ rad/s for the case of fixed-end beam springs and $m = 16.1 \times 10^{-11}$ kg. Equation (b) can thus be expressed in the following form for the present case:

$$\frac{d^2 X(t)}{dt^2} + 0.6435 \frac{dX(t)}{dt} + (1.4786 \times 10^5)^2 X(t) = 0 \tag{c}$$

with the initial conditions

$$X(t)|_{t=0} = 0 \tag{d}$$

and

$$\frac{dX(t)}{dt}\bigg|_{t=0} = 13.8888 \text{ m/s} \tag{e}$$

as given in Example 4.9. Solution of Equation (c) is available as in Equation (4.20c), with

$$\lambda = \frac{c}{2m} = \frac{1.036 \times 10^{-10}}{2 \times 16.1 \times 10^{-11}} = 0.3217$$

We may readily find that $\lambda^2 - \omega^2 \ll 0$, which leads to the solution of Equation (c) in the form

$$X(t) = e^{-\lambda t} \left(c_1 \cos \sqrt{\omega^2 - \lambda^2}\, t + c_2 \sin \sqrt{\omega^2 - \lambda^2}\, t \right) \tag{f}$$

where c_1 and c_2 are two arbitrary constants to be determined by the initial conditions expressed in Equations (d) and (e). We have $c_1 = 0$ and $c_2 = 9.4 \times 10^{-5}$. The complete solution of Equation (c) with numerical values of λ and ω and c_1 and c_2 is

$$X(t) = 9.4 \times 10^{-5} e^{-0.3217t} \sin(1.4786 \times 10^5 t) \tag{g}$$

The solution of $X(t)$ in Equation (g) represents the movement of the beam mass at any instant t in the enclosure after the specified deceleration. The movement of the beam mass at $t = 1$ ms can be evaluated from Equation (g) as

$$X(10^{-3}) = 9.4 \times 10^{-5} e^{-0.3217(10^{-3})} \sin(146.86 \text{ rad})$$

or

$$X(10^{-3}) = 9.4 \times 10^{-5} e^{-0.3217(10^{-3})} \sin(8476°) = -2.58 \times 10^{-5} \text{ m or} -25.8 \,\mu\text{m}$$

The result indicated in this example on the movement of the beam mass with silicone oil as damping medium is approximately 5% less than that without damping as illustrated in Example 4.10.

Example 4.14. Two vehicles with respective masses m_1 and m_2 are traveling in opposite directions at velocities V_1 and V_2, as illustrated in Figure 4.26. Each vehicle is equipped with an inertia sensor (or microaccelerometer) built with a cantilever beam as configured in Example 4.8. Estimate the deflection of the proof mass in the sensor in vehicle 1 with mass m_1 and also the strain in the two piezoresistors embedded underneath the top and bottom surfaces of the beam near the support after the two vehicles collide.

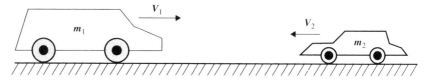

Figure 4.26. Two vehicles in a course of head-on collision.

We have the following information on the microaccelerometer from Example 4.8: The equivalent spring constant $k = 59.39$ N/m and the natural frequency $\omega_n = 2437$ rad/s. The beam has the following properties: Young's modulus $E = 1.9 \times 10^{11}$ N/m^2; moment of inertia I of the cross section 0.1042×10^{-18} m^4; nominal length L of the beam $1000\,\mu$m; distance C from outer surface to the centroid 25×10^{-6} m.

We will have the following conditions for the computations: $m_1 = 12{,}000$ kg; $m_2 = 8000$ kg; $V_1 = V_2 = 50$ km/h.

Solution: We postulate that the two vehicles will tangle together after the collision and the entangled vehicles move at a velocity V as illustrated in Figure 4.27. This final velocity of the entangled vehicles, V, can be obtained by using the principle of conservation of momentum as

$$V = \frac{m_1 V_1 - m_2 V_2}{m_1 + m_2} = \frac{12{,}000 \times 50 - 8000 \times 50}{12{,}000 + 8000} = 10 \text{ km/h}$$

The decelerations of both vehicles can be determined by the expressions

$$\ddot{X} = \begin{cases} \dfrac{V - V_1}{\Delta t} & \text{for vehicle with } m_1 \\[2mm] \dfrac{V - V_2}{\Delta t} & \text{for vehicle with } m_2 \end{cases}$$

where Δt is the time required for the deceleration.

Let us assume that it takes 0.5 s for vehicle 1 to decelerate from 50 to 10 km/h after the collision. Thus the time for deceleration of vehicle m_1 is $\Delta t = 0.5$ s in the above expressions. We may thus compute the deceleration of vehicle m_1 to be

$$\ddot{X} = a_{\text{base}} = \frac{(10 - 50) \times 10^3 / 3600}{0.5} = -22.22 \text{ m/s}^2$$

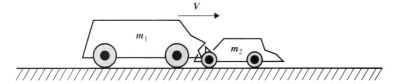

Figure 4.27. Vehicles after the collision.

We may assert that the frequency of vibration, ω, of either vehicle in the case of collision is much smaller than the natural frequency of the microaccelerometer, ω_n; that is, $\omega \ll \omega_n$. We may thus justify the use of the expression in Equation (4.33) for the maximum movement of the proof mass, Z:

$$Z \doteq -\frac{a_{\text{base}}}{\omega_n^2} = \frac{22.22}{(2437)^2} = 3.74 \times 10^{-6} \text{ m or } 3.74 \, \mu\text{m}$$

We thus expect the proof mass to move by $3.74 \, \mu$m during the collision, which means the maximum deflection of the cantilever beam is the same amount at the free end.

The corresponding force for such deflection at the free end is

$$F = \frac{3EIZ}{L^3} = \frac{3(1.9 \times 10^{11})(0.1042 \times 10^{-18})(3.74 \times 10^{-6})}{(1000 \times 10^{-6})^3} = 2.2213 \times 10^{-4} \text{ N}$$

The equivalent maximum bending moment $M_{\text{max}} = FL$, with

$$M_{\text{max}} = 2.2213 \times 10^{-4} \times 10^{-3} = 2.2213 \times 10^{-7} \text{ N-m}$$

from which we can compute the maximum bending stress σ_{max} near the support

$$\sigma_{\text{max}} = \frac{M_{\text{max}}C}{I} = \frac{(2.2213 \times 10^{-7})(25 \times 10^{-6})}{0.1042 \times 10^{-18}} = 532.95 \times 10^5 \text{ Pa} = 53.30 \text{ MPa}$$

The corresponding maximum strains in the piezoresistors in these locations are

$$\varepsilon_{\text{max}} = \frac{\sigma_{\text{max}}}{E} = \frac{53.30 \times 10^5}{190 \times 10^9} = 02.81 \times 10^{-4} = 0.0281\%$$

Example 4.15. Determine if the force-balanced microaccelerometer described in Example 4.9 would be suitable for measuring the movement of the beam mass at a deceleration of vehicle 1 at $\ddot{X} = -22.22 \text{ m/s}^2$ as computed in Example 4.14.

Solution: Figures 4.19 and 4.20 show the dimensions of the microaccelerometer. It is attached to the chassis of the vehicles. The beam mass is attached to two beam springs at each end. Deceleration of the vehicle at 22.22 m/s² produces an inertia force by the beam mass along the longitudinal direction with a magnitude $F = m\ddot{X}$, where m is the mass of the proof beam mass. Each attached spring shares half of this force at its midspan. We will thus have the situation illustrated in Figure 4.28.

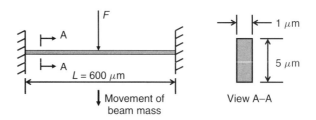

Figure 4.28. Beam spring subject to inertia force.

We computed the following quantities in Example 4.9:

The moment of inertia of the beam spring, $I = 10.42 \times 10^{-24}\,\text{m}^4$.
The mass of the proof beam mass, $m = 16.1 \times 10^{-11}\,\text{kg}$.

We may thus compute the induced inertia force to the beam springs:

$$F = \frac{1}{2}m\ddot{X} = \frac{1}{2}(16.1 \times 10^{-11})(22.22) = 1.7887 \times 10^{-9}\,\text{N}$$

The maximum deflection of the beam spring, which is the corresponding movement of the beam mass in the microaccelerometer, can be computed from formulas available in a handbook for the bending of simply supported beams with both ends fixed (Roark, 1965):

$$\delta_m = \frac{1}{192}\frac{FL^3}{EI} = \frac{(1.7882 \times 10^{-9})(600 \times 10^{-6})^3}{192(1.9 \times 10^{11})(10.42 \times 10^{-24})}$$
$$= 1.012 \times 10^{-9}\,\text{m or }1.012\,\text{nm}$$

We notice the corresponding movement of the beam mass is approximately 1 nm. It is too small to be measured by conventional means. The microaccelerometer described in Example 4.9 is thus not practical for this application.

4.3.6 Resonant Microsensors

We have learned from Equation (4.16) that the natural frequency of a simple mass–spring system is related to the spring constant and the vibrating mass. In reality, a structure or a device has geometry that is far more complicated than that of simple mass–spring systems. Yet the devices of complex geometry are just as vulnerable to resonant vibration as in simple mass–spring systems. Determination of the natural frequency of a device of complex geometry is related to the device's *stiffness matrix* [**K**] and the *mass matrix* [**M**], as described in Section 4.3.2. The natural frequency of various modes of a device can thus be expressed in a form similar to that for a simple mass–spring system:

$$\omega_n = \sqrt{\frac{[\mathbf{K}]}{[\mathbf{M}]}} \tag{4.44}$$

with mode number $n = 1, 2, 3, \ldots$.

Both the stiffness and mass matrices in Equation (4.44) are obtainable from a finite element analysis, as will be described in Chapter 10. The [**K**] matrix is related to both the geometry and material properties of the device. One can thus perceive that any change of the stress and the associated strains in a structure component will result in the change of the component's geometry, which in turn will alter the [**K**] matrix. Consequently, the natural frequency of a device or a component can vary with the change of the states of the stresses in the structure. Shifting of natural frequencies of a device is expected when it is subject to different state of stresses.

As we observe from Figures 4.12 and 4.15, resonant vibrations occur when the casing to which an accelerometer is attached vibrates with a frequency that is equal to any mode of the natural frequency of the device; that is, $\beta = \omega/\omega_n \to 1.0$ with $n = 1, 2, 3, \ldots$. The amplitude of vibration of the proof mass reaches the peak value at this frequency. For a microaccelerometer, the device provides the most sensitive output at this frequency. Thus, the advantage of using the peak sensitivity of a vibration-measuring device at resonant frequency has been adopted in microsensor design. Howe (1987) developed a theory for the natural frequency of a vibrating beam in mode 1 subjected to a longitudinal force as

$$\omega_{n,1}^2 = \frac{\int_0^L (EI/2)\left[d^2 Y_1(x)/dx^2\right]^2 dx + \int_0^L F/2\left[dY_1(x)/dx\right]^2 dx}{\int_0^L (1/2)\rho b h Y_1^2(x)\, dx} \tag{4.45}$$

where

F = applied force to vibrating beam which can be converted to normal stress in beam

ρ = mass density of beam material

E = Young's modulus

I = area moment of inertia of beam cross section

b, h = respective width and thickness of rectangular beam cross section

The function $Y_1(x)$ in Equation (4.45) is the amplitude of the vibrating beam, as illustrated in Figure 4.29. The function $Y_1(x)$ can be obtained from the solution $y(x,t)$ in the form $y(x,t) = Y_1(x) \exp(i\omega t)$, where $i = \sqrt{-1}$ and ω is the frequency of the vibrating beam.

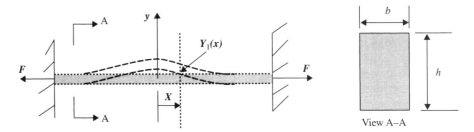

Figure 4.29. Vibrating beam subject to tensile force.

The solution $y(x, t)$ in a vibrating beam can be obtained by solving the following partial differential equation:

$$\alpha^2 \frac{\partial^4 y(x, t)}{\partial x^4} + \frac{\partial^2 y(x, t)}{\partial t^2} = 0 \tag{4.46}$$

where $\alpha = \sqrt{EI/\gamma}$ with γ the mass per unit length of the beam.

The solution of Equation (4.46) obviously requires the specified initial and end conditions. Solutions of several cases with clamped end conditions are available for the design of microbridges (Bouwstra and Geijselaers, 1991). The solution of a special case with the following initial and end conditions is presented below (Trim, 1990).

The situation is for a simply supported beam vibrating in the first mode, as illustrated in Figure 4.30. The following initial and end conditions are used in solving Equation (4.46): The initial deflection and the velocity of the beam in static equilibrium are used as the initial conditions:

$$y(x, 0) = f(x) = x \sin \frac{\pi x}{L} \qquad 0 \le x \le L$$

$$\left. \frac{\partial y(x, t)}{\partial t} \right|_{t=0} = 0 \qquad 0 \le x \le L$$

End, or boundary, conditions for a simply supported beam at $t > 0$ require that

$$y(0, t) = 0 \qquad y(L, t) = 0$$

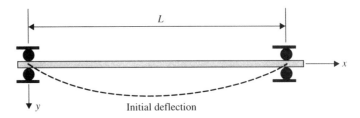

Figure 4.30. Free vibrating Beam.

$$\left.\frac{\partial^2 y(x,t)}{\partial x^2}\right|_{x=0} = 0 \qquad \left.\frac{\partial^2 y(x,t)}{\partial x^2}\right|_{x=L} = 0$$

The solution $y(x,t)$ in Equation (4.46) with the above conditions has the form

$$y(x,t) = \frac{L}{2}\sin\frac{\pi x}{L}\cos\frac{\pi^2 \alpha t}{L^2} - \frac{16L}{\pi^2}\sum_{n=1}^{\infty}\frac{n}{(4n^2-1)^2}\sin\frac{2n\pi x}{L}\cos\frac{4n^2\pi^2\alpha t}{L^2} \qquad (4.47)$$

The number $n = 1,2,3,\ldots$ in Equation (4.47) represents the various modes of vibration of the beam. It is customary to take the first mode (i.e., $n = 1$) for the solution in the design process. One may deduce the maximum amplitude of vibration, that is, the function $Y_1(x)$, for Equation (4.45).

Equation (4.45) has indicated that the natural frequency of a vibrating beam can be changed with the change of the applied axial force (or the normal stress). This principle was used to construct a highly sensitive pressure sensor (Petersen et al., 1991). In this microdevice, a single-crystal p-type silicon beam 40 μm wide $\times$ 6 μm thick $\times$ 600 μm long is attached to the middle of the front-face cavity of the diaphragm in a pressure sensor by means of fusion bonding. The beam is first excited to resonance by applying an AC voltage signal to the diffused deflection electrode beneath the beam. This takes typically a few millivolts. Electrostatic forces from the AC power source intermittently pull the beam downward at the proper frequency to excite the beam to vibrate at its natural frequency. The resonant vibration of the beam is detected by the piezoresistors diffused into the ends of the beam. A schematic of this type of pressure sensor is shown in Figure 4.31.

The application of pressure at the back face of the diaphragm would induce tensile stress in the vibrating beam and thus alter its natural frequency. Adjustment of the supply excitation voltage is necessary to get the beam to vibrate at resonance at the new state. The applied pressure can be calibrated to correspond to the shifting of the resonant frequency of the vibrating beam. This method has proven to offer much higher sensitivity than the micro–pressure sensors that use other forms of signal transduction, such as piezoresistors or capacitors. An overall measurement accuracy of 0.01% was achieved. The corresponding sensitivity of the pressure sensor was presented in Figure 2.14.

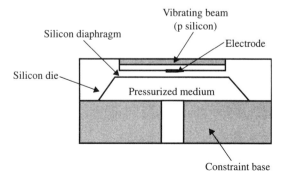

Figure 4.31. Resonant vibrating beam pressure sensor.

Example 4.16. Determine the shift of natural frequency of a fixed-end beam made of silicon with geometry and dimensions shown in Figure 4.32 when subjected to a longitudinal stress at 187 MPa. The dimensions of the beam are width $b = 40 \times 10^{-6}$ m, depth (thickness) $h = 6 \times 10^{-6}$ m, and length $L = 600 \times 10^{-6}$ m. The following properties from Table 7.3 are appropriate for silicon beams:

Young's modulus $E = 190{,}000$ MPa
Mass density $\rho = 2300$ kg/m^3

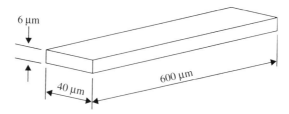

Figure 4.32. Resonant vibrating beam.

Solution: The parameters required for the computation are as follows:

Mass per unit length of the beam, $\gamma = 5.52 \times 10^{-7}$ kg/m
Weight per unit length of the beam, $w = 5.4096 \times 10^{-6}$ N/m
Area moment of inertia, $I = 7.2 \times 10^{-22}$ m^4

The partial differential equation (4.46) is used to compute the amplitude of vibration with the following conditions:
End conditions:

$$\left.\frac{\partial y(x,t)}{\partial x}\right|_{x=0} = 0 \quad y(x,t)|_{x=0} \qquad = 0 \quad \text{at the end } x = 0$$

$$\left.\frac{\partial y(x,t)}{\partial x}\right|_{x=L} = 0 \quad y(x,t)|_{x=L} \qquad = 0 \quad \text{at the other end } x = L$$

Initial conditions:

$$\left.\frac{\partial y(x,t)}{\partial t}\right|_{t=0} = 0$$

for the initial velocity and

$$y(x,t)|_{t=0} = f(x) = \frac{w}{24EI}(x-L)^2 x^2$$

for the initial sag of the beam due to its own weight (from handbooks).

The separation-of-variables technique is used for the solution of the partial differential equation. The solution $y(x, t)$ in Equation (4.46) with the above end and initial conditions can be expressed as

$$y(x, t) = \sum_{n=1}^{\infty} K_n \left[\cosh(\lambda_n x) \right.$$

$$\left. - \cos(\lambda_n x) - \frac{\cosh(\lambda_n L) - \cos(\lambda_n L)}{\sinh(\lambda_n L) - \sin(\lambda_n L)} [\sinh(\lambda_n x) - \sin(\lambda_n x)] \right]$$

$$\times \cos[\alpha (\lambda_n)^2 t]$$

where the coefficient kernel K_n is obtained from the following integrals:

$$K_n = \frac{\int_0^L f(x) X(x) \, dx}{\int_0^L [X(x)]^2 \, dx}$$

where the $X(x)$ function is the part in the above general solution of $y(x, t)$, that is,

$$X(x) = [\cosh(\lambda_n x) - \cos(\lambda_n x)] - \frac{\cosh(\lambda_n L) - \cos(\lambda_n L)}{\sinh(\lambda_n L) - \sin(\lambda_n L)} [\sinh(\lambda_n x) - \sin(\lambda_n x)]$$

The function $f(x)$ is given as one of the two initial conditions.

The number n in the infinite-series solution represents the *mode number* of vibration. For the problem at hand, we are only interested in the mode 1 vibration, that is, $n = 1$. The corresponding eigenvalue λ_1 can be evaluated as $4.73/L$. The natural frequency of the beam in mode 1 is

$$\omega_1 = \lambda_1^2 \sqrt{\frac{EI}{\gamma}} = 9.783 \times 10^5 \text{ rad/s} \quad \text{or} \quad \omega_1 = \frac{9.783 \times 10^5}{2\pi} = 1.557 \times 10^5 \text{ Hz}$$

The function $Y_1(x)$ in Equation (4.45) is used to determine the shift of natural frequencies due to applied stress. It can be obtained from the solution of $y(x,t)$ shown above with $n = 1$:

$$Y_1(x) = K_1 X_1(x)$$

where K_1 and $X_1(x)$ are the respective coefficients K_n and $X(x)$ evaluated at $n = 1$ in the above expressions.

The applied longitudinal force to the beam is $F = A\sigma$, where A is the cross-sectional area of the beam ($240 \times 10^{-12} \text{ m}^2$) and σ is the applied stress at 187 MPa. We can thus compute the shift of the natural frequency of the beam under this stress from Equation (4.45):

$$\omega_{1,s} = 1.932 \times 10^6 \text{ rad/s} \quad \text{or} \quad 3.075 \times 10^5 \text{ Hz}$$

4.4 THERMOMECHANICS

Many microsystems are either fabricated at high temperatures, such as in fusion bonding or oxidation processes, or expected to operate at elevated temperatures in the case of engine cylinder pressure sensors in an automotive application. Some transduction components such as piezoresistors are highly sensitive to the operating temperature. Thermal effects are thus an important factor in the design and packaging of microsystems.

There are generally three serious effects on micromachines and devices exposed to elevated temperatures, which will be discussed below.

4.4.1 Thermal Effects on Mechanical Strength of Materials

Most engineering materials exhibit reductions in the stiffness (e.g., Young's modulus) and the yield and ultimate strengths with increasing temperature, as illustrated in Figure 4.33. These reductions are more drastic with plastics and polymers. Fortunately, materials used for many core elements of microsensors and actuators, including silicon, quartz, and Pyrex glass, are relatively insensitive to the temperature. Significant changes of material properties are not expected in these device components at elevated temperatures. Also from Figure 4.33, we observe that the thermophysical properties of most materials increase with the temperature. Again, these changes are more pronounced in packaging materials such as adhesives and sealing and die protection materials used in most microsystems. Table 4.4 presents the temperature-dependent specific heats and coefficients of thermal expansion of silicon.

4.4.2 Creep Deformation

Creep deformation can occur in materials when the material's temperature exceeds half of the *homologous melting point* of the material (i.e., melting points on the absolute temperature scales). Creep is a form of deformation of the material without being subjected to additional mechanical loads. Creep deformation can develop in some components of microdevices such as adhesives and solder joints over a period of time as illustrated in

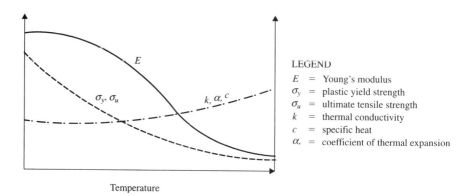

Figure 4.33. Variation of material properties with temperature.

TABLE 4.4. Temperature-Dependent Thermophysical Properties of Silicon

Temperature, K	Specific Heat, J/g-K	Coefficient of Thermal Expansion, 10^{-6}/K
200	0.557	1.406
220	0.597	1.715
240	0.632	1.986
260	0.665	2.223
280	0.691	2.432
300	0.713	2.616
400	0.785	3.253
500	0.832	3.614
600	0.849	3.842

Source: "Properties of Silicon," EMIS Group, INSPEC, NY, 1988.

Figure 4.34. There are generally three stages of creep deformation: primary, steady-state, and tertiary creep. Prolonged exposure of these materials to elevated operating temperature can lead to detrimental tertiary creep with catastrophic failure of the device.

Another important fact about a material's creep deformation is the effect of increasing operating temperature. The creep curves become steeper as temperature increases. We can readily see from Figure 4.34 that the distinction between the three stages of creep

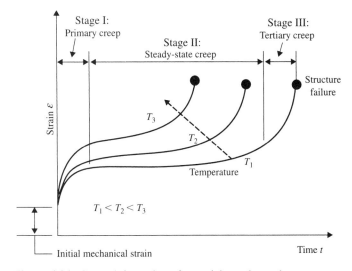

Figure 4.34. Creep deformation of materials at elevated temperature.

becomes less distinct at higher temperatures. The structure takes a shorter time to fail in creep at these temperatures than at lower operating temperatures.

Although creep can be a serious design consideration for plastics, polymers, and bonding materials in microsystems, most sensing and actuating components made of silicon or its compounds typically have very high melting points, which makes creep less of a problem. There are many publications on the creep behavior of solder alloys relating to microelectronic solder joint reliability (e.g., Hsu and Zheng, 1993; Hsu et al., 1993).

4.4.3 Thermal Stresses

Thermal stresses are induced in microdevices operating at elevated temperatures due to either mechanical constraints or mismatch of the coefficient of thermal expansion (CTE) of the mating components. Thermal stresses can also be induced in devices with little or no mechanical constraint as a result of nonuniform temperature distributions in the structure. Since most microsystems are made of components of different materials such as layered thin films, thermal stresses due to mismatch of the CTE need to be accurately assessed during the design stage, as excessive thermal stresses can cause failure of microdevices. Thermal stress analysis is thus an important part of the design of microsystems.

To many engineers, a well-known physical phenomenon is that materials expand when they are heated and contract when they are cooled. The amount of expansion or contraction of a material subjected to the change of the thermal environment is determined by (1) the temperature change $\Delta T = T_2 - T_1$ and (2) the CTE of the material, α. The CTE α is defined as the amount of thermal expansion of a material per unit length per degree of temperature change. It has units of in./in./°F, or m/m/°C, or ppm/K (parts per million per kelvin) as commonly used in industry.

The total expansion (with $+ \Delta T$) or contraction (with $- \Delta T$) of a bar as shown in Figure 4.35b may be determined by $\delta = L\alpha \ \Delta T$, where L is the original length of the bar at the reference temperature, say the room temperature. The associated induced thermal strain $\varepsilon_T = \delta/L = \alpha \ \Delta T$.

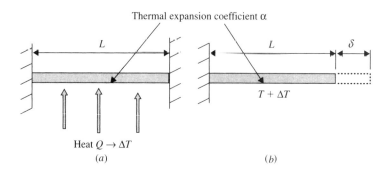

Figure 4.35. Bar with fixed ends subject to a temperature rise: (a) with both ends fixed; (b) free thermal expansion with one endconstraint removed.

Now let us examine the situation in Figure 4.35a. The bar is fixed at both ends. We can expect a compressive stress induced in the bar with an application of a temperature rise $\Delta T = T_2 - T_1$. The magnitude of the stress is

$$\sigma_T = E\varepsilon_T = -\alpha E \, \Delta T \qquad (4.48)$$

The derivation of the above expression can be carried out as follows: By referring to Figure 4.35b, let us first evaluate the amount of thermal expansion of the bar if one end of the bar were set free. The amount of free thermal expansion of the bar is $\delta = L \, (\alpha \, \Delta T)$. However, that end of the bar is fixed; that is, the end is not released, as shown in Figure 4.35a. This situation can be considered to be equivalent to having an equivalent mechanical force "pushing" back the freeexpanded bar with an amount δ. The required "pushing" force would produce an equivalent compressive force, which produces a compressive strain $\varepsilon_T = -\delta/L = -\alpha \, \Delta T$. The corresponding stress is thus equal to a compressive stress $\sigma_T = E\varepsilon_T = -\alpha E \, \Delta T$, as shown in Equation (4.48).

The following case will illustrate how thermal stress and deformation due to mismatch of thermal expansion coefficients can be determined. The reader will recognize the application of this case in the design of a microactuator that involves bilayer strips.

Figure 4.36 shows the schematic of a bilayer beam with two strips bonded together. Bilayer strips of this kind are commonly used in microactuators such as microtweezers actuated by thermal means. The two strips with distinct thermal expansion coefficients can make the strip curve either upward or downward when it is subjected to a temperature rise or drop. This actuated curvature can also result in either opening or closing an electric circuit.

Assume the straight bilayer strip in Figure 4.36 is at an initial temperature T_0. The strip experiences a temperature rise $\Delta T = T - T_0$, resulting in a curvature change with a radius of curvature as illustrated in Figure 4.37. The induced change of the radius of curvature with a temperature change ΔT from its initial temperature T_0 can be computed from the equation (Timoshenko, 1925)

$$\frac{1}{\rho} = \frac{6(1+m)^2(\alpha_1 - \alpha_2)\,\Delta T}{h\left\{3(1+m)^2 + (1+mn)\left[m^2 + 1/(mn)\right]\right\}} \qquad (4.49)$$

where

$m = t_1/t_2$, with t_1 and t_2 the thicknesses of strips 1 and 2, respectively

$n = E_1/E_2$, with E_1 and E_2 the Young's moduli of strips 1 and 2, respectively

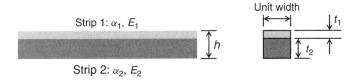

Figure 4.36. Bilayer strip subjected to a temperature change.

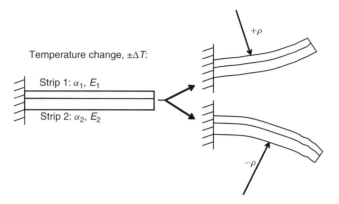

Figure 4.37. Change of curvature of a bilayer strip.

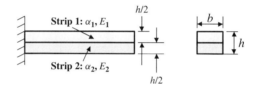

Figure 4.38. Bimaterial strip subject to a temperature change.

Equation (4.49) was derived with unit width of the bilayer strip with respective area moment of inertia of strips 1 and 2 as

$$I_1 = \tfrac{1}{12}t_1^3 \qquad I_2 = \tfrac{1}{12}t_2^3$$

A special case in which the bilayer strip is made of two strips of equal thickness, that is, $m = 1$, is illustrated in Figure 4.38.

Equation (4.49) is used to compute the interface force and the resulting curvature of the strip (Boley and Weiner, 1960):

$$F = \frac{(\alpha_2 - \alpha_1)\,\Delta T}{8}\frac{hb}{1/E_1 + 1/E_2} \tag{4.50}$$

and the curvature ρ of the deformed strip is given as

$$\rho = \frac{2h}{3(\alpha_1 - \alpha_2)\,\Delta T} \tag{4.51}$$

Example 4.17. A microactuator made of a bilayer strip—an oxidized silicon beam—is illustrated in Figure 4.39*a*. A resistant heating film is deposited on the top of the oxide layer. Estimate the interfacial force between the Si and SiO_2 layers

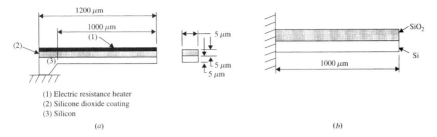

(1) Electric resistance heater
(2) Silicone dioxide coating
(3) Silicon

(a)   (b)

Figure 4.39. Bilayered strip actuator: (a) overall dimensions; (b) bilayer beam for analysis.

and the movement of the free end of the strip with a temperature rise $\Delta T = 10°C$. Use the following material properties:

$$\text{Young's modulus: } E_{SiO_2} = E_1 = 385{,}000 \text{ MPa}$$
$$E_{Si} = E_2 = 190{,}000 \text{ MPa}$$
$$\text{CTE: } \alpha_{SiO_2} = \alpha_1 = 0.5 \times 10^{-6}/°C$$
$$\alpha_{Si} = \alpha_2 = 2.33 \times 10^{-6}/°C$$

Solution: The configuration of the bilayer strip in Figure 4.39 fits what is depicted in Figure 4.38. Consequently, we may use Equations (4.50) and (4.51) to calculate the interfacial force and the radius of curvature of the heated strip with $h = 10 \times 10^{-6}$ m and $b = 5 \times 10^{-6}$ m.

Thus, by substituting the above material properties and the dimensions into Equations (4.50) and (4.51), we have the following values for the interfacial force F and the radius of curvature ρ:

$$F = \frac{(2.33 \times 10^{-6} - 0.5 \times 10^{-6})10}{8} \frac{(10 \times 10^{-6})(5 \times 10^{-6})}{1/(385{,}000 \times 10^6) + 1/(190{,}000 \times 10^6)}$$
$$= 14.55 \times 10^{-6} \text{ N}$$

$$\rho = \frac{2(10 \times 10^{-6})}{3(2.33 \times 10^{-6} - 0.5 \times 10^{-6})10} = 0.3643 \text{ m}$$

We realize that the interfacial force is so small that it can be entirely neglected. The radius of curvature of the bent strip, however, needs to be used to assess the corresponding movement of the free end, which can be approximated from the situation illustrated in Figure 4.40.

Since the radius of curvature of the arc is $\rho = 0.3643$ m, the corresponding perimeter of a complete circle with this radius is $2\pi\rho = 2.2878$ m. The angle θ that corresponds to the arc can be approximated by

$$\frac{\theta}{\text{arc(ac)}} \approx \frac{\theta}{\text{line(ac)}} = \frac{\theta}{1000 \times 10^{-6}} = \frac{360}{2.2878}$$

which leads to $\theta = 0.1574°$.

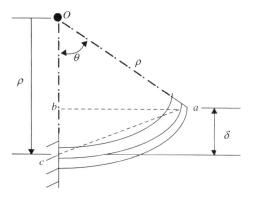

Figure 4.40. Movement of the free end of a bilayer strip actuator.

Consequently, we may use the relationship derived from the triangle ΔOab to obtain the movement of the free end δ to be

$$\delta \approx \rho - \rho \cos \theta = 0.3643 - 0.3643 \cos(0.1574°) = 1.373 \times 10^{-6} \text{ m or } 1.373 \; \mu\text{m}$$

Example 4.18. The bilayer strip described in Example 4.16 is used for the microswitch of a circuit but with the thickness of the SiO_2 film reduced to $2 \; \mu$m and the total thickness h remaining at $10 \; \mu$m, meaning the thickness of the Si strip is increased to $8 \; \mu$m. (1) Estimate what will be the change in the end deflection of the strip. (2) Plot the end deflection–thickness ratio t_1/t_2 with t_1 being the thickness of SiO_2.

Solution: We will use the same length of the strip and material properties presented in Example 4.17 in our computations.

1. Equation (4.49) is used to compute the induced radius of curvature change in the strip with thickness ratio $m = t_1/t_2 = \frac{2}{8} = 0.25$ and $n = E_1/E_2 = 385{,}000/190{,}000 = 2.026$:

$$\frac{1}{\rho} = \frac{6(1 + 0.25)^2(0.5 \times 10^{-6} - 2.33 \times 10^{-6}) \times 10}{10 \times 10^{-6} \left\{ 3(1 + 0.25)^2 + (1 + 0.25 \times 2.026) \left[0.25^2 + 1/(0.25 \times 2.026) \right] \right\}}$$

$$= -2.187 \text{ m}^{-1}$$

from which we obtain the radius of curvature $\rho = -0.4572$ m. The negative radius of curvature means concave-down geometry of the strip, as illustrated in Figure 4.37. By using the procedure in Example 4.17, we may compute the corresponding movement of the strip tip to be $\delta = 1.73 \; \mu$m (a downward movement with negative radius of curvature).

2. The same procedure is followed to plot the movement of the strip tip versus the thickness ratio using Microsoft Excel software. The resultant curve is shown in Figure 4.41.

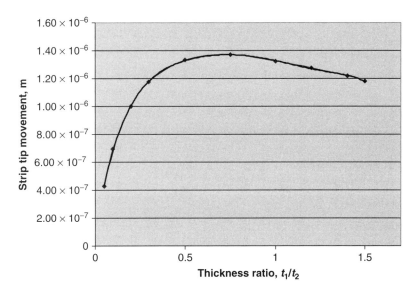

Figure 4.41. Bilayer strip tip movements with variable thickness ratios of strips.

The result shown in Figure 4.41 indicates an interesting phenomenon when designing a microbilayer strip; note that silicon has about five times higher CTE than SiO_2 but the latter material has two times the stiffness of silicon. One may conceive that reversal of the end-point deflection of the bonded strip can occur with increasing value of the thickness ratio or with increasing thickness portion of the stiffer material of SiO_2. One may thus conclude that mismatch of the CTE dominates the geometry of the strip with a smaller portion of SiO_2. The dominance of the CTE is overcome by the stiffness of the material as a thicker portion of SiO_2 is used in the strip, as illustrated in Figure 4.41.

Theories of thermomechanics, which include those for thermoelastic–plastic stress analysis of structures of complex geometry and loading and boundary conditions as well as the associated creep and fracture of these structures due to thermal effects, can be found in a reference book (Hsu, 1986). Finite element formulation for the above thermomechanical analyses is also available in the same book.

As we have learned from Chapter 2, many MEMS components are in the shape of thin plates and beams. We will outline the closed-form solutions of thermal stress distributions in these structure components in the following sections. Detailed derivation of these formulas is available in several sources, including Boley and Weiner (1966).

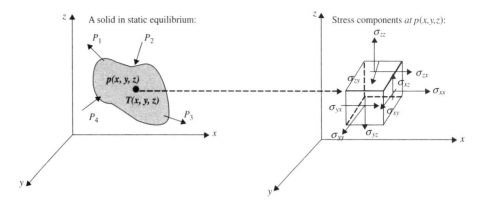

Figure 4.42. Designation of stress components in a solid.

The components of stress and associated strains in a three-dimensional solid under static equilibrium conditions are defined in Figure 4.42. The displacements along the respective x, y, and z directions are designated as $u(x,y,z)$, $v(x,y,z)$, and $w(x,y,z)$.

Thermal Stresses in Thin Plates with Temperature Variation through Thickness. Figure 4.43 illustrates a thin plate of arbitrary shape defined by a cartesian coordinate system. Plane stress is assumed for this thin plate, which means the following stress situations with temperature variation along the thickness direction, or $T = T(z)$:

$$\sigma_{xx}(x, y, z) = f_1(z) \quad \sigma_{yy}(x, y, z) = f_2(z)$$

$$\sigma_{zz} = \sigma_{xz} = \sigma_{yx} = \sigma_{yz} = 0$$

Thermal stresses:

$$\sigma_{xx} = \sigma_{yy} = \frac{1}{1 - \nu} \left[-\alpha E T(z) + \frac{1}{2h} N_T + \frac{3z}{2h^3} M_T \right] \tag{4.52}$$

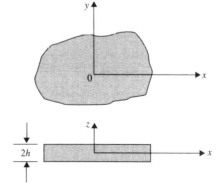

Figure 4.43. Thermal stresses in a thin plate.

Thermal strains:

$$\varepsilon_{xx} = \varepsilon_{yy} = \frac{1}{E} \left(\frac{N_T}{2h} + \frac{3z}{2h^3} M_T \right) \tag{4.53a}$$

$$\varepsilon_{zz} = -\frac{2\nu}{(1-\nu)E} \left(\frac{N_T}{2h} + \frac{3z}{2h^3} M_T \right) + \frac{1+\nu}{1-\nu} \alpha T(z) \tag{4.53b}$$

$$\varepsilon_{xy} = \varepsilon_{yz} = \varepsilon_{zx} = 0$$

Displacement components:

$$u = \frac{x}{E} \left(\frac{N_T}{2h} + \frac{3z}{2h^3} M_T \right) \tag{4.54a}$$

along the x axis,

$$v = \frac{y}{E} \left(\frac{N_T}{2h} + \frac{3z}{2h^3} M_T \right) \tag{4.54b}$$

along the y axis, and

$$w = -\frac{3M_T}{4h^3 E} (x^2 + y^2) + \frac{1}{(1-\nu)E} \left[(1+\nu)\alpha E \int_0^z T(z)dz - \frac{\nu z}{h} N_T - \frac{3\nu z^2}{2h^3} M_T \right] \tag{4.54c}$$

along the z axis.

The *normal thermal force* N_T and *thermal moment* M_T are expressed in terms of the temperature function $T(z)$ as follows:

$$N_T = \alpha E \int_{-h}^{h} T(z)\, dz \tag{4.55a}$$

$$M_T = \alpha E \int_{-h}^{h} T(z)z\, dz \tag{4.55b}$$

where α is the coefficient of thermal expansion, E is Young's modulus, and ν is Poisson's ratio of the material.

Thermal Stresses in Beams due to Temperature Variation through Depth.

This situation is illustrated in Figure 4.44. The beam has a cross-sectional area $A = 2bh$ and the corresponding area moment of inertia $I = 2h^3 b/3$.

The bending stress $\sigma_{xx}(x,z)$ is the dominant stress component. It can be expressed as

$$\sigma_{xx}(x, z) = -\alpha E T(z) + \frac{bN_T}{A} + \frac{z(bM_T)}{I} \tag{4.56}$$

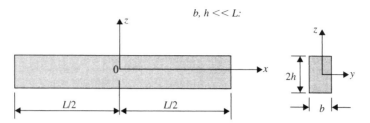

Figure 4.44. Thermal stresses in beam.

with $\sigma_{zz} = \sigma_{xz} = 0$. The associated strain components are

$$\varepsilon_{xx}(x, z) = \frac{1}{E}\left[\frac{bN_T}{A} + \frac{z}{I}(bM_T)\right] \tag{4.57a}$$

$$\varepsilon_{zz}(x, z) = -\frac{\nu}{E}\left[\frac{bN_T}{A} + \frac{z}{I}(bM_T)\right] + \left(\frac{1+\nu}{E}\right)\alpha T(z) \tag{4.57b}$$

with $\varepsilon_{xz} = 0$. The displacement components are

$$u(x, z) = \frac{x}{E}\left[\frac{bN_T}{A} + \frac{z}{I}(bM_T)\right] \tag{4.58a}$$

along the x direction and

$$w(x, z) = -\frac{bM_T}{2EI}x^2 - \frac{\nu}{E}\left[\frac{bN_T}{A}z + \frac{z^2}{2I}(bM_T)\right] + \alpha\left(\frac{1+\nu}{E}\right)\int_0^z T(z)\,dz \tag{4.58b}$$

along the z direction.

The normal force and the bending moment due to thermal force N_T and thermal moment, M_T are expressed in Equations (4.55a) and (4.55b), respectively.

The curvature of the bent beam can be determined by the expression

$$\frac{1}{\rho} \approx -\frac{bM_T}{EI} \tag{4.59}$$

where ρ is the radius of curvature of the bent beam.

Example 4.19. Determine the thermal stresses and strains as well as the deformation of a thin beam at 1 μs after the top surface of the beam is subjected to a sudden heating by the resistance heating of the attached thin copper film. The temperature at the top surface resulting from the heating is 40°C. The geometry and dimensions of the beam are illustrated in Figure 4.45. The beam is made of silicon and has the following material properties:

Mass density $\rho = 2.3 \text{ g/cm}^3$
Specific heat $c = 0.7 \text{ J/g-}^\circ\text{C}$
Thermal conductivity $k = 1.57 \text{ W/cm-}^\circ\text{C}$ (or J/cm-$^\circ$C-s)
Coefficient of thermal expansion $\alpha = 2.33 \times 10^{-6}/^\circ\text{C}$
Young's modulus $E = 190,000 \times 10^6 \text{ N/m}^2$
Poisson's ratio $\nu = 0.25$

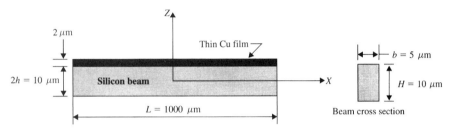

Figure 4.45. Silicon beam subject to heating at top surface.

Solution: Thermal stresses, strains, and displacements induced in the beam by the sudden heating of the top surface can be computed using Equations (4.56), (4.57), and (4.58). However, as in almost all thermal stress analyses, one needs to determine the temperature distribution in the structure first.

In the present case, we need to obtain the temperature variation across the depth of the beam, that is, $T(z)$ at a given time t, as required in these equations.

The temperature $T(z, t)$ in the beam can be obtained either from a heat conduction analysis with appropriate boundary conditions, as will be described in Chapter 5, or from the solution of a similar but approximate case involving the heating of a half-infinite space, as illustrated in Figure 4.46.

As will be described in Chapter 5, the situation in Figure 4.46 can be mathematically modeled with a partial differential equation and appropriate initial and boundary conditions as follows:

$$\frac{\partial^2 T(y, t)}{\partial y^2} = \frac{1}{\alpha} \frac{\partial T(y, t)}{\partial t} \tag{4.60}$$

The initial condition is $T(y,0) = 0$, and the boundary conditions are $T(0,t) = T_0$ and $T(\infty,t) = 0$.

The corresponding solution has the form

$$T(y, t) = T_0 \operatorname{erfc}\left(\frac{y}{2\sqrt{\alpha t}}\right) \tag{4.61}$$

where α is the thermal diffusivity of the solid.

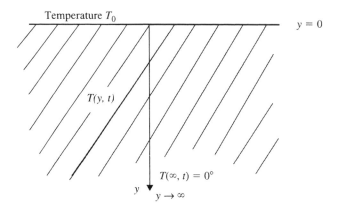

Figure 4.46. Temperature in half space.

The corresponding thermal diffusivity of the silicon beam for the present case can be computed as

$$\alpha = \frac{k}{\rho c} = \frac{1.57 \text{ J/cm-°C-s}}{2.3 \text{ g/cm}^3 \times 0.7\text{J/g-°C}} = 0.9752 \text{ cm}^2/\text{s}$$

or $\alpha = 97.52 \times 10^{-6} \text{ m}^2/\text{s}$ for silicon.

The function erfc(x) in Equation (4.61) is the complimentary error function as described in Chapter 3.

Thus, by substituting the relevant parameters into Equation (4.61), we can obtain the temperature distribution along the depth of the beam at 1 μs after its top surface is heated at 40°C:

$$T(y, 1\,\mu s) = 40 \text{ erfc}\left(\frac{y}{2\sqrt{(97.52 \times 10^{-6})(10^{-6})}}\right) = 40 \text{ erfc}\left(\frac{10^6 y}{19.75}\right) \quad \text{(a)}$$

Application of the above solution to the beam as illustrated in Figure 4.45 requires a transformation of coordinate from the y coordinate in Figure 4.46 to the z coordinate in Figure 4.45 by letting $z = -y + h$, or $y = h - z$, with $h = 5$ μm.

The solution in Equation (a) after the transformation takes the form

$$T(z, 1\,\mu s) = 40 \text{ erfc}\left[\frac{10^6(5 \times 10^{-6} - z)}{19.75}\right] \quad \text{(b)}$$

By substituting the above expression for $T(z)$ into Equation (4.56), we can evaluate the thermal stress at 1 μs and also the corresponding strains and displacements from the respective equations (4.57) and (4.58a).

ALTERNATIVE NUMERICAL SOLUTION

Determination of the thermal force N_T in Equation (4.55a) and the thermal moment M_T in Equation (4.55b) for the present case will require the integration of the function $T(z, 1\ \mu s)$ in Equation (b), which involves the complementary error function erfc(z). This task is manageable with mathematical solution software such as MATHCAD or MATLAB. An alternative approach is to derive a polynomial function that can approximate the temperature distribution expressed in Equation (b). This polynomial function can then be used to obtain the numerical solution for the problem at hand.

We can obtain the numerical solution of the temperature distribution along the depth of the beam by evaluating the complementary error function in Equation (b) using the numerical values of error functions in Figure 3.14. The tabulated values of $T(z)$ at $t = 1\ \mu s$ are given below:

y in Eq. (a)	$z \times 10^{-6}$ m in Eq. (b)	erfc$\left[\dfrac{10^6(5\times 10^{-6}-z)}{19.75}\right]$	$T(z)$, °C at 1 μs
0	5	0	40.00
1	4	0.0506	37.72
2	3	0.1013	35.44
3	2	0.1519	33.20
4	1	0.2025	30.38
5	0	0.2532	28.81
6	-1	0.3038	26.69
7	-2	0.3544	24.63
8	-3	0.4051	22.64
9	-4	0.4557	20.74
10	-5	0.5063	18.92

We will find that a linear polynomial function in the following form fits the temperature distribution in Equation (b) reasonably well:

$$T(z) = 2.1 \times 10^6 z + 28.8 \qquad (c)$$

where z is the coordinate shown in Figure 4.45 in units of micrometers.

We can thus determine the thermal force and thermal moment by using $T(z)$ from Equation (c) in Equations (4.55a) and (4.55b) as follows:

$$N_T = \alpha E \int_{-h}^{h} T(z)\,dz = (2.33 \times 10^{-6})(190{,}000 \times 10^6)$$

$$\times \int_{-5\times 10^{-6}}^{5\times 10^{-6}} (2.1 \times 10^6 z + 28.8)\,dz$$

$$= 127.5\ \text{N}$$

$$M_T = \alpha E \int_{-h}^{h} T(z)z\,dz = (2.33 \times 10^{-6})(190{,}000 \times 10^6)$$

$$\times \int_{-5 \times 10^{-6}}^{5 \times 10^{-6}} (2.1 \times 10^6 z + 28.8)z\,dz$$

$$= 77.4725 \times 10^{-6} \text{ N-m}$$

At this stage, we need to find both the cross-sectional area of the beam, A, and the area moment of inertia, I:

$$A = (5 \times 10^{-6})(10 \times 10^{-6}) = 5 \times 10^{-11} \text{ m}^2$$

$$I = \frac{1}{12}bH^3 = \frac{1}{12}(5 \times 10^{-6})(10 \times 10^{-6})^3 = 4.167 \times 10^{-22} \text{ m}^4$$

We are now ready to compute the induced thermal stresses, strains, and displacements by the temperature distribution in Equation (c). From Equation (4.56)

$$\sigma_{xx}(z, 1\,\mu\text{s}) = -(2.33 \times 10^{-6})(190{,}000 \times 10^6)(2.1 \times 10^6 z + 28.8)$$

$$+ \frac{(5 \times 10^{-6})127.5}{5 \times 10^{-11}} + \frac{z(5 \times 10^{-6})(77.4725 \times 10^{-6})}{4.167 \times 10^{-22}}$$

$$= -4.427 \times 10^5 (2.1 \times 10^6 z + 28.8) + 127.5 \times 10^5$$

$$+ 92.96 \times 10^{10} z \quad \text{Pa}$$

We will have the maximum bending stress $\sigma_{xx,max}$ 1μs after the heating by substituting $z = 5 \times 10^{-6}$ m into the above expression. We will obtain $\sigma_{xx,max} = -500$ Pa in compression.

The associated thermal strain components can be evaluated from Equations (4.57a) and (4.57b) as follows:

$$\varepsilon_{xx}(z) = \frac{1}{190{,}000 \times 10^6} \left[\frac{(5 \times 10^{-6})127.5}{5 \times 10^{-11}} + \frac{z(5 \times 10^{-6} \times 77.4725 \times 10^{-6})}{4.167 \times 10^{-22}} \right]$$

$$= 67.11 \times 10^{-6}(1 + 0.73 \times 10^5 z)$$

$$\varepsilon_{zz}(z) - 0.25(-67.11 \times 10^{-6})(1 + 0.73 \times 10^5 z)$$

$$+ \frac{1 + 0.25}{190{,}000 \times 10^6}(2.33 \times 10^{-6})(2.1 \times 10^6 z + 28.8)$$

$$= (-16.78 \times 10^{-6} - 1.23z) + (3.22 \times 10^{-11} z + 44.15 \times 10^{-17})$$

The maximum strains occur at the top surface. Numerical values of these strains can be obtained by letting $z = 5 \times 10^{-6}$ m in the above expressions:

$$\varepsilon_{xx,max} = \varepsilon_{xx}(5 \times 10^{-6}) = 91.61 \times 10^{-6} = 0.0092\%$$

$$\varepsilon_{zz,max} = \varepsilon_{zz}(5 \times 10^{-6}) = -22.93 \times 10^{-6} = -0.0023\%$$

The displacements at the top free corners with $x = \pm 500 \times 10^{-6}$ m and $z = 5 \times 10^{-6}$ m can be computed from Equations (4.58a) and (4.58b):

$$u = \frac{500 \times 10^{-6}}{190,000 \times 10^6} \left[\frac{(5 \times 10^{-6})127.5}{5 \times 10^{-11}} + \frac{(5 \times 10^{-6})(5 \times 10^{-6})(77.47 \times 10^{-6})}{4.167 \times 10^{-22}} \right]$$

$$= 0.046 \, \mu m$$

$$w = -\frac{(5 \times 10^{-6})(77.47 \times 10^{-6})(500 \times 10^{-6})^2}{2(190,000 \times 10^6)(4.167 \times 10^{-22})}$$

$$-\frac{0.25}{190,000 \times 10^6} \left[\frac{(5 \times 10^{-6})127.5}{5 \times 10^{-11}}(5 \times 10^{-6}) \right.$$

$$\left. + \frac{(5 \times 10^{-6})^2(5 \times 10^{-6})(77.47 \times 10^{-6})}{2(4.167 \times 10^{-22})} \right]$$

$$+ \frac{1 + 0.25}{190,000 \times 10^6}(2.33 \times 10^{-6})$$

$$\times \int_0^{5 \times 10^{-6}} (2.1 \times 10^6 z + 28.8)z \, dz = -0.612 \, \mu m$$

The curvature of the bent beam is estimated from Equation (4.59) to be

$$\frac{1}{\rho} = -\frac{(5 \times 10^{-6})(77.47 \times 10^{-6})}{(190,000 \times 10^6)(4.167 \times 10^{-22})} = -4.892 \text{ m}^{-1}$$

4.5 FRACTURE MECHANICS

Many MEMS devices involve the binding of thin films of distinct materials as illustrated in Chapter 2. These different materials with distinct properties are bonded together by various means, as will be described in Chapter 8 on microfabrication technology and Chapter 11 on the packaging of microsystems. These binding techniques include chemical and physical vapor depositions, thermal diffusion, soldering, and adhesions. Interfaces of bonded structures are likely places where structural failure will take place. Fracture mechanics is a way to assess the structural integrity of these interfaces.

Fracture of bonded interfaces can occur by excessive forces acting normal to the interfaces and/or by the shear forces acting on the interfaces. Some of the linear elastic fracture mechanics (LEFM) principles can be applied in the design analysis to mitigate possible delamination of interfaces in MEMS and microsystems.

Linear elastic fracture mechanics is an engineering discipline on its own. It is not possible to cover this subject in much detail in this section. What we will learn here are the basic concept and the formulation of LEFM that may be used in the microstructure design.

4.5.1 Stress Intensity Factors

Let us consider a small crack existing in a solid subjected to a stress field as illustrated in Figure 4.47. The stress field has a tendency to deform the crack in three distinct modes. These deformation modes, illustrated in Figure 4.48, are mode I, the opening mode; mode II; the edge-sliding mode; and mode III; the tearing mode.

The *opening mode* (mode I) is associated with local displacements in which the crack surfaces tend to move apart in a direction perpendicular to these surfaces (symmetric with respect to the $x-y$ and $z-x$ planes). The *edge-sliding mode* (mode II) is characterized by displacements in which the crack surfaces slide over one another and remain perpendicular to the leading edge of the crack (symmetric with respect to the $x-y$ plane and skew symmetric with respect to the $x-z$ plane). The *tearing mode* (mode III) is defined by the crack surfaces sliding with respect to one another parallel to the leading edge of the crack (skew symmetric with respect to the $x-y$ and $x-z$ planes).

The *stress intensity factor K* is introduced in the LEFM to assess the stress field surrounding a crack. By referring to the situation in Figure 4.47, the stress and displacement

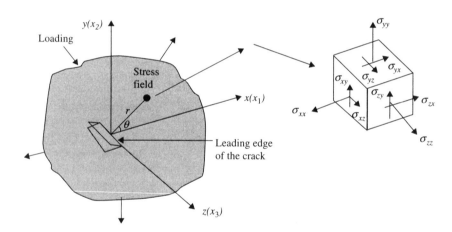

Figure 4.47. Crack in a solid subjected to stress field.

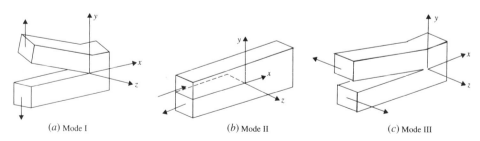

(a) Mode I   *(b)* Mode II   *(c)* Mode III

Figure 4.48. Three modes of fracture.

components around a crack can be expressed as

$$\sigma_{ij} = \frac{K_{\mathrm{I}} \text{ or } K_{\mathrm{II}} \text{ or } K_{\mathrm{III}}}{\sqrt{r}} f_{ij}(\theta) \tag{4.62}$$

$$u_i = K_{\mathrm{I}} \text{ or } K_{\mathrm{II}} \text{ or } K_{\mathrm{III}} \sqrt{r} g_i(\theta) \tag{4.63}$$

where K_{I}, K_{II}, and K_{III} are the stress intensity factors of modes I, II, and III fracture, respectively.

The stress components σ_{ij} in Equation (4.62) with subscripts $i, j = 1,2,3$ denoting the coordinates ($x_1 = x$, $x_2 = y$, and $x_3 = z$ in Figure 4.47) are the same stress components in the figure, for example, $\sigma_{11} = \sigma_{xx}$ and $\sigma_{23} = \sigma_{yz}$ in the (x,y,z) coordinates. The displacements of the small element a distance r away from the crack tip are expressed as u_i ($i = 1, 2, 3$) in Equation (4.63). The functions $f_{ij}(\theta)$ and $g_i(\theta)$ in Equations (4.62) and (4.63) depend on the crack length c and the geometry of the solid. Forms of these functions for solid structures of different geometry are available in textbooks on fracture mechanics.

Notice from Equation (4.62) that the stresses $\sigma_{ij} \to \infty$ as $r \to 0$ (the tip of the crack is located at $r = 0$). In reality, this will not happen. However, we should realize that the stresses rapidly increase to very large magnitudes near the tip of a sharp crack. This situation with $\sigma_{ij} \to \infty$ as $r \to 0$ is called *stress singularity* near the crack tip.

We can determine the values of the stress intensity factors K_{I}, K_{II}, and K_{III} from Equation (4.62) for a structure containing a crack with specific geometry and the stress field surrounding the crack using the theory of LEFM.

4.5.2 Fracture Toughness

Fracture toughness K_C for various modes of fracture are used as criteria for assessing the stability of the crack in a structure. They are expressed as K_{IC}, K_{IIC}, and K_{IIIC} for modes I, II, and III fracture, respectively. The measured *critical load* P_{cr} that causes specimens of given geometry containing a crack to fail establishes these K_C values (K_{IC}, K_{IIC}, and K_{IIIC}). The common geometry of specimens for mode I fracture toughness $K_{I,C}$ includes compact tension specimens and three-point bending beam specimens, as illustrated in Figure 4.49. The respective fracture toughness can be determined by the expressions given below for mode I fracture.

For compact tension (CT) specimens in Figure 4.49a: The specimen is subjected to a tension to open the crack. A notch is machined in the midheight of the specimen to facilitate the fracture. A minute crack with length c is created at the tip of the notch by fatigue load at low magnitudes of loading. This minute crack initiates its growth and subsequent unstable propagation under the applied loading. Fracture toughness of the CT specimen material can be determined by the formula

$$K_{\mathrm{IC}} = \sigma_C \sqrt{\pi c} F\left(\frac{c}{b}\right) \tag{4.64}$$

where c is the crack length and $\sigma_c = P_{cr}/A$, with P_{cr} the applied load P that causes the specimen to fracture with crack propagation, A being the cross section of the bulk specimen.

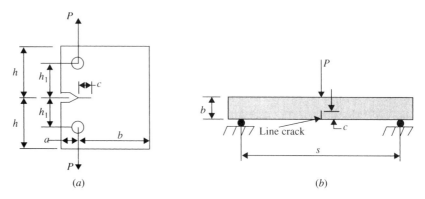

Figure 4.49. Specimens used to determine mode I fracture toughness: (*a*) compact-tension specimen; (*b*) three-point bending specimen.

The function $F(c/b)$ in Equation (4.64) can be found in fracture mechanic handbooks. An approximation of $F(c/b) = 1$ is used if such a reference is not readily available. This value is for the case where the specimen is subjected to a uniform tensile load.

Compact tension specimens used to determine fracture toughness need to satisfy the following specifications in the geometry with reference to Figure 4.49*a*:

$$h = 0.6b \qquad h_1 = 0.275b \qquad D = 0.25b \qquad a = 0.25b$$

The thickness T of the specimen is $b/2$.

For three-point bending specimens in Figure 4.49b: The expression in Equation (4.64) can be used to determine the fracture toughness of the specimen material, except that the function $F(c/b)$ takes the form

$$F\left(\frac{c}{b}\right) = 1.09 - 1.735\left(\frac{c}{b}\right) + 8.2\left(\frac{c}{b}\right)^2 - 14.18\left(\frac{c}{b}\right)^3 + 14.57\left(\frac{c}{b}\right)^4 \qquad (4.65a)$$

for the case $s/b \leq 4$ and

$$F\left(\frac{c}{b}\right) = 1.107 - 2.12\left(\frac{c}{b}\right) + 7.71\left(\frac{c}{b}\right)^2 - 13.55\left(\frac{c}{b}\right)^3 + 14.25\left(\frac{c}{b}\right)^4 \qquad (4.65b)$$

for $s/b \leq 8$.

Fracture toughness so determined from measurements of critical loading P_{cr} on specific specimens is used as a design criterion similar to the allowable stress used in a conventional machine design. In practice, the engineer will conduct a stress analysis to determine the stress field around a hypothetical crack of certain size and orientation in the structure. Equation (4.62) is used to determine the stress intensity factors K_I, K_{II}, and K_{III}. The stress intensity factors so determined are compared with the corresponding fracture toughnesses K_{IC}, K_{IIC}, and K_{IIIC}. Computed K values exceeding the corresponding fracture toughness would mean failure of the structure. In most cases, however, only the dominant mode I fracture is considered. Thus, the case in which the computed K_I is less than the specified K_{IC} is considered a safe case, in which the crack is treated as stable.

4.5.3 Interfacial Fracture Mechanics

Many MEMS devices are made with bonded dissimilar materials, for example, the pressure sensors in Figures 2.8 and 2.9, microactuators in Figures 2.18 and 2.19, and microvalves and pumps in Figures 2.41, 2.42, and 2.43. There are many interfaces between dissimilar materials in these devices. As many of these devices are expected to operate millions of working cycles, the strength of these bonded interfaces has become a serious concern to the design engineer.

Analytical assessment of the fracture strength of interfaces is much more complicated than that for single-mode fracture as presented in the foregoing sections. A major complication is that most of these interfaces are subject to simultaneous modes I and II (and III in some cases) loading. Mixed-mode fracture analysis is significantly different from what was established for single-mode fracture analysis. Here, again, we will only outline the general approach to the solution of this type of problem.

Let us refer to the interface of a bonded bimaterial structure as illustrated in Figure 4.50. The two bonded materials have distinct properties such as Young's moduli E_1 and E_2 and Poisson's ratios ν_1 and ν_2. The interface is subjected to simultaneous loading in both the normal and lateral shearing directions with respective stress fields σ_{yy} and σ_{xy}. These stress fields make the interface vulnerable to coupled opening mode (I) and shearing mode (II) fracture.

The stresses σ_{ij} with $i,j = x,y$ in either material 1 or 2 near the interfacial crack can be expressed by the following expression, which is similar to Equation (4.62):

$$\sigma_{ij} = \frac{K_I \text{ or } K_{II}}{r^\lambda} + L_{ij} \ln(r) + \text{terms}$$

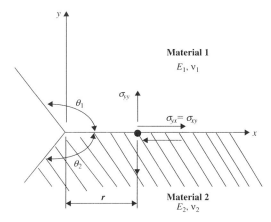

Figure 4.50. Interface of a bimaterial structure.

where K_{I} and K_{II} are the respective stress intensity factors for modes I and II fracture, λ is a *singularity parameter*, and L_{ij} are constants.

For the region that is close to the tip of the interface (i.e., the distance r is very small), the contribution of the term $\ln(r)$ to the stress field can be neglected. We thus have the stresses near the tip of the interface where the stress singularity dominates the stress distribution:

$$\sigma_{ij} = \frac{K_{\mathrm{I}} \text{ or } K_{\mathrm{II}}}{r^{\lambda}} \tag{4.66}$$

from which we can express the stress components along the interface as

$$\sigma_{yy} = \frac{K_{\mathrm{I}}}{r^{\lambda_{\mathrm{I}}}} \tag{4.67a}$$

$$\sigma_{xy} = \frac{K_{\mathrm{II}}}{r^{\lambda_{\mathrm{II}}}} \tag{4.67b}$$

where λ_{I} and λ_{II} are, respectively, singularity parameters for the materials in modes I and II fracture.

The stress intensity factors K_{I} and K_{II} in Equations (4.67a) and (4.67b) can be obtained by linearizing both these equations in logarithm scales:

$$\ln(\sigma_{yy}) = -\lambda_{\mathrm{I}} \ln(r) + \ln(K_{\mathrm{I}}) \tag{4.68a}$$

$$\ln(\sigma_{xy}) = -\lambda_{\mathrm{II}} \ln(r) + \ln(K_{\mathrm{II}}) \tag{4.68b}$$

Equations (4.68a) and (4.68b) can be expressed in graphical forms in Figures 4.51a and 4.51b, respectively. In Figure 4.51, r_0 represents a short distance away from the crack tip, in which the stresses vary linearly, whereas r_e is the closest distance away from the

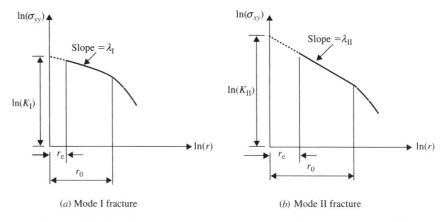

(a) Mode I fracture   (b) Mode II fracture

Figure 4.51. Graphical representation of stress intensities in modes I and II.

crack tip with which numerical values of stresses are available from a finite element analysis.

For a typical interfacial fracture analysis, the stress distributions in both materials along the interface are determined by either a classical method or a finite element analysis. These stress distributions in both mating materials closest to to the interface are then plotted in natural logarithm scales as shown in Figure 4.51. The stress intensity factors K_I and K_{II} of the materials near the interface can be obtained from the intercepting points at the y axes in the respective $\ln(\sigma_{yy})$–$\ln(r)$ and $\ln(\sigma_{xy})$–$\ln(r)$ plots. The slopes of the straight-line portion of the plots represent the singularity parameters λ_I and λ_{II} for both modes of fracture.

The stress intensities K_I and K_{II} obtained for the mating materials from the above procedure will enter the following mixed-mode fracture criterion in the form

$$\left(\frac{K_I}{K_{IC}}\right)^2 + \left(\frac{K_{II}}{K_{IIC}}\right)^2 = 1 \tag{4.69}$$

in which the fracture toughnesses K_{IC} and K_{IIC} are determined by experimental methods for the same mating materials. A database for coupled modes I and II fracture toughness for a limited number of microelectronics materials was reported (Nguyen et al., 1997). Equation (4.69) can be represented graphically by a quarter ellipse, as illustrated in Figure 4.52. An interface of two materials is considered "safe" if the computed K_I and K_{II} by the above procedure are located within the elliptical envelope.

The above analytical technique for the interfacial fracture of layered materials appears to be straightforward. Unfortunately, there are very limited material databases available on this type of coupled fracture toughness for MEMS and microsystems. Application of this method is thus limited.

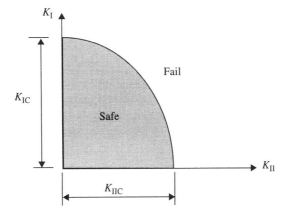

Figure 4.52. Fracture criterion for interfaces.

4.6 THIN-FILM MECHANICS

As we will learn from various microfabrication techniques in Chapter 8, it is customary to deposit thin films made of various materials either on a substrate or on other thin films. These thin films are produced in MEMS and microsystems either to shape the desired three-dimensional structure geometry or to provide necessary functional performances such as thin coatings for reflective surfaces in micro-optical devices or thin-film coating for the passivation of delicate core components. These films, with thickness ranging from submicrometers to a few micrometers, are expected to either carry loads in many microdevices or remain in desired surface conditions with dimensional stability. The mechanistic behavior of thin films is thus a prime concern to design engineers.

Chemical and physical vapor depositions and other processes such as oxidation and epitaxy are common techniques for producing thin films in MEMS and microsystems. Most of these chemical vapor deposition techniques involve elevated ambient temperature and low pressure, whereas many physical deposition techniques such as sputtering involve bombardment of ions of the deposit materials on substrate materials. Almost all these deposition techniques result in curved surfaces after the fabrication.

The principal reason for the curvature change of the thin substrates with deposited thin films is the parasite stresses and strain induced by these deposition processes. There are two common types of these inherent stresses: *residual stresses* and *intrinsic stresses*.

Residual stresses are introduced in thin films as a result of excessive thermal stresses induced to the structure due to mismatch of the CTE of the thin film and the substrate material during the production process at elevated temperature. The unrecoverable strains induced in the thin film after the completion of the process result in the inherent residual stresses and strains that change the flat substrate into a curved surface. A finite element analysis on an oxidation process for a 4-μm-thin silicon oxide film over silicon substrate at $1000°$C indicated a compressive residual stress as high as 2000 MPa in the oxide film (Hsu and Sun, 1998). This falls in the range of 10–5000 MPa residual stresses in thin films according to Madou (1997). The same reference indicated that thin metal films on silicon substrates result in residual stresses after the deposition process that are in the range of 10–100 MPa, but in tension. These tensile stresses can cause cracking of the thin metal films.

The intrinsic stresses are induced in thin films mainly by the reaggregation of crystalline grains of thin-film materials induced mainly by physical deposition processes such as sputtering at low temperature. These stresses can also result in a curved surface of substrates (Zhang and Misra, 2004).

Inherent stresses in thin films induced by microfabrication processes are commonplace in MEMS and microsystems. These stresses (and strains) can seriously affect the intended functionality of the device. There are two ways by which engineers may mitigate the magnitudes of these stresses: annealing and ion beam machining (Bifano et al., 2002). Annealing involves the relaxation of residual thermal stresses/strains in thin-film structures with exposures to specifically designed thermal environments for specific duration—a process that is similar to annealing of metals in traditional macromanufacturing. Ion beam machining relaxes the "locked-in" intrinsic stress in the thin-film structures by orderly removing the material. This process is similar to the residual stress measurements in structures using relaxation methods (Osgood, 1954).

Credible methods for quantifying these forces are not currently available because distribution of material in thin films is not likely to be uniform across the area on which they are deposited. The use of continuum mechanics theory with average material behavior to assess the stresses in thin films is thus not realistic.

The total stress σ_T in a thin film over a thick substrate can be expressed as follows (Madou, 1997):

$$\sigma_T = \sigma_{th} + \sigma_m + \sigma_{int} \tag{4.70}$$

where

$\sigma_{th} =$ thermal stress induced by operating temperature as described in Section 4.4
$\sigma_m =$ stress induced by applied mechanical loads
$\sigma_{int} =$ intrinsic stress

Numerical evaluation of the first two stress components in Equation (4.70) often requires the use of the finite element method, whereas assessment of intrinsic stresses is usually conducted with empirical means developed by measurements as well as by the engineer's own experience (Bifano et al., 2002; Zhang and Misra, 2004).

4.7 OVERVIEW OF FINITE ELEMENT STRESS ANALYSIS

Almost all MEMS and microsystems involve complex three-dimensional geometry. These devices are fabricated under severe thermal and mechanical conditions and are subject to harsh loading during operation. Furthermore, these structures often consist of multilayers of dissimilar thin films bonded together by either physical or chemical means. Credible stress analysis of these complex systems cannot be achieved by conventional methods with closed-form solutions as presented in this chapter. The finite element method (FEM) is the only viable way to attain credible solutions. Following is a brief introduction of this powerful method. Theories and formulation are deliberately omitted. They can be found in a number of references (Hsu, 1986, Hsu and Sinha 1992; Desai, 1979). The fundamental principles of deriving the key *element equations* will be presented in Chapter 10. What the reader will learn from the following is the essential input–output information that is required in using commercial finite element codes such as the popular ANSYS code used by the MEMS industry.

4.7.1 The Principle

In essence, the principle of the FEM is "divide and conquer." The very first step in a finite element analysis (FEA) is to divide the whole solid structure made of continua into a *finite* number of subdivisions of special shapes (called the *elements*) interconnected at the corners or specific points on the edges of the elements (called the *nodes*).

A variety of geometric elements are usually available from element libraries offered by most commercial codes.

Once the structure is subdivided into a great many elements (a process called *discretization*), engineering analysis is performed on these elements instead of the whole structure. Solutions obtained at the element level are assembled to get the corresponding solutions for the whole structure.

Credible results from a FEA are attainable with intelligent discretization of the structure. Two principal rules need to be followed:

1. Place denser and smaller elements in the parts of the structure with abrupt changes of geometry where high stress or strain concentrations are expected.
2. Avoid using elements with high aspect ratio, which is defined as the ratio of the longest dimension to the shortest one in the same element. The user is advised to keep this aspect ratio below 10.

Figure 4.53 shows the discretization of the silicon die/diaphragm/constraint base of a micro–pressure sensor (Schulze, 1998). Only a quarter of the unit is included in the finite element model because of the quarter-plane symmetry of the die/diaphragm/constraint base assembly.

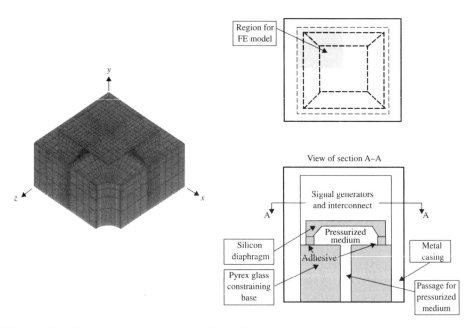

Figure 4.53. Finite element model for the silicon die/constraint base assembly of a micro–pressure sensor.

4.7.2 Engineering Applications

The FEM is used in almost every engineering discipline. Particularly well-developed common applications are in the following three areas, which happen to be necessary for microsystems design:

1. **Stress/Strain Analysis of Solid Structure:** The FEM provides solutions in stresses$\{\sigma\}$, strains $\{\varepsilon\}$, and displacement $\{U\}$ in each element in a discretized model. The displacements are the primary unknown quantities to be solved in this type of analysis.

2. **Heat Conduction Analysis:** It can handle all modes of heat transfer, including conduction, convection, and radiation. Various boundary conditions such as heat flux $\{q\}$ and convective/radiative types can be handled. Expected results include temperature in the elements or at the nodes.

3. **Fluid Dynamics:** The FEM codes for fluid dynamics are built on the basis of Navier–Stokes equations, as will be presented in Chapter 5. Numerical solutions will include mass flow, velocities, and driving pressures in selected control volumes of the moving fluid.

4.7.3 Input Information to FEA

1. General information:
   a. Profile of the structure geometry.
   b. Establish the coordinates: $x-y$ for plane; $r-z$ for axisymmetrical; $x-y-z$ for three-dimensional geometry.

2. Develop the finite element mesh (e.g., in Figure 4.53). For automatic mesh generation, the user usually specifies the desirable density of nodes and elements in specific regions. Information generated for the FEA includes node number, nodal coordinates, nodal conditions (e.g., constraints, applied forces and temperature), element number, element description (e.g., element designations and involved nodes).

3. Material property input: In stress analysis—Young's modulus E, Poisson ratio ν, shear modulus of elasticity G, yield strength σ_y, and ultimate strength σ_u. In heat conduction analysis—mass density ρ, thermal conductivity k, specific heat c, and linear thermal expansion coefficient α.

4. Boundary and loading conditions. In stress analysis—nodes with constrained displacements (e.g., in x, y, or z direction), concentrated forces at specified nodes, or pressure at specified element edge surfaces. In heat conduction analysis—given temperature at specified nodes, or heat flux at specified element edge surfaces, or convective or radiative conditions at specified element surfaces.

4.7.4 Output from FEA

1. Nodal and element information
2. Displacements at nodes
3. Stresses and strains in each element:

a. Normal stress components in x, y, and z directions
b. Shear stress components on xy, xz, and yz planes
c. Normal and shear strain components
d. Maximum and minimum principal stress components
e. The von Mises stress, defined as

$$\overline{\sigma} = \frac{1}{\sqrt{2}}\sqrt{(\sigma_{xx} - \sigma_{yy})^2 + (\sigma_{xx} - \sigma_{zz})^2 + (\sigma_{yy} - \sigma_{zz})^2 + 6(\sigma_{xy}^2 + \sigma_{yz}^2 + \sigma_{xz}^2)}$$

(4.71)

where σ_{xx}, σ_{yy}, and σ_{zz} are respective stress components along the x, y, and z axis and σ_{xy}, σ_{yz}, and σ_{xz} are the shear stress components in the element as defined in Figure 4.42. The von Mises stress in Equation (4.71) is used as the "representative" stress in a multiaxial stress situation. It is compared with the yield strength σ_y for plastic yielding and to σ_u for the prediction of the rupture of the structure.

4.7.5 Graphical Output

All modern commercial FEA codes offer graphical output in the following form:

- A solid model of the structure such as shown in Figure 10.34 for a microgripper
- User input discretized solid structure with finite element mesh such as shown in Figure 4.53 for a micro–pressure sensor die
- A deformed or/and undeformed solid model of the structure
- Color-coded zones to indicate the distribution of stresses, strains, and displacements over the selected plane of the structure
- Color-coded zones to indicate other required output quantities such as temperature or pressure in the structure
- Animated movements or deformation of the structure under specified operating conditions

4.7.6 General Remarks

The FEM is widely used by the MEMS and microsystems industry for design analysis and simulation of the systems. Indeed, because of the inherent complex geometry and loading and boundary conditions of MEMS and microsystems, FEM appears to be the only viable tool for such purposes. However, there are several facts that users need to be aware of. Following are a few specific ones that are relevant to MEMS and microsystems design analyses and simulations:

1. Engineers need to be aware of the fact that FEM offers only approximate solutions to the problem analyzed. Credibility results are obtainable by the intelligent use of the finite element code by the user. As mentioned in Section 4.7.1, the user needs to recognize stress/strain concentrations by allowing denser and smaller elements

in the parts of the structure with abrupt change of geometry. It is also necessary to avoid using elements with high aspect ratio for computational accuracy. The limit on aspect ratio makes discretization of thin-film structural components in MEMS and microsystems an insurmountable task. The user's knowledge and experience in selecting the right elements for special thermomechanical functions of the structural components, such as the spring and contact elements from the available element library, will be desirable. Another user-related issue is the proper establishment of the loading and boundary conditions in the discretized model. In many finite element codes, the pressure loading needs to be properly converted to the concentrated forces at the adjacent nodes on the applicable boundary surfaces. This issue often becomes more complicated in transient thermal analyses.

2. One must be sure of the credibility of the input material properties to the analysis. Because of the relatively short history of the MEMS and microsystems industry, most thermophysical properties of the materials used for these products are lacking. We will present some of these material properties in Chapter 7, but these properties are far from sufficient for proper design analyses. Information on some phenomenological behavior of MEMS structures, such as the interfacial fracture toughness described in Section 4.5 and the intrinsic stress model in Section 4.6, required for modeling and simulations is not currently available. An ongoing effort is being made by scientists and engineers to generate more complete material databases. The user is thus cautioned to be selective in choosing the input material properties for the finite element analyses.

3. The user should be aware of the fact that, unless otherwise specified, most commercial general-purpose finite element codes are developed on the theories of continua. A typical example is the common use of constitutive relations such as the generalized Hooke's law for solid mechanics or the Fourier law for heat conduction. The theories on continuum mechanics and heat conduction are time tested for macrosystems. However, they are not valid for MEMS and microsystems components at submicrometer scales. Commercial finite element codes thus cannot be indiscriminately used for MEMS and microsystems design analyses at submicrometer scales.

4. The reader may have come to the realization from Section 4.6 that it is difficult to predict the mechanistic behavior of MEMS and microsystems, which often consist of components made of layers of thin films or bonded to dissimilar materials. The fabrication of these components introduces unavoidable residual and intrinsic stresses in the structure. As we will learn from later chapters, virtually all microfabrication techniques result in such adverse effects in MEMS and microsystems. The incorporation of these effects in the design analysis requires coupling the FEA and the microfabrication processes. Some developers in computer-aided design (CAD) software packages have made such an effort in coupling, as will be described in Chapter 10. The reader is cautioned to be aware of this fact when using a commercial finite element code in a design analysis.

Despite the aforementioned shortcomings of the current state of the art of commercially available finite element codes, they are still the only viable means by which engineers can acquire a good insight on the MEMS and microsystems they have chosen for their design and simulation. Engineers are merely reminded of the fact that the power of the FEM, like many other tools for craftsmen and professionals, lies in the intelligence of the user. The user needs to have sound knowledge of the tool and be aware of its limitations before he or she can maximize the benefits of the tool.

PROBLEMS

Part 1 Multiple Choice

1. In general, mechanical engineering principles derived for continua can be used for MEMS components of size (a) larger than 1 nm, (b) larger than 1 μm, (c) larger than 1 cm.

2. The theory of thin-plate bending can be used to assess (a) the deflection only, (b) stresses only, (c) both the deflection and stresses in thin diaphragms of micro–pressure sensors.

3. Square diaphragms are (a) most popular, (b) somewhat popular, (c) least popular geometry for micro–pressure sensors.

4. From a mechanics point of view, the most favored geometry of diaphragm in micro–pressure sensors is (a) circular, (b) square, (c) rectangular.

5. The principal theory used in microaccelerometer design is (a) plate bending, (b) mechanical vibration, (c) strength of materials.

6. The natural frequency of a microdevice is determined by its (a) mass, (b) structure stiffness, (c) mass and structure stiffness.

7. Microdevices in theory contain (a) one, (b) several, (c) infinite number of natural frequencies.

8. The analysis that attempts to determine several or all natural frequencies of a microdevice is called (a) modal, (b) vibration, (c) model analysis.

9. The "resonant" vibration of a device made of elastic materials occurs when the frequency of the excitation force (a) approaches, (b) equals, (c) exceeds any of the natural frequencies of the device.

10. The "dashpot" in a "mass–spring" vibration system serves the purpose of including the (a) damping, (b) acceleration, (c) deceleration effect on the system.

11. The damping effect in most microaccelerometer design is (a) very important, (b) somewhat important, (c) not important.

12. The damping effect by compressible fluids (a) increases, (b) decreases, (c) remains unchanged with increase of the input frequency of the vibrating mass.

13. The movement of the beam mass in balanced force microaccelerometers is usually measured by (a) piezoresistors, (b) piezoelectric, (c) capacitance changes.

14. A vibrating beam will have its natural frequency (a) increased, (b) decreased, (c) unchanged with increase of longitudinal stress in tension.

15. Thermal stresses can be induced in mechanically constrained microdevice components by (a) uniform temperature rise, (b) nonuniform temperature rise, (c) any temperature rise.

16. Thermal stresses are induced in a microdevice component made of dissimilar materials due to (a) difference of coefficients of thermal expansion of the materials, (b) the weakness of the bonding interface, (c) the degradation of materials after bonding.

17. Thermal stresses are induced in microdevice components free of mechanical costraints by (a) uniform temperature change, (b) nonuniform temperature change, (c) uniform temperature with time.

18. The creep deformation in a material becomes serious (a) at any temperature, (b) above half the melting point, (c) above half the homologous melting point.

19. The *homologous melting point* of a material is defined as the melting point on the scale of (a) absolute temperature, (b) Celsius temperature, (c) Fahrenheit temperature.

20. The parts of microsystems that are obviously vulnerable to creep failure are (a) solder bonds, (b) epoxy resin bonds, (c) silicone rubber bonds.

21. There are generally (a) two, (b) three, (c) four modes of fracture at the interfaces of microdevices.

22. The most frequently occurring fracture failure in a microstructure is by (a) mode III, (b) mode II, (c) mode I.

23. Interfaces in microdevices are vulnerable to (a) mixed modes I and II, (b) mixed modes I and III, (c) mixed modes II and III fracture.

24. Fracture mechanics analysis of interfaces in microstructures requires the distribution of (a) normal stresses, (b) shear stresses, (c) both normal and shear stresses at the vicinity of the interface.

25. The finite element method is a viable analytical tool for microstructures of (a) simple geometry, (b) complex geometry and loading/boundary conditions, (c) complex loading and boundary conditions.

26. The first step in a finite element analysis is (a) to find the approximate solution, (b) to set the governing equation and boundary condition, (c) to subdivide the continuum into a number of subdivisions called *discretization*.

27. The primary unknown quantity in a finite element analysis is (a) the quantity that appears in the formulation, (b) the most important quantity, (c) the most desirable quantity to be determined.

28. The primary unknown quantity in a stress analysis by using the finite element method is (a) stress, (b) strain, (c) displacement.

29. The constitutive relation in a finite element analysis relates (a) the construction of appropriate formulations, (b) the primary and other essential quantities, (c) the loading and boundary conditions.

30. The von Mises stress represents (a) the stress component following the von Mises principle, (b) the stress for a specific material, (c) stresses in a structure of complex geometry.

Part 2 Computational Problems

1. Use the material properties in Table 7.3 to determine the maximum deflection of a circular diaphragm made of aluminum with conditions specified in Example 4.1.

2. If the stress required to produce measurable signal output in a square diaphragm in a pressure sensor is 350 MPa, what will be the required thickness of the diaphragm? The diaphragm is an integral part of a silicon die that is shaped from a wafer 100 mm in diameter in the (100) plane (i.e., with a 54.74° angle in the slope from the bottom face into the cavity, as shown in Figure 4.6). The die has a plane area of 3 mm × 3 mm. A pressurized medium at 75 MPa pressure is applied at the front side of the silicon die. We realize that the thickness of a 100-mm-diameter wafer is 500 μm. The width of the footprint of the silicon die is assumed to be 250 μm.

3. Derive the expressions for the equivalent spring constants for the two beam springs illustrated in Figure 4.18.

4. A microdevice component 5 g in mass is attached to a fine strip made of silicon as illustrated in Figure 4.54. The equivalent spring constant of the strip is 18,240 N/m. Both the mass and the strip spring are made of silicon. The mass is pulled down by 5 μm initially and is released at rest. Determine (a) the natural frequency of the simulated mass–spring system and (b) the maximum amplitude of vibration.

5. Determine the time required to break the strip spring in problem 4 when a force $F(t) = 5 \cos (1910t)$ newtons is applied to the mass at time $t > 0$, where t is the time into the vibration. Assume the strip breaks at a deflection of 1 mm. The vibration of the strip mass system begins when the system is at rest.

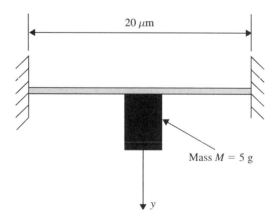

Figure 4.54. Strip spring–mass system.

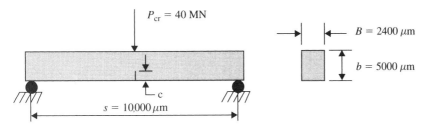

Figure 4.55. Measurements of fracture toughness of silicon using three-point bending Tests.

6. Determine the thickness of the beam spring in Example 4.10 if the maximum allowable deflection of the beams is 5 mm.

7. Assess the effect of damping on the amplitude of vibration of a cantilever beam mass system.

8. Assess the effect of damping on the movement of the beam mass in Example 4.12.

9. Use a balanced-force microaccelerometer such as described in Example 4.12 for the situation described in Example 4.14. The two beam springs have the same dimensions as those given in Example 4.9. Determine the maximum movement of the beam mass. The damping effect can be neglected.

10. The microaccelerometer described in Example 4.17 is expected to operate with temperature rise from 10 to 50°C. Plot the movements of the free end of the actuator with respect to the range of temperature rise.

11. Compute the maximum thermal stresses, thermal strains, and displacements induced in a beam by a temperature variation in the thickness direction, $T(z) = 2 \times 10^6 z + 30$ in degrees Celsius. The geometry and dimensions of the beam are given in Figure 4.45. The beam is made of silicon.

12. Redo Example 4.19 with the width of the beam being $b = 100$ μm. (*Hint:* Use the formulations for thin plates.)

13. Find the maximum thermal stress, strain, and displacements in a beam subjected to heating at its top surface at time t of 0.1, 0.5, and 1.5 μs. The geometry and loading and other conditions are described in Example 4.19. Plot the results versus time during the period of heating.

14. Compute the fracture toughness for a silicon substrate using a three-point bending specimen as illustrated in Figure 4.49b. The dimensions of the beam specimen are shown in Figure 4.55 with $s = 1$ cm, $b = 5$ mm, $c = 100$ μm. The width of the beam, B, is 240 μm. The force required to break the specimen is assumed to be 40 MN.

15. What fracture toughness of the silicon specimen would you get if the width of the specimen, B, is 100 times greater than that used in problem 14. What would you have observed from a comparison of the results obtained from these two cases?

CHAPTER 5

THERMOFLUID ENGINEERING AND MICROSYSTEMS DESIGN

5.1 INTRODUCTION

Many MEMS devices and microsystems, such as the manifold absolute pressure sensor (MAPS) presented in Chapter 2, are expected to handle hot gases. Many others are expected to operate at elevated temperature. Still, there are microvalves, pumps, and fluidics for biosensors that involve moving fluids in both liquid and gaseous forms. The micro–heat pipes for effective heat dissipation when shrinking the size of electronic circuits and devices with increasing power densities, as described in Chapter 2, involve two-phase thermofluid flows in capillary conduits. Mechanical design of these systems requires the application of theories and principles of fluid dynamics and heat transfer. Thermofluids engineering principles are also used in microfabrication process design, as will be described in Chapter 8. For example, gas flow over hot substrate surfaces needs to be properly controlled in many thin-film deposition processes.

In this chapter, we will first review thermofluids engineering principles that are derived from classical continuum theories and how these principles can be used in the design of MEMS and microsystems. Scientists and engineers are making relentless effort in this frontier area of research, as the trend of miniaturization of MEMS and microsystems with many of these devices involving higher power densities not only continues but also will accelerate in the near future.

5.2 OVERVIEW OF BASICS OF FLUID MECHANICS AT MACRO- AND MESOSCALES

Fluid mechanics is the study of fluids in motion (fluid dynamics) or at rest (fluid statics) as well as solid–fluid interactions. For MEMS and microsystems design, fluid dynamics is of primary interest to engineers.

Microsystems design engineers need to deal with two principal types of fluids: (1) noncompressible fluids (e.g., liquids) and (2) compressible fluids (gases). Both types have applications in micro- and nanoscale systems.

Fluids are aggregations of molecules. These molecules are closely spaced in liquids and widely scattered in gases. Unlike solids, fluid molecules are much widely spaced than the molecule size. Further, these molecules are not fixed in the lattice but move freely relative to each other. We may thus view fluids as a substance that has volume but no shape. Unlike solids, fluids cannot resist any shear force or shear stress without moving. All fluids have viscosity that causes friction when they are set in motion. Viscosity is a measure of the fluid's resistance to shear when the fluid is in motion. Thus, it is necessary to have a driving pressure to make a fluid flowing in a conduit, a channel, or, as in the case of fluidics, systems of conduits.

As mentioned in the foregoing section, theories derived from continuum fluids can be used for studying the flow of fluids at macro- and mesoscales. A *continuum fluid* is viewed as a fluid with its properties varying continuously in space. For this reason, differential calculus can be applied to analyze the physical behavior of this substance.

5.2.1 Viscosity of Fluids

Fluids can be put in motion by even a slight shear force. The induced shear strain in a bulk of fluid, as illustrated in Figure 5.1, can be expressed by the change of the right angle at the corners in Figure 5.1a to that of angle θ in Figure 5.1b after deformation in shear. The magnitude of angle θ represents the shear strain of the fluid and has a unit of radians.

The shear deformation illustrated in Figure 5.1 is considered possible by the relative motion of a pair of plates placed at the top and bottom of the bulk fluid, as illustrated in Figure 5.2.

We notice that the deformed fluid shown in Figure 5.2b is produced by the motion of the top plate at a velocity u_0. A *non-slip* boundary condition at the interface of the

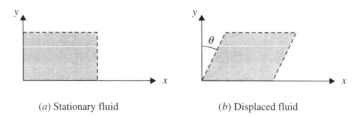

(*a*) Stationary fluid   (*b*) Displaced fluid

Figure 5.1. Shear deformation of a fluid.

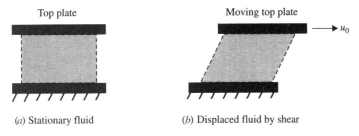

(a) Stationary fluid   (b) Displaced fluid by shear

Figure 5.2. Shear flow of fluid by a moving plate.

fluid at both the top and bottom plates is assumed in the situation. The relative motion of the top and bottom plates in parallel represents a *shear force* that causes the flow of the fluid. The associated *shear stress* τ is considered proportional to the rate of change of the induced shear strain θ. Mathematically, this relationship can be expressed as

$$\tau \propto \frac{\delta\theta}{\delta t}$$

or in the form

$$\tau = \mu \frac{\delta\theta}{\delta t}$$

In the case of a fluid in continuous motion, the increment of the variation of the shear strains is taken as infinitesimally small, that is $\delta\theta \to 0$; we have

$$\tau = \lim_{\delta t \to 0}\left(\mu\frac{\delta\theta}{\delta t}\right) = \mu\frac{d\theta}{dt} \tag{5.1}$$

The proportional constant μ is called the *dynamic viscosity*, or the *viscosity*, of the fluid. It is a very important property of a fluid. We notice the linear relationship between the shear stress and the shear strain rate exhibited in Equation (5.1). Fluids that exhibit such linear relationship are classified as *newtonian fluids*. Other classifications of fluid are qualitatively illustrated in Figure 5.3.

The velocity profile of the fluid flow in Figure 5.2 is illustrated in Figure 5.4, in which we perceive a linear variation of the velocity in the y direction. One may readily prove that $d\theta/dt$ in Equation (5.1) is equal to $du(y)/dy$, where $u(y)$ represents the velocity of the fluid at a distance y from the stationary state at the bottom plate. The following expression for the shear stress τ is obtained:

$$\tau = \mu\frac{du(y)}{dy} \tag{5.2}$$

Numerical values of dynamic viscosity in Equation (5.2) for seven selected fluids are presented in Table 4.3. Another measure of fluid viscosity, called the *kinematic viscosity* v, is used in fluid mechanics analysis. It is defined as $v = \mu/\rho$, where ρ is the mass density of the fluid.

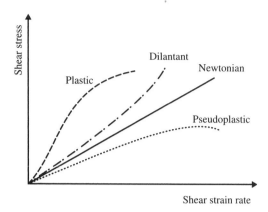

Figure 5.3. Classes of fluids (White, 1994).

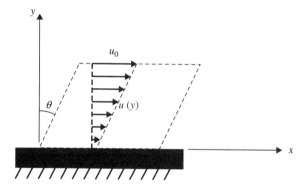

Figure 5.4. Velocity profile in newtonian fluid flow.

5.2.2 Streamlines and Stream Tubes

A *streamline* is a trace of any point in a moving fluid. The direction of the line coincides with the direction of the motion of the moving point in the fluid. Streamlines have the following properties:

1. No flow can cross a streamline.
2. No two streamlines may intersect, except at a point where the velocity of the fluid is zero.
3. Any boundary line of the flow must also be a streamline.
4. In general, a streamline can change the position and its shape from time to time in a steady-state flow.

A *stream tube* is made up of a bundle of streamlines through all points of a closed curve.

5.2.3 Control Volumes and Control Surfaces

As fluids do not have fixed shapes of their own, it is not possible to define their volumes as we do with solids. Consequently the term *control volume* is adopted for computational purposes. The control volume of a bulk of fluid (v) is selected in such a way that the flow properties (e.g., the velocities) can be evaluated at locations where the mass of the fluid crosses the surface of this volume. The surface of a control volume is called *control surface s*.

5.2.4 Flow Patterns and Reynolds Number

The patterns of fluid flow can significantly influence the engineering results. There are generally two patterns of fluid flow: (1) laminar flow and (2) turbulent flow. Laminar flow is a gentle flow of the fluid that follows the streamlines. Turbulent flow, on the other hand, is violent. It does not follow any traceable pattern.

The Reynolds number, defined as follows, usually determines the fluid flow patterns:

$$\text{Re} = \frac{\rho L V}{\mu} \tag{5.3}$$

where ρ, V, and μ represent the corresponding mass density, velocity, and dynamic viscosity of the fluid. The characteristic length L is the dimension that is a principal factor that relates to the flow. For example, L is defined as the diameter d of a pipeline or the length L of a flat plate on which the fluid flows.

Laminar fluid flows occur at Re < 10–100 for compressible fluids and Re < 1000 for incompressible fluids. Turbulent flows result in large drag forces when the fluid flows over a solid surface. Fortunately, in almost all MEMS and microsystems applications, fluid flow is always in the laminar flow regime.

5.3 BASIC EQUATIONS IN CONTINUUM FLUID DYNAMICS

5.3.1 Continuity Equation

The *continuity equation* of fluid flow is used to evaluate volumetric flow rates. With reference to the situation illustrated in Figure 5.5, a control volume of the fluid flowing through a stream tube is selected. This volume is enclosed in the region designated by *MNPQ* in the figure. The two control surfaces *MN* and *PQ* are separated by an infinitesimally small distance ds. The rates of the fluid flow at the control surfaces *MN* and *PQ* are m and $m + \Delta m = m + (\partial m/\partial s)\,ds$ respectively. We may readily account for the total volume of fluid entering and leaving the two control surfaces during a small time interval Δt to be $m\,\Delta t$ at *MN* and $[m + (\partial m/\partial s)\,ds]\,\Delta t$ at *PQ*. The change of mass in the control volume is $(A\,ds)\,\Delta\rho$, where A is the average cross-sectional area of the stream tube of the control volume. By the law of conservation of mass, we have the relation

$$m\,\Delta t - \left(m + \frac{\partial m}{\partial s}\,ds \right) \Delta t = (A\,ds)\,\Delta\rho$$

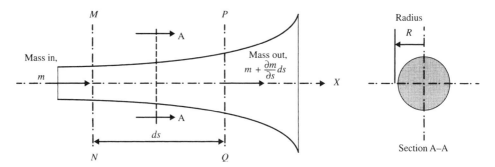

Figure 5.5. Continuous fluid flow in a stream tube.

For the case with $\Delta t \rightarrow 0$ for a fluid in continuous motion, we have $\Delta \rho / \Delta t \rightarrow \partial \rho / \partial t$. Consequently, we have the following equation for the equation of continuity for one-dimensional flow:

$$A\frac{\partial \rho}{\partial t} + \frac{\partial m}{\partial s} = 0 \tag{5.4}$$

For steady-state fluid flows, $\partial \rho / \partial t = 0$, which leads to $\partial m / \partial s = 0$, as in Equation (5.4). We will thus have the *rate of mass flow* in the steady-state flow condition with $\dot{m} = \text{const}$, or

$$\dot{m} = \rho_1 V_1 A_1 = \rho_2 V_2 A_2 \tag{5.5}$$

where A_1 and A_2 are cross-sectional areas through which the fluid flows at the respective velocities V_1 and V_2.

If units of kilograms per cubic meter, meters per second, and meters squared are used for the corresponding mass density ρ, the velocity V and the cross-sectional area A, respectively, then the rate of mass flow $\dot{m}$ in Equation (5.5) will have a unit of kilograms per second.

For incompressible fluids, the above relationship can be used to compute the volumetric flow rate:

$$Q - V_1 A_1 = V_2 A_2 \tag{5.6}$$

where Q has a unit of cubic meters per second.

> **Example 5.1.** An incompressible fluid is used in a microfluidic system. It flows through a tube with a diameter of 1 mm at a rate of 1 microliter (μL) per minute. A reducer is used to connect this tube to the microconduits in the fluidic system. The reducer has an outlet diameter of 20 μm. Determine the velocity of the fluid at the inlet and outlet of the reducer. The system is illustrated in Figure 5.6.

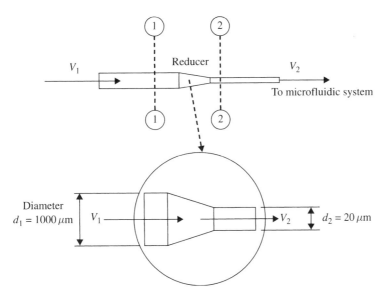

Figure 5.6. Fluid supply to a microfluidic system.

Solution: It is safe to assume a laminar flow situation because of the low flow rate. We further assume that a steady-state flow is maintained. Thus, from the continuity equation (5.6), the rate of fluid flow through the control surface 1–1 in Figure 5.6 should be the same as that through surface 2–2.

From Equation (5.6), the volumetric mass flow rate is given as

$$Q = A_2 V_2 = A_1 V_1$$

where $Q = 1 \times 10^{-6}\,\text{cm}^3/\text{min} = 1.67 \times 10^{-14}\,\text{m}^3/\text{s}$ and the entrance and exit cross-sectional areas of the reducer are

$$A_1 = \frac{1}{4}\pi(1000 \times 10^{-6})^2 = 0.785 \times 10^{-6}\,\text{m}^2$$

$$A_2 = \frac{1}{4}\pi(20 \times 10^{-6})^2 = 314 \times 10^{-12}\,\text{m}^2$$

We can thus compute the velocities of the fluid at both the entrance and exit of the reducer by using the continuity relation as follows:

$$V_1 = \frac{Q}{A_1} = \frac{1.67 \times 10^{-14}}{0.785 \times 10^{-6}} = 0.0217\ \mu\text{m/s}$$

$$V_2 = \frac{Q}{A_2} = \frac{1.67 \times 10^{-14}}{314 \times 10^{-12}} = 53.2\ \mu\text{m/s}$$

5.3.2 Momentum Equation

The momentum equation is used to compute the fluid-induced forces on the solid on which the fluid flows. This equation is derived on the bases of conservation of momentum and Newton's law of dynamic equilibrium.

Consider the two-dimensional steady-state flow situation illustrated in Figure 5.7. The selected control volume of $ABCD$ makes a small movement to a new position designated as $A'B'C'D'$.

The change of momentum in the control volume during this short period dt is given as

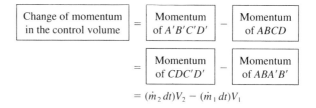

$$= (\dot{m}_2 \, dt)V_2 - (\dot{m}_1 \, dt)V_1$$

where $\dot{m}_1$ and $\dot{m}_2$ are the rates of mass flow across the respective control surfaces 1–1 and 2–2 in Figure 5.7.

In the steady-state flow condition, the rate of mass flow remains constant, that is, $\dot{m}_1 = \dot{m}_2 = \dot{m}$. We can obtain the induced forces using the relationships between the impulse forces and change of momentum as follows:

$$\sum F = \dot{m}(\mathbf{V}_2 - \mathbf{V}_1) \tag{5.7}$$

where $\mathbf{V}_1$ and $\mathbf{V}_2$ are the respective velocity vectors at control surfaces 1–1 and 2–2.

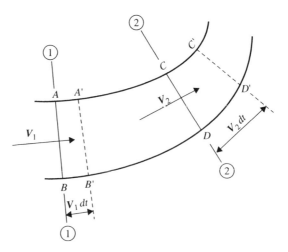

Figure 5.7. Moving fluid in a stream tube.

Example 5.2. A micromachined silicon valve utilizing electrostatic actuation is constructed. The valve unit has a similar configuration as that reported in Ohnstein et al. (1990), as illustrated in Figure 5.8a. The thin closure plate is used as the valve with a dimension of 300 μm wide $\times$ 400 μm long $\times$ 4 μm thick. The plate is bent to open or close by electrostatic actuation to regulate the hydrogen gas flow. The maximum opening of the closure plate is a $15°$ tilt from the horizontal closed position. Determine the force induced by the flow of the gas at a velocity of 60 cm/min and a volumetric rate of 30,000 cm^3/min. Also, calculate the mass flow split over the lower surface of the plate.

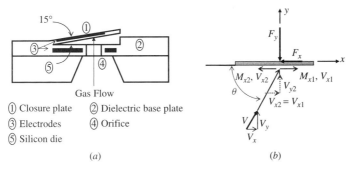

Gas Flow

① Closure plate ② Dielectric base plate
③ Electrodes ④ Orifice
⑤ Silicon die

(a) (b)

Figure 5.8. Fluid-induced force on a microvalve: (a) microvalve unit; (b) fluid-induced forces on closure plate.

Solution: We assume that the gas stream impinges on the plate at an angle $\theta = 75°$ at the maximum opened position. The gas stream splits into two components, that is, M_{x1} induced by velocity V_{x1} and M_{x2} by velocity V_{x2} as shown in Figure 5.8b. We designate M_{x1} and M_{x2} to be the respective components of the rate of mass flow of the gas, $\dot{m}$.

The volumetric flow of the gas, Q, is 30,000 cm^3/min, or 500×10^{-6} m^3/s. The rate of mass flow of the hydrogen gas involves the mass density of the gas, $\rho = 0.0826$ kg/m^3 (Janna, 1993), with $\dot{m} = \rho Q = 0.0826 \times (500 \times 10^{-6}) = 41.3 \times 10^{-6}$ kg/s.

Referring to the force diagram in Figure 5.8b and using Equation (5.7), the induced force components are expressed in the forms

$$\sum F_y = \dot{m}(V_{y2} - V_y) \tag{a}$$

$$\sum F_x = (M_{x1}V_{x1} - M_{x2}V_{x2}) - \dot{m}V_x \tag{b}$$

where

V_{y2} = fluid velocity along y axis at plate surface (zero in this case)

V_x = velocity component along x axis in gas stream $= V\cos\theta$

V_y = velocity component along y axis in gas stream $= V\sin\theta$

Thus, by substituting the values $\theta = 75°$ and $V = 60\,\text{cm/min}$ or $10^{-2}\,\text{m/s}$ into Equation (a), we obtained the force $F_y = 40 \times 10^{-8}\,\text{kg-m/s}^2$, or $40 \times 10^{-8}\,\text{N}$.

The horizontal force component F_x on the plate exists only if the coefficient of friction between the gas and the contacting plate surface is known. However, we may reasonably assume frictionless gas flow at that surface, that is, $\sum F_x = 0$, which, according to Equation (b), will lead to the relationship

$$(M_{x1} V_{x1} - M_{x2} V_{x2}) - \dot{m} V \cos \theta = 0$$

It is further reasonable to assume that $V_{x1} = V_{x2} = V$ in frictionless flow. Consequently, the mass flow split at the lower surface of the plate can be obtained by solving the simultaneous equations

$$M_{x1} - M_{x2} = \dot{m} \cos \theta \qquad (c)$$

$$M_{x1} + M_{x2} = \dot{m} \qquad (d)$$

from which we obtain the split mass flow rates

$$M_{x1} = \frac{1}{2}\dot{m}(1 + \cos \theta) = \frac{1}{2}(41.3 \times 10^{-6})(1 + \cos 75°) = 26 \times 10^{-6}\,\text{kg/s}$$

$$M_{x2} = \frac{1}{2}\dot{m}(1 - \cos \theta) = \frac{1}{2}(41.3 \times 10^{-6})(1 - \cos 75°) = 15.3 \times 10^{-6}\,\text{kg/s}$$

5.3.3 Equation of Motion

The equation of motion in fluid dynamics is used to evaluate the relationship between the motion of the fluid and the required driving force, namely the pressure. It is a very important part of fluid dynamics in assessing the pumping power required to move the fluid. The following derivation of the equation is based on an assumption that the friction between the fluid and the contacting surface of the conduit is negligible.

Figure 5.9 illustrates a fluid element moving along a streamline with a unit thickness (in the direction normal to the figure) in a two-dimensional domain defined by the x–y plane. The small fluid element is situated in a cartesian coordinate system (x, y) with the two sides, ds and dn, along the respective tangential (s) and normal (n) directions that define the streamline. The movement of the fluid element is maintained by the application of pressure p at the backside, that is the left face of the element. The weight of the fluid, dw, is always pointed vertically downward. By summing all forces acting on both the tangential and normal faces of the element, we may establish the relationship of the applied pressure and the velocities of the element in motion by using Newton's second law, expressed as $\sum F_s = m\,a_s$ and $\sum F_n = m\,a_n$, where F_s and F_n are the respective inertia forces along the s and n directions in Figure 5.9, m is the mass of the fluid element, and a_s and a_n are the respective accelerations in both the s and n directions. Both these acceleration components can be related to the tangential velocity of the fluid

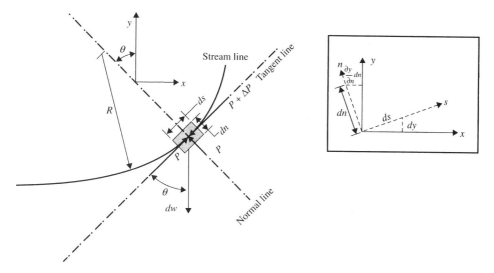

Figure 5.9. A fluid element traveling along a streamline.

element, $V(s, t)$, by the following expressions according to the theory of the kinematics of a moving particle along a curvilinear path:

$$a_s = \frac{\partial V(s, t)}{\partial t} + V(s, t)\frac{\partial V(s, t)}{\partial s}$$

$$a_n = \frac{\partial V_n(s, t)}{\partial t} + \frac{V(s, t)^2}{R}$$

where $V_n(s, t)$ is the component of the tangential velocity $V(s, t)$ in the normal direction.

Thus, using the above relationships in Newton's second law, we may derive the equations of motion in the forms

$$\rho\left(\frac{\partial V}{\partial t} + V\frac{\partial V}{\partial s}\right) = -\frac{\partial P}{\partial s} - \rho g \sin \alpha \tag{5.8a}$$

$$\rho\left(\frac{\partial V_n}{\partial t} + \frac{V^2}{R}\right) = -\frac{\partial P}{\partial n} - \rho g \cos \alpha \tag{5.8b}$$

Equation (5.8) is called *Euler's equation*.

The equations of motion in a three-dimensional cartesian coordinate systems, with the x coordinate coinciding with the streamline, can be expressed as

$$\rho\left(\frac{\partial u}{\partial t} + u\frac{\partial u}{\partial x} + v\frac{\partial u}{\partial y} + w\frac{\partial u}{\partial z}\right) = -\frac{\partial P}{\partial x} + X \tag{5.9a}$$

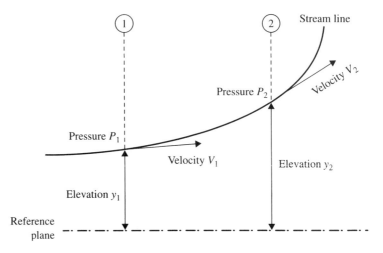

Figure 5.10. Fluid flow with varying state properties.

$$\rho \left(\frac{\partial v}{\partial t} + u\frac{\partial v}{\partial x} + v\frac{\partial v}{\partial y} + w\frac{\partial v}{\partial z} \right) = -\frac{\partial P}{\partial y} + Y \qquad (5.9b)$$

$$\rho \left(\frac{\partial w}{\partial t} + u\frac{\partial w}{\partial x} + v\frac{\partial w}{\partial y} + w\frac{\partial w}{\partial z} \right) = -\frac{\partial P}{\partial z} + Z \qquad (5.9c)$$

where

u = fluid velocity along x direction, $= u(x,y,z)$

v = fluid velocity along y direction, $= v(x,y,z)$

w = fluid velocity along z direction, $= w(x,y,z)$

X, Y, Z = components of body force of fluid (e.g., weight) along respective x, y, and z directions

The above equations of motion can be deduced into the well-known *Bernoulli's equation*

$$\frac{V_1^2}{2g} + \frac{P_1}{\rho g} + y_1 = \frac{V_2^2}{2g} + \frac{P_2}{\rho g} + y_2 \qquad (5.10)$$

which depicts the situation illustrated in Figure 5.10.

Example 5.3. Determine the pressure drop in a minute stream of alcohol flowing through a section of a tapered tube 10 cm in length. The inlet velocity is 600 μm/s. The mass density of alcohol is 789.6 kg/m^3 (Avallone and Baumeister, 1996). The tube is inclined 30° from the horizontal plane, as illustrated in Figure 5.11.

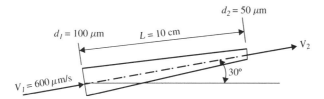

Figure 5.11. Flow in a tapered tube.

Solution: The cross-sectional areas of the tube at the entrance, A_1, and the exit, A_2, are calculated to be 0.784×10^{-8} m^2 and 0.1964×10^{-8} m^2, respectively. We are given the entrance velocity as 600 μm/s, or 600×10^{-6} m/s. The exit velocity of the flow can be determined from Equation (5.6) as

$$V_2 = \frac{V_1 A_1}{A_2} = 2.4 \times 10^{-3} \text{ m/s}$$

By rearranging Bernoulli's equation in Equation (5.10), we have the following expression for the pressure drop ΔP:

$$\Delta P = (P_1 - P_2) = \rho g \left[\frac{V_2^2 - V_1^2}{2g} + (y_2 - y_1) \right] \tag{5.11}$$

The difference of elevation of the tube between the exit and entrance $(y_2 - y_1)$ in Equation (5.11) is equal to $L \sin 30° = 5 \times 10^{-2}$ m.

The corresponding pressure drop is thus calculated to be 387.3 N/m^2, or 387 Pa, from Equation (5.11).

5.4 LAMINAR FLUID FLOW IN CIRCULAR CONDUITS

We made an assumption in deriving the equations of motion in the foregoing section that fluid flows through a conduit with no slipping at the conduit wall. This assumption obviously contradicts our subsequent assumption that fluid is also frictionless between the streamlines as well as between the fluid and the contacting wall. In reality, however, friction does exist in the fluid flowing in a conduit. This friction factor inevitably results in additional pressure drop in these cases. The equivalent *head* loss due to the friction in the fluid and between the fluid and the contacting wall in a circular pipe with diameter d and length L can be accounted for by the Darcy–Weisbach equation (White, 1994):

$$h_f = f \frac{L}{d} \frac{V^2}{2g} \tag{5.12}$$

where f is the *Darcy friction factor*:

$$f = \frac{8\tau_w}{\rho V^2} \tag{5.13}$$

where τ_w is the shear stress at the pipe wall. This quantity can be evaluated using Equation (5.2) with $y = d/2 = a$, where d is the inside diameter of the pipe. We may use the following expression to evaluate τ_w in Equation (5.13):

$$\tau_w = \frac{a}{2}\left|\frac{d}{dx}(P + \rho g y)\right| \quad \text{for laminar flow} \tag{5.14}$$

The Darcy friction factor is deduced to a simple expression as

$$f_\ell = \frac{64}{\text{Re}} \tag{5.15}$$

where Re is the Reynolds number as expressed in Equation (5.3), with the characteristic length L to be replaced by the diameter d of the pipe.

The head loss due to friction, h_f, in Equation (5.12) is included on the right-hand side of the Bernoullis equation (5.10) in evaluating the pressure drop in a pipeline flow.

> **Example 5.4.** Estimate the equivalent head loss due to friction in Example 5.3.
>
> *Solution:* We will approach the problem with average values: average diameter d of tube $= 75$ μm and the average velocity V between entrance and exit of tube $= 1.5 \times 10^{-3}$ m/s. We further obtain the dynamic viscosity of alcohol: $\mu = 1199.87 \times 10^{-6}$ N-s/m^2 from Table 4.3.
>
> Thus, we obtain the Reynolds number Re from Equation (5.3):
>
> $$\text{Re} = \frac{\rho V d}{\mu} = \frac{789.6 \times (1.5 \times 10^{-3}) \times (75 \times 10^{-6})}{1199.87 \times 10^{-6}} = 0.074$$
>
> The friction factor in Equation (5.15) leads to the following:
>
> $$f = \frac{64}{\text{Re}} = \frac{64}{0.074} = 864.86$$
>
> The equivalent head loss h_f can thus be determined using Equation (5.12) as
>
> $$h_f = (864.86)\frac{10^{-1}}{75 \times 10^{-6}}\frac{(1.5 \times 10^{-3})^2}{2 \times 9.81} = 0.1322 \text{ m}$$

Figure 5.12 illustrates the laminar flow of a fluid in a circular tube of radius a. The Hagen–Poiseuille equation for the volumetric flow Q and the pressure drop ΔP can be derived from the equations of motion (5.8) and (5.9) by using a cylindrical polar coordinate system (White, 1994):

$$Q = \frac{\pi a^4}{8\mu}\left[-\frac{d}{dx}(P + \rho g y)\right] \tag{5.16}$$

where y is the elevation of the tube from a reference plane.

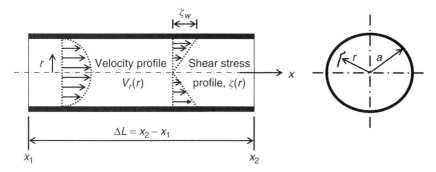

Figure 5.12. Fluid flow in a circular tube.

The pressure drop in the fluid over the tube length L is given as

$$\Delta P = \frac{8\mu L Q}{\pi a^4} \qquad (5.17)$$

The equivalent head loss in relation to the volumetric flow Q is determined as

$$h_{f,\ell} = \frac{128\mu L Q}{\pi \rho g d^4} \qquad (5.18)$$

The parameter d in the above expression stands for the diameter of a circular pipe filled with fluid. It is replaced by the *hydraulic diameter* d_h for conduits of circular cross section with partial fluid flow or channels with geometry other than circular cross sections. A hydraulic diameter is defined as

$$d_h = \frac{4A}{p} \qquad (5.19)$$

In Equation (5.19), A is the cross-sectional area of fluid flow and p is the *wet perimeter*, meaning the perimeter that is in contact with the fluid. We may readily find the hydraulic diameter d_h for a conduit of rectangular cross section with width w and height h to be

$$d_h = \frac{4A}{p} = \frac{4(wh)}{2(w+h)} = \frac{2wh}{w+h}$$

One striking phenomenon that one may observe from Equations (5.17) and (5.18) is that both the pressure drop and the friction head loss for laminar flows are inversely proportional to the fourth power of the conduit diameter. This means that a 16 times higher pumping power would be required to pump the same amount of volumetric flow of fluid with the conduit diameter reduced by half.

5.5   COMPUTATIONAL FLUID DYNAMICS

The governing equations for fluid dynamics are the well-known *Navier–Stokes equations* on which modern CFD codes are built. Derivation of Navier–Stokes equations is beyond the scope of this book. They can be found in many fluid mechanics books (e.g., White, 1994). Here we will present only the equations for the solutions for velocity vectors u, v, and w along the respective x, y, and z coordinates in a moving fluid in a three-dimensional space:

$$\rho g - \frac{\partial P}{\partial x} + \mu \left(\frac{\partial^2}{\partial x^2} + \frac{\partial^2}{\partial y^2} + \frac{\partial^2}{\partial z^2} \right) u(x, y, z) = \rho \frac{\partial u(x, y, z)}{\partial t} \tag{5.20a}$$

$$\rho g - \frac{\partial P}{\partial y} + \mu \left(\frac{\partial^2}{\partial x^2} + \frac{\partial^2}{\partial y^2} + \frac{\partial^2}{\partial z^2} \right) v(x, y, z) = \rho \frac{\partial v(x, y, z)}{\partial t} \tag{5.20b}$$

$$\rho g - \frac{\partial P}{\partial z} + \mu \left(\frac{\partial^2}{\partial x^2} + \frac{\partial^2}{\partial y^2} + \frac{\partial^2}{\partial z^2} \right) w(x, y, z) = \rho \frac{\partial w(x, y, z)}{\partial t} \tag{5.20c}$$

where

ρ = mass density of fluid

g = gravitational acceleration

P = driving pressure

μ = dynamic viscosity of fluid

t = time

The stress components in fluid in a small cubical element of the control volume of the fluid (Figure 5.13) can be obtained by differentiating various velocity components from the Navier–Stokes equations as follows:

$$\tau_{xx} = 2\mu \frac{\partial u(x, y, z)}{\partial x} \qquad \tau_{yy} = 2\mu \frac{\partial v(x, y, z)}{\partial y} \qquad \tau_{zz} = 2\mu \frac{\partial w(x, y, z)}{\partial z}$$

$$\tau_{xy} = \tau_{yx} = \mu \left(\frac{\partial u(x, y, z)}{\partial y} + \frac{\partial v(x, y, z)}{\partial x} \right)$$

$$\tau_{xz} = \iota_{zx} - \mu \left(\frac{\partial w(x, y, z)}{\partial x} + \frac{\partial u(x, y, z)}{\partial z} \right) \tag{5.21}$$

$$\tau_{yz} = \tau_{zy} = \mu \left(\frac{\partial v(x, y, z)}{\partial z} + \frac{\partial w(x, y, z)}{\partial y} \right)$$

All stress components in a moving fluid can be expressed as follows:

$$\sigma_{ij} = \begin{bmatrix} -P + \tau_{xx} & \tau_{yx} & \tau_{zx} \\ \tau_{xy} & -P + \tau_{yy} & \tau_{zy} \\ \tau_{xz} & \tau_{yz} & -P + \tau_{zz} \end{bmatrix} \tag{5.22}$$

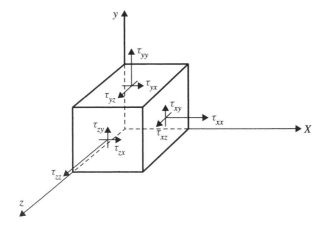

Figure 5.13. Notation for stresses in a fluid element.

Various CFD computer codes are available commercially for solving dynamic behavior of fluids in microsystems.

5.6 INCOMPRESSIBLE FLUID FLOW IN MICROCONDUITS

A well-recognized phenomenon in fluid dynamics is that the velocity profile of a fluid flowing in a conduit has its velocity increasing toward the center of the conduit, as illustrated in Figure 5.12. This is partly due to the nonslip boundary at the conduit wall. Another common phenomenon that has been observed by many is the spherical surface of a droplet of viscous fluid on a flat surface. It has been further observed that such a phenomenon exists only for very small droplets. The formation of these small droplets is due to the fact that there exists a *tensile force* in the spherical surface of the droplet that exceeds the hydrostatic pressure in the contained fluid. This tensile force is called the *surface tension* of incompressible fluids.

5.6.1 Surface Tension

Surface tension in liquids relates to the cohesion forces of molecules. When a liquid is in contact with air or a solid (e.g., a pipe wall), intermolecular forces as described in Chapter 3 bond the molecules beneath the surface of the liquid. The molecules at the interface, on the other hand, are not bonded in the same way as by its contacting neighbor media. For liquid in contact with air, the surface is subjected to a tensile force that causes the contact surface to be concave upward, as observed by many in a capillary.

The nonslip boundary condition is principally due to the maximum *friction* existing between the fluid and the conduit wall. Surface tension also exists at the interface between the fluid and its solid contacting wall of the conduit but is not a dominant force under normal circumstances. The combined effects of the maximum friction between the interface of the fluid/conduit wall and the surface tension in the fluid make the pressure that is required to drive fluid flow in conduits nonuniformly distributed across the lateral section of the conduit. Such nonuniformity is the principal cause for the well-known *capillary effect*. A capillary can affect the dynamics of fluid flowing in minute conduits, as is the case of many microsystems described in Chapter 2. Navier–Stokes equations that are derived on the basis of continuum fluid flow without accounting for the surface tensions of the fluid are no longer applicable in these cases. Demarcation between continuum and capillary flows is less than clear for incompressible fluids. Some discussion is available in Madou (1997).

The surface tension F_s in a liquid can be expressed as

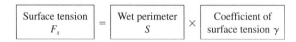

The *coefficient of surface tension* γ, with a unit of newtons per meter, is a measure of the magnitude of the surface tension. As shown in White (1994), the γ value for water can be obtained by the empirical formula

$$\gamma(T) = 0.07615 - 1.692 \times 10^{-4}T \tag{5.23}$$

where T is the temperature in degrees Celsius.

Surface tension has significant effect in liquid flow in microconduits, as additional pressure must be applied to overcome the surface tension of the fluid, which becomes more dominant in the flow at this scale. The following formulation (White, 1994) may be used to assess such additional pressure drop due to surface tension.

Figure 5.14 illustrates the pressure change due to surface tension in a liquid cylinder and in the interior of a spherical droplet. The following relationships are derived on the principles of equilibrium of pressure changes and surface tensions in liquid volumes, as shown in Figure 5.14.

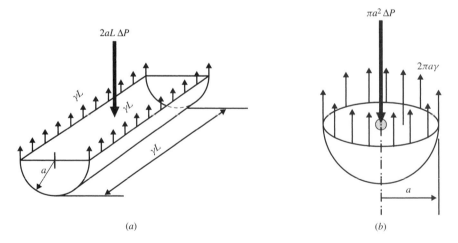

(a)   (b)

Figure 5.14. Pressure change due to surface tension across liquid volumes: (*a*) in a liquid cylinder; (*b*) in a sphere.

In the liquid cylinder in Figure 5.14a, the total force $(2aL)\Delta P$ is equal to the wet perimeter $(2L)$ times the coefficient of surface tension, γ, from which we have

$$\Delta P = \frac{\gamma}{a} \tag{5.24a}$$

In the liquid sphere in Figure 5.14b, the total force $(\pi a^2)\Delta P$ is equal to the wet perimeter $(2\pi a)$ times the coefficient of surface tension, γ, from which we have:

$$\Delta P = \frac{2\gamma}{a} \tag{5.24b}$$

Consequently, we may estimate the total pressure change in a free-standing liquid inside a small tube with diameter $d \approx 2a$, as shown in Figure 5.15, to be the sum of the pressure changes in Equations (5.24a) and (5.24b) as

$$\Delta P = \frac{3\gamma}{a}$$

Example 5.5. Determine the pressure required to overcome the surface tension of water in a small tube of 0.5 mm inside diameter. Assume that the water is at 20°C.

Solution: We first determine the surface tension coefficient of water at 20°C from Equation (5.23) to be $\gamma = 0.073$ N/m. The tube has a radius $a = 250\ \mu$m $= 250 \times 10^{-6}$ m. Following Equations (5.24a) and (5.24b), we have the pressure required to overcome the surface tension:

$$\Delta P = \frac{3\gamma}{a} = \frac{3 \times 0.073}{250 \times 10^{-6}} = 876\ \text{N/m}^2\ \text{or } 876\ \text{Pa}$$

5.6.2 Capillary Effect

The capillary effect of fluid is related to the surface tension of the fluid and the size of the conduit in which the fluid flows. An obvious capillary phenomenon is the rise of fluid in a minute open-ended tube when one of its ends is inserted into a volume of liquid, as illustrated in Figure 5.16.

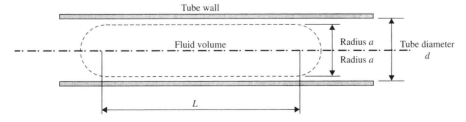

Figure 5.15. Fluid volume in a small tube.

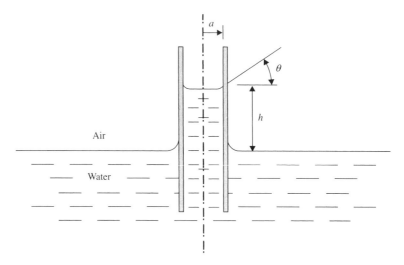

Figure 5.16. Capillary effect on a fluid in a small conduit.

The capillary height of a fluid in a small tube can be computed from the following expression (White, 1994):

$$h = \frac{2\gamma \cos \theta}{wa} \tag{5.25}$$

where $w = \rho g$ is the specific weight of the fluid, θ is the angle between the edge of the free fluid surface and the tube wall, and a is the radius of the capillary tube.

The rise of the liquid in a capillary tube, h, as illustrated in Figure 5.16, is the result of a driving pressure that is induced by the capillary effect. This driving pressure is responsible for the back flow of the condensed liquid along the minute sharp corners of the channel in micro–heat pipes as described in Section 2.6.3.

Example 5.6. Find the height of water rising in a small tube, h, in Figure 5.16. The tube has a diameter of 1 mm.

Solution: We may assume that the angle θ of the water surface is very small: $\theta \approx 0°$. The radius of the tube is 0.5 mm, or 500×10^{-6} m. The surface tension coefficient of water at 20°C is $\gamma = 0.073$ N/m as found in Example 5.5, and the mass density of water is $\rho = 1000$ kg/m³, which leads to the specific gravity of 9810 kg$_f$.

Thus, using the expression in Equation (5.25), we have the height of water column in the tube:

$$h = \frac{2\gamma \cos \theta}{wa} = \frac{2(0.073) \cos(0)}{9810(500 \times 10^{-6})} = 0.02976 \text{ m or } 2.9 \text{ cm}$$

The corresponding static pressure for a column of water at this height is

$$P = \rho g h = (1000 \text{ kg/m}^3)(9.81 \text{ m/s}^2)(0.02976 \text{ m}) = 292 \text{ N/m}^2 \text{ or } 292 \text{ Pa}$$

5.6.3 Micropumping

A serious physical limitation on pumping liquid through microconduits is indicated in Equation (5.17), in which the pressure drop for a liquid flow in circular conduits (ΔP) is inversely proportional to the fourth power of the radius a and thus the diameter d of the conduit. As the diameter reduces to half, the required pumping power is increased by 16 times. To make the situation worse, the surface tension effect also becomes more pronounced in liquid flow in minute conduits, as indicated in Equation (5.24). Conventional pumping methods on the volumetric flow of liquids in microconduits are thus not feasible in an engineering sense.

There are a number of ways that effective pumping of liquids in microscale conduits can be achieved (Madou, 1997). Electrohydrodynamic pumping with electro-osmosis and electrophoretic pumping methods such as presented in Chapter 3 are common practices in the industry. Following is a discussion of another feasible micropumping technique called *piezoelectric pumping*. It uses the surface forces instead of the volumetric pressure to prompt the flow of liquids in minute conduits.

Figure 5.17 illustrates the working principle of piezoelectric pumping for microflow in a minute tube (Nyborg, 1965; Madou, 1997). The tube usually has a thin wall on the order of a few micrometers. The thin membrane walls make these tubes highly flexible. The outside wall is coated with a piezoelectric film such as ZnO with IDTs. A RF voltage is applied to one IDT, which produces mechanical stress in the piezoelectric layer. The mechanical stress so generated can produce flexural acoustic waves in the membrane tube wall (see Section 2.2.1 for a discussion of acoustic wave generation). The wave motion of the tube wall can produce a pumping effect to move the contained fluid, as illustrated in Figure 5.18. The law of physics indicates that the force generated by the surface of the tube wall is proportional to the amplitude of the acoustic wave generated in the wall by the piezoelectric effect, and it decays exponentially toward the center of the tube (see the variation of F in Figure 5.17). This variation of the force results in a more uniform

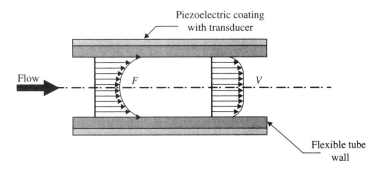

Figure 5.17. Capillary tube for microflow.

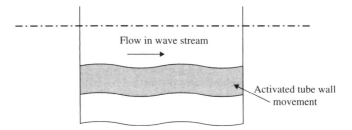

Figure 5.18. Microflow in wave stream in a capillary tube.

velocity of fluid flow inside the tube (see the variation of V in Figure 5.17), which is in contrast to what is illustrated in Figure 5.12 for volumetric fluid flow.

Acoustic streaming using piezoelectric pumping as described above was used by Moroney et al. (1990, 1991) to move air at a speed of 30 mm/s and water at 0.3 mm/s with 5 V supplied by a RF drive.

5.7 OVERVIEW OF HEAT CONDUCTION IN SOLIDS

We have learned from Chapter 2 that many MEMS devices are actuated by thermal means. Key issues involved in the design of these actuators are as follows:

1. The amount of heat required to invoke the desired action
2. The time required for initiating and terminating the action
3. The associated thermal stresses or distortions induced in the device
4. The possible damage to the delicate components of the device due to the heating

These issues are less of a problem for devices at the macroscale, as there are established theories and formulations available to engineers to address these issues. However, the situation at the submicro- and nanoscales can be significantly different, and most theories and formulations derived for the macroscale require significant modifications, as the mechanisms for heat transmission in solids at these scales are radically different. As in the section on fluid mechanics, we will begin the subject of heat transfer with a review of the formulations for the macroscale. Some of these formulations can be used to deal with the aforementioned four issues involving the design and fabrication of microdevices. Necessary modifications of these formulations for submicro- and nanoscales will be presented in Chapter 12.

5.7.1 General Principle of Heat Conduction

Let us first look at heat conduction in a solid slab, as illustrated in Figure 5.19, where a solid slab has temperature at the left side wall maintained at T_a and at the right side wall maintained at T_b, with $T_a > T_b$. The temperature difference between the two walls causes the heat to flow from the left side to the right side of the slab. One would

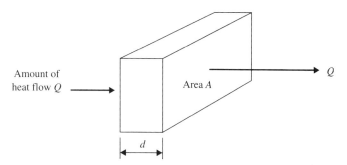

Figure 5.19. Conduction of heat in a solid slab.

envisage that the total amount of heat flow through the slab, Q, is proportional to the size of the cross-sectional area A, the temperature difference between the two faces, and the time t allowed for the heat to flow. However, the amount of heat flow is inversely proportional to the distance that the heat has to travel, that is, the thickness d of the slab. Mathematically, one may express the above qualitative correlations in the form

$$Q \propto \frac{A(T_a - T_b)t}{d}$$

Expressing the above expression in the form of an equation, we have

$$Q = k\frac{A(T_a - T_b)t}{d} \tag{5.26}$$

where the proportionality constant k is the *thermal conductivity* of the solid and has units of Btu/in-s-°F in the imperial system and W/m-°C in the SI system.

Thermal conductivity k is a material property. The value of k for solids generally increases with the temperature. However, for most engineering materials at realistic operating temperature ranges, k can be regarded as constant. The value of k is a measure of how conductive the material is to heat. Thus gold, silver, copper, and aluminum have the heat-conducting capability in this order. These materials have higher k values than many other materials, and they are much better heat-conducting materials than ceramics such as SiC, SiO_2, or Al_2O_3.

5.7.2 Fourier Law of Heat Conduction

Equation (5.26) provides us with a way to account for the total heat flow in a planar slab. A more realistic quantity to account for in heat conduction analysis is the *heat flux q*, defined as the heat flow per unit area and time. Thus, from Equation (5.26), we may express the heat flux in the slab as

$$q = \frac{Q}{At} = k\frac{T_a - T_b}{d} \tag{5.27}$$

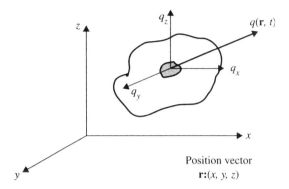

Figure 5.20. Heat flux in a solid.

Heat flux is a vector quantity as illustrated in Figure 5.20. It can be related to the associated temperature gradient by the *Fourier law of heat conduction* as follows: With reference to Figure 5.20, the heat flux in a solid situated in a space defined by **r**: (x,y,z) coordinate system can be expressed as

$$q(\mathbf{r}, t) = -k \, \nabla T(\mathbf{r}, t) \tag{5.28}$$

The negative sign in Equation (5.28) indicates that the heat flux vector $q(\mathbf{r},t)$ is along the outward normal to the surface of the solid. From Equations (5.26) and (5.27), we may assign the unit for heat flux to be W/m^2 or J/m^2-s or N/m-s.

Equation (5.28) is the mathematical expression of the Fourier law of heat conduction. Expanding the above equation in cartesian coordinates will lead to the following expression for heat flux in a solid:

$$q(x, y, z, t) = \sqrt{q_x^2 + q_y^2 + q_z^2} \tag{5.29}$$

where

$$q_x = -k_x \frac{\partial T(x, y, z, t)}{\partial x} \tag{5.30a}$$

$$q_y = -k_y \frac{\partial T(x, y, z, t)}{\partial y} \tag{5.30b}$$

$$q_z = -k_z \frac{\partial T(x, y, z, t)}{\partial z} \tag{5.30c}$$

where q_x, q_y, and q_z are the respective heat flux components in the x, y, and z directions, as shown in Figure 5.20. The terms k_x, k_y, and k_z are the respective thermal conductivities in the solid in the x, y, and z directions. For isotropic materials, $k_x = k_y = k_z$.

5.7.3 Heat Conduction Equation

When a solid is subjected to a heat input from a heat source or dissipates heat to the surrounding medium which acts as a heat sink, the temperature field, that is, the temperature distribution in the solid, will follow a state of nature after reaching a state of equilibrium. The temperature field $T(\mathbf{r},t)$, with $\mathbf{r}$: (x,y,z) in cartesian coordinates or (r, θ, z) in cylindrical polar coordinates, can be obtained by solving the following heat conduction equation:

$$\nabla^2 T(\mathbf{r}, t) + \frac{Q(\mathbf{r}, t)}{k} = \frac{1}{\alpha} \frac{\partial T(\mathbf{r}, t)}{\partial t} \qquad (5.31)$$

where the Laplacian is defined as

$$\nabla^2 = \begin{cases} \dfrac{\partial^2}{\partial x^2} + \dfrac{\partial^2}{\partial y^2} + \dfrac{\partial^2}{\partial z^2} & \text{in cartesian coordinates} \\[2ex] \dfrac{\partial^2}{\partial r^2} + \dfrac{1}{r} \dfrac{\partial}{\partial r} + \dfrac{1}{r^2} \dfrac{\partial^2}{\partial \theta^2} + \dfrac{\partial^2}{\partial z^2} & \text{in cylindrical polar coordinates} \end{cases}$$

The term $Q(\mathbf{r},t)$ in Equation (5.31) is heat generation by the material per unit volume and time. A common heat source in microsystems is electric resistance heating in which the electric current is allowed to pass through an ohmic conductor such as copper or aluminum film. The heat generated by electric resistance heating can be evaluated by the relationship

$$\boxed{\begin{array}{c} \text{Power } P \text{ in} \\ \text{watts (W)} \end{array}} = \boxed{\begin{array}{c} \text{Current } i \\ \text{in amperes } (A) \end{array}}^2 \times \boxed{\begin{array}{c} \text{Resistance } R \\ \text{in ohms } (\Omega) \end{array}}$$

Power in the above expression has a unit of watts, which is equivalent to 1 J/s. It is also equivalent to 1 N-m/s in SI units.

The constant α in Equation (5.31) is called the *thermal diffusivity* of the material with a unit of meters squared per second. It has an important physical meaning as a measure of how fast heat can conduct in solids. Mathematically, it is given as

$$\alpha = \frac{k}{\rho c} \qquad (5.32)$$

where ρ and c are the respective mass density and specific heat of the solid. The unit for ρ is g/cm^3 and the unit for c is J/g - $^\circ$C.

It is apparent that a solid with higher value of α can conduct heat faster than one with a lower value of α. Thus, a thermally actuated microdevice made of materials with higher α values will respond to thermal actuation forces more rapidly than those made of materials with low α values.

5.7.4   Newton's Cooling Law

We have learned that heat transmission in solids is in the mode of conduction and that Fourier's law of heat conduction in Equation (5.28) governs such transmission. Heat transfer in fluid is quite different; convection is the mode for the heat flow in fluids. *Newton's cooling law* is used as the basis for convective heat transfer analysis in fluids.

With reference to the fluid in Figure 5.21, the heat flux between any two points with respective temperatures T_a and T_b is proportional to the temperature difference between these two points, that is, $q \propto T_a - T_b$, from which we can express Newton's cooling law by the expression

$$q = h(T_a - T_b) \tag{5.33}$$

The constant h in Equation (5.33) has several different names. We will call it *heat transfer coefficient* in our analyses.

The value of h is evaluated by dimensional analysis, with analytical parameters determined by experiments. It is usually embedded in the *Nusselt Number* $\mathrm{Nu} = hL/k$, with L the characteristic length and k the thermal conductivity of the fluid. The following relations are normally used for determining the numerical values of the Nusselt number:

$$(Nu) = \begin{cases} \alpha(\mathrm{Re})^{\beta}(\mathrm{Pr})^{\gamma} & \text{for forced convection} \\ \alpha(\mathrm{Re})^{\beta}(\mathrm{Pr})^{\gamma}(\mathrm{Gr})^{\delta} & \text{for free convection with low velocity} \end{cases}$$

The parameters α, β, γ, and δ in the above expressions are determined by experiments. Also involved in the above expressions are the *Reynolds number* Re, *Prandtl number* Pr, and *Grashoff number* Gr. Of these, the Reynolds number has the dominant effect on the value of h. The Reynolds Number was defined in Equation (5.3) as

$$\mathrm{Re} = \frac{\rho L V}{\mu}$$

where ρ and μ are the respective mass density and dynamic viscosity of the fluid and V is the velocity of fluid flow.

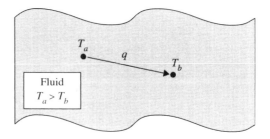

Figure 5.21. Heat flux in a fluid.

The Prandtl number and the Grashoff number are defined as

$$\mathrm{Pr} = \frac{C_p \mu}{k} \tag{5.34a}$$

$$\mathrm{Gr} = \frac{L^3 \rho^2 g}{\mu^2 (\beta \, \Delta t)} \tag{5.34b}$$

where C_p is the specific heat of fluids under constant pressure, β is the volumetric coefficient of thermal expansion, Δt is the duration, and g is gravitational acceleration.

The Grashoff number usually appears in analyses that involve free or natural convection.

For most convective heat transfers, the h value is related to the velocity of the fluid, or $h \propto V^\phi$, where ϕ is a constant determined by experiments.

5.7.5 Solid–Fluid Interaction

As described in Chapter 2, many thermally actuated MEMS devices involve transferring heat from solid members to the surrounding fluids in contact, or vice versa. We will also learn from several microfabrication processes in Chapter 8 (e.g., chemical vapor depositions) that gaseous fluids are made to flow over the surfaces of substrate solids and thereby transfer heat between the two media. Since heat transmission in solids differs from that in fluids, it is necessary to learn how these two modes of heat transfer interact at solid–fluid interfaces.

Let us look at the situation illustrated in Figure 5.22. Heat is being dissipated from the solid with a temperature field $T(\mathbf{r},t)$ into the surrounding fluid at a temperature T_f. The reverse situation, of course, is also possible.

Let us first take a closer look at the interface of the solid and fluid. We realize that a *boundary layer* is built in the fluid immediately adjacent to the solid surface. The thickness of this layer depends on the velocity of the fluid over the solid surface and the property of the fluid. This layer creates a barrier for freer heat transfer between the solid and the fluid. A resistance to heat flow is generated in this layer. Thus, because of this resistance, the surface temperature of the solid is not equal to the bulk fluid

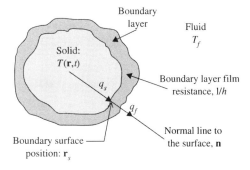

Figure 5.22. Heat flow between solid and fluid.

temperature T_f. Such thermal resistance may be numerically evaluated by $1/h$, where h is the heat transfer coefficient of the fluid.

By equating the heat flux vector q_s entering the interface from the solid side and the heat flux q_f leaving the interface into the bulk fluid, as illustrated in Figure 5.22, we obtain the relation

$$-k \left. \frac{\partial T(\mathbf{r}, t)}{\partial \mathbf{n}} \right|_{\mathbf{r}_s} = h[T(\mathbf{r}, t) - T_f] \tag{5.35}$$

at the interface.

The vector $\mathbf{n}$ is the normal vector to the point of interest on the specific boundary between the solid and the fluid.

5.7.6 Boundary Conditions

Equation (5.31) is used to determine the temperature distribution $T(\mathbf{r},t)$ in a MEMS device component. The temperature distribution is then used to identify the location and the magnitude of the maximum temperature in the component as well as to determine the associated thermal stresses as demonstrated in Section 4.4.3. Both the maximum temperature and the temperature-induced thermal stresses and strains are primary design considerations of these components. They are thus an important part of the procedure in the design of microsystems involving thermal effects.

The solution of Equation (5.31) requires proper formulation of boundary conditions. There are three types of boundary conditions that can be used in thermal analyses:

1. **Prescribed Surface Temperature:** This type of boundary condition is applied to a solid with the temperature at specific locations on the surface specified. With reference to Figure 5.23a, the boundary condition at the location $\mathbf{r}_s$ can be expressed as

$$T(\mathbf{r}, t) \left|_{\mathbf{r}=\mathbf{r}_s} \right. = f(\mathbf{r}_s, t) = F(t) \tag{5.36}$$

   Of course, the function $F(t)$ in Equation (5.36) can be a constant in special cases.

2. **Prescribed Heat Flux at Boundary:** Figure 5.23b illustrates the case in which the heat flux entering or leaving the part of the boundary at $\mathbf{r} = \mathbf{r}_s$ is specified. As we have learned from Equations (5.28) and (5.30), the heat flux in a solid can be expressed by the Fourier law of heat conduction. Thus, by assuming line $n-n$ to be the *outward normal line* to the surface at which the heat flux enters or leaves, we may express the corresponding boundary conditions with q_{in} or q_{out} specified as follows:

$$q_{in} = +k \left. \frac{\partial T(\mathbf{r}, t)}{\partial \mathbf{n}} \right|_{\mathbf{r}=\mathbf{r}_s} \tag{5.37a}$$

   or

$$q_{out} = -k \left. \frac{\partial T(\mathbf{r}, t)}{\partial \mathbf{n}} \right|_{\mathbf{r}=\mathbf{r}_s} \tag{5.37b}$$

   The plus and minus signs attached to the heat flux in Equation (5.37) designate the direction of the heat flux along the outward normal line to the surface. Table 5.1

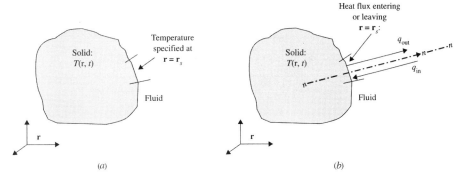

Figure 5.23. Prescribed surface temperature or heat flux on solid surfaces: (a) with prescribed temperature; (b) with prescribed heat flux.

TABLE 5.1. Signs for Specified Heat Flux Boundary Conditions

Sign of Outward Normal, n	Is q Along Same Direction as n?	Sign of q in Equations (5.37)
+	Yes	−
+	No	+
−	Yes	+
−	No	−

will help the reader to assign a proper sign to the temperature gradients in the above equations.

Example 5.7. Express the heat flux boundary conditions at the four faces of a rectangular block as shown in Figure 5.24. Heat transfers in the x–y plane and the block has a temperature distribution $T(x,y)$. The direction of heat fluxes q_1, q_2, q_3, and q_4 crossing these four faces are specified in the figure.

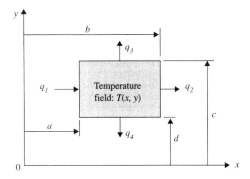

Figure 5.24. Prescribed heat fluxes across four faces of a rectangular object.

Solution: By using Equation (5.37) and Table 5.1, we may specify the following boundary conditions:

At the left face, we have a negative outward normal and heat flux is in the opposite direction to the outward normal line. We thus have the last case in Table 5.1 with a negative sign to Equation (5.37):

$$\left.\frac{\partial T(x, y)}{\partial x}\right|_{x=a} = -\frac{q_1}{k} \tag{a}$$

At the right face, the outward normal is positive and the heat flux is in the same direction as the outward normal, which leads to the following expression for this boundary according to case 1 in Table 5.1:

$$\left.\frac{\partial T(x, y)}{\partial x}\right|_{x=b} = -\frac{q_2}{k} \tag{b}$$

At the top face, the situation is similar to that for the right face. We have

$$\left.\frac{\partial T(x, y)}{\partial y}\right|_{y=c} = -\frac{q_3}{k} \tag{c}$$

At the bottom face, the outward normal is negative and the heat flux is in the same direction as the outward normal line. This situation fits the third case in Table 5.1. We thus have

$$\left.\frac{\partial T(x, y)}{\partial y}\right|_{y=d} = +\frac{q_4}{k} \tag{d}$$

in Equations (a), (b), (c), and (d), k is the thermal conductivity of the solid.

3. **Convective Boundary Conditions:** This type of boundary conditions applies to the situation when the solid boundary is in contact with the fluid at the bulk fluid temperature T_f as shown in Figure 5.25. The boundary conditions for this situation are expressed by rearranging Equation (5.35) to give

$$\left.\frac{\partial T(\mathbf{r}, t)}{\partial \mathbf{n}}\right|_{\mathbf{r}=\mathbf{r}_s} + \frac{h}{k} T(\mathbf{r}, t)|_{\mathbf{r}=\mathbf{r}_s} = \frac{h}{k} T_f \tag{5.38}$$

One will realize from the above formulations that the case $h \to \infty$ is equivalent to the prescribed surface boundary condition in Equation (5.36), whereas the case $h = 0$ leads to the insulated boundary condition in which

$$\left.\frac{\partial T(\mathbf{r}, t)}{\partial \mathbf{n}}\right|_{\mathbf{r}=\mathbf{r}_s} = 0$$

in Equation (5.37).

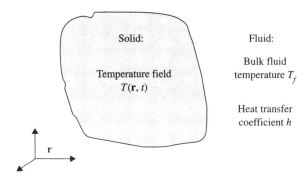

Figure 5.25. Solid surrounded by fluid.

Example 5.8. Show the differential equation and the appropriate initial and boundary conditions for a thermally actuated microbeam as illustrated in Figure 5.26. A thin copper film is attached to the top surface of the silicon beam and used as a resistant heater. The actuator is initially at $20°C$. Consider two cases for the contacting air at the bottom surface of the beam: (1) still air, (2) the air has a bulk temperature of $20°C$ but has a heat transfer coefficient of 10^{-4} W/m²-°C.

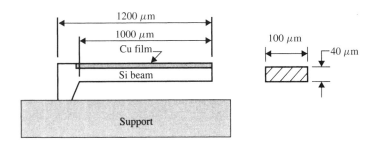

Figure 5.26. Thermally actuated beam.

Solution: We may consider that the induced temperature field in the beam predominantly varies with the thickness of the beam. It is thus reasonable to assume a temperature function $T(x,t)$ in the beam with x being the coordinate in the thickness direction, as illustrated in Figure 5.27.

The governing differential equation from the general form in Equation (5.31) for the present case is

$$\frac{\partial^2 T(x,t)}{\partial x^2} = \frac{1}{\alpha} \frac{\partial T(x,t)}{\partial t} \tag{5.39}$$

The thermal diffusivity α in Equation (5.39) for the silicon beam can be obtained by the thermophysical properties listed in Table 7.3, with the thermal conductivity $k = 1.57$ W/cm-°C, the mass density $\rho = 2.3$ g/cm³, and the specific heat $c = 0.7$ J/g-°C, which gives $\alpha = 0.9752$ cm²/s.

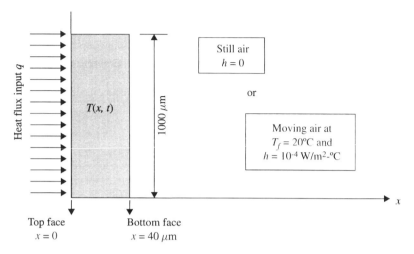

Figure 5.27. Heat conduction in a silicon beam.

The initial condition is

$$T(x, t)|_{t=0} = 20°C \qquad (a)$$

For the boundary condition at the top surface, $x = 0$, we have heat flux input condition

$$\frac{\partial T(x, t)}{\partial x}\bigg|_{x=0} = -\frac{q}{k}$$

where q is the heat flux generated by the copper film and k is the thermal conductivity of the silicon beam.

The magnitude of the heat flux generated by the copper film is $q = i^2 R/A$, where i is the electric current passing through the film in amperes and R is the electric resistance of the film in ohms. The top surface area of the beam, A, is 1000 μm × 100 μm, or 10^{-3} cm^2. The boundary condition in the above expression for the present case is thus

$$\frac{\partial T(x, t)}{\partial x}\bigg|_{x=0} = -\frac{i^2 R}{1.57 \times 10^{-3}} \qquad °C/cm \qquad (b)$$

We may apply a factor of 10^{-2} to the right-hand side of Equation (b) to convert the unit to degrees Celsius per meter for consistency with the units used in other parameters.

Two possible boundary conditions can be applied at the bottom surface of the beam:

1. **The insulated boundary condition with still air (i.e., h = 0) at the bottom surface of the beam (x = 40 μm):** This boundary condition makes the solution of Equation (5.39) relatively easy. However, it is valid on the assumption that the temperature of the entrapped air beneath the bottom surface of the beam has not changed. Thus, the temperature solution $T(x,t)$ derived from using this boundary condition is applicable for the very early stage of the heating process before any significant temperature rise takes place in the entrapped air. Mathematically, we can express this condition in the form

$$\frac{\partial T(x,t)}{\partial x}\bigg|_{x=40\times10^{-6} \text{ m}} = 0 \tag{c}$$

2. **The convective boundary condition at the bottom surface of the beam (x = 40 μm):** The mathematical expression for this boundary condition is represented by Equation (5.38) as

$$\frac{\partial T(x,t)}{\partial x}\bigg|_{x=40\times10^{-6} \text{ m}} + \frac{10^{-4}}{157} T(x,t)|_{x=40\times10^{-6} \text{ m}} = \frac{10^{-4}}{157} \times 20 \tag{d}$$

The reader is reminded that the value $k = 157$ W/m-$^\circ$C is used in Equation (d) for consistency of the units used in the problem.

The temperature field in the beam, $T(x,t)$, can be obtained by solving Equation (5.39) with the conditions in Equations (a), (b), and (c) or the conditions in Equations (a), (b), and (d). Various ways are available for solving Equation (5.39), for example, the separation-of-variables technique or the integral transformation method (Ozisik, 1968). There are also various numerical techniques available for the same solution.

5.8 HEAT CONDUCTION IN MULTILAYERED THIN FILMS

Many MEMS structures are made of multilayered thin films of different materials. These structures are often used in devices fabricated using surface micromachining, as will be described in Chapters 8 and 9. Thin films are typically in sandwich form and are subjected to thermal loads, as in the case of actuators and microvalves. A heat conduction analysis is required not only to assess the sensitivity of the structure to the thermal loading but also as input to thermal stress analysis, as presented in the Chapter 4.

Complete heat conduction analysis of multilayered media requires complex mathematical derivation and is beyond the scope of this book. What we will present below is a simple one-dimensional formulation for such analysis. Layers in the structure are

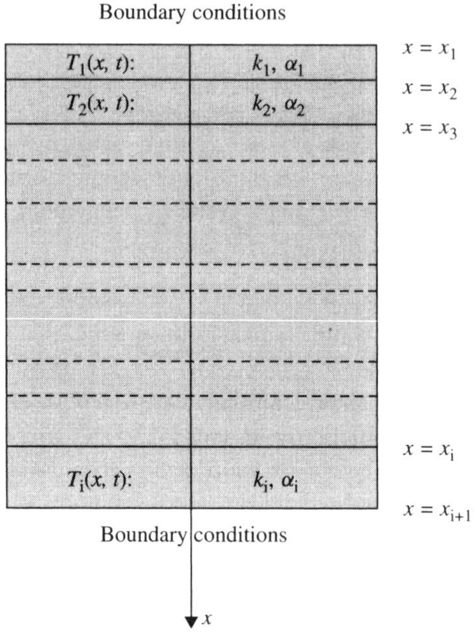

Boundary conditions

Figure 5.28. Temperatures in a multilayered structure.

assumed to be in perfect thermal contact at the interfaces. Detailed derivation of the formulation is available in a few heat conduction books (e.g., Ozisik, 1968).

With reference to Figure 5.28, the temperature $T_i(x,t)$ in each i-layered structure may be obtained by solving the system of equations

$$\frac{\partial^2 T_i(x,t)}{\partial x^2} = \frac{1}{\alpha_i}\frac{\partial T_i(x,t)}{\partial t} \tag{5.40}$$

where the layer designation $i = 1,2,3,\ldots$, $x_i \leq x \leq x_{i+1}$, and $t > 0$, satisfying the following conditions:

1. Prescribed initial conditions in $x_i \leq \text{x} \leq x_{i+1}$ at $t = 0$
2. Prescribed boundary conditions at $x = 0$ and $x = x_{i+1}$ for $t > 0$

In Figure 5.28 and Equation (5.40), k_i and α_i, $i = 1, 2, 3,\ldots$, denote the respective thermal conductivity and thermal diffusivity of the material in layer i.

Other conditions applicable to the problem involve the continuities of temperatures and heat fluxes at the interfaces, that is,

$$T_i(x_{i+1}, t) = T_{i+1}(x_{i+1}, t) \quad \text{for } i = 1, 2, 3, \ldots$$

$$k_i\frac{\partial T_i(x_{i+1}, t)}{\partial x} = k_{i+1}\frac{\partial T_{i+1}(x_{i+1}, t)}{\partial x} \quad \text{for } i = 1, 2, 3, \ldots$$

Example 5.9. The structure of a thermal actuator is made of a compound beam involving silicon and SiO_2 as illustrated in Figure 5.29. A thin copper film is deposited on the top of the SiO_2 layer as the resistant heater. This heater will provide a maximum temperature of $50°C$ at the top surface of the SiO_2 layer. Determine the time required for the silicon beam to reach the input surface temperature of $50°C$. Since heat will predominantly flow through the thickness of the compound beam because of the short distance of the passage, a one-dimensional heat conduction analysis along the thickness direction is justified.

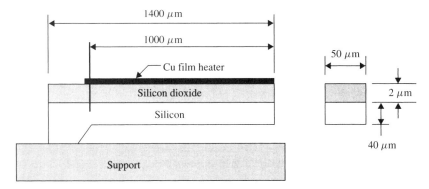

Figure 5.29. Thermal actuator made of a compound beam.

The given material properties are (see Table 7.3):

Thermal conductivities: $k_1 = 1.4\,\text{W/m-}°\text{C}$ for SiO_2 and $k_2 = 157\,\text{W/m-}°\text{C}$ for silicon
Thermal diffusivities: $\alpha_1 = 0.62 \times 10^{-6}\,\text{m}^2\text{/s}$ for SiO_2 and $\alpha_2 = 97.52 \times 10^{-6}\,\text{m}^2\text{/s}$ for silicon

Solution: We assume that the temperatures $T_1(x,t)$ and $T_2(x,t)$ in the respective SiO_2 and silicon layers can be computed from Equation (5.40) and the similar arrangement shown in Figure 5.28. We may use the model illustrated in Figure 5.30 for this problem.

The following system of differential equations and specific conditions are used to solve the problem: The differential equations are given as

$$\frac{\partial^2 T_1(x,t)}{\partial x^2} = \frac{1}{\alpha_1}\frac{\partial T_1(x,t)}{\partial t} \tag{5.41a}$$

for the SiO_2 layer with $0 \le x \le a$ and

$$\frac{\partial^2 T_2(x,t)}{\partial x^2} = \frac{1}{\alpha_2}\frac{\partial T_2(x,t)}{\partial t} \tag{5.41b}$$

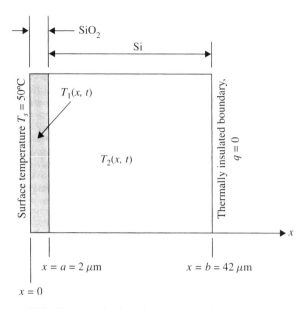

Figure 5.30. Heat conduction through a two-layer compoundbeam.

for the silicon beam with $a \leq x \leq b$. The initial conditions are

$$T_1(x,t)|_{t=0} = F_1(x) = 20°C \tag{5.42a}$$

$$T_2(x,t)|_{t=0} = F_2(x) = 20°C \tag{5.42b}$$

The boundary conditions are

$$T_1(x,t)|_{x=0} = 50°C \tag{5.42c}$$

$$\left. \frac{\partial T_2(x,t)}{\partial x} \right|_{x=b=42\mu m} = 0 \tag{5.42d}$$

The compatibility conditions at the interface are

$$T_1(x,t)|_{x=a=2\mu m} = T_2(x,t)|_{x=a=2\mu m} \tag{5.42e}$$

$$k_1 \left. \frac{\partial T_1(x,t)}{\partial x} \right|_{x=a=2\mu m} = k_2 \left. \frac{\partial T_2(x,t)}{\partial x} \right|_{x=a=2\mu m} \tag{5.42f}$$

There are several ways to solve the temperature distribution functions $T_1(x,t)$ and $T_2(x,t)$ from Equation (5.41) with the conditions specified in Equation (5.42).

The following solution is obtained using the integral transform method presented in Ozisik (1968):

$$T_i(x, t) = \sum_{n=1}^{\infty} \frac{e^{-\beta_n^2 t}}{N_n} \Psi_{in}(x) \left[\frac{k_1}{\alpha_1} \int_0^a \Psi_{1n}(x) F_1(x) \, dx \right.$$

$$\left. + \frac{k_2}{\alpha_2} \int_a^b \Psi_{2n}(x) \, dx \right] \qquad i = 1, 2 \qquad (5.43)$$

where the functions $F_1(x)$ and $F_2(x)$ are the respective initial conditions in SiO_2 and silicon. The norm N_n is defined as

$$N_n = \frac{k_1}{\alpha_1} \int_0^a \Psi_{1n}^2(x) \, dx + \frac{k_2}{\alpha_2} \int_a^b \Psi_{2n}^2(x) \, dx$$

The functions $\Psi_{1n}(x)$, $i = 1, 2$, have the forms

$$\Psi_{1n}(x) = \sin\left(\frac{\beta_n}{\sqrt{\alpha_1}} x\right)$$

$$\Psi_{2n}(x) = A_{2n} \sin\left(\frac{\beta_n}{\sqrt{\alpha_1}} x\right) + B_{2n} \cos\left(\frac{\beta_n}{\sqrt{\alpha_2}} x\right)$$

The coefficients A_{2n} and B_{2n} have the forms

$$A_{2n} = \sin\omega \sin\left(\frac{a\lambda}{b}\right) + \mu \cos\omega \cos\left(\frac{a\lambda}{b}\right)$$

$$B_{2n} = -\mu \cos\omega \sin\left(\frac{a\lambda}{b}\right) + \sin\omega \cos\left(\frac{a\lambda}{b}\right)$$

with

$$\omega = \frac{\beta_n a}{\sqrt{\alpha_1}} \qquad \lambda = \frac{\beta_n b}{\sqrt{\alpha_2}} \qquad \mu = \frac{k_1 \sqrt{\alpha_2}}{k_2 \sqrt{\alpha_1}}$$

The eigenvalues β_n in the above formulations can be obtained by solving the transcendental equation

$$\begin{vmatrix} \sin\omega & -\sin\left(\frac{a\lambda}{b}\right) & -\cos\left(\frac{a\lambda}{b}\right) \\ \mu \cos\omega & -\cos\left(\frac{a\lambda}{b}\right) & \sin\left(\frac{a\lambda}{b}\right) \\ 0 & \cos\lambda & -\sin\lambda \end{vmatrix} = 0$$

Numerical solutions for the temperature distributions in Equation (5.43) for both SiO_2 and silicon layers are obtained by using a commercial software package, MATHCAD. The temperature variations in both layers at selected instances are plotted as shown in Figure 5.31, from which we can determine the time required for the silicon layer to reach the input temperature of $50°C$ to be 600 μs. This information will enable the design engineer to assess the sensitivity of the thermally actuated device. The expression of temperature distributions in Equation (5.43) is also used as input to thermal stresses and interfacial fracture analyses, such as illustrated in Sections 4.4 and 4.5.

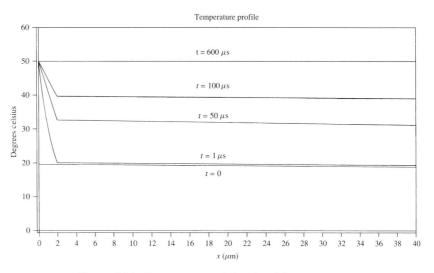

Figure 5.31. Temperature variation in a bilayer actuator.

5.9   HEAT CONDUCTION IN SOLIDS AT SUBMICROMETER SCALE

In Section 5.7 we presented a phenomenological model for heat transmission in a solid block in Figure 5.19, on which the Fourier law of heat conduction in solids is developed. This fundamental law of physics combined with the first law of thermodynamics leads to the heat conduction equation (5.31) for assessing the temperature field induced by input heat fluxes to or from a solid. We can use this equation to assess not only the sensitivity of thermally induced microactuators but also the associated stresses and strains for both the strength and dimensional stability of microactuators in design analyses.

This model for heat conduction in solids is both accurate and reliable for most solid structures. It also works reasonably well for structures at the micrometer scale. However, its validity diminishes as the size of the solids reduces from macro- to submicro- and nanoscales. The principal reason for this diminishing validity is the more pronounced influence of thermal energy carriers on heat transmission in solids of drastically reduced sizes.

Heat is a form of energy. As such, the flow of heat requires a carrier, as in the case of other energy carriers. The carriers for thermal energy and thus heat vary from substance to substance. For example, the primary carriers for heat in metals are electrons and phonons, whereas the carrier in dielectric and semiconducting materials is predominantly phonons. We will learn from Chapter 12 that phonons may be viewed as a group of virtual mass particles that characterize the state of energy of the lattice in molecules. The vibration of the lattices that bond the atoms supplies the energy to phonons and electrons, and thus the heat in the substance. Consequently, the transmission of heat in a substance relies on the movement of phonons in semiconductors or the movement of both electrons and phonons in metals.

The study of movements of electrons and phonons associated with heat transmission in a substance is an extremely complicated subject and is beyond the scope of this book. However, one may well imagine the situation in which zillions of phonons and electrons are present. These energy carriers, traveling from the point of energy supply in a solid (e.g., a heat source) to other parts of the solid, would inevitably encounter millions of collisions with other phonons and electrons that already exist in the solid. The time required for those traveling energy carriers to reach other parts of the solid will thus become a significant parameter in heat transmission. The change of paths due to multiple collisions and the time required for energy carriers to travel in heat transmission are not critical factors in the case of solids at the macroscale as the time involved is usually too short to significantly affect the overall behavior of heat transmission. However, in the case of solids of extremely small in sizes (e.g., less than a micrometer in characteristic length) the time required for the energy carriers to travel becomes more significant. Consequently, all heat conduction analysis of solids of submicrometer size needs to be treated as a transient phenomenon. The phenomenological properties, such as the thermal conductivity of a material, also become size dependent. A linear heat conduction equation such as Equation (5.31) can no longer be used for the design analysis of solids of this size without proper modifications. Detailed description of the analytical model for heat conduction in solids at the submicrometer scale will be presented in Chapter 12.

PROBLEMS

Part 1 Multiple Choice

1. Viscosity of a fluid is a measure of the fluid's resistance to motion created by (a) pressure, (b) driving forces, (c) shear stress.

2. Newtonian fluids are defined by their relationships between the shear stress and shear strain rates, which exhibit (a) linear, (b) nonlinear, (c) mixed linear and nonlinear characteristics.

3. The Reynolds number is related to characteristics of (a) a stationary, (b) a moving, (c) any state of the fluid.

4. The Reynolds number is proportional to (a) the traveling distance, (b) the velocity, (c) the pressure of a fluid.

5. Control volume in a fluid dynamic analysis means (a) a conveniently selected volume of the fluid for the analysis, (b) the volume of the fluid in which the Reynolds number is constant, (c) the volume of the fluid in which the fluid properties are constant.

6. A laminar fluid flow means a (a) low velocity, (b) high velocity, (c) quasi-stagnant fluid flow.

7. Laminar flow of compressible fluids normally takes place with Reynolds numbers in the range of (a) 0–10, (b) 10–100, (c) 100–1000.

8. In general, fluid flows in microsystems are (a) laminar, (b) turbulent, (c) neither laminar nor turbulent.

9. The continuity equation is used to evaluate (a) volumetric flow rate, (b) relationship between the motion and the driving forces, (c) the induced forces in a moving fluid.

10. The momentum equation is used to evaluate (a) volumetric flow rate, (b) relationship between the motion and the driving forces, (c) induced forces in a moving fluid.

11. We will use (a) continuity equation, (b) momentum equation, (c) equation of motion to assess the fluid induced forces on microsystem components.

12. The hydraulic diameter is used to evaluate the cross-sectional area of fluid flowing in (a) circular, (b) rectangular, (c) any shape of conduits.

13. CFD stands for (a) critical fluid dynamics, (b) computational fluid dynamics, (c) computerized fluid dynamics.

14. Navier–Stokes equations relate (a) pressure–velocity, (b) pressure–density change, (c) pressure–viscosity in a moving fluid.

15. The surface tension in a fluid is a form of (a) applied tension at the surface of the fluid, (b) an existing tension at the fluid surface, (c) tension that makes the surface of the fluid.

16. The surface tension is the principal cause for (a) capillary, (b) newtonian, (c) laminar flow of a fluid.

17. The coefficient of surface tension of a fluid is a measure of the (a) inherent strength, (b) the surface tension, (c) topology of a fluid surface.

18. The capillary height of a fluid in a small tube is (a) equal to, (b) directly proportional to, (c) inversely proportional to the diameter of the tube.

19. Pressure drop at two points in a fluid drives the flow of the fluid in a circular conduit. It is inversely proportional to the (a) second, (b) third, (c) fourth power of the diameter of the conduit.

20. The driving force in the piezoelectric pumping of fluids in minute conduits is of (a) linear, (b) surface, (c) volumetric nature.

21. The thermal conductivity of a material is a measure of its (a) conductance to heat, (b) resistance to conducting heat and electricity, (c) speed of heat conduction.

22. Metals have (a) better than, (b) worse than, (c) about the same as semiconductors and ceramics in conducting heat.

23. The thermal diffusivity of a material is a measure of its (a) conductance to heat, (b) resistance to conducting heat and electricity, (c) speed of heat conduction.

24. Heat flux is a measure of heat conduction in a solid per unit (a) length, (b) area, (c) volume for a given period of time.

25. Heat flux is a (a) scalar, (b) vector, (c) tensor quantity.

26. Heat generation in a solid by electric resistance is related to (a) current and resistance, (b) voltage and resistance, (c) inductance and resistance.

27. For micro–thermal actuators, one would choose the actuating materials with (a) high thermal conductivity, (b) high thermal diffusivity, (c) neither of the above.

28. Newton's cooling law is used for microsystem components in contact with (a) fluids, (b) another solid component, (c) any substance.

29. The heat transfer coefficient of a fluid in contact with a solid is (a) equal, (b) directly proportional, (c) inversely proportional to the velocity of the fluid flow.

30. The natural (or free) convective heat transfer is prompted by (a) the driving forces of the fluid, (b) the pressure drop in the fluid, (c) the change of density of the fluid due to heating or cooling of the fluid.

31. The surface of the solid becomes virtually impermeable to heat if the surrounding fluid is moving at (a) high, (b) low, (c) stagnant velocity.

32. The larger the Nusselt number, the (a) larger, (b) smaller, (c) same the heat transfer coefficient in the fluid.

33. In thermally actuated micropumps, one needs to pay attention to the (a) conductive, (b) convective, (c) radiative heat transfer between the actuating elements and the contacting fluid.

34. When a thermally actuated element is in contact with a working fluid, the interface temperature is normally (a) lower than, (b) about the same as, (c) higher than the bulk fluid temperature.

35. There are generally (a) three, (b) four, (c) five types of boundary conditions involved in heat conduction analysis of solids.

36. When designing a thermally actuated beam element, one will be primarily concerned with (a) the weight of the beam, (b) the mechanical strength of the beam, (c) the thermal response of the beam.

37. Heat conduction analysis is necessary for the subsequent (a) thermal stresses, (b) heat dissipation, (c) safety analysis in the design of a thermally actuated microdevice.

38. One needs to be concerned with the validity of using continuum heat transfer theories when the size of the solids is (a) greater than, (b) about equal to, (c) less than 1 μm.

39. In general, heat transportation in solids is by (a) vibration of the lattices of molecules, (b) the movement of phonons in the solids, (c) the movement of phonons and electrons in the solids.

40. Heat transportation in dielectric and semiconducting solids at submicrometer- and nanoscales is dominated by (a) the vibration of the lattices of molecules, (b) the movement of phonons in the solids, (c) the movement of phonons and electrons in the solids.

Part 2 Computational Problems

1. List the dynamic viscosity, in newton-seconds per square meter, for the following fluids at standard conditions, 1 atm and $20°C$: hydrogen, argon, helium, oxygen, nitrogen, water, hydrogen peroxide, and silicone oil.

2. Repeat the problem in Example 5.1 with $d_1 = 500$ μm and $d_2 = 50$ μm.

3. Estimate the opening of the check valve in Example 5.2 as illustrated in Figure 5.8 and the velocity of gas flow in that microdevice. Will this valve allow the gas to flow at a rate of 30,000 cm³/min?

4. Estimate the power requirement and tilt angle θ of a microvalve to allow for a gas flow of 1 cm³/min at a velocity of 1 m/s. The valve is actuated by a set of embedded plate electrodes as illustrated in Figure 5.32a. The width of the electrodes is the same as that of the valve, as illustrated in Figure 5.32b. We may assume that the dielectric material between the electrodes is the gas that gives the same dielectric constant as air. The dimension of the plate valve is shown in Figure 5.32b. (*Hint:* You must first determine the appropriate value of gap H for the gas passage, as shown in Figure 5.32b. Use beam theory with a linearly varying distributed load for the answer.)

5. Estimate the pressure drop in the situation described in Example 5.3 using the Hagen–Poiseuille equation (5.17) instead and compare the results.

6. Water is to flow through a capillary tube 20 μm in diameter over a length of 1000 μm. Use the Hagen–Poiseuille equation to determine the pressure drop in the flow and estimate the additional pressure drop due to the surface tension in the water. Laminar

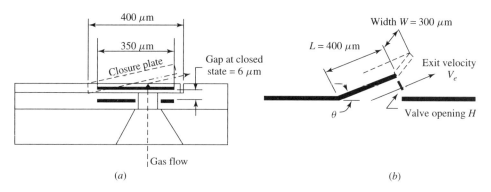

Figure 5.32. Dimensions of electrodes for electrostatic forces on a microvalve: (*a*) passage of gas flow; (*b*) valve opening.

flow with Reynolds number Re ≤ 1000 is assumed for the capillary flow. (*Hint:* Use the two extreme values of dynamic viscosity of water given in Table 4.3 to determine the range of volumetric flow of water in the capillary tube.)

7. When one end of the tube in problem 6 is inserted vertically into a pot of water, what will be the rise of water level in the tube due to the capillary effect?

8. Rank all the materials common to MEMS as listed in Table 7.3 for their sensitivity to be thermally actuated materials. This sensitivity is measured by how fast they can be heated or cooled under the influence of thermal forces.

9. Prove that the convective heat transfer boundary condition in Equation (5.38) includes both the other boundary conditions of prescribed temperature and heat flux, for example, by letting $h = 0$ for a thermally insulated boundary and the case $h \to \infty$ approaches the prescribed boundary temperature condition.

10. Specify appropriate boundary conditions for a thermal analysis in a thermally actuated beam as illustrated in Figure 5.33. Electric current i and voltage V are applied to the copper film to generate the heat for actuation. Use the material properties in Table 7.3 if necessary. Heat flow in the normal direction to the $x-y$ plane is negligible.

11. Show the differential equations and appropriate boundary conditions for the temperature rise in the thermally actuated beam system in Figure 5.33 initially at a uniform temperature of 20°C. You are not required to solve the system of equations.

12. Show the differential equations required to find the temperature fields in the components of the system in Figure 5.33. The surrounding air carries away heat generated in the copper film to the system. The surrounding air has a bulk temperature of 20°C and a heat transfer coefficient h by natural convection.

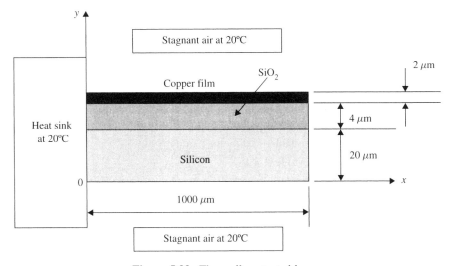

Figure 5.33. Thermally actuated beam.

CHAPTER 6

SCALING LAWS IN MINIATURIZATION

6.1 INTRODUCTION TO SCALING

The needs for and the advantages of miniaturization of machines and devices have been described in Chapter 1. Typically, a successful industrial product requires meeting consumer expectations to be intelligent and multifunctional, among many others. Consequently, a great many sensors, actuators, and microprocessors have to be systematically integrated and packaged in many of these products. The constraints on the size and geometry of the product require substantial miniaturization of these sensors, actuators, and microprocessors. It is thus essential that engineers make increasing effort in miniaturizing new industrial products for reasons of physical appearances, volume, weight, and economy. The objective of this chapter is to provide engineers with a few selected scaling laws that will make them aware of the physical consequences of downscaling machines and devices. It should be kept in mind that some miniaturization either may be physically unfeasible or may not make economical sense.

There are generally two types of scaling laws that are applicable to the design of microsystems. The first type is dependent on the size of physical objects, such as the scaling of geometry. Also in this category is the behavior of the objects that are governed by the law of physics. Examples include the scaling of rigid-body dynamics and electrostatic and electromagnetic forces. The second type involves the scaling of the phenomenological behavior of microsystems. Both the size and material properties of systems are involved in the second type of scaling laws, which deal mostly with thermofluids in microsystems. The scaling of material properties at the submicrometer- and nanoscale is described in Chapter 12.

6.2 SCALING IN GEOMETRY

Volume and surface are two physical quantities that are frequently involved in machine design. Volume leads to the mass and weight of device components and is, for example, related to both mechanical and thermal inertia. Thermal inertia is related to the heat capacity of the solid, which is a measure of how fast we can heat or cool a solid. Such a characteristic is important in the design of a thermally actuated device, as described in Chapter 5. Surface properties, on the other hand, are related to pressure and the buoyant forces in fluid mechanics as well as heat absorption or dissipation by a solid in convective heat transfer. We have learned in Chapter 5 that surface pumping is a more practical method in microfluidics than the conventional volumetric pumping in macrosystems. When the physical quantity is to be miniaturized, the design engineer must weigh the possible consequences of the reduction of both the volume and surface of the particular device. Equal reduction of volume and surface of an object is not normally achievable in a scaled-down process, as will be shown below.

Let us look at the example of a solid of rectangular geometry illustrated in Figure 6.1. The rectangular solid has three sides, $a > b > c$. It is readily seen that the volume $V = abc$ and the surface area $S = 2 \times (ac + bc + ab)$. If we let l represent the linear dimension of a solid, then the volume $V \propto l^3$ and the surface S $\propto l^2$, which leads to the relationship

$$\frac{S}{V} = l^{-1} \tag{6.1}$$

Figure 6.2 shows an interesting contrast of two living objects, an elephant and a dragonfly, with respective approximate S/V ratios of 10^{-4} and 10^{-1} mm^{-1}. These distinct S/V ratios explain why a dragonfly requires little energy and power, and thus little consumption of food and water, to fly, whereas an elephant has a huge appetite for food in order to generate sufficient energy (thus power) to make even slow movements. Further, the much higher S/V ratio of the dragonfly makes it better suited for flying because of its larger surface area for aerodynamic buoyant forces over the smaller weight that relates to body volume.

One may thus conclude from the scaling formula in Equation (6.1) that a reduction of size 10 times (i.e., $l = 0.1$) will mean a $10^3 = 1000$ reduction in volume but only $10^2 = 100$ reduction in surface area. A reduction of volume by 1000, of course, means a 1000 times reduction in weight. One may readily prove that the same scaling relations apply to other geometries of solids.

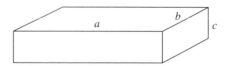

Figure 6.1. Solid rectangle.

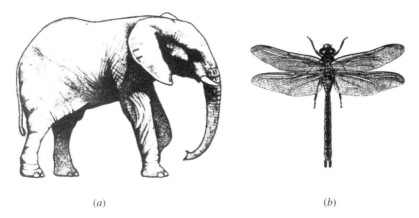

(a)   *(b)*

Figure 6.2. Distinct surface-to-volume ratios of two solids: (*a*) an elephant ($S/V \cong 10^{-4}\,\mathrm{mm}^{-1}$); (*b*) a dragonfly ($S/V \cong 10^{-1}\,\mathrm{mm}^{-1}$).

Example 6.1. As mentioned in Chapter 1, micromirrors are essential parts of microswitches used in fiber-optic networks in telecommunication. These mirrors are expected to rotate in a tightly controlled range at high rates. The angular momentum is a dominating factor in both the rotation control and the rate of rotation. In this example, we will estimate the reduction of torque required to turn a micromirror with a 50% reduction in the dimensions. The arrangement and the dimensions of the mirror are illustrated in Figure 6.3.

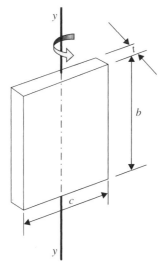

Figure 6.3. Micromirror.

Solution: The torque that is required to turn the mirror about the $y-y$ axis is related to the mass moment of inertia of the mirror, I_{YY}, which can be expressed as (Beer and Johnston, 1988)

$$I_{yy} = \tfrac{1}{12} M c^2$$

where M is the mass of the mirror and c the width of the mirror.

Since the mass of the mirror is $M = \rho V = \rho(bct)$, with ρ the mass density of the mirror material, the mass moment of inertia of the mirror is thus

$$I_{yy} = \tfrac{1}{12} \rho b c^3 t \tag{6.2}$$

The mass moment of inertia of the mirror with a 50% reduction in the size becomes

$$I'_{yy} = \tfrac{1}{12} \rho \left[\left(\tfrac{1}{2}b\right) \left(\tfrac{1}{2}c\right)^3 \left(\tfrac{1}{2}t\right) \right] = \tfrac{1}{32} \left[\tfrac{1}{12} \rho b c^3 t \right] = \tfrac{1}{32} I_{yy}$$

It is clear from the above simple calculation that a reduction in the factor of 32 is achieved in the mass moment of inertia, and thus the required torque for rotating the mirror with a 50% percent reduction of the dimension.

6.3 SCALING IN RIGID-BODY DYNAMICS

In Chapter 4 we demonstrated the important role that engineering mechanics plays in the design of MEMS devices and microsystems. Forces are required to make parts move, whereas power is the source for the generation of forces. The amount of force required to move a part and how fast the desired movements can be achieved, as well as how readily a moving part can be stopped, depend on the inertia of the part. The inertia of a solid is related to its mass and the acceleration or deceleration that is required to initiate or stop the solid motion. In minimizing the solid's inertia, and thus the required force, for the desired motion, one needs to understand the effect of reduction in size on the power P, force F, or pressure p and the time t required to deliver the motion.

6.3.1 Scaling in Dynamic Forces

For a rigid solid traveling from one position to another, the distance that the solid travels, s, can be shown to be $s \propto l$, where l stands for the linear scale. The velocity $v = s/t$, and hence $v \propto (l)t^{-1}$, where t is the time required for travel.

From particle kinematics, we have

$$s = v_0 t + \tfrac{1}{2} a t^2 \tag{6.3}$$

where

$v_0 = $ initial velocity

$a = $ acceleration

By letting $v_0 = 0$, we may express the acceleration using Equation (6.3):

$$a = \frac{2s}{t^2} \tag{6.4}$$

The dynamic force F, from Newton's second law, can be expressed as

$$F = Ma = \frac{2sM}{t^2} \propto (l)(l^3)t^{-2} \tag{6.5}$$

The reader will notice that the particle mass M is proportional to l^3, which is the linear scale of volume, as in Example 6.1.

6.3.2 Trimmer Force Scaling Vector

Trimmer (1989) proposed a unique matrix to represent force scaling with related parameters of acceleration a, time t, and power density P/V_0 that is required for the scaling of systems in motion. This matrix is called *force scaling vector* **F**.

The force scaling vector is defined as

$$\mathbf{F} = [l^F] = \begin{bmatrix} l^1 \\ l^2 \\ l^3 \\ l^4 \end{bmatrix} \tag{6.6}$$

We will now derive other quantities based on the above vector.

Acceleration a. Let us use the first part of Equation (6.5), that is, $F = Ma$, where $a = F/M$. The following scaling is obtained:

$$\mathbf{a} = [l^F][l^3]^{-1} = [l^F][l^{-3}] = \begin{bmatrix} l^1 \\ l^2 \\ l^3 \\ l^4 \end{bmatrix} [l^{-3}] = \begin{bmatrix} l^{-2} \\ l^{-1} \\ l^0 \\ l^1 \end{bmatrix} \tag{6.7}$$

Time t. Now if we take a look at the second part of Equation (6.5) and express the transient time as the subject of the formula, the following relation is obtained:

$$t = \sqrt{\frac{2sM}{F}} \propto \left([l^1][l^3]\right)^{1/2} [l^F]^{-1/2} = [l^2][l^F]^{-\frac{1}{2}}$$

$$= \begin{bmatrix} l^1 \\ l^2 \\ l^3 \\ l^4 \end{bmatrix}^{-1/2} [l^2] = \begin{bmatrix} l^{-1/2} \\ l^{-1} \\ l^{-1.5} \\ l^{-2} \end{bmatrix} [l^2] = \begin{bmatrix} l^{1.5} \\ l^1 \\ l^{0.5} \\ l^0 \end{bmatrix} \tag{6.8}$$

Power Density P/V₀. It is apparent that no substance, whether it is a solid or a fluid, can move without a power supply. Power is a very important parameter in the design of microsystems. Insufficient power supply to a microsystem can result in inactivity of the system. On the other hand, the system may suffer structural damage such as overheating with excessive power supply. Excessive power requirements by microsystems will increase the operational cost as well as reduce the operational lives of biomedical devices that are implanted in human bodies. Here, we will deal with *power density* rather than power supply per se, which is defined as the power supply P per unit volume V_0.

We begin our derivation of power supply with the work required to move a solid with mass M a distance s. Mathematically, the work done is equal to force times distance traveled, that is, $W = F \times s$. Thus, power, defined as the work done per unit time, is $P = W/t$, and power density can be expressed as

$$\frac{P}{V_0} = \frac{Fs}{tV_0} \tag{6.9}$$

We can thus relate power density to the force scaling vector as follows:

$$\frac{P}{V_0} = \frac{[l^F][l^1]}{\{[l^1][l^3][l^{-F}]\}^{\frac{1}{2}}[l^3]} = [l^{1.5F}][l^{-4}] = [l^F]^{1.5}[l^{-4}]$$

$$= \begin{bmatrix} l^1 \\ l^2 \\ l^3 \\ l^4 \end{bmatrix}^{1.5} [l^{-4}] = \begin{bmatrix} l^{-2.5} \\ l^{-1} \\ l^{0.5} \\ l^2 \end{bmatrix} \tag{6.10}$$

By combining Equations (6.6), (6.7), (6.8), and (6.10), we can establish a set of scaling laws for rigid-body dynamics as presented in Table 6.1.

Table 6.1 is very useful in scaling down devices in a design process. The order number in the table refers to the order of reduction in relation to the force scaling vector, shown in the second column from the left.

TABLE 6.1. Scaling Laws for Rigid-Body Dynamics

Order	Force scale, $\mathbf{F}$	Acceleration, a	Time, t	Power density, P/V_0
1	l^1	l^{-2}	$l^{1.5}$	$l^{-2.5}$
2	l^2	l^{-1}	l^1	l^{-1}
3	l^3	l^0	$l^{0.5}$	$l^{0.5}$
4	l^4	l^1	l^0	l^2

Example 6.2. Estimate the associated changes in acceleration a and time t and the power supply to actuate a MEMS component if its weight is reduced by a factor of 10.

Solution: Since the weight of a solid is equal to the mass times gravitational acceleration, and the mass is proportional to the cubic power of the linear scale, we will have the weight $W \propto l^3$, which means of order 3 in Table 6.1. The same table provides the following information:

1. There will be no reduction in acceleration (l^0).
2. There will be $l^{0.5} = 10^{0.5} = 3.16$ reduction in the time to complete the motion.
3. There will be $l^{0.5} = 3.16$ times reduction in power density (P/V_0). The reduction in power consumption is $P = 3.16V_0$. Since the volume of the component is reduced by a factor of 10, power consumption after scaling down reduces by $P = 3.16/10 = 0.3$ times.

6.4 SCALING IN ELECTROSTATIC FORCES

The mathematical expressions for electrostatic forces have been presented in Section 2.3.4. Here, let us revisit the configuration of a parallel-plate capacitor, as illustrated in Figure 6.4.

The electric potential energy induced in the parallel plates is

$$U = -\frac{1}{2}CV^2 = -\frac{\varepsilon_0 \varepsilon_r W L}{2d}V^2 \quad \text{[Equation (2.7)]}$$

where ε_0 and ε_r are respectively the permittivity and relative permittivity of the dielectric medium between the two electrodes and V is the applied voltage. This applied voltage is often called the *electric breakdown voltage* of a capacitor.

The breakdown voltage V in Equation (2.7) varies with the gap between the two plates, following the Paschen effect (Madou, 1997). This effect is illustrated in Figure 6.5.

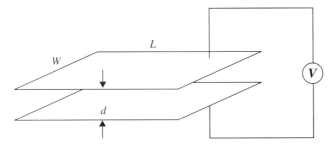

Figure 6.4. Electrostatic in electrically charged parallel plates.

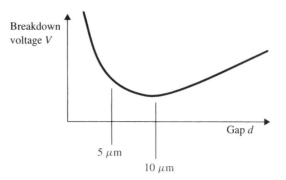

Figure 6.5. Paschen's effect.

We can see from Figure 6.5 that the breakdown voltage V drops drastically with an increase in the gap of $d < 5$ μm. The trend, however, is slowed significantly after the gap widens from $d > 5$ μm. Variation of the voltage reverses at $d \approx 10$ μm. The breakdown voltage continues to increase linearly with further increase of the gap.

We can thus assert that applied voltage $V \propto d$, or in scaling $V \propto l^1$, for the working range of $d > 10$ μm. The scaling of ε_0 and ε_r in Equation (2.7) is neutral, that is, $\varepsilon_0, \varepsilon_r \propto l^0$. We can thus express the scaling of the electrostatic potential energy in Equation (2.7) in the form

$$U \propto \frac{(l^0)(l^0)(l^1)(l^1)(l^1)^2}{l^1} = (l^3) \tag{6.11}$$

The scaling in Equation (6.11) means that a factor of 10 decrease in a linear dimension (i.e., W, L, and d simultaneously) will decrease the potential energy by a factor of $10^3 = 1000$.

Now, let us look at the scaling laws for electrostatic forces. As we learned in Chapter 2, electrostatic forces can be produced in three directions in parallel-plate arrangements. Expressions for these forces were shown in Equations (2.8), (2.10), and (2.11):

$$F_d = -\frac{1}{2} \frac{\varepsilon_0 \varepsilon_r W L V^2}{d^2} \quad \text{[Equation (2.8)]}$$

$$F_W = \frac{1}{2} \frac{\varepsilon_0 \varepsilon_r L V^2}{d} \quad \text{[Equation (2.10)]}$$

$$F_L = \frac{1}{2} \frac{\varepsilon_0 \varepsilon_r W V^2}{d} \quad \text{[Equation (2.11)]}$$

The directions of these forces are illustrated in Figure 6.6. Below we show how to readily derive the scaling laws for these forces.

From Equations (2.8), (2.10), and (2.11) the three force components are F_d, F_W and $F_L \propto (l^2)$, which means that electrostatic forces are of order 2 in the force scaling in

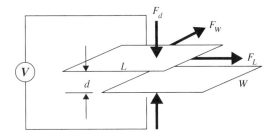

Figure 6.6. Electrostatic forces in charged parallel plates.

Table 6.1. Physically it indicates, for example, that a 10 times reduction in the size of the parallel plates will mean a 100 times decrease in induced electrostatic forces.

> **Example 6.3.** Find the reduction of electrostatic forces generated by a pair of parallel-plate electrodes as illustrated in Figure 6.6 if both the length L and width W of these plates are reduced by a factor of 10.
>
> *Solution:* The gap d between the plate electrodes remains unchanged. Consequently, the following relationships are obtained from Equations (2.8), (2.10), and (2.11) for the respective electrostatic force components with gap d unchanged:
>
> $F_d \propto l^2$ for the normal force component,
>
> $F_W \propto l$ for the force component along the width, and
>
> $F_L \propto l$ for the force component along the length.
>
> One will thus find a reduction of $10^2 = 100$ in the normal direction and $10^1 = 10$ times in the width and length directions.

6.5 SCALING OF ELECTROMAGNETIC FORCES

Although electromagnetic forces are the principal means of actuation in most machines at the macroscale, the principles of electromagnetic force–actuated devices were not included in Chapter 2 because electromagnetic actuation is not scaled down nearly as favorably as electrostatic forces, as will be demonstrated in this section. However, the reader should not rule out the use of electromagnetic actuation. It is used in a few mesoscale machine tools. One such application is in driving a miniaturized car, as shown in Figure 1.19 (Teshigahara et al., 1995).

Classical electromagnetic theory has shown that electromagnetic force **F** can be induced in a conductor or a conducting loop in a magnetic field **B** by passing current i in the conductor. In the following mathematical formulation, **B** is the magnitude of the magnetic field with a unit of webers (Wb/m^2) and i is the passing current in amperes. The electric current is the flow of electrons in a conductor per unit area and time, and it can be mathematically expressed as $i = Q/t$, where Q represents the charge per unit area of the conductor and t is the time. The electromotive force (emf) is the "force" that drives the electrons through the conductor. The energy that is required to drive these charges can be expressed as

$$U = \int dU = \int e\,dQ \tag{6.12}$$

For a current-carrying conductor in a magnetic field with a magnetic flux ϕ, the well-known Faraday's law can be used to express the induced emf in the conductor in the form of a coil with N turns as

$$e = N\frac{d\phi(t)}{dt}$$

By substituting the above relationship into Equation (6.12), along with $Q = it$, we have

$$U = \int \frac{d\phi(t)}{dt} i\,dt = \int i\,d\phi(t) = \frac{1}{L}\int \phi(t)\,d\phi(t) \tag{6.13}$$

The inductance $L = \phi/i$ can be derived with $N = 1$ in the above equation.

By integrating Equation (6.13), we can obtain the relationship

$$U = \frac{1}{2}\frac{\phi^2}{L} \tag{6.14a}$$

or

$$U = \tfrac{1}{2}Li^2 \tag{6.14b}$$

The induced electromagnetic force will change with the relative position of the conductor in the magnetic field. We can thus derive the expression of these forces as

$$\mathbf{F} = \left.\frac{\partial U}{\partial x}\right|_{\phi = \text{constant}} \tag{6.15a}$$

or

$$\mathbf{F} = \left.\frac{\partial U}{\partial x}\right|_{i = \text{constant}} \tag{6.15b}$$

If we consider the case with constant current flow, that is, Equation (6.15b), the induced electromagnetic force is expressed as

$$F = \frac{1}{2}i^2\frac{\partial L}{\partial x} \tag{6.16}$$

A close examination of the expression in Equation (6.16) indicates that the current i, that is, the number of electrons that flow through a conductor, depends on the cross-sectional area of the conductor: $i \propto l^2$. The term $\partial l/\partial x$ is dimensionless. We can thus conclude that the scaling of the electromagnetic force **F** is given as

$$\mathbf{F} \propto (l^2)(l^2) = l^4 \tag{6.17}$$

We can see from the above scaling that a 10 times reduction in size (l) will lead to $10^4 = 10{,}000$ reduction in the electromagnetic force. This is in sharp contrast to the scaling of the electrostatic force with l^2, which means a 100 times reduction in magnitude with the same reduction of linear dimensions. One will thus conclude that electromagnetic forces are 100 times less favorable in scaledown than electrostatic forces. Contrasting the scaling law in Equation (6.17) with that in Section 6.4 for electrostatic forces explains why almost all micromotors and actuators have been built based on electrostatic actuation, rather than the electromagnetic actuation commonly used in macrosized motors and actuators. Another obvious reason is that there is just not sufficient space in microdevices to accommodate the coil inductance needed to generate sufficient magnetic field for actuation power.

Despite what has been said above, engineers should not overlook the disadvantage of using electrostatic actuation forces. One obvious weakness of electrostatic actuation is its inherent low magnitudes. However, not much force is normally required to actuate most microdevices.

6.6 SCALING IN ELECTRICITY

Electricity is a dominant source of power to many MEMS and microsystems. Principal applications of electricity include the actuation of many microsystems by electrostatic, piezoelectric, and thermal resistance heating as presented in Chapter 2. Electrodynamic pumping is currently widely used pumping methods for microsystems. Electromechanical transduction is another common application of electricity in microsystems. Scaling of electricity thus is a very important design issue.

Some of the scaling laws related to electricity may be derived from simple laws of physics as shown below:

$$\text{Electric resistance:} \quad R = \frac{\rho L}{A} \propto (l)^{-1} \tag{6.18}$$

where

$\rho =$ electric resistivity of material

$L =$ length

$A =$ across-sectional area of conductor

$$\text{Resistive power loss:} \quad P = \frac{V^2}{R} \propto (l)^1 \tag{6.19}$$

where

$V =$ applied voltage

$$\text{Electric field energy:} \quad U = \tfrac{1}{2}\varepsilon E^2 \propto (l)^{-2} \qquad (6.20)$$

where

ε = permittivity of dielectric, $\propto (l)^0$

E = electric field strength, $\propto (l)^{-1}$

The above scaling law proves useful when miniaturizing devices. The central issue, however, is the scaling of electric power supply for miniaturization. For example, the power loss due to resistivity of the material in Equation (6.19) follows the first-order law: $P \propto l^1$. For a system that carries its own power supply such as an electrostatic actuation circuit, the available power is directly related to the system's volume: $E_{av} \propto (l)^3$. The ratio of power loss to available energy or power for performing the designed functions can be expressed as

$$\frac{P}{E_{av}} = \frac{(l)^1}{(l)^3} = (l)^{-2} \qquad (6.21)$$

The relationship in Equation (6.21) indicates a significant disadvantage of scaling down power supply systems. It means, for example, that a 10 times reduction in the size (l) of the power supply system (e.g., the linear dimension of the material for passing electric current for the power supply) would lead to a 100 times power loss due to the increase of resistivity.

6.7   SCALING IN FLUID MECHANICS

We have learned from Chapter 5 that fluids flow under the influence of shear forces (or shear stresses) and the continuum fluid mechanics based on Navier–Stokers equations breaks down in flows at submicrometer- and nanoscales. The capillary effect is the major reason for this breakdown. Here, we will learn why capillary flow does not scale down favorably and what are good alternatives for microflow.

Let us look at a volume of fluid as illustrated in Figure 6.7. The volume of the fluid, V, can be expressed as the area (A) times the height (h). Assume that the fluid is originally in a rectangular shape between two imaginary parallel plates. Moving the top plate in the x direction to the right induces the motion of the fluid. The original rectangular-shaped fluid volume is now a parallelogram, as shown in the figure. The shear force that is applied to the fluid volume for this motion is represented by the shear stress τ. The action of the shear force on the fluid results in an instantaneous velocity profile that varies from the maximum value along the top wall to a zero value at the bottom wall. For newtonian flow, the variation of the velocity between the top and bottom walls follows a linear relation, as illustrated in Figure 6.7. The viscosity of the fluid, which was defined in

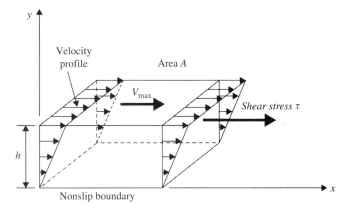

Figure 6.7. Velocity profile of a volume of moving fluid.

Equations (5.1) and (5.2), causes the variation of the velocities in the moving fluid. Let us express the viscosity in a different way:

$$\mu = \frac{\tau}{R_s} \qquad (6.22)$$

where

μ = dynamic viscosity of fluid, Pa-s

R_s = shear rate, = V_{max}/h.

The shear stress $\tau = F_s/A$, where F_s is the shear force.

The average velocity and the cross-sectional area of the passage are used to obtain the rate of volumetric fluid flow:

$$Q = A_s V_{ave} \qquad (6.23)$$

where

A_s = cross-sectional area of flow

V_{ave} = average velocity of fluid

We have learned from Chapter 5 that almost all fluid flow at the microscale is in the laminar flow regime. Consequently, formulations in Section 5.4 for laminar fluid flow in circular conduits could be used to derive the scaling law for liquid flow at the microscale.

Figure 6.8 illustrates fluid flowing through a small circular conduit of length L and radius a. The pressure drop ΔP over the length L of the conduit can be computed by using the Hagen–Poiseuille law expressed in Equation (5.17). The rate of volumetric

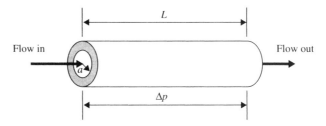

Figure 6.8. Fluid flow in a small circular conduit.

flow of the fluid, Q, can be expressed by the same equation but in a different form:

$$Q = \frac{\pi a^4}{8\mu L}\frac{\Delta P}{}$$ (6.24)

where

a = radius of tube

ΔP = pressure drop over length of tube L

From Equation (6.23), the average velocity V_{ave} can be shown to be

$$V_{\text{ave}} = \frac{Q}{\pi a^2}$$

and the pressure gradient is

$$\frac{\Delta P}{\Delta x} = -\frac{8\mu V_{\text{ave}}}{a^2}$$

We may compute the pressure drop for a section of a capillary tube of length L as

$$\Delta P = -\frac{8\mu V_{\text{ave}} L}{a^2}$$ (6.25)

The scaling laws for fluid flow in capillary tubes are thus derived as $Q \propto a^4$ for volumetric flow from Equation (6.24) and $\Delta P/L \propto a^{-2}$ for pressure drop per unit length from Equation (6.25), where a is the radius of the tube.

> **Example 6.4.** Use the scaling laws to estimate the variations of volumetric flow and pressure drop in a circular tube if the radius of the tube is reduced by a factor of 10. Make an observation on this scaling practice.
>
> *Solution:* From Equation (6.24), volumetric flow $Q \propto a^4$, where a is the radius of the tube. From this scaling law, the volumetric flow is reduced by

$10^4 = 10,000$ times. Likewise, the scaling law in Equation (6.25) indicates a pressure drop per unit length increase of $10^2 = 100$ times.

This is obviously a highly undesirable situation for a scale-down device. Alternative mechanisms would thus be desirable for microfluid flow.

In regard to scaling of fluid flow in capillary tubes, when the radius of the conduit is very small (microscale), a more complete scaling for microfluid flow would include the capillary effect, which occurs to liquid flow in minute conduits, as described in Section 5.6. This effect is attributed to the dominance of the surface tension of fluids at a small scale.

The pressure required to overcome the surface tension is expressed in Equations (5.24a) and (5.24b), from which we have $\Delta P \propto a^{-1}$, where a is the radius of the tube. Thus, the following relation can be used to scale the pressure drop per unit length of a liquid in a microscale tube:

$$\frac{\Delta P}{L} \propto l^{-3} \qquad (6.26)$$

Example 6.5. What will happen to the pressure drop in the fluid in Example 6.4 if the tube radius is at the microscale?

Solution: The pressure drop per unit length of the tube, from the scaling law in Equation (6.26), increases 1000 time with a 10 times reduction of the tube radius. The situation is thus one order of magnitude more severe than the case at the meso- or macroscale in Example 6.4.

The significantly adverse effect in scaling down fluid flow at the micrometer- and submicrometerscal has prompted engineers to search for alternative fluid propulsion mechanisms to that used in conventional volumetric pumping. These mechanisms include piezoelectric, electro-osmotic, electrowetting, and electrohydrodynamic pumping (Madou, 1997). Many of these special pumping techniques are based on surface pumping forces, which scale more favorably in microdomains than conventional volumetric pumping forces. A typical pumping technique based on surface forces is use of a piezoelectric pump.

The principle of piezoelectric pumping is to use the forces generated on tube wall to drive the fluid flow instead of the conventional pressure difference. This way of moving fluid is similar to squeezing toothpaste from a plastic tubular container. The surface force F, which is proportional to the surface area of the inner wall of the tube, scales much more favorably than the backpressure, such as given in Equation (6.25), required to drive the equivalent volume of fluid. The surface force is favored over the volumetric pumping pressure because the surface area of the inner wall of the section of the small tube in Figure 6.8 is $S = 2\pi aL$ and the equivalent volume of the fluid is $V = \pi a^2 L$. These relations result in surface area–volume ratio $S/V = 2/a$. Since the surface force that is required to pump a volume of the fluid is proportional to the surface area, we may readily see from the S/V ratio that $F \propto l^{-1}$. Consequently, scaling down the tube radius will result in an increase of the surface force available for pumping the unit volume of fluid. The feasibility of the piezoelectric pumping technique for microflow was demonstrated in Section 5.6.3.

The working principles of the two principal electrohydrodynamic pumping methods, electro-osmosis and electrophoresis, were presented in Section 3.8.2. These pumping techniques are widely used in microfluidic systems designed and constructed by biotechnology and pharmaceutical industries.

6.8 SCALING IN HEAT TRANSFER

Heat transfer is an essential part of design for many microsystems. For most cases, heat transmission in microsystems is in the modes of conduction and convection. In some special cases, heat also transfers in radiation, such as in laser treatments involved in some manufacturing techniques for microsystems. Formulation of heat conduction and convection in solids and fluids was presented in the second part of Chapter 5. Here, we will take an overview of the scaling of heat transmission in these two modes and offer scaling laws at the microscale. Modified scaling laws for device components at the submicrometer- and nanoscale will be dealt with in Chapter 12, where thermophysical properties that vary with the size of the solids are considered. In addition to the size parameter, these properties should be included in the scaling law.

6.8.1 Scaling in Heat Conduction

Scaling of Heat Flux. Heat conduction in solids is governed by Fourier's law, expressed in a general form in Equation (5.28). For one-dimensional heat conduction along the x coordinate, we have

$$q_x = -k \frac{\partial T(x, y, z, t)}{\partial x}$$

where q_x is the heat flux along the x coordinate, k is the thermal conductivity of the solid, and $T(x,y,z,t)$ is the temperature field in the solid in a cartesian coordinate system at time t. A more generic form for the rate of heat conduction in a solid is

$$Q = qA = -kA \frac{\Delta T}{\Delta x} \tag{6.27}$$

It is readily seen from Equation (6.27) that the scaling law for heat conduction for solids at the meso- and macroscales is

$$Q \propto (l^2)(l^{-1}) = (l^1) \tag{6.28}$$

It is easy to see from this simple scaling law that reduction in size leads to the decrease of total heat flow in a solid.

Scaling in Effect of Heat Conduction in Solids at Meso- and Microscales.

A dimensionless number, called the *Fourier number* Fo, is frequently used to determine the time increments in a transient heat conduction analysis:

$$\text{Fo} = \frac{\alpha t}{L^2} \qquad (6.29)$$

where α is the thermal diffusivity of the material as defined in Equation (5.32) and t is the time for heat to flow across the characteristic length L.

Physically, the Fourier number is the ratio of the rate of heat transfer by conduction to the rate of energy storage in the system (Kreith and Bohn, 1997). By rearranging Equation (6.29), we will obtain the scaling of the time for heat conduction in a solid:

$$t = \frac{\text{Fo}}{\alpha} L^2 \propto (l^2) \qquad (6.30)$$

We assign both Fo and α in Equation (6.29) to be constants.

> **Example 6.6.** Estimate the variation of total heat flow and the time required to transmit heat in a solid with a reduction in size by a factor of 10.
>
> *Solution:* Using the scaling laws in Equations (6.28) and (6.30), we estimate that total heat flow is reduced by a factor of 10, and the time required for heat transmission is reduced by $10^2 = 100$ times with a reduction in size by a factor of 10.

6.8.2 Scaling in Heat Convection

We have seen from Figure 5.22 that boundary layers are present at the interfaces of solids and fluids. Heat transfer in a fluid is in the mode of convection, which is governed by Newton's cooling law as expressed in Equation (5.33) or in a more generic form as

$$Q = qA = hA\,\Delta T \qquad (6.31)$$

where

Q = total heat flow between two points in fluid

q = corresponding heat flux

A = cross-sectional area for heat flow

h = heat transfer coefficient,

ΔT = temperature difference between these two points

As described in Section 5.7.4, the heat transfer coefficient h depends primarily on the velocity of fluid, which does not play a significant role in the scaling of heat flow. Consequently, total heat flow from Equation (6.31) primarily depends on the cross-sectional area A, which is of order l^2. We thus have the scaling of heat transfer in convection to be $Q \propto l^2$ for fluids in the meso- and microregimes.

The scaling of heat convection for fluids in the submicrometer regime is not quite as simple as that described above. The scaling of heat transmission in gases will be presented in Chapter 12.

PROBLEMS

1. The application of the scaling laws in miniaturization is to assess the consequences on (a) the physical effect, (b) the economical effect, (c) the market effect on miniaturized products.

2. Scaling laws are derived from (a) design engineer's experience, (b) the laws of physics, (c) the market demands.

3. Scaling in geometry is critical in miniaturizing (a) moving components, (b) sensing components, (c) overall dimensions of MEMS products.

4. Trimmer's force scaling vector is used to assess miniaturization relating to (a) heat flow in solids, (b) fluid flow, (c) rigid-body dynamics in the design of MEMS.

5. For order 1 scaling such as surface-to-volume scaling, the acceleration varies (a) linearly, (b) in square power, (c) in cubic power.

6. The reason electrostatic forces are favored over electromagnetic forces in microactuation is that electrostatic forces scale (a) better, (b) worse, (c) about the same as electromagnetic forces.

7. Electromagnetic forces scale (a) two, (b) three, (c) four orders of magnitude worse than electrostatic forces.

8. The power loss due to electric resistivity is (a) much more severe, (b) less severe, (c) about the same in small-sized systems.

9. Pressure drop in a fluid flowing through a smaller circular conduit is (a) much greater, (b) about equal to, (c) much less than that in a larger conduit.

10. The volumetric flow of fluid in a smaller circular conduit is (a) much greater, (b) about equal to, (c) much smaller than that in a larger conduit.

11. The effect of surface tension on fluid flowing in a capillary tube makes the pressure drop (a) much greater, (b) about equal to, (c) much less than the same flow in mesosize tubes.

12. The effect of surface tension on fluid flowing in a capillary tube makes the volumetric flow (a) much greater than, (b) about equal to, (c) much less than the same flow in mesosize tubes.

13. Heat flows (a) faster, (b) slower, (c) about the same in a smaller solid than in a larger solid.

14. A reduction in size of 10 times would result in reduction of total heat flow in a solid by (a) 10, (b) 100, (c) 1000 times.

15. A reduction in size of 10 times would result in reduction of total heat flow in a fluid by (a) 10, (b) 100, (c) 1000 times.

CHAPTER 7

MATERIALS FOR MEMS AND MICROSYSTEMS

7.1 INTRODUCTION

In Chapter 1, we maintained that the current technologies used in producing MEMS and microsystems are inseparable from those of microelectronics. This close relationship between microelectronics and microsystems fabrications often misleads engineers to a common belief that the two are interchangeable. It is true that many of the current microsystem fabrication techniques are closely related to those used in microelectronics. Design of microsystems and their packaging, however, is significantly different from that for microelectronics. Many microsystems use microelectronics materials such as silicon and gallium arsenide (GaAs) for the sensing or actuating elements. These materials are chosen mainly because they are dimensionally slable and their microfabrication and packaging techniques are well established in microelectronics. However, there are other materials used for MEMS and microsystems products—such as quartz and Pyrex, polymers and plastics, and ceramics—that are not commonly used in microelectronics. Plastics and polymers are also used extensively in the case of microsystems produced by the LIGA processes, as will be described in Chapter 9.

7.2 SUBSTRATES AND WAFERS

The frequently used term *substrate* in microelectronics means a flat macroscopic object on which microfabrication processes take place (Ruska, 1987). In microsystems, a substrate serves an additional purpose: It acts as a signal transducer besides supporting other transducers that convert mechanical action to electrical output or vice versa. For example, in Chapter 2, we saw pressure sensors that convert the applied pressure to the deflection of a thin diaphragm that is an integral part of a silicon die cut from a silicon substrate. The same applies to microactuators, in which the actuating components, such as microbeams made of silicon in microaccelerometers, are also called substrates.

In semiconductors, the substrate is a single crystal cut in slices from a larger piece called a *wafer*. Wafers can be of silicon or other single crystalline materials such as quartz or gallium arsenide. Substrates in microsystems, however, are somewhat different. Two types of substrate materials are used in microsystems: (1) active substrate materials and (2) passive substrate materials, as will be described in detail in the subsequent sections.

Table 7.1 presents a group of materials that are classified as electric *insulators* (or dielectrics), *semiconductors*, and *conductors* (Sze, 1985). The same reference classifies insulators as having electrical resistivity ρ in the range of $\geq 10^8$ Ω-cm; semiconductors with ρ in the range of 10^{-3}–10^8 Ω-cm; and conductors with $\rho < 10^{-3}$ Ω-cm. We will find that common substrate materials used in MEMS such as silicon (Si), germanium (Ge), and gallium arsenide (GaAs) all fall in the category of semiconductors. One major reason for using these materials as principal substrate materials in both microelectronics and microsystems is that these materials can be used as either conductors and insulators as needs arise. Indeed, the doping techniques that were described in Chapter 3 can

TABLE 7.1. Typical Electrical Resistivities of Insulators, Semiconductors, and Conductors

Materials	Approximate Electrical Resistivity, $\rho(\Omega\text{-cm})$
Conductors	
Silver (Ag)	10^{-6}
Copper (Cu)	$10^{-5.8}$
Aluminum (Al)	$10^{-5.5}$
Platinum (Pt)	10^{-5}
Semiconductors	
Germanium (Ge)	10^{-3}–$10^{1.5}$
Silicon (Si)	10^{-3}–$10^{4.5}$
Gallium arsenide (GaAs)	10^{-3}–10^8
Gallium phosphide (GaP)	10^{-2}–$10^{6.5}$
Insulators	
Oxide	10^9
Glass	$10^{10.5}$
Nickel (pure)	10^{13}
Diamond	10^{14}
Quartz (fused)	10^{18}

be used to convert the most commonly used semiconducting material, silicon, to an electrically conducting material by doping it with a foreign material to form either p- or n-type silicon. All semiconductors are amenable to such doping. Another reason for using semiconductors is that the fabrication processes, such as etching and thin-film deposition, and the equipment required for these processes have already been developed for these materials.

A checklist of factors that can help the designer in selecting substrate materials for microsystems is available in Madou (1997).

7.3 ACTIVE SUBSTRATE MATERIALS

Active substrate materials are primarily used for sensors and actuators in a microsystem (Figure 1.5) or other MEMS components (Figure 1.7). Typical active substrate materials for microsystems include silicon, gallium arsenide, germanium, and quartz. All these materials except quartz are classified as semiconductors in Table 7.1. These substrate materials have basically a cubic crystal lattice with a tetrahedral atomic bond (Sze, 1985). They are selected as active substrates primarily for their dimensional stability, which is relatively insensitive to environmental conditions. Dimensional stability is a critical requirement for sensors and actuators with high precision.

As indicated in the periodic table (Figure 3.3), each atom of these semiconductor materials carries four electrons in the outer orbit. Each atom also shares these four electrons with its four neighbors. The force of attraction for electrons by both nuclei holds each pair of shared atoms together. They can be doped with foreign materials to alter their electric conductivity, as described in Section 3.5.

7.4 SILICON AS SUBSTRATE MATERIAL

7.4.1 Ideal Substrate for MEMS

Silicon is the most abundant material on earth. However, it almost always exists in compounds with other elements. Single-crystal silicon is the most widely used substrate material for MEMS and microsystems. The popularity of silicon for such application is primarily for the following reasons:

1. It is mechanically stable and can be integrated into electronics on the same substrate. Electronics for signal transduction, such as a p- or n-type piezoresistive, can be readily integrated with the silicon substrate.
2. Silicon is almost an ideal structure material. It has about the same Young's modulus as steel (about 2×10^5 MPa) but is as light as aluminum, with a mass density of about 2.3 g/cm^3. Materials with higher Young's modulus can better maintain a linear relationship between applied load and the induced deformations.
3. It has a melting point at 1400°C, which is about twice as high as that of aluminum. This high melting point makes silicon dimensionally stable even at elevated temperatures.

4. Its coefficient of thermal expansion is about 8 times smaller than that of steel and more than 10 times smaller than that of aluminum.

5. Above all, silicon shows virtually no mechanical hysteresis. It is thus an ideal candidate material for sensors and actuators. Moreover, silicon wafers are extremely flat and accept coatings and additional thin-film layers for building microstructural geometry or conducting electricity.

6. There is a greater flexibility in design and manufacture with silicon than with other substrate materials. Treatments and fabrication processes for silicon substrates are well established and documented.

7.4.2 Single-Crystal Silicon and Wafers

To use silicon as a substrate material, it has to be pure silicon in a single-crystal form. The Czochralski (CZ) method appears to be the most popular method among several methods that have been developed for producing pure silicon crystal. The raw silicon, in the form of quartzite, is melted in a quartz crucible with carbon (coal, coke, wood chips, etc.). The crucible is placed in a high-temperature furnace, as shown in Figure 7.1 (Ruska, 1987). A "seed" crystal, which is attached at the tip of a puller, is brought into contact with the molten silicon to form a larger crystal. The puller is slowly pulled up along with the continuous deposition of silicon melt onto the seed crystal. As the puller

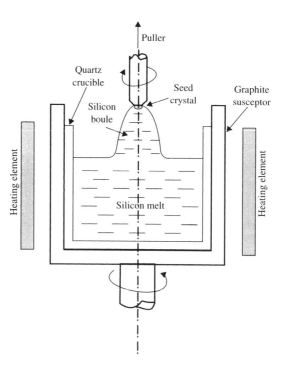

Figure 7.1. Czochralski method for growing single crystals (Ruska, 1987).

Figure 7.2. A 300-mm single-crystal silicon boule cooling on a material-handling device. (Courtesy of MEMC Electronic Materials, St. Peters, MO.)

is pulled up, the deposited silicon melt condenses and a large bologna-shaped boule of single-crystal silicon several feet long is formed. Figure 7.2 shows a boule produced by this method. The diameter of the boule ranges from 100 to 300 mm.

The silicon crystal boule produced by the CZ method is then ground into a perfect circle in its outside surface, then sliced to form thin disks of the desired thickness by fine diamond saws. These thin disks are then chemically lap polished to form the finished wafers.

Principal materials in the silicon melt are silicon oxide and silicon carbide. These materials react at high temperature to produce pure silicon, along with other gaseous by-products, as shown by the chemical reaction

$$SiC + SiO_2 \rightarrow Si + CO + SiO$$

The gases produced by the above reaction escape to the atmosphere and the liquid Si is left and solidifies to pure silicon. Circular pure-crystal silicon boules are produced by this technique in three standard sizes: 100 mm (4 in.), 150 mm (6 in.), and 200 mm (8 in.) in diameter. A larger boule size, 300 mm (12 in.) in diameter, is the latest addition to the standard wafer sizes. Current industry standard on wafer sizes and thickness are as follows:

100 mm (4 in.) diameter $\times$ 500 μm thick
150 mm (6 in.) diameter $\times$ 750 μm thick
200 mm (8 in.) diameter $\times$ 1 mm thick
300 mm (12 in.) diameter $\times$ 775 μm thick

Figure 7.3. Size difference between a 200-mm wafer and a 300-mm wafer. (Courtesy of MEMC Electronic Materials, St. Peters, MO.)

The size difference between a 200- and a 300-mm wafer is shown in Figure 7.3. The latter size wafer has 2.25 times more surface area than the 200-mm wafer and thus provides significant economic advantage for accommodating many more substrates on a single wafer.

Silicon substrates often are expected to carry electric charges, either in certain designated parts or in the entire area, as in the resonant frequency pressure sensors described in Section 4.3.6 as well as the silicon solar photovoltaic cells described in Section 2.2.4. Substrates thus often require p or n doping of the entire wafers. The doping of p- and n-type impurities, as described in Section 3.5, can be done by either ion implantation or diffusion, as will be described in detail in Chapter 8. Common n-type dopants of silicon are phosphorus, arsenic, and antimony, whereas boron is the most common p-type dopant for silicon.

7.4.3 Crystal Structure

As indicated in Figure 3.2b, a silicon atom has four electrons in its outermost orbit. Electrons at the outermost orbit of an atom are called *valence electrons*. Large numbers of silicon atoms are bonded with each other by means of their valence electrons to form a crystal. Each silicon atom normally shares one of its four valence electrons in a *covalent bond* with each of four neighboring silicon atoms. A covalent bond is the result of binding two atoms together by shared electrons. The solid thus consists of basic units of five silicon atoms: the original atom plus the four atoms with which it shares valence

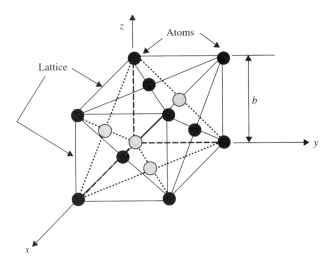

Figure 7.4. Typical FCC unit cell.

electrons. The silicon crystal is thus made up of a regular series of units of five silicon atoms. This regular, fixed arrangement of silicon atoms is known as the crystal lattice.

The atoms of silicon have an uneven lattice geometry and are arranged in a face-centered-cubic (FCC) unit cell, as illustrated in Figure 7.4. A unit cell consists of atoms situated at fixed locations defined by imaginary lines called *lattices*. The dimension b of the lattice that forms the edges of a cubic unit cell is called the *lattice constant* in the figure. In a typical FCC crystal, atoms are situated at the eight corners of the cubic lattice structure as well as at the center of each of the six faces. We show those "visible" atoms in black and the "invisible" or "hidden" ones in gray in Figure 7.4. For silicon crystals, the lattice constant, $b = 0.543$ nm. In an FCC lattice, each atom is bonded to 12 nearest-neighbor atoms.

The crystal structure of silicon, however, is more complex than that of regular FCC structure as illustrated in Figure 7.4. It can be considered the result of two interpenetrating FCC crystals, FCC A and FCC B, illustrated in Figure 7.5a (Angell et al., 1983). Consequently, the silicon crystal contains an additional four atoms, as shown in Figure 7.5b.

Figure 7.6 shows a three-dimensional model of the crystal structure of silicon. A closer look at this structure will reveal that these four additional atoms in the interior of the FCC (the white balls in Figure 7.6) form a subcubic cell of the diamond lattice type, as illustrated in Figure 7.7 (Ruska, 1987). A silicon unit cell thus has 18 atoms with 8 atoms at the corners plus 6 atoms on the faces and another 4 interior atoms. Many perceive the crystal structure of silicon to be a *diamond lattice* at cubic lattice spacing of 0.543 nm (Kwok, 1997).

The spacing between adjacent atoms in the diamond subcell is 0.235 nm (Bryzek et al., 1991; Sze, 1985; Ruska, 1987). Four equally spaced nearest-neighbor atoms that lie at the corners of a tetrahedron make the diamond lattice, as shown in the inset in Figure 7.7. One may also perceive the silicon crystal as being stacked layers of repeating cubes.

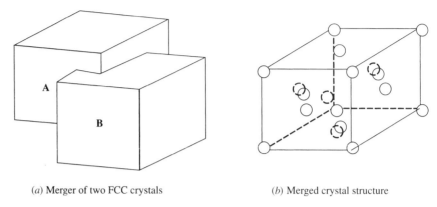

(*a*) Merger of two FCC crystals   (*b*) Merged crystal structure

Figure 7.5. Structure of silicon crystal.

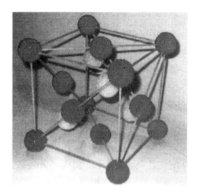

Figure 7.6. Photograph of silicon crystal structure.

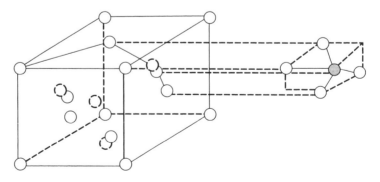

Figure 7.7. Subcubic cell of diamond lattice in silicon crystal.

Each cube has an atom at each corner and at the center of each face (FCC structure). These cubes are interlocked with four neighboring cubes in bulk single-crystal silicon boules, and the wafers are sliced from the boules.

Because of the asymmetrical and nonuniform lattice distance between atoms, single-crystal silicon exhibits anisotropic thermophysical and mechanical characteristics that need to be understood for the benefits of handling and manufacturing. These orientation-dependent material characteristics can be better expressed using the Miller indices (Ruska, 1987; Sze, 1985).

Example 7.1. Estimate the number of atoms per cubic centimeter of pure silicon.

Solution: Since the lattice constant $b = 0.543 \, \text{nm} = 0.543 \times 10^{-9} \, \text{m}$ and there are 18 atoms in each stand-alone cubic cell, the number of atoms in a cubic centimeter, with $1 \, \text{cm} = 0.01 \, \text{m}$, is given as

$$N = \left(\frac{V}{v}\right)n = \left(\frac{0.01}{0.543 \times 10^{-9}}\right)^3 \times 18 = 1.12 \times 10^{23} \text{ atoms/cm}^3$$

where V and v represent the bulk volume of silicon in question and the volume of a single crystal, respectively, and n is the number of atoms in each stand-alone single-silicon crystal.

7.4.4 Miller Indices

Because of the skew distribution of atoms in a silicon crystal, material properties are by no means uniform in the crystal. It is important to be able to designate the principal orientations as well as planes in the crystal on which the properties are specified. A popular method of designating crystal planes and orientations is by the *Miller indices*. These indices are effectively used to designate planes of materials in cubic crystal families. We will briefly outline the principle of these indices below.

Consider a point $P(x,y,z)$ in an arbitrary plane in a space defined by the cartesian coordinate system $x-y-z$. The equation that defines the point P is

$$\frac{x}{a} + \frac{y}{b} + \frac{z}{c} = 1 \tag{7.1}$$

where a, b, and c are the intercepts formed by the plane with the respective x, y, and z axes.

Equation (7.1) can be rewritten as

$$hx + ky + mz = 1 \tag{7.2}$$

It is apparent that $h = 1/a$, $k = 1/b$, and $m = 1/c$ in Equation (7.2).

Now, if we let (hkm) designate the plane and $<hkm>$ designate the direction that is normal to the plane (hkm), then we may designate the three planes that apply to a cubic

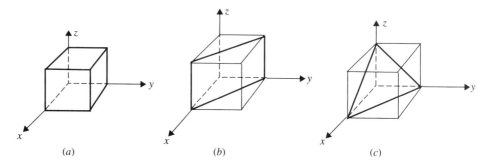

Figure 7.8. Designation of plane of cubic silicon crystal: (*a*) normal planes; (*b*) diagonal plane; (*c*) inclined plane.

silicon crystal. These three planes are illustrated in Figure 7.8. We will assume that the cubic structure has unit length with intercepts $a = b = c = 1.0$.

We can designate various planes in Figure 7.8 using Equations (7.1) and (7.2) as follows:

Top face in Figure 7.8*a*: (001)
Right face in Figure 7.8*a*: (010)
Front face in Figure 7.8*a*: (100)
Diagonal face in Figure 7.8*b*: (110)
Inclined face in Figure 7.8*c*: (111)

The orientations of these planes can be represented by <100>, which is perpendicular to plane (100); <111>, which is normal to the plane (111); and so on. Figure 7.9 shows these three planes and orientations for a unit cell in a single-crystal silicon.

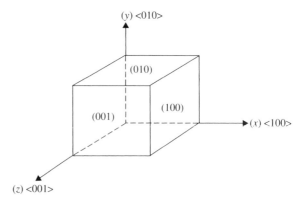

Figure 7.9. Silicon crystal structure and planes and orientations.

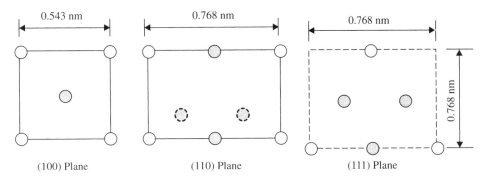

Figure 7.10. Silicon atoms on three designated planes.

The silicon atoms in the three principal planes, (100), (110), and (111), are illustrated in Figure 7.10. The atoms shown in open circles are those at the corners of the cubic, solid circles in gray are the ones at the center of the faces, and the gray circles in dotted lines are the atoms in the interior of the unit cell.

As shown in Figures 7.6 and 7.10, the lattice distances between adjacent atoms are shortest for those atoms on the (111) plane. These short lattice distances between atoms make the attractive forces between atoms stronger on this plane than those on the other two planes. Also, this plane contains three of the four atoms that are situated at the center of the faces of the unit cell (in gray). Thus, the growth of crystal in this plane is the slowest and the fabrication processes, for example, etching, as we will learn in Chapters 8 and 9, will proceed slowest.

Because of the importance of the orientation-dependent machinability of silicon substrate, wafers that are shipped by suppliers normally indicate in which directions, that is, <110>, <100>, or <111>, the cuts have been made by "flats," as illustrated in Figure 7.11. The edge of silicon crystal boules can be ground to produce single

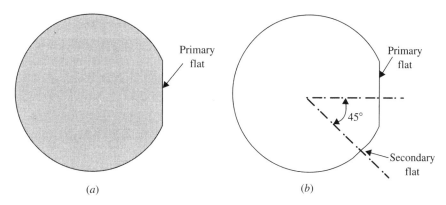

Figure 7.11. Primary and secondary flats in silicon wafers: (*a*) *n*-type; (*b*) *p*-type.

primary flats, in some cases, with additional single *secondary flats*. The wafers that are sliced from these crystal boules may thus contain one or two flats, as shown in Figure 7.11. The primary flats are used to indicate the crystal orientation of the wafer structure, whereas the secondary flats are used to indicate the dopant type of the wafer. For example, Figure 7.11a indicates a flat that is normal to the <111> orientation with *p*-type silicon crystal. The wafer with an additional second flat at 45° from the primary flat in Figure 7.11b indicates *n*-type doping in the wafer. Other arrangements of flats in wafers for various designations can be found in the literature (van Zant, 1997; Madou, 1997).

7.4.5 Mechanical Properties of Silicon

Silicon is mainly used as an IC carrier in microelectronics. For microsystems, it is used as the prime candidate material for sensors and actuators as well as substrates for microfluidics. The technology that is used in implementing the IC on silicon dies is used in a similar way for microsystems. However, silicon, as the material of components of generally three-dimensional geometry, needs to withstand often-severe mechanical and thermal loads in addition to accommodating electrical instruments such as piezoresistives integrated into it. It is thus important to have a good understanding of the mechanical characteristics of silicon as a structural material.

Silicon is an elastic material with no plasticity or creep below 800°C. It shows virtually no fatigue failure under all conceivable circumstances. These unique characteristics make it an ideal material for sensing and actuating in microsystems. However, it is a brittle material. Therefore, undesirable brittle fracture behavior with weak resistance to impact loads needs to be considered in the design of such microsystems. Another disadvantage of silicon substrate is that it is anisotropic. This makes accurate stress analysis of silicon structures tedious, since directional mechanical properties must be included. For example, Table 7.2 indicates the different Young's moduli and shear moduli of elasticity of silicon crystals in different orientations (Madou, 1997). For most cases in microsystem design, the bulk material properties of silicon, silicon compounds, and other active substrate materials presented in Table 7.3 are used.

TABLE 7.2. Young's Modulus and Shear Modulus of Elasticity of Silicon Crystals

Miller Index for Orientation	Young's Modulus, E (GPa)	Shear Modulus, G (GPa)
<100>	129.5	79.0
<110>	168.0	61.7
<111>	186.5	57.5

TABLE 7.3. Mechanical and Thermophysical Properties of MEMS Materials

Materials	σ_y (10^9 N/m^2)	E (10^{11} N/m^2)	ρ (g/cm^3)	C (J/g-°C)	k (W/cm-°C)	α (10^{-6} °C)	T_M °C
Si	7.00	1.90	2.30	0.70	1.57	2.33	1400
SiC	21.00	7.00	3.20	0.67	3.50	3.30	2300
Si$_3$N$_4$	14.00	3.85	3.10	0.69	0.19	0.80	1930
SiO$_2$	8.40	0.73	2.27	1.00	0.014	0.50	1700
Aluminum	0.17	0.70	2.70	0.942	2.36	25	660
Stainless steel	2.10	2.00	7.90	0.47	0.329	17.30	1500
Copper	0.07	0.11	8.9	0.386	3.93	16.56	1080
GaAs	2.70	0.75	5.30	0.35	0.50	6.86	1238
Ge		1.03	5.32	0.31	0.60	5.80	937
Quartz	0.5–0.7	0.76–0.97	2.66	0.82–1.20	0.067–0.12	7.10	1710

Note: Source for semiconductor material properties: *Fundamentals of Microfabrication,* by M. Madou, CRC Press, Boca Raton, FL, 1997.
[a]*Legend*: σ_y = yield strength, E = Young's modulus, ρ = mass density, C = specific heat, k = thermal conducticity, α = coefficient of thermal expansion, T_M = melting point.

Example 7.2. As indicated in Chapter 5, the thermal diffusivity of a material is a measure of how fast heat can flow in the material. List the thermal diffusivity of silicon, silicon dioxide, aluminum, and copper and make an observation on the results.

Solution: The thermal diffusivity α is a function of several properties of the material as shown in Equation (5.32):

$$\alpha = \frac{k}{\rho C}$$

where the properties k, ρ, and C for the four materials are given in Table 7.3. They are listed with slightly different units in Table 7.4.

By substituting the material properties tabulated in the three left columns in Table 7.4 into Equation (5.32), we can compute the thermal diffusivities of the materials as indicated in the right column in the same table. It is not surprising to observe that copper has the highest thermal diffusivity, whereas silicon and aluminum have about the same value. Useful information from this exercise is that silicon oxide conducts heat more than 150 times slower than silicon and aluminum. We may thus conclude that copper films are the best material for fast heat transmission in microsystems, whereas silicon dioxide can be used as an effective thermal barrier. Materials with high thermal diffusivities are preferred for microactuators by thermal forces, as described in Chapter 2.

TABLE 7.4. Thermal Diffusivities of Selected Materials for Microsystems

Materials	K (J/s-m-°C)	ρ(g/m^3)	C (J/g-°C)	Thermal Diffusivity, α(m^2/s)
Si	157	2.3×10^6	0.7	97.52×10^{-6}
SiO$_2$	1.4	2.27×10^6	1.0	0.62×10^{-6}
Aluminum	236	2.7×10^6	0.94	93×10^{-6}
Copper	393	8.9×10^6	0.386	114.4×10^{-6}

7.5 SILICON COMPOUNDS

Three silicon compounds are often used in microsystems: silicon dioxide, SiO_2; silicon carbide, SiC; and silicon nitride, Si_3N_4. We will take a brief look at the roles these compounds play in microsystems.

7.5.1 Silicon Dioxide

There are three principal uses of silicon oxide in microsystems: (1) as a thermal and electric insulator (see Table 7.1 for the low electric resistivity of oxides), (2) as a mask in the etching of silicon substrates, and (3) as a sacrificial layer in surface micromachining, as will be described in Chapter 9. Silicon oxide has much stronger resistance to most etchants than silicon. Important properties of oxide are listed in Table 7.5.

Silicon dioxide can be produced by heating silicon in oxidant such as oxygen with or without steam. The reactions for such processes are

$$Si + O_2 \rightarrow SiO_2 \qquad\qquad \text{for "dry" oxidation} \qquad\qquad (7.3)$$

$$Si + 2H_2O \rightarrow SiO_2 + 2H_2 \qquad\qquad \text{for "wet" oxidation in steam} \qquad (7.4)$$

TABLE 7.5. Properties of Silicon Dioxide

Properties	Value
Density (g/cm^3)	2.27
Resistivity (Ω-cm)	$\geq 10^{16}$
Dielectric constant	3.9
Melting point (°C)	~ 1700
Specific heat (J/g-°C)	1.0
Thermal conductivity (W/cm-°C)	0.014
Coefficient of thermal expansion (ppm/°C)	0.5

Source: Ruska, 1987.

Oxidation is a diffusion process, as described in Chapter 3. Therefore, the rate of oxidation can be controlled by similar techniques used for most other diffusion processes. Typical diffusivities of silicon dioxide at $900°C$ in dry oxidation are 4×10^{-19} cm²/s for arsenic doped (n-type) and 3×10^{-19} cm²/s for boron doped (p-type) (Sze, 1985). The process can be accelerated to much faster rates by the presence of steam; the highly activated H_2O molecules enhance the process. As in all diffusion processes, the diffusivity of the substance to be diffused into the base material is the key parameter for the effectiveness of the diffusion process.

7.5.2 Silicon Carbide

The principal advantage of silicon carbide (SiC) in microsystems is its dimensional and chemical stability at high temperature. It has very strong resistance to oxidation even at very high temperatures. Thin films of silicon carbide are often deposited over the MEMS components to protect them from extreme temperature. Another attraction of using SiC in MEMS is that dry etching (to be described in Chapters 8 and 9) with aluminum masks can easily pattern the thin SiC film. The patterned SiC film can further be used as a passivation layer (protective layer) in micromachining for the underlying silicon substrate, as SiC can resist common etchants, such as KOH and HF.

Silicon carbide is a by-product in the process of producing single-crystal silicon boules as described in Section 7.4.2. As silicon exists in the raw materials of carbon (coal, coke, wood chips, etc.), intense heating of these materials in the electric arc furnace results in SiC sinking to the bottom of the crucible. Silicon carbide films can be produced by various deposition techniques. Pertinent thermophysical properties of SiC are given in Table 7.3.

7.5.3 Silicon Nitride

Silicon nitride (Si_3N_4) has many superior properties that are attractive for MEMS and microsystems. It provides an excellent barrier to diffusion of water and ions such as sodium. Its ultrastrong resistance to oxidation and many etchants makes it suitable in a mask for deep etchings. Applications of silicon nitride include optical waveguides and encapsulants to prevent diffusion of water and other toxic fluids into the substrate. It is also used in high-strength electric insulators and ion implantation masks.

Silicon nitride can be produced from silicon-containing gases and NH_3 according to the chemical reaction

$$3SiCl_2H_2 + 4NH_3 \rightarrow Si_3N_4 + 6HCl + 6H_2 \tag{7.5}$$

Selected properties of silicon nitride are listed in Table 7.6. Both chemical vapor deposition processes [*low-pressure chemical vapor deposition* (LPCVD) and *plasma-enhanced chemical vapor deposition* (PECVD)] given in Table 7.6 will be described in detail in Chapter 8. Additional material properties are given in Table 7.3.

TABLE 7.6. Selected Properties of Silicon Nitride

Properties	LPCVD[a]	PECVD[b]
Deposition temperature (°C)	700–800	250–350
Density (g/cm³)	2.9–3.2	2.4–2.8
Film quality	Excellent	Poor
Dielectric constant	6–7	6–9
Resistivity (Ω-cm)	10^{16}	10^6–10^{15}
Refractive index	2.01	1.8–2.5
Atom % H	4–8	20–25
Etch rate in concentrated HF (Å/min)	200	—
Etch rate in BHF (Å/min)	5–10	—
Poisson's ratio	0.27	—
Young's modulus (GPa)	385	—
Coefficient of thermal expansion, ppm/°C	1.6	—

[a]Low-pressure chemical vapor deposition.
[b]Plasma-enhanced chemical vapor deposition
Source: Madou, 1997.

7.5.4 Polycrystalline Silicon

Silicon in polycrystalline form can be deposited onto silicon substrates by chemical vapor deposition (CVD), as illustrated in Figure 7.12. It has become a principal material in surface micromachining, as will be described in Chapter 9.

The LPCVD process is frequently used for depositing polycrystalline silicon onto silicon substrates. The temperature involved in this process is about 600–650°C. Polysilicon (an abbreviation of polycrystalline silicon) is widely used in the IC industry, for examples, for resistors, gates for transistors, and thin-film transistors. Highly doped polysilicon (with arsenic and phosphorus for *n*-type or boron for *p*-type) can drastically reduce the resistivity of polysilicon to produce conductors and control switches. They are thus ideal

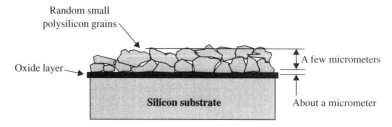

Figure 7.12. Polysilicon deposits on silicon substrate.

TABLE 7.7. Comparison of Mechanical Properties of Polysilicon and Other Materials

Materials	Young's Modulus (GPa)	Poisson's Ratio	Coefficient of Thermal (ppm/°C)
As substrates			
Silicon	190	0.23	2.6
Alumina	415	—	8.7
Silica	73	0.17	0.4
As thin films			
	160	*0.23*	*2.8*
Polysilicon	70	0.2	0.35
Thermal SiO$_2$	270	0.27	1.6
LPCVD SiO$_2$		—	2.3
PACVD SiO$_2$	70	0.35	25
Aluminum	410	0.28	4.3
Tungsten	3.2	0.42	20–70
Polymide			

Source: Madou, 1977.

materials for microresistors as well as easy ohmic contacts. Doped polysilicon wafers are preferred materials for solar photovoltaic cells with improved conversion efficiencies. A comparison of some key properties of polysilicon and other materials is presented in Table 7.7. Being a congregation of single-silicon crystals in random sizes and orientations, polysilicon can be treated as isotropic material in thermal and structural analyses.

7.6   SILICON PIEZORESISTORS

Piezoresistance is defined as the change in electrical resistance of solids when subjected to stress fields. Silicon piezoresistors are widely used in microsensors and actuators.

We have learned from Chapter 3 and Section 7.4.2 that doping boron to the silicon lattice produces *p*-type silicon crystal while doping arsenic or phosphorus results in *n*-type silicon. Both *p*- and *n*-type silicon exhibit excellent piezoelectric effect. Charles Smith in 1954 discovered the piezoresistance of *p*- or *n*-type silicon.

The fact that silicon crystal, whether it is *p*-type or *n*-type, is anisotropic has made the relationship between the change of resistance and the existent stress field more complex. This relationship is shown below:

$$\{\Delta \mathbf{R}\} = [\pi]\{\sigma\} \tag{7.6}$$

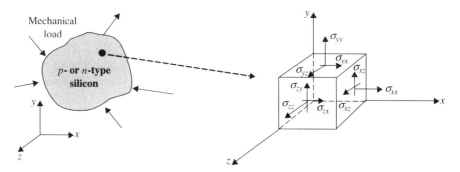

Figure 7.13. Silicon piezoresistance subjected to a stress field.

where the vector $\{\Delta \mathbf{R}\} = \{\Delta R_{xx} \quad \Delta R_{yy} \quad \Delta R_{zz} \quad \Delta R_{xy} \quad \Delta R_{xz} \quad \Delta R_{yz}\}^T$ represents the change of resistances in an infinitesimally small cubic piezoresistive crystal element with corresponding stress components $\{\boldsymbol{\sigma}\} = \{\sigma_{xx} \quad \sigma_{yy} \quad \sigma_{zz} \quad \sigma_{xy} \quad \sigma_{xz} \quad \sigma_{yz}\}^T$, as shown in Figure 7.13. Of the six independent stress components in the stress tensor $\{\boldsymbol{\sigma}\}$, there are three normal stress components, σ_{xx}, σ_{yy}, and σ_{zz}, and three shearing stress components, σ_{xy}, σ_{xz}, and σ_{yz}. The matrix $[\pi]$ in Equation (7.6) is called the *piezoresistive coefficient matrix* and has the form

$$[\pi] = \begin{bmatrix} \pi_{11} & \pi_{12} & \pi_{12} & 0 & 0 & 0 \\ \pi_{12} & \pi_{11} & \pi_{12} & 0 & 0 & 0 \\ \pi_{12} & \pi_{12} & \pi_{11} & 0 & 0 & 0 \\ 0 & 0 & 0 & \pi_{44} & 0 & 0 \\ 0 & 0 & 0 & 0 & \pi_{44} & 0 \\ 0 & 0 & 0 & 0 & 0 & \pi_{44} \end{bmatrix} \tag{7.7}$$

We notice from Equation (7.7) that only three coefficients, π_{11}, π_{12}, and π_{44}, appear in the matrix. Expanding the matrix equation (7.6) with the appropriate piezoresistive coefficients in Equation (7.7), we have the relations

$$\Delta R_{xx} = \pi_{11}\sigma_{xx} + \pi_{12}(\sigma_{yy} + \sigma_{zz})$$

$$\Delta R_{yy} = \pi_{11}\sigma_{yy} + \pi_{12}(\sigma_{xx} + \sigma_{zz})$$

$$\Delta R_{zz} = \pi_{11}\sigma_{11} + \pi_{12}(\sigma_{xx} + \sigma_{yy})$$

$$\Delta R_{xy} = \pi_{44}\sigma_{xy}$$

$$\Delta R_{xz} = \pi_{44}\sigma_{xz}$$

$$\Delta R_{yz} = \pi_{44}\sigma_{yz}$$

It is thus apparent that the coefficients π_{11} and π_{12} are associated with the normal stress components, whereas the coefficient π_{44} is related to the shearing stress components.

TABLE 7.8. Resistivities and Piezoresistive Coefficients of Silicon at Room Temperature in <100> Orientation

Materials	Resistivity (Ω-cm)	$\pi_{11}{}^a$	$\pi_{12}{}^a$	$\pi_{44}{}^a$
p-Silicon	7.8	+6.6	−1.1	+138.1
n-Silicon	11.7	−102.2	+53.4	−13.6

^{a}In 10^{-12} cm^2/dyne or 10^{-11} m^2/N (or Pa^{-1}).
Source: French and Evans, 1988.

The actual values of these three coefficients depend on the angles of the piezoresistor with respect to the silicon crystal lattice. The values of these coefficients at room temperature are given in Table 7.8.

Equation (7.6) and the situation illustrated in Figure 7.13, of course, represent general cases of piezoresistive crystals in a three-dimensional geometry. In almost all applications in MEMS and microsystems, silicon piezoresistors exist in the form of thin strips such as illustrated in Figure 7.14. In such cases, only the in-plane stresses in the x and y directions need to be accounted for.

We will realize from Table 7.8 that the maximum piezoresistive coefficient for the p-type silicon is $\pi_{44} = +138.1 \times 10^{-11}$ Pa^{-1}, and the maximum coefficient for the n-type silicon is $\pi_{11} = -102.2 \times 10^{-11}$ Pa^{-1}. Thus, many silicon piezoresistives are made of p-type material with boron as the dopant. Table 7.9 presents piezoresistive coefficients for p-type silicon piezoresistives made of various crystal planes (Bryzek et al., 1991).

The values of π_L denote the piezoresistive coefficient along the longitudinal direction, that is, along the $<x>$ direction in Figure 7.14, whereas π_T represents the piezoresistive coefficients in the tangential direction, that is, along the $<y>$ direction in the same figure.

The change of electric resistance in a silicon piezoresistance gage can thus be expressed as

$$\frac{\Delta R}{R} = \pi_L \sigma_L + \pi_T \sigma_T \tag{7.8}$$

where ΔR and R are the change of resistance and the original resistance of the silicon piezoresistive, respectively. The value of the original resistance R in Equation (7.8) can be obtained either by direct measurement or using the formula $R = \rho L/A$, where ρ is the

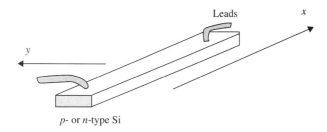

Figure 7.14. Silicon strain gages.

TABLE 7.9. Piezoresistive Coefficients of p-Type Silicon Piezoresistance in Various Directions

Crystal Planes	Orientation $<x>$	Orientation $<y>$	π_L	π_T
(100)	$<111>$	$<211>$	$+0.66\pi_{44}$	$-0.33\pi_{44}$
(100)	$<110>$	$<100>$	$+0.5\pi_{44}$	~ 0
(100)	$<110>$	$<110>$	$+0.5\pi_{44}$	$-0.5\pi_{44}$
(100)	$<100>$	$<100>$	$+0.02\pi_{44}$	$+0.02\pi_{44}$

Source: Bryzek et al., 1991.

resistivity of the piezoresistor, such as given in Figure 3.8, and L and A are the length and cross-sectional area, respectively, of the piezoresistor. The piezoresistive coefficients in both longitudinal and tangential directions are given in Table 7.9 for p-type silicon piezoresistors. The stress components in the longitudinal and tangential directions, σ_L and σ_T, are the stresses that cause the change of the resistance in the piezoresistor.

> *Example 7.3.* Estimate the change of resistance in piezoresistors attached to the diaphragm of a pressure sensor described in Example 4.4.
>
> *Solution:* Let us reillustrate the situation in Example 4.4 in Figure 7.15. There are four identical piezoresistors, A, B, C, and D, diffused in the locations at the top face of the silicon die, as shown in the figure. Resistors A and D are subjected predominantly to the transverse stress component σ_T that is normal to the horizontal edges, whereas resistors B and C are subjected to the longitudinal stress σ_L that is normal to the vertical edges.
>
> Because of the square plane geometry of the diaphragm that is subjected to uniform pressure loading at the top surface, the bending moments normal to all edges are equal in magnitude. Thus, $\sigma_L = \sigma_T = \sigma_{max} = 186.8\,\text{MPa} = 186.8 \times 10^6$ Pa (from Example 4.4).
>
> We assume that the diaphragm is on the (100) plane and both stresses are along the $<100>$ directions (why?); we thus obtain the piezoresistive coefficient $\pi_L = \pi_T = 0.02\pi_{44}$ from Table 7.9 for p-type piezoresistors. The value of the piezoresistive coefficient π_{44}, 138.1×10^{-11} Pa^{-1}, is available from Table 7.8.
>
> We can thus estimate the change of electric resistance in the piezoresistors to be
>
> $$\frac{\Delta R}{R} = \pi_L \sigma_L + \pi_T \sigma_T = 2 \times 0.02\pi_{44}\sigma_{max} = 2 \times 0.02(138.1 \times 10^{-11})(186.8 \times 10^6)$$
>
> $$= 0.01032\ \Omega/\Omega$$
>
> Determining the net change of resistance in the resistors requires knowledge of the length of the resistor at the stress-free state and the resistivity of the resistor material as given in Figure 3.8.

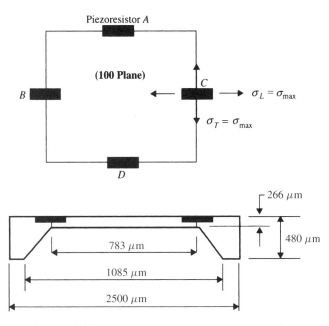

Figure 7.15. Piezoresistors in a pressure sensor die.

One major drawback of silicon piezoresistors is the strong temperature dependence of their piezoresistivity. The sensitivity of piezoresistors to the applied stress deteriorates rapidly with increases of temperature. Table 7.10 presents the variation of the piezoresistive coefficients with reference to those at room temperature.

Take, for example, a p-type silicon piezoresistor with a doping concentration of $10^{19}/cm^3$; the loss of piezoresistivity is 0.27% per degree Celsius. The same

TABLE 7.10. Temperature Dependence of Resistivity and Piezoresistivity of Silicon Piezoresistors

Doping Concentration $(10^18/cm^3)$	p-Type TCR[a] $(\%/^\circ C)$	p-Type TCP[b] $(\%/^\circ C)$	n-Type TCR $(\%/^\circ C)$	n-Type TCP $(\%/^\circ C)$
5	0.0	-0.27	0.01	-0.28
10	0.01	-0.27	0.05	-0.27
30	0.06	-0.18	0.09	-0.18
100	0.17	-0.16	0.19	-0.12

[a]TCR = Temperature coefficients of resistance.
[b]TCP = Temperature coefficient of piezoresistivity.
Source: French and Evans, 1988.

piezoresistor operating at $120°C$ would have lost $(120 - 20) \times 0.27\% = 27\%$ of the value of the piezoresistivity coefficient. Appropriate compensation for this loss must be considered in the design of a signal conditioning system.

The doping concentration for piezoresistives normally should be kept below $10^{19}/cm^3$ because the piezoresistive coefficients drop considerably above this dose, and reverse breakdown becomes an issue.

7.7 GALLIUM ARSENIDE

Gallium arsenide (GaAs) is a compound semiconductor made of equal numbers of gallium and arsenic atoms. Because it is a compound, it is more complicated in lattice structure, with atoms of both constituents, and hence is more difficult to process than silicon. However, GaAs is an excellent material for monolithic integration of electronic and photonic devices on a single substrate. The main reason for GaAs to be a prime candidate material for photonic devices is its high mobility of electrons in comparison to other semiconducting materials, as shown in Table 7.11.

As we see from the table, GaAs has about 7 times higher electron mobility than silicon. The high electron mobility in this material means it is easier for the electric current to flow in the material. The photoelectronic effect, as illustrated in Section 2.2.4, describes electric current flow in a photoelectric material when it is energized by incoming photons. Being a material with high mobility of electrons, GaAs can thus better facilitate electric current flow when it is energized by photon sources.

Gallium arsenide is also a superior thermal insulator, with excellent dimensional stability at high temperatures. The negative aspect of this material is its low yield strength, as indicated in Table 7.3. Its yield strength at 2700 MPa is only one-third that of silicon. This makes GaAs less attractive as a substrate in microsystems. Because of its relatively low use in the microelectronics industry, GaAs is much more expensive than silicon.

In addition to the differences in thermophysical properties indicated in Table 7.3, a good comparison of these two substrate materials in microsystems is given in Table 7.12.

TABLE 7.11. Electron Mobility of Selected Materials at 300 K

Materials	Electron Mobility, m^2/V-s
Aluminum	0.00435
Copper	0.00136
Silicon	0.145
Gallium arsenide, GaAs	*0.850*
Silicon oxide	≈ 0
Silicon nitride	≈ 0

Source: Kwok, 1997.

TABLE 7.12. Comparison of GaAs and Silicon in Micromachining

Properties	GaAs	Silicon
Optoelectronics	Very good	Not good
Piezoelectric effect	Yes	No
Piezoelectric coefficient (pN/°C)	2.6	Nil
Thermal conductivity	Relatively low	Relatively high
Cost	High	Low
Bonding to other substrates	Difficult	Relatively easy
Fracture	Brittle, fragile	Brittle, strong
Operating temperature	High	Low
Optimum operating temperature (°C)	460	300
Physical stability	Fair	Very good
Hardness (GPa)	7	10
Fracture strength (GPa)	2.7	6

Source: Madou, 1997.

7.8 QUARTZ

Quartz is a compound of SiO_2. The single-unit cell for quartz is in the shape of tetrahedron with three oxygen atoms at the apexes at the base and one silicon atom at the other apex of the tetrahedron, as shown in Figure 7.16. The axis that is normal to the base plane is called the *Z axis*. The quartz crystal structure is made up of rings with six silicon atoms.

Quartz is close to being an ideal material for sensors because of its near absolute thermal dimensional stability. It is used in many piezoelectric devices in the market, as will be described in Section 7.9. Commercial applications of quartz crystals include wristwatches, electronic filters, and resonators. Quartz is a desirable material in microfluidics applications in biomedical analyses. It is inexpensive and works well in electrophoretic fluid transportation, as described in Chapters 3 and 5, because of its excellent electric insulation properties. It is transparent to ultravoilet light, which often is used to detect the various species in the fluid.

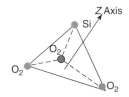

Figure 7.16. Cellular structure of quartz.

TABLE 7.13. Selected Properties of Quartz

Properties	Value‖Z	Value⊥Z	Temperature Dependency
Thermal conductivity, cal/cm/s/°C	29×10^{-3}	16×10^{-3}	↓ With T
Dielectric constant	4.6	4.5	↓ With T
Density, kg/m³	2.66×10^3	2.66×10^3	
Coefficient of thermal expansion, ppm/°C	7.1	13.2	↑ With T
Electrical resistivity, Ω/cm	0.1×10^{15}	20×10^{15}	↓ With T
Fracture strength, GPa	1.7	1.7	↓ With T
Hardness, GPa	12	12	

Source: Madou, 1997.

Quartz is a material that is hard to machine. Diamond cutting is a common method, although ultrasonic cutting has been used for more precise geometric trimming. It can be chemically etched into the desired shape using HF/NH$_4$F. Quartz wafers up to 75 mm in diameter by 100 μm thick are available commercially.

Quartz is even more dimensionally stable than silicon, especially at high temperatures. It offers more flexibility in geometry than silicon despite the difficulty in machining. Some key properties are presented in Table 7.13.

7.9 PIEZOELECTRIC CRYSTALS

One of the most commonly used nonsemiconducting materials in MEMS and microsystems is the piezoelectric crystal. Piezoelectric crystals are the solids of ceramic compounds that can produce a voltage when a mechanical force is applied between their faces. The reverse situation, the application of voltage to the crystal, can also change its shape. This conversion of mechanical energy to electronic signals (i.e., voltage) and vice versa is illustrated in Figure 7.17. This unique behavior is called the *piezoelectric effect*. Jacques and Pierre Curie discovered the piezoelectric effect in 1880. This effect exists in a number of natural crystals, such as quartz, tourmaline, and sodium potassium tartrate, and quartz has been used as an electromechanical transducer for many years. There are many other synthesized crystals, such as Rochelle salt (N$_a$KC$_4$H$_4$O$_6$–4H$_2$O), barium titanate (BaTiO$_3$), and lead zirconate titanate (PZT).

For a crystal to exhibit the piezoelectric effect, its structure should have no center of symmetry. A mechanical load applied to such a crystal will alter the separation between the positive and negative charge sites in each elementary cell, leading to a net polarization at the crystal surface (Waanders, 1991). An electric field with voltage potential is thus created in the crystal because of such polarization.

The most common use of the piezoelectric effect is for high-voltage generation through the application of high compressive forces. The generated high-voltage field can be used

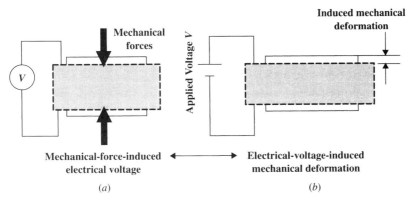

Figure 7.17. Conversion of mechanical and electrical energies using piezoelectric crystals: (*a*) from mechanical to electrical; (*b*) from electrical to mechanical. (From Kasap, 1997.)

as an impact detonation device. It can also be used to send signals for depth detection in a sonar system. The principal applications of the piezoelectric effect in MEMS and microsystems, however, are in actuations, as described in Chapter 2, and dynamic signal transducers for pressure sensors and accelerometers. The piezoelectric effect was also used in pumping mechanisms, as described in Chapters 5 and 6, for microfluid flows, as well as for inkjet printer heads.

The effectiveness of the conversion of mechanical to electrical energies and vice versa can be assessed by the electromechanical conversion factor K defined as follows (Kasap, 1997):

$$K^2 = \frac{\text{output of mechanical energy}}{\text{input of electrical energy}} \tag{7.9a}$$

or

$$K^2 = \frac{\text{output of electrical energy}}{\text{input of mechanical energy}} \tag{7.9b}$$

The following simple mathematical relationships between the electromechanical effects can be used in the design of piezoelectric transducers in a unidirectional loading situation (Askeland, 1994):

1. The electric field produced by stress:

$$V = f\sigma \tag{7.10}$$

where V is the generated electric field in volts per meter and σ, in pascals, is the stress in the piezoelectric crystal induced by the applied mechanical load. The coefficient f is a constant.

TABLE 7.14. Piezoelectric Coefficients of Selected Materials

Piezoelectric Crystals	Coefficient, d $(10^{-12}$ m/V)	Electromechanical Conversion Factor, K
Quartz (crystal SiO_2)	2.3	0.1
Barium titanate ($BaTiO_3$)	100–190	0.49
Lead zirconate titanate, ($PbTi_{1-x}Zr_xO_3$)	480	0.72
$PbZrTiO_6$	250	
$PbNb_2O6$	80	
Rochelle salt ($NaKC_4H_4O6\text{-}4H_2O$)	350	0.78
Polyvinylidene fluoride, PVDF	18	

Sources: Kasap, 1997; Askeland, 1994.

2. The mechanical strain produced by the electric field:

$$\varepsilon = Vd \tag{7.11}$$

where ε is the induced strain and V is the applied electric field in volts per meter. The piezoelectric coefficient d for common piezoelectric crystals is given in Table 7.14.

The coefficients f and d in Equations (7.10) and (7.11) have the relationship

$$\frac{1}{fd} = E \tag{7.12}$$

where E is Young's modulus of the piezoelectric crystal.

Two examples are given below to illustrate the application of piezoelectric crystals as either a signal transducer or an actuator in microdevices.

Example 7.4. A thin piezoelectric crystal film of PZT is used to transduce the signal in a microaccelerometer with a cantilever beam made of silicon as described in Example 4.8. The accelerometer is designed for maximum acceleration/deceleration of $10g$. The PZT transducer is located at the support base of the cantilever beam where the maximum strain exists during the bending of the beam, as illustrated in Figure 7.18. Determine the electrical voltage output from the PZT film at the maximum acceleration/deceleration of $10g$. The mass of the cantilever beam is negligible.

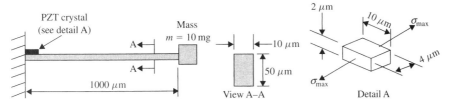

Figure 7.18. Piezoelectric transducer in a beam accelerometer.

Solution: The solution can be obtained by determining the maximum bending stress and thus the maximum bending strain in the beam due to the dynamic load of the attached mass accelerated to $10g$. The maximum bending strain in the beam is assumed to give the same magnitude of strain to the attached thin PZT film. The voltage generated in the PZT can be computed from Equation (7.11).

We will first determine the equivalent bending load P_{eq} that is equivalent to that of the 10-mg mass accelerated or decelerated to $10g$:

$$P_{eq} = ma = (10 \times 10^{-6}) \times (10 \times 9.81) = 981 \times 10^{-6} \text{ N}$$

The beam accelerometer is equivalent to a statically loaded cantilever beam subjected to the equivalent force acting at its free end, as illustrated in Figure 7.19.

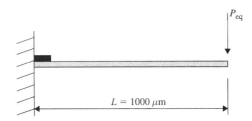

Figure 7.19. Equivalent static bending of a cantilever beam in an accelerometer.

The maximum bending moment $M_{max} = P_{eq}L = (981 \times 10^{-6})(1000 \times 10^{-6}) = 0.981 \times 10^{-6}$ N-m.

We will need the moment of inertia I of the beam cross section to calculate the equivalent maximum bending stress σ_{max}. The value of I was computed in Example 4.8 as 0.1042×10^{-18} m^4. We thus have the maximum bending stress at the support as

$$\sigma_{max} = \frac{M_{max}C}{I} = \frac{(0.981 \times 10^{-6})(25 \times 10^{-6})}{0.1042 \times 10^{-18}} = 235.36 \times 10^6 \text{ Pa}$$

where C is the half-depth of the beam cross section.

The associate maximum bending strain $\varepsilon_{\max}$ in the beam is determined as

$$\varepsilon_{\max} = \frac{\sigma_{\max}}{E} = \frac{235.36 \times 10^6}{1.9 \times 10^{11}} = 123.87 \times 10^{-5} \text{ m/m}$$

We used Young's modulus $E = 1.9 \times 10^{11}$ Pa as given in Table 7.3 for the silicon beam in the above calculation.

We assumed that the strain in the beam will result in the same strain in the attached PZT film. Thus with a strain of 123.87×10^{-5} m/m in the PZT, the induced voltage per meter in the crystal is

$$V = \frac{\varepsilon}{d} = \frac{\varepsilon_{\max}}{d} = \frac{123.87 \times 10^{-5}}{480 \times 10^{-12}} = 0.258 \times 10^7 \text{ V/m}$$

We used the piezoelectric coefficient $d = 480 \times 10^{-12}$ m/V obtained from Table 7.14 in the above computation.

Since the actual length l of the PZT crystal attached to the beam is 4 μm, we will expect the total voltage generated by the transducer at $10g$ load to be

$$v = V\ell = (0.258 \times 10^7)(4 \times 10^{-6}) = 10.32 \text{ V}$$

Example 7.5. Determine the electric voltage required to eject a droplet of ink from an inkjet printer head with a PZT piezoelectric crystal as a pumping mechanism. The ejected ink will have a resolution of 300 dots per inch (dpi). The ink droplet is assumed to produce a dot with a film thickness of 1 μm on the paper. The geometry and dimensions of the printer head are illustrated in Figure 7.20. Assume that the ink droplet takes the shape of a sphere and the inkwell is always refilled after each ejection.

Solution: We first have to determine the diameter of the ejection nozzle, d, corresponding to ink dots with 300 dpi resolution on the paper. We will first determine the diameter of the dot, D, on the paper. With 300 such dots in a linear space of 1 in., we can readily find the corresponding diameter to be $D = 1$ in./300 = 25.4 mm/300 = 0.084666 mm, or 84.66 μm.

Since the dots are produced from spherical droplets of diameter d, by letting the volume of the ink droplet from the nozzle be equal to that of the ink dot on the paper, the following relation can be used:

$$\tfrac{4}{3}\pi r^3 = \left(\tfrac{1}{4}\pi D^2\right)(t)$$

where r is the radius of the spherical ink droplet and t is the thickness of the ink dot on the paper. We thus find the radius of the spherical ink droplet to be $r = 11.04 \times 10^{-6}$ m, with $D = 84.66$ μm and $t = 1$ μm.

Next, we assume that the volume of an ink droplet leaving the inkwell is equivalent to the volume created by vertical expansion of the PZT piezoelectric cover at

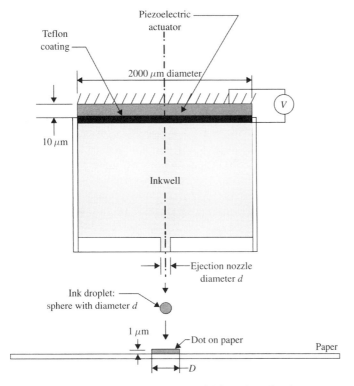

Figure 7.20. Schematic of an inkjet printer head.

the back of the inkwell. Let the vertical expansion of the piezoelectric cover be W and the corresponding volume of the displaced ink be $(\pi/4)\Delta^2 W$, where Δ is the diameter of the piezoelectric cover at 2000 μm.

By equating the above displaced ink volume and the volume of the ink dot, V_{dot}, we can compute the required piezoelectric expansion W to be

$$W = \frac{4V_{\text{dot}}}{\pi\Delta^2} = \frac{4 \times 5629.21 \times 10^{-18}}{3.1416(2000 \times 10^{-6})^2} = 1791.83 \times 10^{-12} \text{ m}$$

The corresponding strain in the piezoelectric cover is given as

$$\varepsilon = \frac{W}{L} = \frac{1791.83 \times 10^{-12}}{10 \times 10^{-6}} = 179.183 \times 10^{-6} \text{ m/m}$$

From Equation (7.11) and the piezoelectric coefficient of PZT crystals from Table 7.14, we have the required applied voltage per meter:

$$V = \frac{\varepsilon}{d} = \frac{179.183 \times 10^{-6}}{480 \times 10^{-12}} = 0.3733 \times 10^6 \text{ V/m}$$

Since the cover has a thickness of 10 μm, the required applied voltage is

$$v = LV = (10 \times 10^{-6})(0.3733 \times 10^6) = 3.733 \text{ V}$$

7.10 POLYMERS

Polymers, which include such diverse materials as plastics, adhesives, Plexiglas, and Lucite, have become increasingly popular materials for MEMS and microsystems. For example, plastic cards approximately 150 mm wide containing over 1000 microchannels have been adopted in microfluidic electrophoretic systems by the biomedical industry (Lipman, 1999), as described in Chapter 2. Epoxy resins and adhesives such as silicone rubber are customarily used in MEMS and microsystems packaging.

This type of material is made up of a long chain of organic (mainly hydrocarbon) molecules. The combined molecules (i.e., polymer molecules) can be a few hundred nanometers long. Low mechanical strength, low melting points, and poor electrical conductivity characterize polymers. Thermoplastics and thermosets are two groups of polymers that are commonly used for industrial products. Thermoplastics can be easily formed to desirable shapes for the specific products, whereas thermosets have better mechanical strength and temperature resistance up to 350°C. Because of the rapid increase of applications in industrial products, polymers and polymerization, the process of producing various kinds of polymers, constitute a distinct engineering subject. It is not realistic to offer a complete list of available polymers and plastics as well as the many polymerization processes in this chapter. What will be presented in this section is information on the application of polymers that is relevant to the design and packaging of MEMS and microsystems. Much of the material presented here is available in greater detail in a special reference on polymers for electronics and optoelectronics (Chilton and Goosey, 1995).

7.10.1 Polymers as Industrial Materials

Traditionally, polymers have been used as insulators, sheathing, capacitor films in electric devices, and die pads in ICs. A special form of polymer, plastic, has been widely used for machine and device components. Following is a summary of the many advantages of polymers as industrial materials:

Light weight
Ease in processing
Low cost of raw materials and processes for producing polymers
High corrosion resistance
High electric resistance
High flexibility in structures
High dimensional stability

Perhaps the most intriguing fact about polymers is their variety of molecular structures. This unique feature has offered scientists and engineers great flexibility in developing "organic alloys" by mixing various ingredients to produce polymers for specific applications. Consequently, there are a great variety of polymers available for industrial applications in today's marketplace.

7.10.2 Polymers for MEMS and Microsystems

Polymers have become increasingly important materials for MEMS and microsystems. Some applications are as follows:

1. Photoresist polymers are used to produce masks for creating desired patterns on substrates by photolithography, as described in Chapter 8.

2. The same photoresist polymers are used to produce the prime mold with the desired geometry of MEMS components in the LIGA process for manufacturing microdevice components, as described in Chapter 9. These prime molds are plated with metals such as nickel for subsequent injection molding for mass production of microcomponents.

3. As will be described later in this section, conductive polymers are used as organic substrates for MEMS and microsystems.

4. Ferroelectric polymers, which behave like piezoelectric crystals, can be used as a source of actuation in microdevices, such as those for micropumping, as described in Section 5.6.3.

5. Thin Langmuir–Blodgett (LB) films can be used to produce multilayer microstructures, similar to the micromachining technique presented in Chapter 9.

6. Polymers with unique characteristics are used as a coating substance for capillary tubes to facilitate electro-osmotic flow in microfluidics, as described in Section 3.8.2.

7. Thin polymer films are used as electric insulators in microdevices and as a dielectric substance in microcapacitors.

8. Polymers are widely used for electromagnetic interference (EMI) and radio frequency interference (RFI) shielding in microsystems.

9. Polymers are ideal materials for encapsulation of microsensors and packaging of other microsystems.

10. Special polymers, such as SU-8 negative photoresists, can be used to create permanent microstructures with high aspect ratios, as described in Section 7.10.5.

7.10.3 Conductive Polymers

For polymers to be used in certain applications in MEMS and microsystems, they have to be made electrically conductive with superior dimensional stability. Polymers have been used extensively in the packaging of MEMS, but they have also been used as substrate for some MEMS components in recent years with the successful development of techniques for controlling the electric conductivity of these materials.

By nature, polymers are poor electric conductors. Table 7.15 shows the electric conductivity of various materials. One will readily see that polymers, represented by polyethelene, have the lowest electric conductance of all the listed materials.

TABLE 7.15. Electric Conductivity of Selected Materials

Materials	Electric Conductivity, S/m[a]
Conductors	
Copper	10^6–10^8
Carbon	10^4
Semiconductors	
Germanium	10^0
Silicon	10^{-4}–10^{-2}
Insulators	
Glass	10^{-10}–10^{-8}
Nylon	10^{-14}–10^{-12}
SiO_2	10^{-16}–10^{-14}
Polyethylene	10^{-16}–10^{-14}

[a] S/m = siemens per meter; siemens = Ω^{-1} = m^{-2} kg^{-1} s^3 A^2.

Polymers can be made electrically conductive by the following three methods:

1. **Pyrolysis:** A pyropolymer based on phthalonitrile resin can be made electrically conductive by adding an amine heated above $600°C$. The conductivity of the polymer produced by this process can be as high as 2.7×10^4 S/m, which is slightly better than that of carbon.

2. **Doping:** Doping with the introduction of inherently conductive polymer structure, such as by incorporating a transition metal atom into the polymer backbone, can result in electrically conductive polymers. Doping of polymers depends on the dopants and the individual polymer. Following are examples of dopants used in producing electrically conductive polymers:

   For polyacetylenes (PAs): Dopants such as Br_2, I_2 AsF_5, $HClO_4$, and H_2SO_4 are used to produce the p-type polymer, and sodium naphthalide in tetrahydrofuran (THF) is used for the n-type polymers.

   For polyparaphenylenes (PPPs): AsF_5 is used for the p-type and alkali metals are used for the n-type polymers.

   For polyphenylene sulfide (PPS): The dopant used is AsF_5.

3. **Insertion of Conductive Fibers:** Incorporating conductive fillers into both the thermosetting and thermoplastic of polymer structures can result in electrically conductive polymer. Fillers include such materials as carbon, aluminum flakes, stainless steel, gold and silver fibers. Other inserts include semiconducting fibers (e.g., silicon and germanium). Fibers are on the order of nanometers in length.

7.10.4 Langmuir–Blodgett Film

A special process developed by Langmuir as early as 1917 and refined by Blodgett 15 years later can be used to produce thin polymer films at the molecular scale. This process is generally known as the *LB process*. It involves the spreading of volatile solvent over surface-active materials. The LB process can produce more than one single monolayer by depositing films of various compositions onto a substrate to create a multilayer structure. The process closely resembles that of the micromachining manufacturing technique presented in Chapter 9. It is thus an alternative micromanufacturing technique.

The LB films are good candidate materials for exhibiting ferro- (magnetic), pyro- (heat-generating), and piezoelectric properties. They may also be produced with controlled optical properties such as refractive index and antireflectivity. They are thus ideal materials for microsensors and optoelectronic devices. Following are a few examples of LB film applications in microsystems:

1. **Ferroelectric Polymer Thin Films:** Useful in particular is polyvinylidene fluoride (PVDF). Applications of this type of film include sound transducers in air and water, tactile sensors, biomedical applications such as tissue compatibility implants, cardiopulmonary sensors, and implantable transducers and sensors for prosthetics and rehabilitation devices. The piezoelectric coefficient of PVDF is given in Table 7.14.

2. **Coating Materials with Controllable Optical Properties:** These are widely used in broadband optical fibers, which can transmit laser light at different wavelengths.

3. **Microsensors:** The sensitivity of many electrically conducting polymeric materials to gases and other environmental conditions makes this material suitable for microsensors. Its ability to detect specific substances relies on the reversible and specific absorption of species of interest on the surface of the polymer layer and the subsequent measurable change of conductivity of the polymer. Figure 7.21 shows a schematic of such a sensor. These sensors work on the principle that the electric conductivity of the polymer sensing element will change when it is exposed to a specific gas. This principle is similar to that of chemical sensors, as described in Chapter 2.

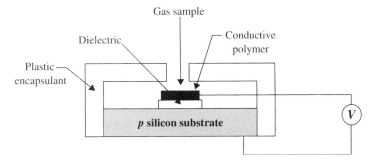

Figure 7.21. Microsensor using polymers.

Conductive polymers are also used in electrochemical biosensors, such as microamperometric glucose and galactose biosensors.

7.10.5 SU-8 Photoresists

SU-8 is a negative epoxy-based polymer photoresist used for thin-film production with thickness ranging from 1 μm to 2 mm (Conradie and Moore, 2002). This material was originally developed by IBM-Watson Research Center and patented in 1989. It has been widely used in microfabrication and for MEMS structures since 1996. Being a photoresist and sensitive to UV light at wavelengths in the range of 350–400 nm, SU-8 can be used to produce the masks for photolithography for micropatterning, as described in Chapter 8, for electroplating in a LIGA micromanufacturing process described in Chapter 9, or for the permanent structure of a microdevice. It is chemically stable and has good mechanical and optical properties. Microscale components such as gears, channels, and valves in microfluidic networks and optical waveguides have been produced using this material. Detailed description of the design and performance of a micro–check valve in a microchannel and a microgripper made of SU-8 polymer is available in Seidemann et al. (2002).

The popularity of SU-8 polymer for MEMS and microsystems application is primarily because of its two unique advantages: (1) the thick functional SU-8 films can be used to create microstructures with aspect ratio as high as 50 (Wright et al., 2004) and (2) low production costs of the material and production of microstructures over silicon-based microdevice components. A comprehensive source of reference for SU-8 polymer is available on the Web (Guerin, 2005). Mechanical properties of SU-8 are presented in Table 7.16.

SU-8 is commercially available in liquid form. Micro Chem Corporation is a major supplier of this material (Micro Chem, 2007). Its viscosities vary with the composition of the solid SU-8 resin and the solvent, as shown in Table 7.16. Thin films of SU-8 polymer can be produced by a spin-coating process, as described in Section 8.2.2. Figure 7.22 shows the relationship between the thickness of thin films produced by two commercial-grade SU-8 polymers and the spin speed.

Processes for producing SU-8 vary with supplier's specifications and by application. Figure 7.23 presents a typical process flow as offered by a major supplier of SU-8 photoresist (Micro Chem, 2007).

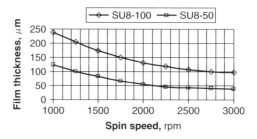

Figure 7.22. Spin speed versus selected SU-8 film thickness. (*Source:* Micro Chem, 2007.)

TABLE 7.16. Mechanical Properties of SU-8 Polymer

Young's modulus	4400 MPa
Poisson's ratio	0.22
Viscosity	0.06 Pa-s (40% SU-8–60% solvent)
	1.50 Pa-s (60% SU-8–40% solvent)
	15.0 Pa-s (70% SU-8–30% solvent)
Coefficient of thermal expansion[a]	0.183 ppm/°C
Thermal conductivity	0.073 W/cm-°C
Glass transition temperature	200°C
Reflective index	1.8 at 100 GHz
	1.7 at 1.6 THz
Absorption coefficient	2/cm at 100 GHZ
	40/cm a 1.6 THz
Relative dielectric constant	3 at 10 MHz

[a]In comparison to 2.33 ppm/°C for silicon.
Source: Guerin, 2005.

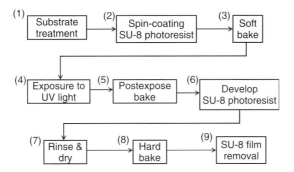

Figure 7.23. Typical process flow for constructing SU-8 films. (*Source:* Micro Chem, 2007.)

With respect to the process flow chart in Figure 7.23, the substrate, such as a silicon wafer, should be dry and clean prior to being coated with SU-8 photoresist (step 1). A thorough cleaning can be accomplished by having the surface subject to, for example, piranha etches that contain 75% H_2SO_4 and 25% H_2O_2. The wafer, with a clean surface, is baked on a flat hot plate at 200°C for 5 min. The dry and clean wafer is then secured on top of a spinning table, as illustrated in Figure 8.3. Approximately 1 mL per inch of substrate diameter of liquid SU-8 is dispensed onto the center of the wafer followed by a preset spinning speed according to the desired film thickness, such as illustrated in Figure 7.22. Spinning usually takes about 30 s. The wafer, with a thin coat of SU-8, is then subjected to a soft bake (step 3) to drive out the inherent solvent in the coating. This can be done in a well-vented oven. Exposure of the photoresist in step 4 is carried out by placing the wafer with a thin SU-8 coating under a UV lamp with a wavelength in the range of 350–400 nm. The required exposure time relates to the exposure energy by the specific SU-8 supplied by the manufacturer. Conditions for postbake of the exposed photoresist in step 5 are normally specified by the suppliers. The exposed photoresist is now ready for development, as specified in step 6. Solvent for developing SU-8 may include ethyl lactate and diacetone alcohol or as specified by the supplier. The equipment shown in Figure 8.3 can be used for spraying the development solvent in this process. This development solvent will remove the parts of the SU-8 film that were not exposed to UV light and thereby creates the desired pattern of the film over the substrate. After the development, the substrate should be rinsed briefly with chemicals such as isopropyl alcohol and then dried by gentle streams of hot air or inert gas, as in step 7. The developed SU-8 film is subjected to hard baking for a brief period at moderate temperature ranging from 150 to 200°C on a hot plate with well vent of dry air. Lift-off of the patterned SU-8 films from the substrates in step 9 is a major effort in many cases. Depending on the film thickness and the crosslink density of the materials, there are several ways that can be used to remove the film from the substrate: immersing the substrate in a special liquid bath containing oxidizing acid solutions, reactive ion etching as described in Chapter 8, laser ablation, and pyrolosis. Many of these techniques are recommended by suppliers in data sheets included with the products they sell.

7.11 PACKAGING MATERIALS

We learned about the evolution of micromachining technology from the IC industry in Chapter 1. Consequently, many techniques that were developed for IC packaging are now used for microsystems packaging. However, a major distinction between the two types of packaging is that in microelectronics packaging it is usually sufficient to protect the IC die and the interconnects from the often hostile operating environment. In microsystems packaging, however, not only do the sensing or actuating elements need to be protected, but also they should be in contact with the media that are the sources of actions. Many of these media are hostile to these elements. Special technologies that have been developed to package delicate aerospace device components are also used frequently in the packaging of microsystems. Technical issues and technologies relating to microsystems packaging will be presented in Chapter 11.

Materials used in microsystem packaging include those used in IC packaging—wires made of noble metals at the silicon die level, metal layers for lead wires, solders for die/constraint base attachments, and so on. Microsystems packaging also includes metals and plastics. Figure 7.24 shows the schematic of a typical pressure sensor. At the die level, we use aluminum or gold metal films serving as the ohmic contacts to the piezoresistors that are diffused in the silicon diaphragm. Similar materials are used for the lead wires to the interconnects outside the casing. The casing can be made of plastic or stainless steel, depending on the severity of the operating environment. Glass such as Pyrex or ceramics such as alumina are often used as constraint bases. The adhesives that enable the silicon

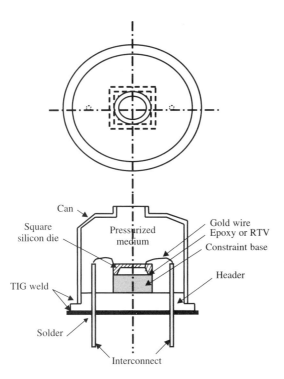

Figure 7.24. Typical packaged micropressure sensor.

die to be attached to the constraint base can be tin–lead solder alloys or epoxy resins or adhesives such as room temperature vulcanizing (RTV) silicone rubber. In the case of soldered attachments, thin metal layers need to be sputtered at the joints to facilitate the soldering. Copper and aluminum are good candidate materials for this purpose. We will learn in Chapter 11 that these adhesive materials for die/base attachments are very important in isolating the die from undesired interference from other components in the package. When part of a microsystem package, such as a silicon diaphragm, needs to be in contact with the hostile medium, silicone gel or silicone oil is used to shield the part.

PROBLEMS

Part 1 Multiple Choice

1. A substrate is (a) a sublayer in MEMS, (b) a flat microscopic object, (c) a flat macroscopic object in microelectronics.

2. A semiconducting material can be made to become an electrically conducting material by (a) applying high electric voltage, (b) applying strong electric current, (c) introducing the right kind of foreign atoms into the semiconducting material.

3. Silicon has a Young's modulus similar to that of (a) aluminum, (b) stainless steel, (c) copper.

4. Silicon has a mass density similar to that of (a) aluminum, (b) stainless steel, (c) copper.

5. The principal reason why silicon is an ideal material for MEMS is (a) its dimensional stability over a wide range of temperatures, (b) it is light and strong, (c) it is readily available.

6. Silicon has a coefficient of thermal expansion (a) higher than, (b) lower than, (c) about the same as that of silicon dioxide.

7. The 300-mm wafers offer (a) 2, (b) 2.25, (c) 2.5 times more area for substrates than that by 200-mm wafers.

8. The length of the lattice of a silicon crystal is (a) 0.543, (b) 0.643, (c) 0.743 nanometer.

9. Miller's indices are used to designate (a) the length, (b) the plane, (c) the volume of a face-centered-cubic crystal.

10. The (100) plane in a silicon crystal consists of (a) five, (b) eight, (c) six atoms.

11. The (110) plane in a silicon crystal consists of (a) five, (b) eight, (c) six atoms.

12. The (111) plane in a silicon crystal consists of (a) five, (b) eight, (c) six atoms.

13. The growth of silicon crystals is slowest in the (a) <100>, (b) <110>, (c) <111> direction.

14. Silicon conducts heat (a) 50, (b) 150, (c) 200 times faster than silicon oxide.

15. Silicon carbide films are used to protect (a) the underlying substrates, (b) the integrated circuits, (c) the electric interconnects in a microsystem.

16. Silicon nitride is (a) tougher than, (b) weaker than, (c) about the same as silicon in strength.

17. Pure and single-crystal silicon (a) exists in nature, (b) is grown by special processes, (c) is made by electrolysis.

18. Wafers used in MEMS and microelectronics (a) are the products of a single-crystal silicon boule, (b) are synthesized from silicon compounds, (c) exist in nature.

19. MEMS design engineers are advised to adopt (a) any size, (b) a custom-specified size, (c) an industrial standard size of wafer.

20. The total number of atoms in a single stand-alone silicon crystal is (a) 18, (b) 16, (c) 14.

21. The toughest plane for processing in a single-silicon crystal is (a) the (100) plane, (b) the (110) plane, (c) the (111) plane.

22. The 54.74° slope in the cavity of a silicon die for a pressure sensor is (a) designed by choice, (b) a result of the crystal's resistance to etching in the (111) plane, (c) a result of the crystal's resistance to etching in the (110) plane.

23. Polysilicon is popular because it can easily be made as a (a) semiconductor, (b) insulator, (c) electric conductor.

24. Polysilicon films are used in microsystems as (a) dielectric material, (b) substrate material, (c) electrically conducting material.

25. The electrical resistance of silicon piezoresistors varies in (a) all directions, (b) in only the preferred directions, (c) neither of the above applies.

26. It is customary to relate silicon piezoresistance change to (a) deformations, (b) strains, (c) stresses induced in the piezoresistors in MEMS and microsystems.

27. There are (a) three, (b) four, (c) six piezoresistive coefficients in silicon piezoresistors.

28. The single most serious disadvantage of using silicon piezoresistor is (a) the high cost of producing such resistors, (b) its strong sensitivity to signal transduction, (c) its strong sensitivity to temperature.

29. Gallium arsenide has (a) 6, (b) 7, (c) 8 times higher electron mobility than silicon.

30. Gallium arsenide is chosen over silicon for use in micro-optical devices because of its (a) optical reflectivity, (b) dimensional stability, (c) high electron mobility.

31. Gallium arsenate is not as popular as silicon in MEMS applications because of (a) its higher cost in production, (b) difficulty of mechanical work, (c) low mechanical strength.

32. Quartz crystals have the shape of (a) a cube, (b) a tetrahedron, (c) a body-centered cube.

33. It is customary to relate the voltage produced by a piezoelectric crystal to the (a) deformations, (b) temperature, (c) stresses induced in the crystal.

34. Application of mechanical deformation to a piezoelectric crystal can result in the production of (a) electric resistance change, (b) electric current change, (c) electric voltage change in the crystal.

35. Most piezoelectric crystals (a) exist in nature, (b) are made by synthetic processes, (c) are made by doping the substrate.

36. A polymer is a material that is made up of many (a) small-size, (b) large-size, (c) long-chain molecules.

37. In general, polymers are (a) electric conductive, (b) semi–electrically conductive, (c) insulating.

38. Polymers (a) can, (b) cannot, (c) may never be made electrically conductive.

39. The LB process is used to produce (a) thin films, (b) dies, (c) piezoelectric polymers in MEMS and microsystems.

40. SU-8 is a (a) negative-type, (b) positive-type, (c) neutral-type photoresistive polymer.

41. SU-8 is widely used in MEMS and microsystems because it (a) is cheap in production, (b) is resistant to high temperature, (c) can produce high-aspect-ratio microstructures.

42. SU-8 is sensitive to light from (a) UV source, (b) mercury lamp, (c) any white light source.

43. The most challenging part in producing SU-8 microstructures is (a) to apply thick films on the substrate, (b) to develop the exposed films, (c) to remove the developed films from the substrate.

44. In developing the exposed SU-8 films, the parts are (a) not exposed to, (b) exposed to, (c) not penetrated by the lights dissolved in the development solvent.

45. MEMS and microsystems packaging materials are (a) restricted to microelectronics packaging materials, (b) just about all engineering materials, (c) semiconducting materials.

Part 2 Computational Problems

1. Estimate the maximum number silicon dies of size 2 mm wide × 4 mm long that can be accommodated in the four standard-sized wafers given in Section 7.4.2. The dies are laid out in parallel on the wafer with 0.25-μm gaps between dies. Make your observation on the results.

2. Prove that the 300-mm wafers have 2.25 times larger area for substrates than 200-mm wafers.

3. Estimate the number of atoms per cubic millimeter and cubic micrometer of pure silicon.

4. Find the change of electric resistance if a piezoresistor made of p-type silicon replaces the piezoelectric crystal in Example 7.4. The piezoresistor has a length of 4 μm. Assume that the piezoresistor of (100) plane in the <100> direction is used in this case, and the resistivity of the material is 7.8 Ω-cm as from Table 7.8.

5. What would be the voltage output of the piezoelectric transducer in Example 7.4 if PVDF polymer films were used instead?

6. Determine the length of lattices that bond the atoms in the three principal planes of a silicon crystal shown in Figure 7.8.

7. Determine the spacing between atoms in the three planes in a single-silicon crystal as illustrated in Figure 7.10.

8. Determine the angle between the orientation <100> to the (111) plane in a single-silicon crystal cell.

9. For homogeneous and isotropic solids, the relation between the three elastic constants is $E = 2(1 + v)G$, where E is the Young's modulus, v is Poisson's ratio, and G is the shear modulus of elasticity. Use this relationship to calculate the Poisson's ratio of silicon crystal in three directions in Table 7.2 using the tabulated values of E and G.

10. Determine the electric voltage required to pump a droplet of ink from the well in Example 7.5 with a resolution of 600 dpi. All other conditions in Example 7.5 remain unchanged.

CHAPTER 8

MICROSYSTEMS FABRICATION PROCESSES

8.1 INTRODUCTION

We have mentioned throughout this book that MEMS technology has evolved from microelectronics fabrication technology. Consequently, many of the fabrication techniques used in producing ICs have been adopted to create the complex three-dimensional shapes of many MEMS and microsystems. As we will learn, the microfabrication techniques presented in this chapter are radically different from those used in producing traditional machines and devices. Furthermore, almost all microfabrication techniques or processes involve physical and chemical treatment of materials whose effects are relatively unknown to many engineers with traditional engineering background. It is thus necessary for engineers who are involved in the design of these products to acquire good knowledge in solid-state physics and the associated microfabrication techniques. Such knowledge and experience not only are required for the analyses in the design process but also are necessary to ensure the manufacturability of the products designed.

This chapter will provide the reader with an overview of various fabrication techniques involved in MEMS and microsystems manufacturing; each technique will be treated separately and will be integrated in micromanufacturing processes as described in Chapter 9.

8.2 PHOTOLITHOGRAPHY

Among many distinct characteristics and features of microelectronics and microsystems as presented in Table 1.1, a fundamental difference between these two advanced technologies is that microsystems almost always involve complex three-dimensional structural geometry at the microscale. Patterning of geometry with extremely high precision at this scale thus becomes a major challenge to engineers. *Photolithography* or *microlithography* appears to be the only viable way for producing high-precision patterning on substrates at the microscale at the present time.

8.2.1 Overview

The word *lithography* is a derivation from two Greek words: *litho* (stone) and *graphein* (to write) (Madou, 1997). According to Ruska (1987), the photolithography process involves the use of an optical image and a photosensitive film to produce a pattern on a substrate. Photolithography is one of the most important steps in microfabrication. It appears to be the only reliable technique that is available at present to create patterns on substrates with submicrometer resolutions. In microelectronics, these patterns are necessary for the *p–n* junctions, diodes, capacitors, and so on, for ICs. In microsystems, however, photolithography is used to set patterns for masks for cavity etching in bulk micromanufacturing or for thin-film deposition and etching of sacrificial layers in surface micromachining as well as for the primary circuitry of electrical signal transduction in sensors and actuators.

The procedure of photolithography is outlined in Figure 8.1. Let us begin with the substrate at the top left of the figure. This substrate can be a silicon wafer, as in microelectronics, or other substrate materials, such as silicon dioxide or silicon nitride in MEMS or microsystems. A photoresist is first coated onto the flat surface of the substrate. The substrate with photoresist is then exposed to a set of lights through a transparent mask with the desired opaque patterns. Masks used for this purpose are often made of quartz. Patterns on the mask were photographically reduced from macro- or mesosizes to the desired microscales. The solubility of photoresist materials changes when they are exposed to light. Photoresists that become more soluble under light are classified as *positive photoresists*, whereas *negative photoresists* become more soluble under the shadow. The exposed substrate after development with solvents will have opposite effects between these two types of photoresists, as illustrated in the right column of Figure 8.1. The retained photoresist materials create the imprinted patterns after the development [see step (a) in the figure]. The portion of the substrate under the shadow of the photoresists is protected from the subsequent etching [step (b)]. A permanent pattern is thus created in the substrate after removal of the photoresist [step (c)].

Photolithography for MEMS and microsystems needs to be performed in a class 10 or better clean room. The class number of a clean room is a designation of the air quality in it. A class 10 clean room means that the number of dust particles 0.5 μm or larger in a cubic foot of air in the room is less than 10. Most other microfabrication processes presented in this chapter can tolerate a clean room of class 100. These requirements for clean-room air quality are in sharp contrast to those of the air quality of class 5 million in a typical urban environment!

8.2.2 Photoresists and Application

There are a number of photoresists that can be used in photolithography. Photoresists are classified as polymers as described in Chapter 7. Different types of photoresists can result in significantly different results. Commercially available photoresists of both positive and negative types are listed in several microelectronics process books (Sze, 1985; Ruska, 1987; van Zant, 1997).

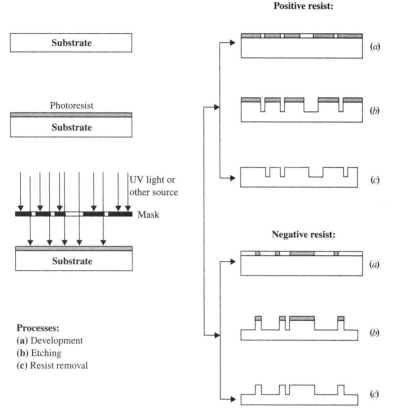

Figure 8.1. General procedure of photolithography.

- **Positive Resists:** In general, there are two kinds of positive photoresists: (1) the PMMA (polymethyl methacrylate) resists and (2) the two-component DQN resist involving diazoquinone ester (DQ) and phenolic novolak resin (N). In the latter kind, the first component accounts for about 20–50% by weight in the compound.

  Positive resists are sensitive to ultraviolet (UV) light, with the maximum sensitivity at a wavelength of 220 nm. The PMMA resists are also used in photolithography involving electron beams, ion beams, and x-rays. Most positive resists can be developed in alkaline solvents such as KOH (potassium peroxide), TMAH (tetramethylammonium hydroxide), ketones, or acetates.

- **Negative Resists:** Two popular negative resists are (1) two-component bis(aryl)-azide rubber resists and (2) Kodak KTFR (azide-sensitized polyisotroprene rubber). Negative resists are less sensitive to optical and x-ray exposure but more sensitive to electron beams. Xylene is the most commonly used solvent for developing negative resists.

In general, positive photoresists provide more clear edge definitions than the negative resists, as illustrated in Figure 8.2. The better edge definition by positive resists makes these resists a better option for high-resolution patterns for microdevices.

(a) By negative resists   (b) By positive resists

Figure 8.2. Profiles of lines from photolithography.

Application of photoresist onto the surface of substrates begins with securing the substrate wafer on the top of a vacuum chuck, as illustrated in Figure 8.3a. A resist puddle is first applied to the center portion of the wafer from a dispenser. The wafer is then subjected to high-speed spinning at a rotational speed that varies from 1500 to 8000 rpm for 10–60 s. The spinning speed depends on the type of the resist, the desired thickness, and the uniformity of the resist coating. The centrifugal forces applied to the resist puddle spread the fluid over the entire surface of the wafer. Typically the thickness is between 0.5 and 2 μm with $\pm$5-nm variation. For some microsystem applications, the thickness had been increased to as large as 1 cm. However, that is hardly a normal practice.

The spinner is stopped once the desired thickness of the coating is reached. A common problem is the bead of resist that occurs at the edge of the wafer, as shown in Figure 8.3b. These edge beads can be several times the thickness of the resist. The uniformity of the coating can be increased and the thickness of the edge beads can be reduced by controlling the spin speed, often by spinning slowly at first, followed by high-speed spinning.

8.2.3 Light Sources

Most photoresists are sensitive to light with wavelength ranging from 300 to 500 nm. The most popular light source for photolithography is the mercury vapor lamp. This light source provides a wavelength spectrum from 310 to 440 nm. Deep UV light has a wavelength of 150–300 nm, while normal UV light has wavelengths between 350 and 500 nm. These are all suitable light sources for photolithography purposes. In special

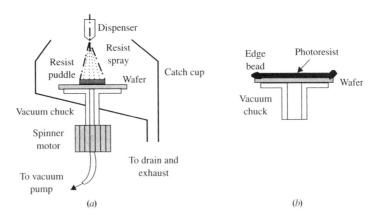

Figure 8.3. Schematic of photoresist application: (a) vacuum chuck; (b) photoresist coating with edge beads.

applications for extremely high resolutions, x-rays are used. The wavelength of x-rays is in the range from 4 to 50 Å (1 angstrom = 0.1 nm or 10^{-4} μm).

8.2.4 Photoresist Development

The same chamber used for the photoresist application in Figure 8.3 may be used for the development of negative photoresists. In such applications, the exposed wafer substrate is secured on the vacuum chuck and is spun at a high speed with a spray of solvent (e.g., xylene) from the overhead nozzle. A rinsing process with distilled water follows the development.

The development of positive photoresists is a little more complicated than for negative photoresists. Wafers with positive photoresists are usually developed in batches in a tank. This development process requires more controlled chemical reaction than just washing. A developer agent such as KOH or TMAH is projected onto the wafer surface in streams of mist at low velocity. After the development, the wafer is rinsed with a spray of distilled water, as with negative photoresists.

8.2.5 Photoresist Removal and Postbaking

After the photoresist is developed and the desired pattern is created on the substrate, a process called *descumming* takes place. It involves the application of mild oxygen plasma treatment on the developed wafers. This process removes the bulk of the resist.

Postbaking is necessary to remove the residue solvent used in the development. It frequently takes place at $120°C$ for about 20 min. Subsequent etching will remove all residual resists over the patterned areas.

8.3 ION IMPLANTATION

As described in Chapters 3 and 7, it is customary to use *p*- or *n*-type silicon for signal transduction in MEMS. There we also learned that doping boron atoms into silicon crystals can produce *p*-type silicon piezoresistors, whereas doping arsenic or phosphorus atoms can produce the *n*-type silicon. There are generally two ways of doping a semiconductor such as silicon with foreign substances, and *ion implantation* is one of these two methods.

Ion implantation involves "forcing" free atoms, such as boron or phosphorus, with charged particles (i.e., ions) into a substrate, thereby achieving an imbalance between the number of protons and electrons in the resulting atomic structure. The ions, whether they are boron ions or phosphorus ions, must carry sufficient kinetic energy to be implanted (that is, to penetrate) into the silicon substrate. For this reason, the ion implantation procedure always involves the acceleration of the ions in order to gain sufficient kinetic energy for the implantation.

As we see from the schematic of ion implantation in Figure 8.4, ions to be implanted are created in a ion source, in which an ion beam is formed. Descriptions of two ion-generating sources are presented in Chapter 3. Figure 3.4 illustrates how

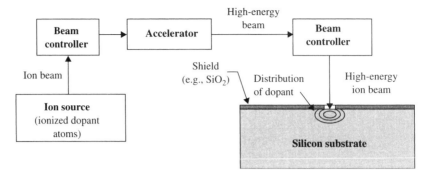

Figure 8.4. Ion implantation on a substrate.

ions can be produced using electron beams. Ions are extracted from the substance in the gaseous state, that is, the plasma. The ion beam is then led into a beam controller in which the size and direction of the beam can be adjusted. The ions in the beam are then energized in the acceleration tube or *accelerator*, in which the final energy with which the ions will impact the substrate surface is attained. The ions in the beam are focused onto the substrate, which is protected by a shield, or a mask, usually made of SiO_2. The highly energized ions enter the substrate and collide with electrons and nuclei of the substrate. The ions will transfer all their energy to the substrate upon collision and finally come to a stop at a certain depth inside the substrate. The distribution of the implanted ions in silicon substrate is illustrated in Figure 8.5. The energy required for *p*- or *n*-type doping is presented in Table 8.1.

Unlike the diffusion technique presented in Section 8.4, ion implantation does not require high temperature. This offers a major advantage, as little thermal stress or strain will be introduced between the substrate and the shield. The disadvantage, however, is that the dopant distribution in the substrate is less uniform, as illustrated in Figure 8.4, with an enlarged view in Figure 8.5.

In Figure 8.5, we observe that the highest concentration of implanted dopant in the substrate appears to be under the substrate surface, not on the surface. The distribution of the dopant into the substrate tends to follow a Gaussian (or normal) distribution, as

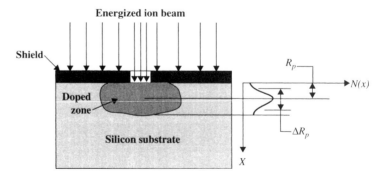

Figure 8.5. Distribution of dopant through a shield (Ruska, 1987).

TABLE 8.1. Ionization Energy for Dopants in Silicon

Dopants	*p*- or *n*-Type	Ionization Energy, eV
Phosphorus (P)	n	0.044
Arsenic (As)	n	0.049
Antimony (Sb)	n	0.039
Boron (B)	p	0.045
Aluminum (Al)	p	0.057
Gallium (Ga)	p	0.065
Indium (In)	p	0.160

Source: Ruska, 1987.

illustrated at the right of the figure. According to Ruska (1987), the concentration of dopant at an x distance below the substrate surface, $N(x)$, may be determined from the equation

$$N(x) = \frac{Q}{\sqrt{2\pi}\ \Delta R_p} \exp\left[\frac{-(x - R_p)^2}{2\ \Delta R_p^2}\right] \qquad (8.1)$$

where

R_p = projected range, μm

ΔR_p = scatter or "straggle," μm

Q = dose of ion beam, atoms/cm^2

The doping ranges for selected dopants in silicon substrate are given in Table 8.2.

TABLE 8.2. Ion Implantation of Common Silicon Dopants

Ion	Range, R_p (nm)	Straggle, Delta R_p(nm)
At 30 keV Energy Level		
Boron (B)	106.5	39.0
Phosphorus (P)	42.0	19.5
Arsenic (As)	23.3	9.0
At 100 keV Energy Level		
Boron (B)	307.0	69.0
Phosphorus (P)	135.0	53.5
Arsenic (As)	67.8	26.1

Source: Ruska, 1987.

Equation (8.1), with the data in Table 8.2, can assist engineers in determining the distribution of dopant concentration, the maximum density of the dopant, and the depth of doping inside a silicon substrate.

> **Example 8.1.** A silicon substrate is doped with boron ions at 100 keV. Assume the maximum concentration after the doping is 30×10^{18}/cm^3 (refer to Figure 3.8). Find (1) the dose Q in Equation (8.1), (2) the dopant concentration at a depth of 0.15 μm, and (3) the depth at which the dopant concentration is 0.1% of the maximum value.
>
> *Solution:*   From Table 8.2, we find the projected range for boron ion is 307 nm at 100 keV, or $R_p = 307 \times 10^{-7}$ cm at 100 keV, and the straggle $\Delta R_p = 69 \times 10^{-7}$ cm.
>
> 1. We are given the maximum concentration $N_{max} = 30 \times 10^{18}$ cm^{-3} at $x = R_p$ in Equation (8.1), so from the same equation, we have
>
> $$N_{max} = \frac{Q}{\sqrt{2\pi} \, \Delta R_p}$$
>
> from which we have the dose $Q = (2\pi)^{0.5} \, (\Delta R_p) N_{max} = (6.28)^{0.5} \, (69 \times 10^{-7}$ cm$)$ $(30 \times 10^{18}$ cm$^{-3}) = 5.2 \times 10^{14}$/cm^2.
>
> 2. To find the concentration at $x = 0.15$ μm, we use the following relationship derived from Equation (8.1):
>
> $$N(0.15\mu m) = N_{max} \exp\left[-\frac{(0.150 - 0.307)^2}{2(0.069)2}\right] = (30 \times 10^{18}) \exp\left(\frac{-0.0246}{0.009522}\right)$$
> $$= 2.27 \times 10^{18} \text{ cm}^{-3}$$
>
> 3. To find $x = x_0$ at which the concentration $N(x_0) = (0.1\%)N_{max} = 3 \times 10^{16}$/cm^3, we solve for x_0 in the equation
>
> $$N(x_0) = \frac{5.2 \times 10^{14}}{\sqrt{2 \times 3.14} \times 69 \times 10^{-7}} \exp\left[-\frac{(x_0 - 307 \times 10^{-7})^2}{2(69 \times 10^{-7})^2}\right] = 3 \times 10^{16}$$
>
> from which we solve for $x_0 = 563.5 \times 10^{-7}$ cm or 0.5635 μm.

8.4 DIFFUSION

The principle of diffusion was presented in Section 3.6. The diffusion process is often used in microelectronics for the introduction of a controlled amount of foreign materials (dopants) into selected regions of another material (substrates).

Unlike ion implantation, diffusion is a slow doping process. It is also used as thin-film buildings in microelectronics and microsystems. Diffusion takes place at elevated temperatures. As illustrated in Figure 8.6, the atoms of the dopant gases move (diffuse) into the

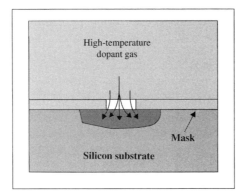

Figure 8.6. Doping of silicon substrate by diffusion.

crystal vacancies or interstitials of the silicon substrate. The profile of the doped region in the substrate as shown in Figure 8.6 is also different from that of ion implantation.

The physics of diffusion is similar to that of heat conduction in solids as described in Chapter 5. The mathematical model for diffusion is Fick's law, represented in Equation (3.3), which is similar to the Fourier law for heat conduction in solids in Equation (5.28). Fick's law gives the dopant flux in the substrate in the x direction during a diffusion process. We express Fick's law, but in different notation, for the diffusion of foreign atoms into the substrates of microsystems:

$$F = -D\frac{\partial N(x)}{\partial x} \tag{8.2}$$

where

F = dopant flux, number of dopant atoms passing through unit area of substrate in unit time, atoms/cm^2-s

D = diffusion coefficient, or diffusivity, of dopant to substrate, cm^2/s

N = dopant concentration in substrate per unit volume, atoms/cm^3

The distribution of dopant in the substrate at any given time during the diffusion process can be obtained by the solution of the following diffusion equation, which is similar to the heat conduction equation (5.31) in the direction along the x coordinate:

$$\frac{\partial N(x,t)}{\partial t} = D\frac{\partial^2 N(x,t)}{\partial x^2} \tag{8.3}$$

The solution of Equation (8.3) depends on the initial and boundary conditions involved in the process. One set of such conditions is as follows:

Initial Condition: $N(x,0)=0$, that is, there is no impurity in the substrate when the diffusion process begins.

TABLE 8.3. Constants for Equation (8.6)

Dopants	Constant a	Constant b
Boron	−19.9820	13.1109
Arsenic	−26.8404	17.2250
Phosphorus ($N_s = 10^{21}$/cm^3)	−15.8456	11.1168
Phosphorus ($N_s = 10^{19}$/cm^3)	−20.4278	13.6430

Boundary Conditions: (1) $N(0,t) = N_s$, which is the concentration at the surface exposed to the gaseous dopant. (2) $N(\infty,t) = 0$, that is, the diffusion of foreign substance is localized, and the concentration far from the exposed surface is negligible.

The solution of Equation (8.3) with the above initial and boundary conditions has the form

$$N(x, t) = N_s \, \text{erfc} \left[\frac{x}{2\sqrt{Dt}} \right] \tag{8.4}$$

where erfc(x) is the complementary error function, which is equal to

$$\text{erfc}(x) = 1 - \text{erf}(x) = 1 - \frac{2}{\sqrt{\pi}} \int_0^x e^{-y^2} \, dy \tag{8.5}$$

The solutions expressed in Equations (8.3) and (8.5) are similar to those in Equations (3.4) and (3.5) for the general cases of diffusion.

Numerical values of the complimentary error function in Equation (8.4) can be obtained either from the above integral or from Figure 3.14. The term $\sqrt{Dt}$ in the solution in Equation (8.4) is the diffusion length.

The diffusivity D for a silicon substrate with common dopants such as boron, arsenic, and phosphorus can be found in the references (e.g., Ruska, 1987) or as given in Figure 3.12. We may use the following expression in Equation (8.6), derived from the graphs provided in the above-cited reference, to estimate the values of D for three common dopants, boron, arsenic, and phosphorus:

$$\ln(\sqrt{D}) = aT' + b \tag{8.6}$$

where D is the diffusivity in micrometers per hours. The temperature $T' = 1000/T$ with T the diffusion temperature, K. The values of constants a and b in Equation (8.6) are given in Table 8.3.

> **Example 8.2.** A silicon substrate is subjected to diffusion of boron dopant at 1000°C with a dose of 10^{11}/cm^2. Find (1) the expression for estimating the concentration of the dopant in the substrate and (2) the concentration at 0.1 μm beneath the surface after 1 h into the diffusion process. The substrate is initially free of impurity.

Solution: We first determine the thermal diffusivity of boron in silicon. This can be done using Equation (8.6) with the constants a and b obtained from Table 8.3. The temperature T' in Equation (8.6) is determined as $T' = 1000/(1000 + 273) = 0.7855$ and $a = -19.982$ and $b = 13.1109$ from Table 8.3. We will thus find the square root of the diffusivity of boron in silicon, $\sqrt{D} = 0.07534$, or $D = 0.005676$ $\mu m^2/h = 1.5766 \times 10^{-6}$ $\mu m^2/s$. This value is in the "ballpark" range $D = 2 \times 10^{-6}$ $\mu m^2/s$ from Figure 3.12.

1. Since the initial condition $N(x,0) = 0$ and the boundary conditions are $N(0,t) = N_s = 10^{11}/cm^2$ and $N(\infty, t) = 0$, the solution presented in Equation (8.4) can be used to express the concentration of dopant in the substrate:

$$N(x,t) = N_s \, \text{erfc}\left[\frac{x}{2\sqrt{Dt}}\right] = 10^{11} \, \text{erfc}\left[\frac{x}{2\sqrt{1.5766 \times 10^{-6}t}}\right]$$

$$= 10^{11} \, \text{erfc}\left(\frac{398.21x}{\sqrt{t}}\right) = 10^{11}\left[1 - \text{erf}\left(\frac{398.21x}{\sqrt{t}}\right)\right]$$

where x is in micrometers and t is in seconds.

2. For $x = 0.1$ μm and $t = 1\,h = 3600\,s$

$$N(0.1\,\mu m, 1\,h) = 10^{11}\left[1 - \text{erf}\left(\frac{398.21 \times 0.1}{\sqrt{3600}}\right)\right] = 10^{11}[1 - \text{erf}(0.6637)]$$

$$= 10^{11}(1 - 0.6518) = 3.482 \times 10^{10} \text{ cm}^{-3}$$

The value of the function erf(0.6637) is obtained from Figure 3.14.

8.5 OXIDATION

8.5.1 Thermal Oxidation

Oxidation is a very important process in both microelectronics and microsystems fabrication. According to Sze (1985), there four types of thin films are frequently used in microelectronics: (1) thermal oxidation for electrical or thermal insulation media, (2) dielectric layers for electrical insulation, (3) polycrystalline silicon for local electrical conduction, and (4) metal films for electrical (ohmic) contact and junctions. All these types of thin films are also widely used in MEMS and microsystems for similar purposes. We have learned that dielectric layers are used to separate electrodes in capacitance transducers or electrically insulate the metal lead wires from the embedded transducers. Materials for dielectric films involve ceramics such as alumina and quartz or those grown over the substrate's surface, such as silicon dioxide or silicon nitride. The polycrystalline silicon films are necessary in providing localized electric conduction, as described in Chapter 7. The metal films are used as leads for piezoresistive transducers as well as the pads for soldering silicon die to the constraint bases, for example, at the footprint in a pressure sensor. We will focus our attention on the thermal oxidation process commonly used for the production of silicon dioxide films in microsystems.

8.5.2 Silicon Dioxide

Silicon dioxide is used as an electric insulator as well as for etching masks for silicon and sacrificial layers in surface micromachining, as will be described in detail in Chapter 9. There are several ways silicon dioxide can be produced on substrate surfaces. The least expensive way to produce SiO_2 film over the silicon substrate is by thermal oxidation. The chemical reactions in this process are as follows:

$$Si_{(solid)} + O_{2(gas)} \rightarrow SiO_{2(solid)} \tag{8.7}$$

or

$$Si_{(solid)} + 2H_2O_{(steam)} \rightarrow SiO_{2(solid)} + 2H_{2(gas)} \tag{8.8}$$

Silicon oxide is produced by thermal oxidation in an electric resistance furnace. A typical furnace consists of a large fused quartz tube, as illustrated in Figure 8.7. Resistance heating coils surround the tube to provide the necessary high temperature in the tube. The furnace tube used in industry is on the order of 30 cm in diameter and 3 m in length.

In the thermal oxidation process, wafers are placed in fused quartz cassettes that are pushed into the preheated furnace tube at a temperature in the range of 900–1200°C. Oxygen is blown into the tubular furnace for the oxidation of wafer surfaces. Often, steam is used instead of oxygen for accelerated oxidation. The timing, temperature, and gas flow are strictly controlled in order to achieve the desired quality and thickness of the SiO_2 film.

8.5.3 Thermal Oxidation Rates

The growth of an oxide layer in a silicon substrate is primarily a thermal diffusion process, as presented in Section 8.4. However, because of the coupling of chemical reactions that

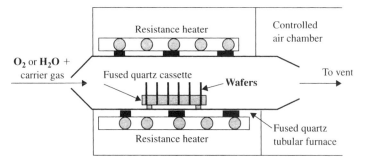

Figure 8.7. Facility for thermal oxidation of silicon dioxide.

takes place simultaneously with the diffusion [see Equations (8.7) and (8.8)], the theory of diffusion as presented in Section 8.4 cannot be applied without significant modifications. The analysis is further complicated by the fact that both heat and mass transfer with moving silicon–SiO$_2$ boundaries need to be considered in the analysis. The so-called *kinetics of thermal oxidation* has to be used in assessing the growth of the oxidized SiO$_2$ layer in the substrate (Ruska, 1987). The growth of an SiO$_2$ layer in a silicon substrate is simulated in Figure 8.8.

The silicon substrate is first exposed to oxidizing species such as oxygen in dry air or wet steam in a hot furnace, as illustrated in Figure 8.7. The oxygen molecules in the oxidizing species begin to diffuse into the surface of the "fresh" silicon wafer, as shown in Figure 8.8a, and a SiO$_2$ layer (shown as the darker gray area in Figure 8.8b) is formed, with SiO$_2$ molecules in solid circles in the figure. While the drive from the oxidizing species continues, the SiO$_2$ molecules that are already in the oxide layer diffuse and chemically react with the silicon molecules, crossing the layer's boundary, as illustrated in Figure 8.8c. Thus the process can be viewed as one that begins with simple diffusion of oxidizing species into silicon substrate. It is soon coupled with the diffusion of the same species into SiO$_2$. Meanwhile, there is a simultaneous diffusion of SiO$_2$ molecules and a chemical reaction between the SiO$_2$ molecules and the silicon substrate, which creates new SiO$_2$–silicon boundaries.

Due to the highly complex oxidation process described above, prediction of the growth of the oxide layer in the silicon substrate is by no means easy. However, simple one-dimensional coupled diffusion/heat and mass transfer models were derived and they can be found in several publications (Sze, 1985; Ruska, 1987). We will use the following simple expressions to estimate the growth of oxide layers in silicon substrates

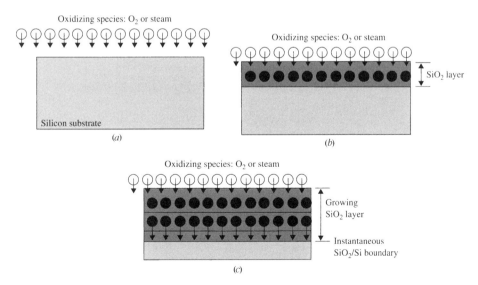

Figure 8.8. Kinetics of SiO$_2$ in silicon substrates: (a) at inception of oxidation; (b) formation of oxide layer; (c) growth of oxide layer.

(in micrometers):

$$x = \begin{cases} \dfrac{B}{A}(t + \tau) & \text{for small time } t \qquad\qquad (8.9) \\[2em] \sqrt{B(t + \tau)} & \text{for large time } t \qquad\qquad (8.10) \end{cases}$$

where x is the thickness of the oxide layer in the silicon substrate in micrometers at time t in hours, A and B are constants, and the parameter τ can be obtained as

$$\tau = \frac{[(d_0{}^2 + 2Dd_0)/k_s]N_1}{2DN_0} \qquad\qquad (8.11)$$

where

D = diffusivity of oxide in silicon, e.g., $D = 4.4 \times 10^{-16}\ \mathrm{cm^2/s}$ at $900°\mathrm{C}$

d_0 = initial oxide layer (~ 200 Å in dry oxidation, 0 for wet oxidation)

k_s = surface reaction rate constant

N_0 = concentration of oxygen molecules in carrier gas: 5.2×10^{16} molecules/cm^3 in dry O_2 at $1000°\mathrm{C}$ and 1 atm; 3000×10^{16} molecules/cm^3 in water vapor at same temperature and pressure

N_1 = number of oxidizing species in oxide: 2.2×10^{22} SiO_2 molecules/cm^3 with dry O_2; 4.4×10^{22} SiO_2 molecules/cm^3 in water vapor

It is readily seen that τ in Equation (8.11) represents the time coordinate shifts to account for the initial oxide layer d_0 that may be in existence before the oxidation process begins. For wet oxidation, or in the case $d_0 = 0$, $\tau = 0$.

Equations (8.9) and (8.10) can be used to estimate the thickness of the oxide layer at time t.

The constant ratio B/A in Equation (8.9) is called the *linear rate constant*, whereas the constant B in Equation (8.10) is called the *parabolic rate constant*. Numerical values of these constants depend on the activation energies used in the oxidation as well as the orientations of the silicon crystal surfaces on which the oxidation takes place. We will offer the following equations to fit the graphical representations available in Sze (1985):

$$\log\left(\frac{B}{A}\right) = aT' + b \quad \text{for linear rate constants} \qquad\qquad (8.12)$$

$$\ln(B) = aT' + b \quad \text{for parabolic rate constants} \qquad\qquad (8.13)$$

TABLE 8.4. Coefficients for Determining Rates of Oxidation in Silicon

Constants	Coefficient a	Coefficient b	Conditions
Linear rate	-10.4422	6.96426	Dry O_2, $E_a = 2\,eV$, (100) silicon
constant,	-10.1257	6.93576	Dry O_2, $E_a = 2\,eV$, (111) silicon
Equation (8.12)	-9.905525	7.82039	H_2O vapor, $E_a = 2.05\,eV$, (110) silicon
	-9.92655	7.948585	H_2O vapor, $E_a = 2.05\,eV$, (111) silicon
Parabolic rate	-14.40273	6.74356	Dry O_2, $E_a = 1.24\,eV$, 760 Torr vacuum
constant,	-10.615	7.1040	H_2O vapor, $E_a = 0.71\,eV$, 760 Torr vacuum
Equation (8.13)			

where $T' = 1000/T$, with the temperature T in kelvin. The coefficients a and b in Equations (8.12) and (8.13) can be obtained from Table 8.4.

A question arises as to which of the two equations (8.9) or (8.10) is to be used for estimating the thickness of the oxide layer in the time scale. The linear nature of Equation (8.9) reflects the growth of the oxide layer in the early stage of oxidation. However, when the layer thickness increases, the coupling effect of diffusion and the chemical reaction becomes dominant and the growth of the layer becomes nonlinear. Consequently, Equation (8.10) is a more realistic way for estimating the thickness of the oxide layer. No clear demarcation between the validity of either equation is available, as the definition of "smaller" and "larger" times for these equations depends on the conditions involved in the oxidation processes.

Example 8.3. Estimate the thicknesses of the SiO_2 layer over the (111) plane of a clean silicon wafer resulting from both dry and wet oxidation at $950°C$ for 1.5 h.

Solution: Since the wafer surface is free from initial oxidation, there is no need to consider the time coordinate shift; that is, $\tau = 0$ in Equations (8.9) and (8.10). Consequently, the oxide thicknesses from the dry and wet oxidation cases can be estimated using the expressions

$$x = \begin{cases} \dfrac{B}{A}\,t & \text{for small time } t \quad\quad\text{(a)} \\[2mm] \sqrt{Bt} & \text{for large time } t \quad\quad\text{(b)} \end{cases}$$

The constants A and B in the above expressions can be obtained from Equations (8.12) and (8.13), with the coefficients a and b given in Table 8.4. Our selection of these coefficients is tabulated below:

	a	b	Conditions
Equation (8.12)	-10.1257	6.9357	Dry O_2
Equation (8.12)	-9.9266	7.9486	Wet steam
Equation (8.13)	-14.4027	6.7436	Dry O_2
Equation (8.13)	-10.6150	7.1040	Wet steam

We will find the temperature $T' = 1000/(950 + 273) = 0.8177$. On substituting T' and the coefficients into Equations (8.12) and (8.13), we have the constants B/A and B as follows:

	Dry Oxidation	Wet Oxidation
B/A, μm/h	0.04532	0.6786
B, μm^2/h	0.006516	0.2068

Using Equations (a) and (b), we may estimate the thicknesses of the oxide layers in the respective dry and wet processes as follows:

	Dry Oxidation Thickness, μm	Wet Oxidation Thickness, μm
Equation (a) for small time	0.068	1.018
Equation (b) for larger time	0.0989	0.5572

We observe from the above summary of results that wet oxidation is more effective in the depth of penetration into the substrate than dry oxidation.

For the cases that involve initial oxide layers, prediction of the growth of the oxide layer becomes much more complex, as it will require determination of the time parameter τ in Equation (8.11) with the value of k_s, the *surface reaction rate constant* in the process. In this case, it is also referred to as the *silicon oxidation rate constant*. This constant is a function of temperature, oxidant, crystal orientation, and doping. We will deal with the evaluation of k_s in Section 8.6.

8.5.4   Oxide Thickness by Color

Both SiO_2 and Si_3Ni_4 layers have a color distinct from that of the silicon substrates on which they grow. The SiO_2 layers are essentially transparent but with a different light

TABLE 8.5. Color of Silicon Dioxide Layers of Selected Thickness

SiO_2 layer thickness, μm	0.050	0.075	0.275 0.465	0.310 0.493	0.50	0.375	0.390
Color	Tan	Brown	Red-violet	Blue	Green to yellow green	Green-yellow	Yellow

refraction index from that of the silicon substrate. Consequently, when the surface is illuminated by white light, one can view the different colors on the surface corresponding to the layer thickness. The color of the surface of a SiO_2 layer is the result of interference of reflected light rays. It is thus not a surprise that the same color may repeat with different layer thickness. Table 8.5 offers a partial list of the colors of the surface of SiO_2 layers of selected thickness. A more complete color chart can be found in several prominent references (Ruska, 1987; Madou, 1997; van Zant, 1997).

8.6 CHEMICAL VAPOR DEPOSITION

Depositing thin films over the surface of substrates and other MEMS and microsystem components is a common and necessary practice in micromachining. Unlike the diffusion and thermal oxidation processes that we learned about in previous sections, deposition adds thin films to, instead of consuming, the substrates. There are abundant circumstances in which the need to add thin films of a wide range of materials on the substrate surfaces arises. These thin-film materials can be organic or inorganic. They include a variety of metals; common ones are Al, Ag, Au, Ti, W, Cu, Pt, and Sn. These materials also include compounds such as the common shapememory alloys NiTi and piezoelectric ZnO, used for coating tubes for microfluidics, as described in Chapter 5. The application of shapememory alloys in microdevices was covered in Chapter 2.

Generally two classes of deposition are used in microelectronics and micromachining: (1) *physical vapor deposition* (PVD) and (2) *chemical vapor deposition* (CVD). Physical vapor deposition involves the direct impingement of particles on hot substrate surfaces. On the other hand, CVD involves convective heat and mass transfer as well as diffusion with chemical reactions at the substrate surfaces. It is a much more complex process than PVD but a great deal more effective in terms of the rate of growth and the quality of deposition. Most CVD processes involve low gas pressures, whereas others that are carried out in high vacuum. We will focus on CVD in the subsequent sections.

8.6.1 Working Principle of CVD

The working principle of CVD involves the flow of a gas with diffused reactants over a hot substrate surface. The gas that carries the reactants is called *carrier gas*. While the gas flows over the hot solid surface, the energy supplied by the surface temperature provokes

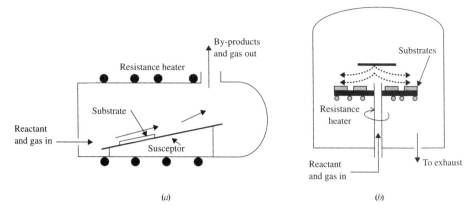

Figure 8.9. Two typical CVD reactors: (a) horizontal reactor; (b) vertical reactor.

chemical reactions of the reactants that form films during and after the reactions. The by-products of the chemical reactions are then vented. Thin films of desired compositions can thus be created over the surface of the substrate. Various types of CVD reactors are built to perform the CVD processes. Two are illustrated in Figure 8.9.

In both types of reactors, the substrate surface is exposed to the flowing gas with diffused reactants. Resistance heaters either surround the chamber, as in Figure 8.9a, or lie directly under the susceptor that holds the substrates, as in Figure 8.9b.

8.6.2   Chemical Reactions in CVD

We will show the chemical reactions that are used in depositing three common thin films over silicon substrates: (1) silicon dioxide, (2) silicon nitride, and (3) polycrystalline silicon.

Silicon Dioxide. Thin silicon dioxide films can be produced at the surface of silicon substrate by the diffusion process as described in Section 8.4. Here, we will learn how SiO_2 thin films can be deposited on the surface of silicon substrates by chemical reaction. A number of chemical reactants could be used to deposit silicon dioxide films on silicon substrates, including $SiCl_4$, $SiBr_4$, and SiH_2Cl_2 (Ruska, 1987). The carrier gases that can be used in these processes are O_2, NO, NO_2, and CO_2 with H_2. The most common reactant used for CVD is silane (SiH_4) together with oxygen. The chemical reaction in this process can be expressed as

$$SiH_4 + O_2 \rightarrow SiO_2 + 2H_2 \tag{8.14}$$

The chemical reaction takes place in a temperature range of 400–500°C with an activation energy E_a around 0.4 eV.

Silicon Nitride (Si₃N₄). Ammonia is a common carrier gas for depositing silicon nitride on silicon substrates. Three reactants can produce the thin silicon nitride films (Ruska, 1987):

$$3SiH_4 + 4NH_3 \rightarrow Si_3N_4 + 12H_2 \tag{8.15a}$$

$$3SiCl_4 + 4NH_3 \rightarrow Si_3N_4 + 12HCl \tag{8.15b}$$

$$3SiH_2Cl_2 + 4NH_3 \rightarrow Si_3N_4 + 6HCl + 6H_2 \tag{8.15c}$$

The activation energy E_a required for the above reactions is 1.8 eV. The temperature range for the reactions is 700– 900°C for silane in Equation (8.15a); 850°C for silicon tetrachloride in Equation (8.15b); and 650–750°C for dichlorosilane in Equation (8.15c).

Polycrystalline Silicon. As described in Section 7.5.4, polycrystalline silicon films consist of single-silicon crystals of different sizes. Deposition of polycrystalline silicon is a *pyrolysis* process, which is a decomposition process using heat, as we see from the chemical reaction

$$SiH_4 \rightarrow Si + 2H_2 \tag{8.16}$$

The process takes place in a temperature range of 600–650°C with activation energy of 1.7 eV.

8.6.3   Rate of Deposition

We learned from Sections 8.6.1 and 8.6.2 that CVD is a process by which films of desired substance are produced by chemical reactions of the reactants and carrier gas at the hot substrate surface. The rate of buildup of these thin films obviously is a concern to process design engineers, as in many cases the production of three-dimensional MEMS structure relies on these buildups.

Let us now look at the situation illustrated in Figure 8.10*a*, where the reactant and the carrier gas flow over the hot substrate (silicon) surface with a velocity. In a close-up view, we will observe the velocity profile of the gas–reactant mixture, $V(x)$, over the substrate surface, as shown in Figure 8.10*b*. As we learned from Section 5.7.5, a boundary layer with a thickness $\delta(x)$ exists at the solid–fluid interface whenever a fluid flows over a solid surface. This boundary layer is of great importance in heat transfer, as it acts as an additional thermal barrier. A similar effect occurs in a diffusion process, as additional time is required for the reactant to diffuse across this layer. Thus, the boundary layer is a retarding factor to the chemical reactions. Consequently, this layer affects the rate of deposition of the desired thin film on the substrate surfaces.

The approximate thickness of the boundary layer, $\delta(x)$, a distance x away from the leading edge of the silicon substrate can be obtained from the expression

$$\delta(x) = \frac{x}{\sqrt{Re(x)}} \tag{8.17}$$

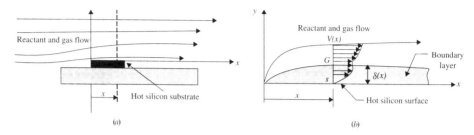

Figure 8.10. Thermal/fluid aspect of CVD: (*a*) gas flow over substrate surface; (*b*) velocity profile and boundary layer.

where $Re(x)$ is the Reynolds number of the gas mixture. By definition in Equation (5.3), the Reynolds number can be expressed as

$$Re(x) = \frac{\rho L V(x)}{\mu} \tag{8.18}$$

where ρ is the mass density of the gas mixture, L is the characteristic length of the flow, and the dynamic viscosity of the gas, μ, can be obtained from Table 8.6.

The diffusion flux of the reactant, N across the boundary layer thickness, with units of atoms or molecules per unit time per area (atoms or molecules per meters squared per second) through the boundary layer can be expressed by Fick's law, described in Section 3.6. The expanded mathematical expression of Fick's law for the present case is

$$\mathbf{N} = \frac{D}{\delta}(N_G - N_S) \tag{8.19}$$

where D is the diffusivity of the reactant in the carrier gas with units of centimeters squared per second and N_G and N_S are the respective concentrations of the reactant at the top of the boundary layer and at the surface of the substrate. Both N_G and N_S have units of molecules per volume, or molecules per cubic meter.

TABLE 8.6. Dynamic Viscosity of Gases Used in CVD Processes

Gas	Viscosity, μP			
	0 °C	490 °C	600 °C	825 °C
Hydrogen, H_2	83	167	183	214
Nitrogen, N_2	153	337	—	419
Oxygen, O_2	189	400	—	501
Argon, Ar	210	448	—	563

1 poise (P) = 1 dyne-s/cm^2 = 0.1 N-s/m^2 = 0.1 kg/m-s.
Source: Ruska, 1987.

Determination of the concentrations N_G and N_S can be carried out by the following procedure:

1. Use *Avogadro's theory*, which states that the volume occupied by all gases at standard conditions by a mole of gas is $22.4 \times 10^{-3}\,m^3$, which leads to a molar density of $44.643\,mol/m^3$. Also, *Avogadro's number* is defined as the number of molecules contained in 1 mol of any gas or substance, or 6.022×10^{23}.
2. Find the molar mass of commonly used gases in Table 8.6:

Gas	Hydrogen, H_2	Nitrogen, N_2	Oxygen, O_2	Argon, Ar
Molar mass, g	2	28	32	40

3. Use the ideal gas law to determine the molar density of the gas at its temperature and pressure if they are different from the standard conditions.
4. The concentrations N_G and N_s can be determined by dividing Avogadro's number, given in step 1, by the molar density obtained in step 3.

Example 8.4. A CVD process involves a reactant being diluted at 2% in the carrier oxygen gas at 490°C. Find the number of molecules in a cubic meter volume of the carrier gas. The pressure variation in the process is negligible.

Solution: The ideal gas law follows the relationship

$$\frac{P_1 V_1}{P_2 V_2} = \frac{T_1}{T_2}$$

where V_1 and V_2 are the respective volumes of the gas at state 1 with pressure P_1 and temperature T_1 and state 2 with P_2 and T_2.

Since $P_1 = P_2$, and V_2, the molar volume of the gas at room temperature, 293 K, is $22.4 \times 10^{-3}\,m^3/mol$, we have the molar density of the gas at 490°C (763 K) as

$$d_2 = \left(\frac{T_1}{T_2}\right) d_1 = \left(\frac{293}{763}\right)(44.643) = 17.1433\,mol/m^3$$

The reader will realize that $d_1 = V_1 = 44.643\,mol/m^3$ is used in the above computation.

The concentration N_G (molecules of gas per cubic meter) is thus given as

$$N_G = \text{Avogadro's number } d_2 = (6.022 \times 10^{23})(17.1433)$$

$$= 103.24 \times 10^{23}\,molecules/m^3$$

Because most CVD processes take place at very low gas velocity, the corresponding Reynolds number Re is at the low value of about 100. The low velocity of the gas flow allows significant amount of reactant to diffuse through the boundary layer and form the film by chemical reaction at the hot surface of the substrate. The following relationship exists:

$$\mathbf{N} = k_s N_s \tag{8.20}$$

where k_s is the surface reaction rate constant as shown in Equation (8.11). This rate can be expressed as

$$k_s = k' \exp\left(-\frac{E_a}{kT}\right) \tag{8.21}$$

where

$k' = $ a constant whose value depends on reaction and reactant concentration

$E_a = $ activation energy

$k = $ Boltzmann's constant

$T = $ absolute temperature

The flux of the carrier gas and the reactant, N in Equation (8.19), may be expressed in terms of the surface reaction rate k_s, with the substitution of Equation (8.20), as

$$\mathbf{N} = \frac{DN_G k_s}{D + \delta k_s} \tag{8.22}$$

where δ is the mean thickness of the boundary layer shown in Figure 8.10 as expressed in Equation (8.17).

The rate of growth of the thin film over the substrate surface, r, in meters per second, can be estimated by the expressions (Ruska, 1987)

$$r = \begin{cases} \dfrac{DN_G}{\gamma^\delta} & \text{for } \delta k_s \geq D \tag{8.23a} \\[2em] \dfrac{N_G k_S}{\gamma} & \text{for } \delta k_s \ll D \tag{8.23b} \end{cases}$$

In both Equations (8.23a) and (8.23b), γ is the number of atoms or molecules per unit volume of the thin film. The value of γ may be estimated by a postulation that the thin film is "densely" packed by atoms or molecules in spherical shapes with the radius according to the materials listed in Table 8.7.

TABLE 8.7. Atomic and Ionic Radii of Selected Materials

Reactant Materials	Atomic Radius, nm	Ionic Radius, nm
Hydrogen	0.046	0.154
Helium	0.046	0.154
Boron	0.097	0.02
Nitrogen	0.071	0.02
Oxygen	0.060	0.132
Aluminum	0.143	0.057
Silicon	0.117	0.198
Phosphorus	0.109	0.039
Argon	0.192	—
Iron	0.124	0.067
Nickel	0.125	0.078
Copper	0.128	0.072
Gallium	0.135	0.062
Germanium	0.122	0.044
Arsenic	0.125	0.04

Source: Kwok, 1997.

The value of γ can thus be determined by the expression

$$\gamma = \frac{1}{v} = \frac{1}{\left(\frac{4}{3}\right)\pi a^3} \tag{8.24}$$

where a is the radius of atoms in meters based on Table 8.7. The units for γ are atoms or molecules per cubic meter.

Example 8.5. A CVD process is used to deposit SiO_2 film over a silicon substrate. Oxygen is used as the carrier gas in the chemical process shown in Equation (8.14). The CVD is carried out in a horizontal reactor as illustrated in Figure 8.11. Other conditions are identical to those given in Example 8.4.

Determine the following:

1. Density of carrier gas
2. Reynolds number of gas flow
3. Thickness of boundary layer over substrate surface

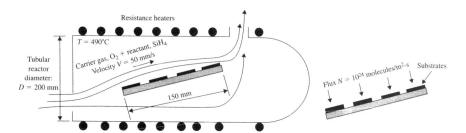

Figure 8.11. CVD over silicon substrates.

4. Diffusivity of carrier gas and reactant in silicon substrate
5. Surface reaction rate
6. Deposition rate

Solution:

1. The density of the carrier gas, ρ, may be obtained by taking the product of the molar mass of the carrier gas and the molar density from Example 8.4. The molar mass of oxygen is 32 g. Hence the density of the oxygen carrier gas is

$$\rho = (32 \text{ g/mol})(17.1433 \text{ mol/m}^3) = 548.586 \text{ g/m}^3$$

2. The Reynolds number Re is computed from Equation (8.18):

$$D = 200 \text{ mm} = 0.2 \text{ m} \quad V = 50 \text{ mm/s} = 0.05 \text{ m/s}$$

$$\mu = 400 \times 10^{-6} \text{ P(Table 8.6)} = 0.04 \text{ g/m-s}$$

$$\text{Re} = \frac{\rho D V}{\mu} = \frac{(548.586)(0.2)(0.05)}{0.04} = 137.147$$

3. The mean thickness of the boundary layer between the substrate surface and the carrier gas, δ, is, from Equation (8.17) with the length of the sled that carries the substrate, $L = 150 \text{ mm} = 0.15 \text{ m}$:

$$\delta = \frac{L}{\sqrt{\text{Re}}} = \frac{0.15}{\sqrt{137.147}} = 0.01281 \text{ m}$$

4. The diffusivity of the carrier gas and the reactant, D, is computed from Equation (8.19) with a modification to account for the dilution factor of the reactant, η, in percent of dilution (2% for the present case):

$$D = \frac{\delta N}{\eta (N_G - N_s)}$$

where

$\mathbf{N}$ = given carrier gas flux, $=10^{24}$ molecules/m^2-s

N_G = equivalent density of gas molecules, $=103.24 \times 10^{23}$ molecules/m^3 from Example 8.4

$N_s = 0$ (assumed value for complete diffusion within the film)

Thus, we have the diffusivity D:

$$D = \frac{(0.01281)(10^{24})}{(0.02)(103.24 \times 10^{23})} = 0.062 \text{ m}^2/\text{s}$$

5. The following expression is used to compute the surface reaction rate k_s, derived from Equation (8.22):

$$k_s = \frac{D\mathbf{N}}{DN_G - \delta\mathbf{N}} = \frac{(0.062)(10^{24})}{(0.062)(103.24 \times 10^{23}) - (0.01281)(10^{24})} = 0.09884 \text{ m/s}$$

6. The approximate rate of deposition r can be obtained by Equation (8.23a) or (8.23b), depending on the value of δk_s. This value is computed as

$$\delta k_s = (0.01281)(0.09884) = 0.0013 \ll D = 0.062$$

Hence Equation (8.23b) is used with a slight modification to account for the dilution factor of the reactant:

$$r = \eta \frac{N_G k_s}{\gamma}$$

From Table 8.7, we get the radii of atoms of silicon, a_{Si}, and oxygen, a_{O_2}:

$$a_{Si} = 0.117 \text{ nm} = 0.117 \times 10^{-9} \text{ m} \quad a_{O_2} = 0.06 \text{ nm} = 0.06 \times 10^{-9} \text{ m}$$

We will make a bold assumption that the radius of the SiO$_2$ molecule is the summation of the radii of the silicon and oxygen atoms, that is, $a_{SiO_2} = 0.177 \times 10^{-9}$ m. The corresponding number of SiO$_2$ molecules per unit film volume can be computed from Equation (8.24) as

$$\gamma = \frac{1}{(4/3)\pi (a_{SiO_2})^3} = \frac{1}{(4/3)(3.14)(0.177 \times 10^{-9})^3} = 4.3074 \times 10^{28}$$

This value leads to the following deposition rate:

$$r = \frac{N_G k_s}{\gamma} = (0.02)\frac{(103.24 \times 10^{23})(0.09884)}{4.3074 \times 10^{28}} = 0.4738 \times 10^{-6} \text{ m/s, or } 0.47 \ \mu\text{m/s}$$

It is useful for process engineers to be aware of the fact that the rate of CVD is affected by the following parameters:

Temperature $T^{3/2}$

Pressure of carrier gas, P^{-1}

Velocity of gas flow, V^{-1}

Distance in direction of gas flow, $x^{1/2}$, where x is as shown in Figure 8.10b

8.6.4 Enhanced CVD

The CVD process we have described involves elevated temperature but near atmospheric pressure. It is called *APCVD*, which stands for *atmospheric pressure CVD*. Several other CVD processes are used with better results for either higher rate of growth or better quality of deposit films. We will mention two such popular CVD processes here.

1. **Low-Pressure CVD (LPCVD):** As we see from Equation(8.23), the rate of growth of the deposited film is inversely proportional to the thickness of the boundary layer (δ) but directly proportional to the diffusivity of the reactant in the carrier gas (D). We have also learned from Equation (8.17) that the thickness δ of the boundary layer is inversely proportional to the square root of the Reynolds number Re. The diffusivity, on the other hand, varies inversely with pressure. The Reynolds number in Equation (8.18) depends on gas velocity V, viscosity μ, and density ρ.

Consider now a 1000 times reduction in gas pressure in the process. We may expect an increase in D by 1000 times, but with the same 1000 times reduction in density. The velocity of the gas with 1000 times reduction of pressure will reduce by 10–100 times in the process. These variations will result in a net reduction of Reynolds number by 10–100 times, which means a net increase of δ from $\sqrt{10} \approx 3$ to $\sqrt{100} = 10$ times. Despite the increase of δ by 3–10 times, the 1000 times increase of D will result in an increase in the diffusion flux **N** in Equation (8.19) by 10–30 times. It is therefore clear that reduction of gas pressure will increase the rate of deposition. Consequently, the deposition rate is inversely proportional to the gas pressure, as mentioned in the previous section.

The LPCVD operates in vacuum at about 1 Torr (1 mm of Hg). It uses a reactor chamber that is not much different from that used for APCVD. However, the chamber must be leakproof for the vacuum and structurally strong to withstand vacuum pressure in operation. The end product of the deposit film is typically more uniform. This technique allows the use of stacked wafers in the process, which is attractive from a mass production point of view.

2. **Plasma-Enhanced CVD (PECVD):** The CVD processes that we have learned thus far require the substrates and the carrier gases to be at elevated temperature for sufficient activation energies to allow for chemical reactions to take place. This high temperature can damage substrates, especially the metallized ones. *Plasma-enhanced CVD* utilizes the *RF plasma* to transfer energy into the reactant gases, which allows the substrates to remain at lower temperature than that in APCVD or LPCVD. A RF source can be electromagnetic radiation in the frequency band between 3 kHz and 300 GHz, or alternating currents in the same frequency range. Precise temperature control on

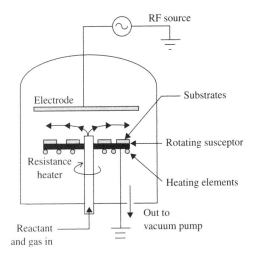

Figure 8.12. A PECVD reactor.

the substrate surfaces is necessary to ensure the quality of the deposit films. A typical PECVD reactor is illustrated in Figure 8.12.

A comprehensive summary and comparison of the three principal CVD processes are compiled in Table 8.8. It provides a useful guideline for process engineers in selecting a suitable CVD process for their microsystems.

TABLE 8.8. Summary and Comparison of Three Principal CVD Processes

CVD Process	Pressure and Temperature	Normal Depostion Rates, 10^{-10} m/min	Advantages	Disadvantages	Applications
APCVD	100–10 kPa, 350–400°C	700 for SiO_2	Simple, high rate, low temperature	Poor step coverage, particle contamination	Doped and undoped oxides
LPCVD	1–8 Torr, 550–900°C	50–180 for SiO_2 30–80 for Si_3N_4 100–200 for polysilicon	Excellent purity and uniformity, large wafer capacity	High temperature and low deposition rates	Doped and undoped oxides, silicon nitride, polysilicon, and tungsten.
PECVD	0.2–5 Torr, 300–400°C	300–350 for Si_3N_4	Lower substrate temperature; fast, good adhesion	Vulnerable to chemical contamination	Low-temperature insulators over metals, passivation

Source: Madou, 1997.

8.7   PHYSICAL VAPOR DEPOSITION: SPUTTERING

Sputtering is a process that is often used to deposit thin metallic films on the order of 100 Å thick ($1 \text{ Å} = 10^{-10}$ m) on substrate surfaces. Metallic films (or layers) are required to conduct electricity from signal generators in sensors or for the supply of electricity to an actuator. For example, metallic layers are required to transmit the signals generated in a piezoresistor in a micropressure sensor as illustrated in Figure 8.13a. A detailed arrangement of the electrical connection is shown in Figure 8.13b.

The sputtering process is carried out with plasma under very low pressure (i.e., in high vacuum at around 5×10^{-7} Torr). This process involves low temperature, which is contrary to CVD as presented in the previous section. At this temperature, little chemical reaction can take place. The process is thus regarded as physical deposition.

We have learned from Chapter 3 that plasma is made of positively charged gas ions, and plasma can be produced by either high-voltage DC sources or RF sources. Whichever

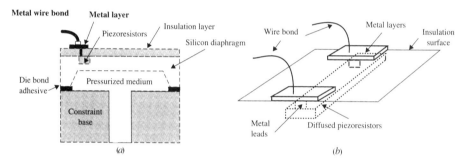

Figure 8.13. Metallic layers for signal transmission in a micropressure sensor: (*a*) metal layers for piezoresistors; (*b*) derailed arrangement for metal layers.

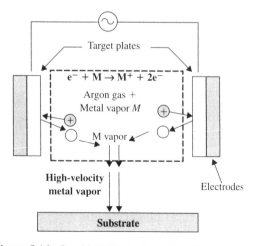

Figure 8.14. Graphical illustration of sputtering process.

the method, the positive ions of the metal in the inert argon gas carrier bombard the surface of the target at such a high velocity that the momentum transfer on impingement causes the metal ions to evaporate. The metal vapor is then led to the substrate surface and is deposited after condensation. This process is illustrated in Figure 8.14.

8.8   DEPOSITION BY EPITAXY

Epitaxy is the extension of a single-crystal substrate by growing a film of the *same* single-crystal material. For example, one may use this process to deposit silicon films over the desired parts of the silicon substrate in order to build the thickness of the microstructure. This process is frequently used in the microelectronics industry in the production of silicon diodes and transistors. For MEMS and microsystems, this technique is used to build the three-dimensional geometry of the devices.

Epitaxial deposition is very similar to the CVD processes in, for instance, the use of carrier gases with reactants involving the same substrate material. The main difference, however, is its ability to deposit not only the same substrate materials, such as silicon, but also compounds such as GaAs over the surface of the same material. Because most MEMS and microsystems use silicon as the substrate material, we will focus our attention only on the epitaxial deposition of silicon over silicon substrates.

There are several methods available for epitaxial deposition in microelectronics technology:

1. Vapor-phase epitaxy (VPE)
2. Molecular beam epitaxy (MBE)
3. Metal organic CVD (MOCVD)
4. Complementary metal oxidation of semiconductors (CMOS) epitaxy

The VPE technique appears to be the most popular in the IC industry, although CMOS is also frequently used in fabricating MEMS components. The VPE technique involves the use of reactant vapors containing silicon, such as those listed in Table 8.9, diluted in hydrogen carrier gas.

The production of silicon film over silicon substrates using silane vapor in Table 8.9 is the simplest of all. Silicon can be produced by simple pyrolysis at about $1000°C$ as follows:

$$SiH_4 \rightarrow Si_{(solid)} + 2H_{2(gas)} \tag{8.25}$$

All the other three reactant vapors in Table 8.9 with the hydrogen carrier gas will react with the silicon substrate surface to produce more silicon on the surface but also with by-products such as $SiC\ell_2$ and $HC\ell$. The silicon dichloride $SiC\ell_2$ in the by-products in turn releases more silicon and $SiC\ell_4$. The hydrogen chloride ($HC\ell$) is highly erosive, and it has an etching effect on the newly produced silicon film. We thus can envision a picture in which the process continuously produces silicon crystals while some of the produced silicon is etched by the by-product $HC\ell$. Delicate control of the process is thus

TABLE 8.9. Reactant Vapors for Epitaxial Deposition

Reactant Vapors	Normal Process Temperature, °C	Normal Deposition Rate, μm/min	Required energy Supply, eV	Remarks
Silane (SiH$_4$)	1000	0.1–0.5	1.6–1.7	No pattern shift
Dichlorosilane (SiH$_2$Cl$_2$)	1100	0.1–0.8	0.3–0.6	Some pattern shift
Trichlorosilane (SiHCl$_3$)	1175	0.2–0.8	0.8–1.0	Large pattern shift
Silicon tetrachloride (SiCl$_4$)	1225	0.2–1.0	1.6–1.7	Very large pattern shift

Source: Ruska, 1987.

critically important in order to ensure that the rate of silicon production exceeds the rate of silicon etching.

Epitaxy deposition is carried out in reactors (or chambers) that have similar arrangement to some of the reactors used for CVD (see Figure 8.9). A typical reactor is illustrated in Figure 8.15.

Epitaxial deposition has high risk of explosion as a result of the use of hydrogen as the carrier gas at high temperature. Therefore, the substrate surfaces must be thoroughly cleaned with solvents and remain clean during and after they are loaded and placed in the reactor. Once the substrates are in place in the reactor, purging of nitrogen gas takes

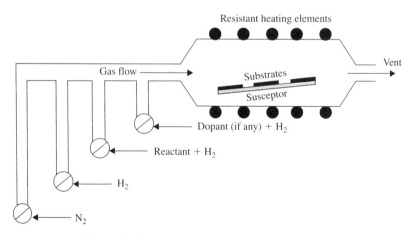

Figure 8.15. Horizontal epitaxy deposition reactor.

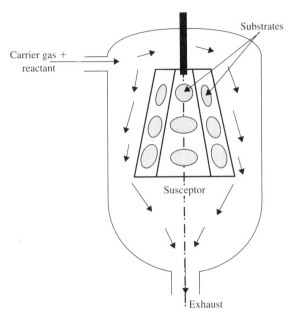

Figure 8.16. Vertical epitaxial deposition reactor.

place. The purging of this inert gas is necessary to drive out all the oxygen inside the chamber in order to avoid explosion from the mixing of oxygen and the hydrogen carrier gas at a high operating temperature. Hydrogen carrier gas is supplied to the chamber at an intermediate temperature of approximately $500°C$. This step is followed by the supply of selected reactant from Table 8.9 and more hydrogen gas for the epitaxy deposition of silicon on the silicon substrates. As mentioned earlier, the by-product HCℓ can etch the silicon film product. Thus, careful control of the process is maintained to assure that the produced silicon outpaces the silicon that is etched away by the hydrogen chloride. The substrates with deposited silicon must be thoroughly cleaned before they are removed from the chamber.

Figure 8.16 illustrates a vertical reactor for epitaxial deposition. The multiple valves to control the mixing of the gases and reactant vapors shown in Figure 8.15 are used for this type of reactor.

8.9 ETCHING

Etching is one of the most important processes in microfabrication. It involves the removal of materials in desired areas by physical or chemical means. It is a way to establish permanent patterns developed at the substrate surfaces by photolithography, as described in Section 8.2. In micromachining, etching is used to shape the geometry of microcomponents in MEMS and microsystems. For example, the cavity of the silicon die for a micropressure sensor such as illustrated in Figure 2.8 is produced by etching. A similar technique can be used to produce the silicon membranes and diaphragms of microvalves, as illustrated in Figures 2.41 and 2.42.

Of the two common types of etching techniques mentioned above, physical etching is usually referred to as *dry etching* or *plasma etching*, whereas chemical etching is referred to as *wet etching*. We will only present the working principles of both techniques in this chapter. Much of the detailed application of these techniques will be explained in Chapter 9.

8.9.1 Chemical Etching

Chemical etching involves using solutions with diluted chemicals to dissolve substrates. For instance, diluted hydrofluoric (HF) solution is used to dissolve SiO_2, Si_3N_4, and polycrystalline silicon, whereas KOH is used to etch the silicon substrate. The rates of etching vary, depending on the substrate materials to be etched, the concentration of the chemical reactants in the solution, and the temperature of the solution.

There are generally two types of etching available for shaping the geometry of MEMS components: (1) isotropic etching and (2) anisotropic etching. Isotropic etching is a process in which the etching of substrate takes place uniformly in all directions at the same rate. Anisotropic etching, on the other hand, etches away substrate material at faster rates in preferred directions.

The chemical solutions used in etching, or *etchants*, attack the parts of the substrate that are not protected by the mask. The masking used in micromachining may be either the photoresists for SiO_2 substrates in HF solutions as shown in Figure 8.1 or masks made of SiO_2 for the protection of the silicon substrate in KOH etchants as shown in Figure 8.17.

Wet etching is easy to apply and involves inexpensive equipment and facility for the process. It is also a faster etching process than dry etching. The etching rates in wet etching range from a few micrometers to several tens of micrometers per minute for isotropic etchants and about 1 μm/min for anisotropic etchants, whereas only. 0.1 μm/min is achievable in typical dry etching. Unfortunately, there are several disadvantages associated with wet etching. It often results in poor quality of etched surfaces due

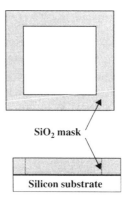

Figure 8.17. SiO_2 masking for etching cavity in micropressure sensors.

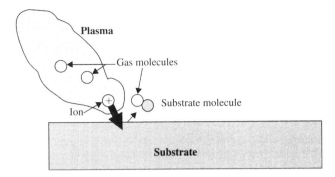

Figure 8.18. Plasma-assisted etching.

to bubbles and flow patterns of the solutions. No effective wet etching is available for some substrates, such as silicon nitrides.

8.9.2 Plasma Etching

The plasma used for etching is a stream of positive-charge-carrying ions of a substance with a large number of electrons diluted by inert carrier gas such as argon. It can be generated by continuous applications of high-voltage electric charge, or by RF sources. Plasmas are usually generated in a low-pressure environment, or a vacuum.

As illustrated in Figure 8.18, the high-energy plasma containing gas molecules, free electrons, and gas ions bombards the surface of the target substrate and knocks off the substrate material from its surface. This process is like a reversed sputtering process at low temperature in the range of 50–100°C. It takes place in high vacuum.

8.10 SUMMARY OF MICROFABRICATION

This chapter has provided an overview of available microfabrication techniques for MEMS and microsystems. These techniques are used to produce the primarily three-dimensional geometry of most microdevices and microsystems.

Other than the special LIGA micromanufacturing technology, which will be described in Chapter 9, most silicon-based MEMS and microsystems are produced by adding material in the form of thin films to the silicon substrates or by removing portions of silicon from the substrates.

One popular technique for adding materials to silicon substrates is CVD. The intended materials, called *reactants*, are diluted in certain carrier gases. The mixture of the depositing material and the carrier gas is then led to a reactor in which the substrates are situated. The reactants are deposited on the surface of the substrates by means of combined diffusion and chemical reactions. A similar process called *epitaxial deposition* is used to deposit thin films of reactant over the surface of a substrate of the same material. It is widely used in depositing silicon crystals on the surface of silicon substrates. Another deposition technique is referred to as *physical deposition*, or *sputtering*. This process is used to deposit mostly thin metal films on silicon substrates by means of ionization of the depositing materials, energized by plasma with RF sources.

Removing portions of silicon from the substrate can also create the three-dimensional geometry of MEMS components. The principal process used for this purpose is the *etching* process. There are generally two types of etching processes in microfabrication: dry etching and wet etching. The former process involves the use of plasma as a carving force for material removal, whereas wet etching uses a chemical reaction to dissolve a portion of the material from the substrate. Dry etching is slow in removing material but usually produces better edge definition of the cavity. The design engineer will choose the etching process according to the needs for the intended MEMS product.

In all microfabrication processes, the geometry of the workpieces is defined and controlled by appropriate masks, which are produced by a photolithographic process. The layout of the masks is usually drawn at a macroscale. They are then reduced to the desired microscale by means of photoreduction procedures. The reduced microscale layout is then projected onto the base mask coated with photoresist materials. A permanent print of the layout is produced at the surface of the mask material after processes that are similar to those in photodevelopment. An etching process is used to remove the unwanted portion of the mask material, and finished masks are thus produced.

Microfabrication processes also include the control of local material properties and characteristics by ion implantation, diffusion, and oxidation. Ion implantation and diffusion processes are frequently used in doping silicon substrate to alter its electrical conductance, either in selected portions or for the entire volume. These processes produce most piezoresistors for signal transduction in microsensors.

PROBLEMS

Part 1 Multiple Choice

1. Microfabrication technologies are developed specifically to shape structures in (a) macro-, (b) meso-, (c) microscale.

2. Established microfabrication techniques primarily involve (a) electromechanical, (b) electrochemical, (c) physical–chemical means.

3. A class 1000 clean room is defined as one in which the number of dust particles smaller than 0.5 μm is less than 1000 per cubic (a) inch, (b) foot, (c) meter.

4. The dust particle size used in defining the clean-room air quality is (a) 0.05 μm, (b) 0.5 μm, (c) 5 μm.

5. The higher the class number of a clean room, the (a) cleaner, (b) dirtier, (c) neither cleaner nor dirtier will the air be in the room.

6. The air quality in a typical urban environment is equivalent to a clean room of class (a) 50,000, (b) 500,000, (c) 5,000,000.

7. Photolithography is used in microfabrication because we need to (a) take a photograph of the microdevice, (b) create microscale patterns on substrates, (c) create pictures at the microscale.

8. The photoresist that, after exposure to light, dissolves in the development is (a) the positive type, (b) the negative type, (c) either positive or negative type.

9. Photolithography using positive-type photoresists results in (a) better, (b) poorer, (c) about the same effect as using the negative photoresists.

10. Typical thickness of photoresists in a photolithographic process is (a) 0.1–1.0 μm, (b) 0.5–2.0 μm, (c) 1–2 μm.

11. Common light sources used in the photolithographic process have wavelengths in the range of (a) 100–300 nm, (b) 300–500 nm, (c) 500–700 nm.

12. The development of positive photoresists is (a) more complex than, (b) less complex than, (c) about equally as complex as that of negative photoresists.

13. Ion implantation is one of (a) two, (b) three, (c) four techniques frequently used for doping semiconductors.

14. Ion implantation is implanting foreign substances in silicon substrates by (a) melting, (b) insertion by force, (c) slow diffusion.

15. The ion implantation process takes place at (a) high temperature, (b) room temperature, (c) low temperature.

16. A common energy source used for ion implantation involves (a) an ion beam, (b) intense heating, (c) high-energy electromagnetic fields.

17. The implanted foreign substance beneath the substrate's surface exhibits (a) uniform distribution in density, (b) nonuniform distribution in density with less near the surface, (c) a distribution that depends on the temperature in the process.

18. Diffusion is used for doping semiconductors. It is (a) slower than, (b) faster than, (c) about the same speed as the ion implantation process.

19. The diffusion process takes place at (a) high temperature, (b) room temperature, (c) low temperature.

20. The diffused foreign substance beneath the substrate's surface exhibits (a) uniform distribution in density, (b) nonuniform distribution in density with the highest near the surface, (c) a distribution that depends on the temperature in the process.

21. A mathematical model of the diffusion process is based on (a) Fourier's law, (b) Newton's law, (c) Fick's law.

22. Wet oxidation of silicon is often preferred because of (a) better quality of SiO_2, (b) faster oxidation, (c) lower cost.

23. Oxidation of silicon substrates is (a) desired for protection of the substrate surface, (b) needed for local electric and thermal insulation, (c) an unavoidable phenomenon.

24. The kinetics of thermal oxidation is used to assess the (a) growth, (b) erosion, (c) plating of silicon oxide layers in silicon substrates.

25. One torr is equal to (a) 1 in. of H_2O, (b) 1 cm of Hg, (c) 1 mm of Hg pressure.

26. The color of oxidized silicon observed under while light represents (a) only one, (b) two, (c) several specific thicknesses of the oxide layer.

27. The deposition process in microfabrication can deposit (a) only organic, (b) only inorganic, (c) any materials onto substrate surfaces.

28. There are generally (a) two, (b) three, (c) four types of deposition methods in microelectronics and micromachining.

29. CVD is effective in depositing foreign materials over silicon substrates because it is a process that (a) is thermally activated, (b) combines mechanical and chemical diffusion, (c) combines thermal diffusion and chemical reactions.

30. The necessary ingredients in CVD are (a) plasma and chemical reactants, (b) carrier gas and chemical reactants, (c) chemical reactants and charge-carrying ions.

31. CVD processes require the substrate's surface to be (a) cold, (b) moderately hot, (c) very hot.

32. Better results in CVD are achievable by (a) increasing the pressure, (b) decreasing the pressure, (c) maintaining high constant pressure in the process.

33. The boundary layer created between the flowing carrier gas and the substrate surface in a CVD process (a) retards, (b) enhances, (c) has no effect on the CVD process.

34. The thickness of the boundary layer in a CVD process (a) increases, (b) decreases, (c) remains constant with a decreased velocity of the carrier gas.

35. Avogadro's number of 6.022×10^{23} is defined as the number of (a) electrons, (b) atoms, (c) molecules contained in 1 mol of any gas.

36. The rate of CVD is (a) proportional to, (b) inversely proportional to, (c) independent of temperature.

37. The rate of CVD is (a) proportional to, (b) inversely proportional to, (c) independent of the carrier gas pressure

38. The rate of CVD is (a) proportional to, (b) inversely proportional to, (c) independent of the carrier gas velocity.

39. The rate of CVD is (a) proportional to, (b) inversely proportional to, (c) independent of thickness of the boundary layer between the carrier gas and the substrate surface.

40. PECVD is popular because it offers (a) good adhesion, (b) a simple process, (c) relatively low operating temperature.

41. The process engineer would choose (a) APCVD, (b) LPCVD, (c) PECVD for lower process temperature.

42. The process engineer would choose (a) APCVD, (b) LPCVD, (c) PECVD for higher rate of deposition.

43. Sputtering is processed at (a) low, (b) elevated, (c) high temperature.

44. Sputtering is normally used for depositing (a) organic, (b) inorganic, (c) metal films over silicon substrates.

45. Epitaxy involves the growth of (a) single-crystal films, (b) organic films, (c) metallic films over a substrate made of the same material.

46. CMOS is a (a) CVD, (b) PVD, (c) epitaxy thin-film growth process.

47. A common carrier gas used in epitaxy deposition is (a) oxygen, (b) nitrogen, (c) hydrogen.

48. Wet etching involves the use of (a) distilled water, (b) mineral water, (c) chemical solutions to dissolve the materials intended for removal.

49. Dry etching involves the use of (a) dry air, (b) dry toxic gas, (c) plasma to remove the substrate material.

50. In general, wet etching is (a) 10, (b) 100, (c) 1000 times faster in removing materials from silicon substrate than dry etching.

Part 2 Computational Problems

1. Solve Example 8.1, but change the doping substance to phosphorus at 30 keV energy level.

2. Solve Example 8.1, but change the doping substances to phosphorus and arsenic. What observations would you make on the relative merits of doping of silicon substrates with these three common dopants using the ion implantation method?

3. Solve Example 8.2, but with diffusion temperatures ranging from 800 to 1100°C. What observation, if any, will you make from this exercise?

4. In Example 8.2, estimate the time required to reach a concentration of the dopant at 0.2 μm beneath the substrate surface.

5. In Example 8.3, estimate the required time to achieve a 1-μm-thick SiO_2 layer over the substrate surface in both wet and dry processes.

6. A CVD process involves using a reactant diluted at 1% in hydrogen gas at 800°C to deposit SiO_2 on silicon substrate. The horizontal tubular reactor has a diameter of 20 cm, and the susceptor that holds the silicon substrate is 20 cm long. Pressure variation in the reactor is negligible. All other operating conditions are identical to those specified in Example 8.5. Find the following:

    a. Number of molecules in a cubic meter volume of gas mixture

    b. Molar density of gas mixture

    c. Reynolds number of gas flow

    d. Thickness of boundary layer over substrate surface

    e. Diffusivity of carrier gas with reactant to silicon substrates

    f. Surface reaction rate

    g. Deposition rate

7. Determine the required time for depositing a 0.5-μm-thick film over silicon substrate with conditions described in problem 6.

8. What will be the deposition rate in problem 6 if the process temperature is dropped to 490°C?

9. What will be the deposition rate in problem 6 if the process pressure is dropped to 1 Torr?

10. What will be the deposition rate in problem 6 if the velocity of the carrier gas is reduced to 25 mm/s?

CHAPTER 9

OVERVIEW OF MICROMANUFACTURING

9.1 INTRODUCTION

For mechanical engineers, a major effort in manufacturing a product involves proper selection and application of fabrication techniques such as machining, drilling, milling, forging, welding, casting, molding, stamping, and penning. We will quickly realize that none of the aforementioned traditional fabrication techniques can be used in manufacturing MEMS and microsystems because of the extremely small size of these products. Some of these traditional fabrication techniques, however, are used in the packaging of MEMS and microsystems products.

The microfabrication techniques that we have learned in the last chapter are process related. What we will learn in this chapter is how these processes can be used either individually or in an integrated nature in the manufacture of MEMS and microsystems products such as microsensors, accelerometers, and actuators as described in Chapter 2. The technique used to produce these products is called *micromachining* or *micromanufacturing.*

Generally speaking, three micromachining techniques are being used: (1) *bulk micromanufacturing*, (2) *surface micromachining*, and (3) the *LIGA* process. The term LIGA is an acronym for the German term for lithography, electroforming, and plastic molding. A number of excellent articles have been published in recent years on micromachining and the MEMS products produced by these three micromanufacturing techniques (O'Connor, 1992; Bryzek et al., 1994; Pottenger et al., 1997).

Of the other process-related micromachining techniques developed in recent years, laser drilling and "machining" appear to be gaining popularity. However, in this chapter we will focus on the aforementioned three principal micromanufacturing techniques.

9.2 BULK MICROMANUFACTURING

Bulk micromanufacturing is widely used in the production of microsensors and accelerometers. It was first used in microelectronics in the 1960s. Further improvement for producing three-dimensonal microstructures took place in the 1970s.

Bulk micromanufacturing or micromachining involves the removal of materials from the bulk substrates, usually silicon wafers, to form the desired three-dimensional geometry of the microstructures. The technique is thus similar to that used by sculptors in shaping sculptures. Neomicrosculpts by handicraft was achievable and was well documented in ancient Chinese history. For instance, a vivid scene of a tea party involving a famous Chinese poet and his friend and two servants inside a small boat was carved out from an olive nut a little over 1 cm long by a microsculptor in 1737. The display of this minute sculpture can be seen at the Imperial Museum of Arts in Taipei, Taiwan. There are similar microsculptures displayed in other museums in China. However, all these sculptures are handicrafts, not industry products, and hardwood was exclusively used in them. Shaping of microsystems components of the size between 0.1 μm and 1 mm made of tough materials such as silicon is beyond any existing mechanical means. Physical or chemical techniques, either by dry or wet etching, are the only practical solutions. Substrates that can be treated this way include silicon, SiC, GaAs, and quartz, as mentioned in Chapter 8. Etching, using either the orientation-independent isotropic etching or the orientation-dependent anisotropic etching, is thus the key technology used in bulk micromachining.

9.2.1 Overview of Etching

We will deal with wet (chemical) etching in this section. As we learned from Section 8.9, etching involves the exposure of a substrate covered by an etchant protection mask exposed to chemical etchants, as illustrated in Figure 9.1a.

The part of the substrate that is not covered by the protective mask is dissolved in the etchants and removed. However, as we will see from Figure 9.1b in an exaggerated way, the etching can undercut the part that is immediately under the protective mask after a lengthy period of time. Additionally, it may also damage the protective mask.

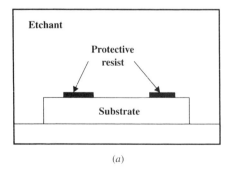
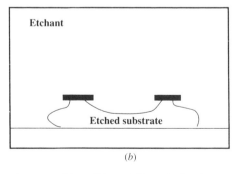

Figure 9.1. Wet etching of substrates: (a) substrate in wet etching; (b) partially etched substrate.

9.2.2 Isotropic and Anisotropic Etching

For substrates made of homogeneous and isotropic materials, the chemical etchants will attack the material uniformly in all directions, as illustrated in Figure 9.1*b*. This orientation-independent etching is referred to as *isotropic etching*.

Isotropic etching is hardly desirable in micromanufacturing due to the lack of control of the finished geometry of the workpiece. Fortunately, most substrate materials are not isotropic in their crystalline structures, as described in Chapter 7. For example, silicon has a diamond cubic crystal structure. Therefore, some parts in the crystal are stronger, and thus more resistant to etching than others. Again, as we have learned from Chapter 7, three planes of silicon crystals are of particular importance in micromachining, the (100), (110), and (111) planes, as illustrated in Figure 9.2.

The three orientations <100>, <110>, and <111> are the respective normal lines to the (100), (110), and (111) planes. The two most common orientations used in the IC industry are the <100> and <111> orientations. However, in micromachining, the <110> orientation is favored. This is because in this orientation the wafer breaks or cleaves more cleanly than in the other orientations. The (110) plane is the only plane in which one can cleave the crystal in vertical edges. The (111) plane, on the other hand, is the toughest plane to treat. Thus, the <111> orientation is the least-used orientation in micromachining. This nonuniformity in mechanical strength also reflects the degree of readiness for etching. The material on the (111) plane obviously is the hardest to be etched. A 400 : 1 ratio in etching rates for silicon with <100> to <111> orientations is possible.

Referring to the arrangement of atoms in silicon crystals illustrated in Figures 7.5 and 7.7, we find that the (111) plane intersects the (100) plane at a steep angle of 54.74° (Bean, 1978). Thus, when a wafer whose face coincides with the (100) plane is exposed to the etchants, we can expect different etching rates in different orientations. A pyramid with sidewall slope at 54.74° exists in the finished product (Angell et al., 1983), as illustrated in Figure 9.3.

Despite the many advantages of anisotropic etching in controlling the shape of the etched substrates, there are several disadvantages: (1) It is slower than isotropic etching; the rate rarely exceeds 1 μm/min. (2) The etching rate is temperature sensitive. (3) It usually requires an elevated temperature around 100°C in the process, which precludes the use of many photoresistive masking materials.

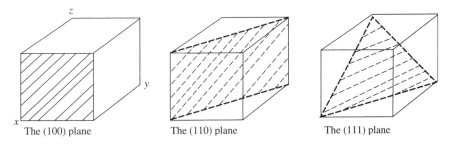

Figure 9.2. Three principal planes in silicon crystal.

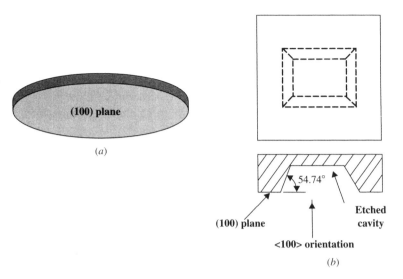

Figure 9.3. Anisotropic etching of silicon substrate: (*a*) unetched wafer; (*b*)wafer etched in <100> orientation.

9.2.3 Wet Etchants

There are a number of different types of etchants that can be used to etch different substrate materials. The common isotropic etchant for silicon is called HNA, which designates acidic agents such as $HF/HNO_3/CH_3COOH$. These etchants can be used effectively at room temperature. Alkaline chemicals with pH > 12, on the other hand, are used for anisotropic etching. Popular anisotropic etchants for silicon include potassium hydroxide (KOH), ethylene-diamine and pyrocatecol (EDP), tetramethyl ammonium hydroxide (TMAH), and hydrazine. Most etchants based on the above chemicals are diluted with water, normally 1 : 1 by weight. Typical ranges for etching rates for common substrate materials with these etchants are given in Table 9.1 (Wise, 1991; Kovacs, 1998).

TABLE 9.1. Typical Etch Rates for Silicon and Silicon Compounds

Materials	Etchants	Etch Rates
Silicon in <100>	KOH	0.25–1.4 μm/min
	EDP	0.75 μm/min
Silicon dioxide	KOH	40–80 nm/h
	EDP	12 nm/h
Silicon nitride	KOH	5 nm/h
	EDP	6 nm/h

TABLE 9.2. Selectivity Ratios of Etchants to Two Silicon Substrates

Substrates	Etchants	Selectivity Ratios
Silicon dioxide	KOH	10^3
	TMAH	10^3–10^4
	EDP	10^3–10^4
Silicon nitride	KOH	10^4
	TMAH	10^3–10^4
	EDP	10^4

Source: Kovacs, 1998.

We may observe from Table 9.1 that the etching rate for silicon dioxide with KOH is 1000 time slower than that for silicon, and silicon nitride is another order slower than that for silicon dioxide. Table 9.2 shows the selectivity ratio of etchants in different substrates. The *selectivity ratio* of a material is defined as the ratio of the etching rate of silicon to the etching rate of another material using the same etchant. For example, silicon dioxide has a selectivity ratio of 10^3 in Table 9.2, meaning this material has an etching rate in KOH that is 10^3 times slower than the etching rate for silicon. Thus, the higher the selectivity ratio of the material, the better masking material it is.

Consequently, the high selectivity ratio of silicon dioxide and silicon nitride makes these materials suitable candidates for masks for etching silicon substrates. However, the timing of etching and the agitated flow patterns of the etchants over the substrate surfaces need to be carefully controlled in order to avoid serious underetching and undercutting, as illustrated in Figure 9.4.

One also has to take special caution in selecting the masking materials. A common practice is to use a SiO_2 layer as a mask for silicon substrate in KOH etchants for trenches of modest depth. SiO_2 masking is relatively inexpensive in an etching process. However, even though etching is a slow process, the SiO_2 mask can be attacked by the etchants if the system is left in the etchant for a long period of time, as in the case of deep etching. In such cases, silicon nitride should be used as the mask.

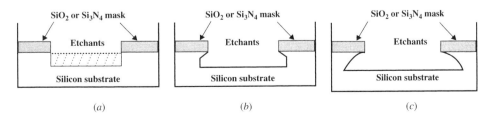

Figure 9.4. Definition of etched geometry: (*a*) ideal etching; (*b*) underetching; (*c*) undercutting.

An effective way to control the shape of the etched silicon substrate as well as achieve acceptably clean and accurate edge definition is to apply etch stop, as will be presented below.

9.2.4 Etch Stop

There are two popular techniques used in etch stop:

Dopant-Controlled Etch Stop   A peculiar phenomenon that can be used to control the etching of silicon is that doped silicon substrates, whether they are doped with boron for p-type silicon or phosphorus or arsenic for n-type silicon, show a different etching rate than pure silicon. When isotropic HNA etchants are used, the p- or n-doped areas are dissolved significantly faster than the undoped regions. The etchant HNA is a mixture of hydrofluric acid, nitric acid, and acetic acid. However, excessive doping of boron in silicon for fast etching can introduce lattice distortion in the silicon crystal and thus produce undesirable internal (residual) stresses.

Electrochemical Etch Stop   This technique is popular in controlling anisotropic etching. As illustrated in Figure 9.5, a lightly doped $p–n$ junction is first produced in the silicon wafer by a diffusion process. The n-type is phosphorus doped at $10^{15}/cm^3$ and the p-type is boron doped at 30 Ω-cm (refer to Figure 3.8 for the corresponding doping). The doped silicon substrate is then mounted on an inert substrate container made of a material such as sapphire. The n-type silicon layer is used as one of the electrodes in an electrolyte system with a constant voltage source as shown in Figure 9.5 (Madou, 1997).

As we may observe from the arrangement in Figure 9.5, the unmasked part of the p-type substrate face is in contact with the etchant. Etching thus takes place as usual until it reaches the interface of the $p–n$ junction, at which point etching stops because of

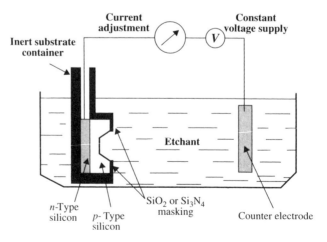

Figure 9.5. Arrangement for electrochemical etch stop.

the rate difference in *p*- and *n*-doped silicon. Consequently, one can effectively control the depth of etching simply by establishing the *p*–*n* silicon boundaries at the desired locations in a doped silicon substrate.

9.2.5 Dry Etching

Dry etching involves the removal of substrate materials by gaseous etchants without wet chemicals or rinsing. There are three dry etching techniques: *plasma, ion milling, and reactive ion etch (RIE)* (van Zant, 1997). In this section we will focus on plasma etching and a relatively new technique called *deep reactive ion etching (DRIE)*.

Plasma Etching. As we learned in Chapter 3, plasma is a neutral ionized gas carrying a large number of free electrons and positively charged ions. A common source of energy for generating plasma is the RF source. The process involves adding a chemically reactive gas such as $CC\ell_2F_2$ to the plasma, one that contains ions and has its own carrier gas (inert gas such as argon). As illustrated in Figure 9.6, the reactive gas produces reactive neutrals when it is ionized in the plasma. The reactive neutrals bombard the target on both sidewalls as well as the normal surface, whereas the charged ions bombard only the normal surface of the substrate. Etching of the substrate materials is accomplished by high-energy ions in the plasma bombarding the substrate surface with the simultaneous chemical reactions between the reactive neutral ions and the substrate material. This high-energy reaction causes local evaporation and thus results in the removal of the substrate material. One may envisage that the etching front moves more rapidly in the depth direction than in the direction of the sidewalls. This is due to the larger number of high-energy particles involving both the neutral ions and the charged ions bombarding the normal surface, while the sidewalls are bombarded by neutral ions only.

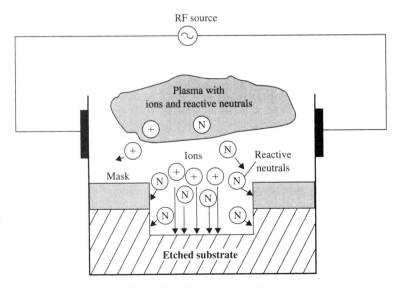

Figure 9.6. Plasma ion etching.

Conventional dry etching is a very slow process, at a rate of about 0.1 μm/min, or 100 Å/min.

Plasma etching can increase the etching rates to on the order of 2000 Å/min. This increase in etching rate is primarily due to the increased mean free path of the reacting gas molecules in the depth to be etched. The working principle of plasma etching was presented in Section 8.9.2. In many ways, it can be viewed as a reversed sputtering process. Plasma etching is normally performed in high vacuum. Table 9.3 offers a number of etchants for selected substrate materials.

Dry etching of silicon substrates, such as by plasma, typically is faster and cleaner than wet etching. A typical dry etching rate is 5 μm/min, which is about five times that by wet etching. Like wet etching, dry etching suffers the shortcoming of being limited to producing shallow trenches. Consequently, both wet and dry etching processes are limited to producing MEMS with low *aspect ratios*. The aspect ratio (*A/P*) of a MEMS component is defined as the ratio of its dimension in depth to those on the surface. For dry etching, *A/P* is less than 15. Another problem with dry etching relates to the contamination of the substrate surface by residues.

TABLE 9.3. Plasma Gas Etchants for Selected Substrate Materials

Substrate Materials	Conventional Chemicals	New Chemicals
Silicon and silicon dioxide, SiO_2	CCl_2F_2	CCl_2F_2
	CF_4	CHF_2/CF_4
	C_2F_6	CHF_3/O_2
	C_3F_8	CH_2CHF_2
Silicon nitride, Si_3N_4	CCl_2F_2	CF_4/O_2
	CHF_3	CF_4/H_2
		CHF_3
		CH_3CHF_2
Polysilicon	Cl_2 or BCl_3/CCl_4	$SiCl_4/Cl_2$
	Cl_2 or BCl_3/CF_4	BCl_2/Cl_2
	Cl_2 or $BCl_3/CHCl_3$	$HBr/Cl_2/O_2$
	Cl_2 or BCl_3/CHF_3	HBr/O_2
		Br_2/SF_6
		SF_6
		CF_4
Gallium arsenide	CCl_2F_2	$SiCl_4/SF_6$
		$SiCl_4/HF_3$
		$SiCl_4/CF_4$

Source: van Zant, 1997.

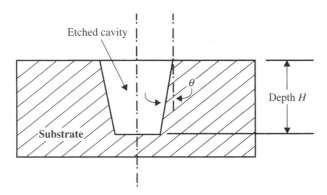

Figure 9.7. Sidewall angle in an etched cavity.

Deep Reactive Ion Etching. Despite the significant increase in the etching rate and the depth of the etched trench or cavity that can be achieved with the use of plasma, the etched walls in the trenches remain at wide angles (θ) to its depth, as illustrated in Figure 9.7. The cavity angle θ is critical in many MEMS structures, such as the plate electrodes and springs in microresonators, as shown in Figure 9.8. These structures require the faces of the electrodes, or "the fingers" in Figure 9.8a, and the curved "leaves" in the spring in Figure 9.8b to be parallel to each other. Etching processes produced most of these comb electrodes. It is highly desirable that the angle θ be kept at a minimum in deep-etched trenches that separate the plate electrodes. Obtaining deep trenches with vertical walls has been a major impediment of bulk manufacturing for a long time. Consequently, the bulk manufacturing technique has been generally regarded as suitable only for MEMS with low aspect ratio and, in many cases, with tapered cavity walls.

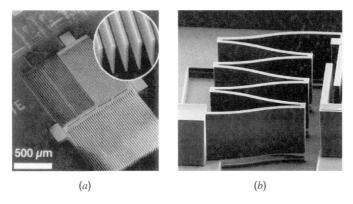

(*a*)   (*b*)

Figure 9.8. High-aspect-ratio plate components in MEMS. (*a*) a plate electrode; (*b*) plate spring. (*Source*: Kovacs, 1998.)

Deep reactive ion etching can overcome the problem described above. The DRIE process has since extended the use of the bulk manufacturing technique to the production of MEMS of high aspect ratio with virtually vertical walls; that is, $\theta \approx 0$.

The DRIE process differs from dry plasma etching in that it produces thin protective films of a few micrometers on the sidewalls during the etching process. It involves the use of a high-density plasma source, which allows alternating between plasma (ion) etching of the substrate material and deposition of etching-protective material on the sidewalls, as illustrated in Figure 9.9. Suitable etching-protective materials (shown in black in the figure) are materials with high selectivity ratio, such as silicon dioxide in Table 9.2. Polymers are also frequently used for this purpose. Polymeric materials such as photoresists are produced by polymerization during the plasma etching process.

The DRIE process with polymeric sidewall protection has been used to produce MEMS structures with $A/P = 30$ with virtually vertical walls of $\theta = \pm 2°$ for several years. Recent developments have substantially improved the performance of DRIE with better sidewall protecting materials. For example, silicon substrates with A/P over 100 with $\theta = \pm 2°$ at a depth of up to 300 μm were achieved by using photoresist sidewall protecting materials as presented in Table 9.4. The DRIE rate, however, was reduced to 2–3 μm/min. A more recent report (Williams, 1998) indicated that a trench depth of up to 380 μm was obtained by this technique.

A number of reactant gases could be used in DRIE. One is fluropolymers (nCF_2) in the plasma of argon gas ions. This reactant can produce a polymer protective layer

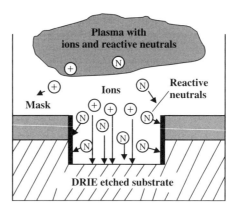

Figure 9.9. Deep reactive ion etching process.

TABLE 9.4. Popular Sidewall Protection Materials for DRIE Process

Sidewall Protection Materials	Selectivity Ratio	Aspect Ratio, A/P
Polymer		30 : 1
Photoresists	50 : 1	100 : 1
Silicon dioxide	120 : 1	200 : 1

TABLE 9.5. Comparison of Wet versus Dry Etching

Parameters	Dry Etching	Wet Etching
Directionality	Good for most materials	Only with single-crystal materials (aspect ratio up to 100)
Production-automation	Good	Poor
Environmental impact	Low	High
Masking film adherence	Not as critical	Very critical
Selectivity	Poor	Very good
Materials to be etched	Only certain materials	All
Process scaleup	Difficult	Easy
Cleanliness	Conditionally clean	Good to very good
Critical dimensional control	Very good (<0.1 μm)	Poor
Equipment cost	Expensive	Less expensive
Typical etch rate	Slow (0.1 μm/min) to fast (6 μm/min)	Fast (1 μm/min and up)
Operational parameters	Many	Few
Control of etch rate	Good in case of slow etch	Difficult

Source: Madou, 1997.

on the sidewalls while etching takes place. The rate of etching is in the range of 2–3 μm/min, which is higher than what wet etching can accomplish. Selectivity ratios of 100 for photoresists and 200 for silicon dioxide have been recorded.

9.2.6 Comparison of Wet versus Dry Etching

Etching is such an important process in bulk micromachining that engineers need to make intelligent choices on which of the two types of etching to use for shaping the micromachine components. Table 9.5 will be a useful reference for this purpose.

9.3 SURFACE MICROMACHINING

9.3.1 Description

In contrast to bulk micromachining in which substrate material is removed by physical or chemical means, the surface micromachining technique builds the microstructure by adding materials layer by layer on top of the substrate. Deposition techniques, in particular LPCVD, such as those described in Section 8.6.4 are used for such buildups, and polycrystalline silicon (polysilicon) is a common material for the layer material. *Sacrificial layers*, usually made of SiO$_2$, are used in constructing the MEMS components but are later removed to create necessary void space in the depth, that is, in the thickness direction. Wet etching is the common method used for this purpose.

We thus see that, although we will still deal with single-crystal silicon as the substrate in most cases, the added layers need not be single crystals or silicon compounds. The overall height of the structure therefore is no longer limited by commercially available wafer thickness. Layers that are being added in surface micromachining are typically 2–5 μm thick each. In special applications, this range can be extended to 5–20 μm. They are thus regarded as thin films in micromanufacturing. We realize that there will be problems associated with structures built with thin films.

Figure 9.10 illustrates the difference between bulk micromanufacturing and surface micromachining. In Figure 9.10*a* we see a microcantilever beam that can be used as either a microaccelerometer (Figure 2.34) or an actuator (Figures 2.19 and 2.21). The cantilever is made of single-crystal silicon with a significant amount of material etched away, as illustrated in Figure 9.11. The same cantilever beam structure can be produced by polysilicon with the surface micromachining technique illustrated in Figure 9.10*b*. We can see from this figure that surface micromachining not only saves material but also eliminates the need for a die attach, as the polysilicon can be built on top of the constraint base directly.

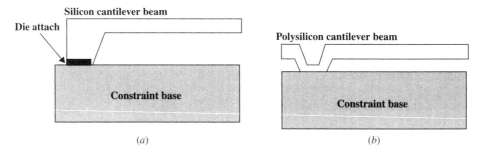

Figure 9.10. Micro–cantilever beams produced by two micromachining techniques: (*a*) bulk micromachining; (*b*) surface micromachining.

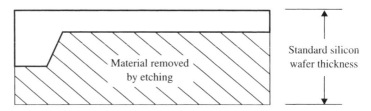

Figure 9.11. Waste of material in bulk micromanufacturing.

9.3.2 Process

Surface-micromachined devices are typically made of three types of components: (1) a sacrificial component (also called spacer layer), (2) a microstructural component, and (3) an insulator component.

The sacrificial components are usually made of phosphosilicate glass (PSG) or SiO_2 deposited on substrates using LPCVD. The PSG can be etched more rapidly than SiO_2 in HF etchants. These components in the form of films can be as long as 1–2000 μm by 0.1–5 μm thick. Both microstructural and insulator components can be deposited in thin films. Polysilicon is a popular material. The etching rates for the sacrificial components must be much higher than those for the other two components.

In Figure 9.12, we demonstrate how a microcantilever beam such as the one shown in Figure 9.10*b* is produced using surface micromachining. We begin in step 1 with a

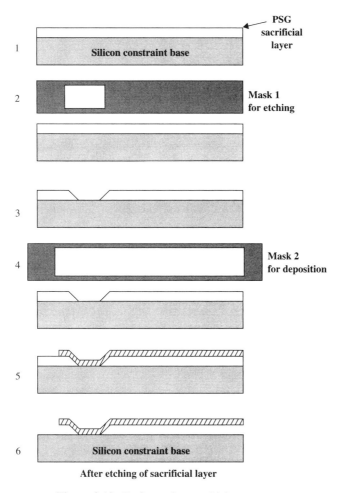

Figure 9.12. Surface micromachining process.

silicon substrate base with a PSG deposited on its surface. A mask (mask 1) is made in step 2 to cover the surface of the PSG layer for the subsequent etching to allow for the attachment of the cantilever beam, as shown in step 3. Another mask (mask 2) is made for the deposition of polysilicon microstructural material in step 4. The PSG that remains in step 5 is subsequently etched away to produce the desired cantilever beam, as shown in step 6. The most suitable etchant used in the last step for removing the sacrificial PSG layer is $1:1$ HF, which is made of $1:1$ HF–$H_2O + 1:1$ HCℓ–H_2O. After etching, the structure is thoroughly rinsed in deionized water followed by drying under infrared lamps. The etching rates of these etchants for various sacrificial materials are available in Table 9.6.

9.3.3 Mechanical Problems Associated with Surface Micromachining

Three major problems of a mechanical nature result from surface micromachining: (1) adhesion of layers, (2) interfacial stresses, and (3) stiction.

Adhesion of Layers. Whenever two layers of materials, whether similar or dissimilar, are bonded together, a possibility of delamination exists. A bilayer structure can delaminate at the interface either by peeling of one layer from the other or by shear that causes the severing of the interfaces locally along the interface. Figure 9.13 illustrates both failures.

Of the many causes for interfacial failures, excessive thermal and mechanical stress is the main cause. However, other causes, including surface conditions (e.g., cleanliness, roughness, and adsorption energy), could also contribute to the weakening of the interfacial bonding strength. Fracture mechanics theories presented in Section 4.5.3 may be used to assess the fracture strength of the bonded structure. However, fracture toughness K_{IC} for the opening mode and K_{IIC} for the shearing mode of the materials must be available for such analysis to be meaningful.

Interfacial Stresses. Typically three types of stresses exist in bilayer structures. The most obvious is the thermal stress resulting from a mismatch of the coefficient of thermal expansion (CTE) of the component materials. This phenomenon was described in Section 4.4.3. For example, we will find that the CTE for silicon is about 5 times that of SiO_2 (see Table 7.3). Severe thermal stress can cause the delamination of SiO_2

TABLE 9.6. Etch Rate in HF/HCl for Sacrificial Oxides

Thin Oxide Films	Lateral Etch Rate (μm/min)
CVD SiO_2 (densified at $1050°C$ for 30 min)	0.6170
Ion-implanted SiO_2 (at 8×10^{15}/cm^2, 50 kev)	0.8330
Phoshosilicate (PSG)	1.1330
5%–5% Boronphosphosilicate (BPSG)	4.1670

Source: Madou, 1997.

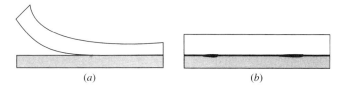

Figure 9.13. Interfacial failure of bilayer materials: (*a*) peeling off; (*b*) severing of interfaces by shear.

layer from the silicon substrate when the bilayer structure is subjected to high enough operating temperature. The same can happen in other combinations of materials in multilayer structures produced by surface micromachining.

The second type of interfacial stress is the residual stress inherent in the microfabrication processes. Take, for example, a SiO_2 layer grown on the top surface of a silicon substrate beam at $1000°C$ by a thermal oxidation process as illustrated in Figure 9.14. The oxidation process was described in Section 8.5. The resultant shape of the bilayer beam at room temperature will be that shown in Figure 9.14*b* because of the significant difference in CTE for both materials. It is not difficult to appreciate the fact that associated with the residual strain are significant residual tensile stress in the SiO_2 layer after it is cooled down to the room temperature at $20°C$. Excessive tensile residual stress in SiO_2 layer can cause multiple cracks in the layer. Hsu and Sun (1998) reported the analysis of residual stresses in the oxide diaphragm of a pressure sensor.

The third type of stress that could be introduced in thin-film structures is the intrinsic stress due to local change of atomic structures during microfabrication processes. Excessive doping, for instance, could introduce substantial residual stresses in the structure after surface micromachining. The exact causes and the quantitative assessment of the intrinsic stresses in thin films are far from being clear to engineers. A qualitative description of this stress was presented in Section 4.6.

Stiction. Many have experienced the difficulty in separating two transparencies after the thin dividing paper is pulled out. A similar phenomenon occurs in surface micromachining. This phenomenon of two separated pieces sticking together is called *stiction*.

Stiction is the most serious problem for engineers to deal with in surface micromachining. It often occurs when the sacrificial layer is removed from the layers of the material that it once separated (e.g., step 6 in the case illustrated in Figure 9.12). The thin structure

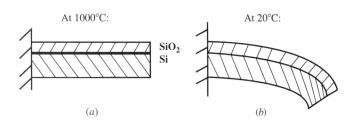

Figure 9.14. Residual stress and strain in a bilayer beam: (*a*) during oxidation; (*b*) after oxidation.

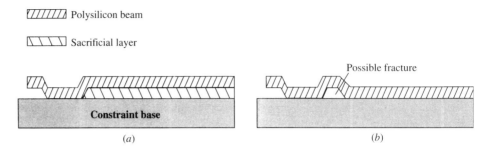

Figure 9.15. Collapse of a thin cantilever beam due to stiction: (*a*) with sacrificial layer in place; (*b*) after removal of sacrificial layer.

that was supported by the sacrificial layer may collapse on the other material. Take, for example, the production of a thin cantilever beam as illustrated in Figure 9.15; stiction could happen with the thin polysilicon beam dropping onto the top surface of the silicon substrate (the constraint base) after removal of the sacrificial PSG layer (Figure 9.15*b*). The two materials would stick together after the joint. Considerable mechanical forces are required to separate the two stuck layers again and these excessive forces can break the delicate microstructure. Stiction is the main cause for the large number of scraps incurred in surface micromachining.

Stiction occurs presumably as a result of hydrogen bonding of surfaces during rinsing of the interface after the etching of the PSG sacrificial layer or by such forces as the van der Waals forces described in Chapter 3. Various ways of avoiding stiction have been suggested by several researchers, as presented in Madou (1997). Among the proposed remedial actions are temporary spacers using polysilicon and sacrificial polymer columns that can be removed by etching with oxygen plasma afterward. Whatever the remedial action, the cost and the time required for production are major concerns to the industry.

9.4 LIGA PROCESS

Both micromanufacturing techniques—bulk micromanufacturing and surface micro machining—involve microfabrication processes evolved from microelectronics technology. Consequently, much of the developed knowledge and experience as well as equipment used for the production of microelectronics and ICs can be adapted for MEMS and microsystems manufacturing with little modification. Unfortunately, these inherited advantages are overshadowed by two major drawbacks: (1) the low geometric aspect ratio and (2) the use of silicon-based materials. The geometric aspect ratio of a microstructure is the ratio of the dimension in depth to that of the surface. Most silicon-based MEMS and microsystems use wafers of standard sizes and thickness as substrates, on which etching or thin-film deposition takes place to form the desired three-dimensional geometry. Severe limitations of the depth dimension is thus unavoidable. The other limitation is on the materials. Silicon-based MEMS preclude the use of conventional materials such as polymers and plastics as well as metals for the structures and thin films.

The LIGA process for manufacturing MEMS and microsystems is radically different from these two manufacturing techniques. This process does not have the two afore-mentioned major shortcomings in silicon-based micromanufacturing techniques. It offers great potential for manufacturing non-silicon-based microstructures. The single most important feature of this process is that it can produce "thick" microstructures that have extremely flat and parallel surfaces, such as microgear trains (Figure 1.9), micromotors (Figure 1.10), and microturbines (Figure 1.11) made of metals and plastics. These unique advantages are the primary reasons for its increasing popularity in the MEMS industry.

The term LIGA is an acronym for the German terms meaning Lithography (*Lithographie*), electroforming (*Galvanoformung*), and molding (*Abformung*), and these terms represent the three major steps in the process as outlined in Figure 9.16. The technique was first developed at the Karlsruhe Nuclear Research Center in Karlsruhe, Germany.

9.4.1 Description

As shown in Figure 9.16, the LIGA process begins with deep x-ray lithography that sets the desired patterns in a thick film of photoresist. X-rays are used as the light source in photolithography because of their short wavelength, which provide higher penetration power into the photoresist materials. This high penetration power is necessary for high resolution in lithography and for a high aspect ratio in the depth. The short wavelength of x-rays allows a line width of 0.2 μm and an aspect ratio of more than $100:1$ to be achieved. The x-rays used in this process are provided by a synchrotron radiation source, which allows high throughput because the high flux of collimated rays shortens the exposure time.

The LIGA process outlined in Figure 9.16 may be demonstrated by a specific example as illustrated in Figure 9.17. The desired product in this example is a microthin-wall metal tube of square cross section. We begin the process by depositing a thick film of photoresist material on the surface of a substrate, as shown in Figure 9.17a. A popular photoresist material that is sensitive to x-ray is PMMA. Masks are used in x-ray lithography. Most masking materials are transparent to x-ray, so it is necessary to apply a thin film of gold to

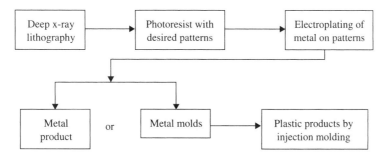

Figure 9.16. Major fabrication steps in LIGA process.

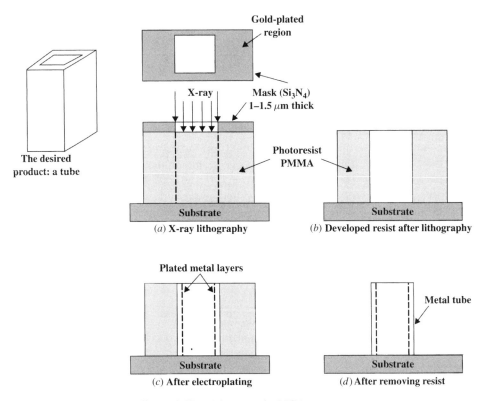

Figure 9.17. Major steps in LIGA process.

the area that will block x-ray transmission. The thin mask used for this purpose is silicon nitride with a thickness varying from 1 to 1.5 μm. The deep x-ray lithography causes the exposed area to be dissolved in the subsequent development of the resist material (see Figure 9.17*b*). The PMMA photoresist after the development will have the outline of the product, that is, the outside profile of the tube. This is followed by electroplating the PMMA photoresist with a desired metal, usually nickel, to produce a tubular product of required wall thickness (see Figure 9.17*c*). The desired tubular product is produced after removal of the photoresist materials (i.e., PMMA in this case) by oxygen plasma or chemical solvents.

For most applications the desired product is metal molds for subsequent injection molding of microplastic products, as shown in Figure 9.16.

9.4.2 Materials for Substrates and Photoresists

Substrate Materials for LIGA Process. The substrate used in the LIGA process is often called the *base plate*. It must be an electrical conductor or an insulator coated with electrically conductive materials. Electrical conduction of the substrate is necessary in order to facilitate electroplating, which is a part of the LIGA process. Suitable materials for the substrates include austenite steel; silicon wafers with a thin titanium or Ag/Cr top layer; and copper plated with gold, titanium, and nickel. Glass plates with thin metal plating could also be used as the substrate.

Photoresist Materials. Basic requirements for photoresist materials for the LIGA process include the following:

- It must be sensitive to x-ray radiation.
- It must have high resolution as well as high resistance to dry and wet etching.
- It must have thermal stability at up to 140°C.
- The unexposed resist must be insoluble during development.
- It must exhibit very good adhesion to substrate during electroplating.

Based on these requirements, PMMA is considered to be an optimal choice of photoresist material for the LIGA process at the present time. However, its low lithographic sensitivity makes the lithographic process extremely slow. According to Madou (1997), at a short wavelength of 5 Å, over 90 min of irradiation is needed for a PMMA resist 500 μm thick with a power consumption of 2 MW at 2.3 GeV by the ELSA synchrotron in Bonn, Germany. Another shortcoming of PMMA is its vulnerability to crack due to stress. For these reasons, other resist materials have been considered and used. Madou (1997) provides a qualitative comparison of the properties of various resist materials, such as polyoxymethylene (POM), polyalkensulfone (PAS), polymethacrylimide (PMI), and poly(lactide-co-glycolide) (PLG). See Table 9.7.

9.4.3 Electroplating

Electroplating is an important step in the LIGA process. Electroforming of metal films onto the surface of the cavities in the photoresist after x-ray lithography has been performed as illustrated in Figure 9.17c. Nickel is the common metal to be electroplated on the photoresist walls. Other metals and metallic compounds that could be used for electroplating include Cu, Au, NiFe, and NiW. The conductive substrate and the carrying photoresist structure form the cathode in an electroplating process illustrated in Figure 9.18.

TABLE 9.7. Properties of Resists for Deep X-ray Lithography

	PMMA	POM	PAS	PMI	PLG
Sensitivity	Bad	Good	Excellent	Reasonable	Reasonable
Resolution	Excellent	Reasonable	Very bad	Good	Excellent
Sidewall smoothness	Excellent	Very bad	Very bad	Good	Excellent
Stress corrosion	Bad	Excellent	Good	Very bad	Excellent
Adhesion on substrate	Good	Good	Good	Bad	Good

Source: Madou, 1997.

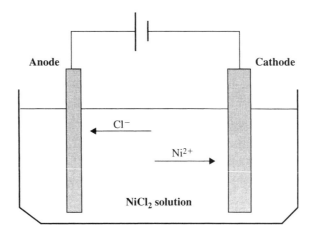

Figure 9.18. Electroplating process of nickel (Schulze, 1998).

Electroplating works on the principle that the nickel ions, Ni^{2+}, from the nickel chloride ($NiC\ell_2$) solution react with the electrons at the cathode to yield nickel:

$$Ni^{2+} + 2e^- \rightarrow Ni$$

However, we should be aware of the competing H_2 on the surface of the cathode that may cause nonuniform Ni plating. The H_2 gas is the product of the H_2^+ ions produced by electrolysis of the solution. The chemical reaction of H_2^+ ions to H_2 gas at the cathode is

$$2H^+ + 2e^- \rightarrow H_2$$

To avoid damage of the plated surfaces by hydrogen bubbles, we need to control the pH of the solution, the temperature, and the current density in the electrolysis.

9.4.4 SLIGA Process

We see from Figures 9.16 and 9.17 that the finished product, whether it is a microstructure or a metal mold, is attached to the substrate, or base plate. Attachment to the electrically conductive substrate is necessary for the electroplating process. However, this attachment is considered a redundancy in the LIGA process. For instance, the hollow square tube produced in the LIGA process as described in Figure 9.17 would not be separated after electroplating of the metal film on the inner walls. A modified process called sacrificial LIGA (SLIGA) has been developed to solve this problem. The principle of SLIGA is to introduce a sacrificial layer between the PMMA resist and the substrate, thereby allowing the separation of the finished mold from the substrate after the electroplating. The separation is achieved by removal of the sacrificial layer by etching. Polyimide with a metal film coating is used as a common sacrificial layer material for that purpose.

9.5 SUMMARY OF MICROMANUFACTURING

This chapter has presented the three principal micromachining techniques currently available. Following is a summary of these techniques which will be useful to engineers in selecting the optimal manufacturing technique for their specific design projects.

9.5.1 Bulk Micromanufacturing

- Straightforward, involving well-documented fabrication processes.
- Less expensive, but material loss is high.
- Suitable for simple geometry, for example, micropressure sensor dies and some actuating elements.
- Limited to low aspect ratio in geometry; that is, the surface dimensions are much greater than that of the depth. This is because the overall height of the microstructure is limited by the thickness of commercially available silicon wafers.

9.5.2 Surface Micromachining

- Requires the building of layers of materials on the substrate.
- Complex masking design and productions.
- Etching of sacrificial layers is necessary.
- The process is tedious and more expensive.
- There are serious engineering problems, such as interfacial stresses and stiction.
- Major advantages: (1) not constrained by the thickness of silicon wafers; (2) wide variety of thin-film materials to be used; (3) suitable for complex geometries such as microvalves and actuators.

9.5.3 LIGA Process

- The most expensive process.
- Requires a special synchrotron radiation facility for deep x-ray lithography.
- Requires the development of microinjection molding technology and a facility for mass production purposes.
- Major advantages: (1) virtually unlimited aspect ratio of the microstructure geometry; (2) flexible in microstructure configurations and geometry; (3) the only technique that allows the production of metallic microstructures; (4) the best manufacturing process for mass production, with a provision for injection molding.

PROBLEMS

Part 1 Multiple Choice

1. *Micromanufacturing* is (a) synonymous to, (b) antonymous to, (c) unrelated to *microfabrication*.

2. In general, there are (a) two, (b) three, (c) four distinct micromanufacturing techniques.

3. Bulk manufacturing involves primarily (a) adding, (b) subtracting, (c) both adding and subtracting portions of material from the substrate.

4. The principal microfabrication process used in bulk manufacturing is (a) etching, (b) deposition, (c) diffusion.

5. Isotropic etching is hardly desirable in micromanufacturing because (a) the etching rate is too low, (b) the cost is too high, (c) it is hard to control the direction of etching.

6. The most favored orientation for micromachining of silicon wafers is the (a) <100>, (b) <110>, (c) <111> orientation.

7. The least-used orientation for micromachining of silicon wafers is the (a) <100>, (b) <110>, (c) <111> orientation.

8. A ratio of 400 : 1 etching rate was observed for silicon between the (a) <100> and <111>, (b) <110> and <111>, (c) <110> and <100> orientations.

9. The (111) plane intersects the (100) plane in a silicon crystal with an angle of (a) $50.74°$, (b) $54.74°$, (c) $57.47°$.

10. Anisotropic etching is (a) faster, (b) slower, (c) about the same speed in comparison to isotropic etching.

11. Silicon dioxide with KOH etchant is (a) 100, (b) 1000, (c) 20,000 times slower than silicon.

12. Silicon dioxide with EDP etchant is (a) 100, (b) 1000, (c) 20,000 times slower than silicon.

13. Silicon nitride is (a) a stronger, (b) a weaker, (c) about the same in etching resistance as in silicon oxide.

14. The higher the selectivity ratio of a material, (a) the better, (b) the worse, (c) neither better nor worse is the material as an etching mask.

15. Doped silicon has resistance to etching (a) stronger than, (b) weaker than, (c) about the same as that of undoped silicon.

16. Excessive doping of silicon introduces (a) residual stress, (b) residual strains, (c) fracture of atomic bonds in a silicon substrate.

17. Wet etching can be stopped at the boundaries of (a) *p–n* doped silicon, (b) *p*-silicon-silicon, (c) *n*-silicon-silicon.

18. There are (a) one, (b) two, (c) three dry etching techniques available for microelectronics.

19. Isotropic etching is (a) 2, (b) 5, (c) 10 times faster than anisotropic etching.

20. When selecting materials for masks in the deep etching process, one would select materials with (a) high, (b) low, (c) medium selectivity ratio.

21. DRIE stands for (a) dry etching, (b) dry reactive ion etching, (c) deep reactive ion etching.

22. DRIE is the best means of (a) dry etching, (b) fast etching, (c) deep etching.

23. PSG stands for (a) polysilicon glass, (b) phosphosilicate glass, (c) phosphorus silicon glass.

24. PSG is a common material for (a) active substrates, (b) passive substrate, (c) sacrificial layers.

25. Sacrificial layers in surface micromachining are used to (a) strengthen the microstructure, (b) create necessary geometric voids in the microstructure, (c) be part of the structure.

26. PSG is a more popular sacrificial layer material than silicon dioxide because it can be etched (a) more rapidly, (b) more slowly, (c) more cheaply in HF etchant.

27. The most popular structural material in surface micromachining is (a) PSG, (b) polysilicon, (c) silicon dioxide.

28. In surface micromachining, the etch rate for sacrificial layers must be (a) much slower, (b) about the same, (c) mush faster than etch rates for other layers.

29. Stiction occurs in (a) bulk micromanufacturing, (b) surface micromachining, (c) the LIGA process.

30. Stiction in finished microstructures made by surface micromachining is a result of (a) layers of dissimilar materials, (b) thin films, (c) atomic forces between layers.

31. The geometric aspect ratio in MEMS structures is defined as the ratio of dimensions in (a) depth to surface, (b) surface to depth, (c) width to length.

32. The LIGA process produces MEMS with materials (a) limited to silicon, (b) limited to ceramics, (c) virtually no limitation on materials.

33. Synchrotron x-rays are used in photolithography in the LIGA process because (a) they are more effective with the photoresist, (b) they are a cheaper source of light, (c) they can penetrate deep into the photoresist material.

34. The photoresist materials commonly used in the LIGA process are (a) any photoresist material, (b) positive photoresist materials, (c) negative photoresist materials.

35. One of the principal advantages of the LIGA process is its ability to produce (a) microstructures with high aspect ratio, (b) microstructures with low cost, (c) microstructures with precise dimensions.

36. X-ray lithography in a LIGA process produces (a) the outline, (b) the actual geometry, (c) the duplicate of a MEMS component.

37. X-ray lithography photoresists provide (a) the outline, (b) the actual geometry, (c) the duplicate of a MEMS component.

38. An electrically conductive base plate must be used in a LIGA process because of the need for (a) signal transduction, (b) electroplating of metals, (c) required electric heating of the mold.

39. Micromolds for injection molding of MEMS are produced from (a) the x-ray lithography photoresist, (b) metals that are electroplated to the photoresist outline, (c) the removal of the base plate in the process.

40. An optimal choice of photoresist in a LIGA process is (a) POM, (b) PMI, (c) PMMA.

41. The photoresist that is most sensitive to x-radiation is (a) POM, (b) PAS, (c) PMMA.

42. SLIGA is an improvement over the LIGA process with the provision of (a) slicing the base plate from the finished product, (b) a sacrificial layer for ready separation of the base plate from the product, (c) a clean product.

43. The least expensive micromanufacturing technique is (a) bulk micromanufacturing, (b) surface micromachining, (c) the LIGA process.

44. The most flexible micromanufacturing technique is (a) bulk micromanufacturing, (b) surface micromachining, (c) the LIGA process.

45. The most costly micromanufacturing technique is (a) bulk micromanufacturing, (b) surface micromachining, (c) the LIGA process.

Part 2 Descriptive Problems

Use no more than 20 words to solve the following problems:

1. What are the limitations of the height (depth) of microstructures that can be produced by bulk micromanufacturing technique?

2. What is your major criterion in selecting materials for the masks used in etching, for example, in etching silicon substrate with moderate depth and also for deeper etching?

3. Describe the DRIE process. How can DRIE achieve virtually perfect vertical etching?

4. What is the principal advantage of DRIE? Why is it necessary to have near perfectly vertical walls in microstructures?

5. What is principal difference between bulk micromanufacturing and surface micromachining?

6. Give one advantage and one disadvantage of using surface micromachining.

7. Describe the role of sacrificial layers in surface micromachining.

8. Describe the phenomenon of stiction and possible ways to avoid it.

9. List the principal advantages and disadvantages of the LIGA process.

10. Why is electroplating necessary in a LIGA process?

Part 3 Design of Micromanufacturing Processes

1. Following the illustrative examples, such as Figure 9.17 offered in this chapter, design micromanufacturing processes for the production of a thick gear made of silicon or plastic with pitch a root diameter of 500 μm and pitch diameter of 750 μm. The gear is 1000 μm thick. Use (a) bulk micromanufacturing, (b) the LIGA process. Draw a conclusion on what you have learned from this exercise.

2. Following the illustrative example in Figure 9.12, design a surface micromachining process for the production of a comb-drive actuator as outlined in Figure 2.27.

CHAPTER 10

MICROSYSTEMS DESIGN

10.1 INTRODUCTION

The working principles of many MEMS devices were presented in Chapter 2. Examples of how various MEMS components can be designed were given in Chapters 4, 5, 6, and 7. The many ways that these components can be manufactured were described in Chapters 8 and 9. However, it requires reliable electromechanical engineering design to configure the optimum geometry and dimensions components that are to be assembled into microsystems. Structural integrity and reliability of these products are primary requirements, but manufacturability of the systems is also a critical consideration in the design process.

A major difference between the engineering design of microsystems and that of other products is that the design of microsystems requires the integration of the related manufacturing and fabrication processes. Mechanical engineering design of traditional products and systems rarely requires the consideration of the consequences of the manufacturing process. For example, components such as gears, bearings, and fasteners in a mechanical system can be purchased from suppliers without the knowledge of how these components are produced. In microsystems, which involve MEMS components, however, the situation is quite different. Components for MEMS are fabricated by various physical–chemical means as described in Chapters 8 and 9. These fabrication and manufacturing processes often involve high temperature and harsh physical and chemical treatments of delicate materials used for the components. These processes can have serious repercussions in the performance of microsystems and hence must be taken into design considerations. Tolerance of the finished components and intrinsic effects such as residual stresses and strains inherent in microfabrication processes are just two obvious examples of such repercussions.

In general, microsystems design involves three major interrelated tasks: (1) *process flow design*, (2) *electromechanical and structural design*, and (3) *design verifications* that include assembly, packaging, and testing. Material selection in microsystems design is also much more complex than in traditional products. Selection of materials for microsystems involves not only the materials for the basic structure of the systems but also the materials in the process flow, such as proper etchants and thin films for depositions.

The intricacy of microsystems design as described above is the main reason for the long design cycle time that has been a major stumbling block for wider acceptance of MEMS and microsystems in the marketplace. Few engineers have the knowledge and experience in this multidisciplinary practice. Later in the chapter we will introduce computer-aided design (CAD) specifically developed for MEMS and microsystems that appears to be a viable way to alleviate such problems.

10.2 DESIGN CONSIDERATIONS

Before we begin with specific design considerations, let us take an overview of the necessary ingredients that will be involved in microsystem design as outlined in Figure 10.1. The diagram indicates that once the product is properly defined with a specification, the following items need to be considered:

1. Design constraints
2. Selection of materials

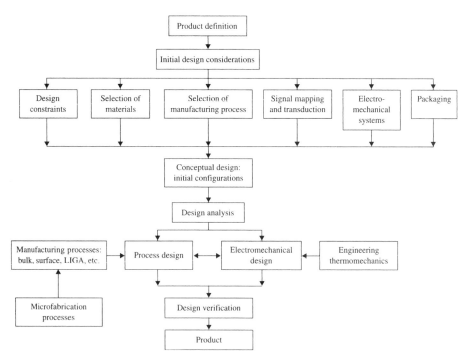

Figure 10.1. Overview of engineering design of microsystems.

3. Selection of manufacturing process
4. Signal mapping and transduction
5. Electromechanical systems
6. Packaging of the product

After these considerations, engineers will be in the position to develop the initial configuration of the product in terms of geometry, dimensions, materials, and fabrication and packaging methods. A more detailed and in-depth design of both the manufacturing process and the electromechanical systems design follows to confirm the feasibility of the initial configuration. Design verification on the prototype usually can be carried out by computer simulation to ensure the structure's integrity and, more importantly, the desired functions of the microsystem.

10.2.1 Design Constraints

Design constraints can be many or only a few and vary from case to case. Many of these constraints are nontechnical and may be related to the marketing of the product. Following are a few typical constraints for microsystems:

1. **Customer Demands:** These include the special requirements that are not specifically included in product specifications. They may include special features that the microdevice must provide in extraordinary circumstances. It is not hard to conceive that a microdevice, such as a sensor or actuator installed in a toy for young children, requires very different considerations in terms of safety and rough-handling than the same device installed in a product for mature office and laboratory workers.

2. **Time to Market (TTM):** This factor is critical, as most high-tech products have what is called a "window for marketing," and these windows are becoming narrower and narrower as technology advances. The shrinking window for marketing specific products is also due to increasingly stiff competition in the marketplace. A microsystem product needs to enter the market at a critical time to capture that market and hence maximize the profit. Usually TTM dictates how much time the design engineer has to design and produce the product. In view of the intricate design process mentioned at the beginning of Section 10.2, special-purpose CAD packages for microsystems appear to be a viable way to alleviate the TTM problem. We will present the application of CAD in MEMS design in Section 10.8.

3. **Environmental Conditions:** Three critical conditions are involved: thermal, mechanical, and chemical. A device operating at elevated temperature requires much more attention in terms of thermal stresses and strains, material deterioration, and degradation of signal transduction. A micropressure sensor designed for monitoring cylinder pressure in an internal combustion engine obviously requires more sophisticated design analysis and careful selection of material than the ones used to monitor the tire pressure in an automobile. Mechanical environment relates to the mechanical stability of the support of the microsystem. A vibratory support

may shake joints loose and also result in breakdown of electric leads in the system. Last, a chemical working medium can degrade both the MEMS and packaging materials. Chemical and moisture contents in fluid media can lead to undesirable oxidation and corrosion of the components in contact. Moisture is also the main reason for the stiction of microswitches in optoelectronic network systems. Clogging of microchannels in microvalves and pumps in microfluidics is a possibility if the system is not properly designed and manufactured.

4. **Physical Size and Weight Limitations:** These constraints are normally covered in the product specification. They can affect the overall configurations of the product with imposition of limitations on some key design parameters.

5. **Applications:** It is important to know whether the microsystem is intended for once-only application or for repeated usage. If the latter is the case, then one needs to design the life expectancy of the product as well as the possibility of creep and fatigue failure of the components.

6. **Fabrication Facility:** This relates to the selection of manufacturing methods for the product. The availability of fabrication facility for the intended product is a critical factor in meeting TTM as well as the cost for the production of the product.

7. **Costs:** This factor can dictate the overall direction of the design. In today's competitive marketplace, cost is a critical factor for the marketability of the product. In this early design stage of the product, engineers should be seriously involved in the cost analysis of the product, which will be translated into constraints on many design parameters such as selection of materials and fabrication methods.

10.2.2   Selection of Materials

We have learned from Chapter 7 about the many materials that can be used in microsystems. Following is a summary of the unique characteristics of these materials. Design engineers may use it in selecting appropriate materials for various parts of the microsystem. As *process flow* is an integrated part of the design process, materials such as etchants and thin films for depositions need to be carefully evaluated and selected for the systems design.

Principal Substrate Materials.   There are two types of substrate materials: (1) *passive substrate* materials for support only, which include polymers, plastics, and ceramics, and (2) *active substrate* materials such as silicon, GaAs, and quartz for the sensing or actuating components in a microsystem.

Silicon

- Mechanically stable, inexpensive, and ready machinability.
- An excellent candidate material for microsensors and accelerometers.

Gallium Arsenide (GaAs)

- Has fast response to externally applied influence, such as photons in light rays even at elevated temperature.
- Can be used as thermal insulation.
- Its high piezoelectricity makes this material suitable for precision microactuation.
- Suitable for surface micromachining.
- A good candidate material for optical shutters, choppers, and actuators.
- A desirable material for both microdevices and microcircuits.
- Unfortunately, it is more expensive than other substrate materials such as silicon.

Quartz

- More mechanically stable than silicon or silicon compounds even at high temperature.
- Virtually immune to thermal expansion. It is thus the ideal material for high-temperature applications.
- Excellent resonance capability for precision microactuation.
- Unfortunately, it is hard to be shaped into desirable configurations.

Polymers

- Used primarily as passive substrate material.
- Special polymers such as SU-8 can be used to produce permanent microstructures with high aspect ratios.
- Low cost in both materials and production processes.
- Easily formed into desired shapes.
- Has flexibility in "alloying" for specific purposes.
- Sensitive to environmental conditions such as temperature and moisture.
- Vulnerable to chemical attacks.
- Most polymers age, that is, they deteriorate with time.

Other Substrates in Silicon Family

Silicon Dioxide (SiO_2)

- Can be easily grown on silicon substrate surface or by deposition as described in Chapter 8.
- Excellent for both thermal and electrical insulation.
- Can be used as good masking material for wet etching of silicon substrates.

Silicon Carbide (SiC)

- Dimensionally and chemically stable even at high temperature.
- Dry etching using aluminum masks can easily pattern it.
- An excellent passivation material for deep etching.

Silicon Nitride (Si_3N_4)

- An excellent barrier for water and sodium ions in diffusion processes.
- A good masking material for deep etching and ion implantation.
- An excellent material for optical wave guidance.
- A good protective material for high-strength electric insulation at high temperature.

Polycrystalline Silicon

- Widely used as resistors, gates for transistors, and thin-film transistors.
- A good material for controlling electrical characteristics of substrates.
- Doped polysilicon makes excellent electrodes in solar photovoltaic cells.

Packaging Materials

- Ceramics (alumina, silicon carbide)
- Glasses (Pyrex, quartz)
- Adhesives (eutectic solder alloys, epoxy resins, silicone rubbers)
- Wire bonds (gold, silver, aluminum, copper, and tungsten)
- Headers and casings (plastics, aluminum, stainless steel)
- Die protections (silicone gel, silicone oil)

10.2.3   Selection of Manufacturing Processes

In Chapter 9, we presented three principal manufacturing processes that are available for producing microdevices and systems. Following is a summary of these manufacturing processes and their advantages and disadvantages.

Bulk Micromachining

- Relatively straightforward in operation. It involves well-documented fabrication processes, mainly the etching processes.
- The least expensive of the three manufacturing techniques.
- Suitable for simple geometry, for example, the dies for micropressure sensors.
- A major drawback is its low aspect ratio. (The ratio of the dimension in depth to that of the plane defines the *aspect ratio* in MEMS industry. The height of microstructures in bulk micromachining is limited by standard silicon wafer thickness.)

- The process involves removal of material from bulk substrates, resulting in high material consumption.

Surface Micromachining

- Requires the building layers of materials over the substrate.
- Requires the design and fabrication of complex masks for deposition and etching in the processes.
- Etching of sacrificial layers is necessary after layer building—a wasteful practice.
- More expensive than the bulk micromanufacturing technique because of its complex fabrication procedures.
- Major advantages: (1) It is less constrained by the thickness of silicon wafers than that in bulk micromanufacturing. (2) It provides wide choices of materials to be used in layer buildings. (3) It is suitable for complex geometries such as microvalves and micromotors.

LIGA and SLIGA and Other High-Aspect-Ratio Processes

- The most expensive micromanufacturing process.
- Both the LIGA and SLIGA processes require a special synchrotron radiation facility for deep x-ray lithography. This facility is not readily accessible to most MEMS industries.
- Require the development of microinjection molding technology and facilities.
- Major advantages: (1) They offer great flexibility in aspect ratio of structure geometry. A high aspect ratio of 200 is achievable by the LIGA process. (2) They offer the most flexibility in microstructure configurations and geometry. (3) There is virtually no restriction on the materials for the microstructure, including metals, by the LIGA process. (4) They are the best of the three manufacturing processes for mass production.

10.2.4 Selection of Signal Transduction

Signal transduction is necessary in both microsensors and actuators. In either case, there is a need to convert biological, chemical, optical, thermal, or mechanical energy such as motion or other physical–chemical behavior of MEMS components into an electrical signal, or vice versa. Figure 10.2 illustrates such conversions of signals as well as the options available for signal transduction for both types of microdevices.

A brief description of each signal transduction technique is presented below. Engineers will select specific techniques that will best serve the purpose for the intended product.

Another critical consideration in the design of a transduction system is signal mapping, which involves a strategy that selects the optimal locations for the transducers and the circuits that transmit the signals. One example is the choice of proper location for signal transduction for pressure sensors, as illustrated in Figures 2.9 and 7.15. The four

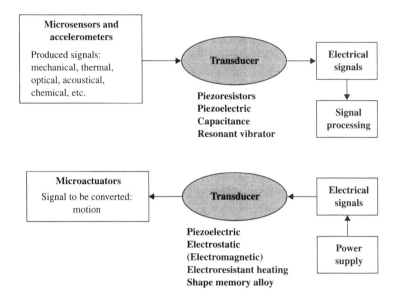

Figure 10.2. Options for signal transduction in microsystems.

piezoresistors for signal transduction appear to be in optimal positions. However, should the pressurized medium be present on the top face of the pressure sensor die, the engineer would be faced with two options:

1. Leave the transduction arrangement as it is in Figure 2.9 but protect the piezoresistors, the wire bond, and the metal circuit and film pads from direct contact with the pressurized medium.
2. Place the transduction system at the bottom face of the die. In this case, a much more complex process for implanting piezoresistors and the wiring needs to be arranged.

Neither option is attractive from a manufacturing point of view.

Following is a summary of materials and signal transduction techniques that are available for engineers in their design of microsystems.

1. **Piezoresistors:** Silicon piezoresistors are most commonly used in microsensors because of the minute sizes and high sensitivity in signal transduction. Piezoresistors can be produced on substrate other than silicon—GaAs and polymers, for example. The characteristics of silicon piezoresistors were presented in Section 7.6. A major disadvantage of using piezoresistors is the stringent control of the doping process required to achieve good quality, and an even more serious drawback is the strong temperature dependence of resistivity (see Table 7.10). The sensitivity of piezoresistors deteriorates rapidly with the increase of temperature. Proper temperature compensation in signal processing is required for applications at elevated temperatures.

2. **Piezoelectric:** Piezoelectric materials are made of crystals, as described in Section 7.9. Common piezoelectric crystals are presented in Table 7.14, by which engineers can select the proper crystal for the specific application. For example, PZT ($PbTi_{1-x}Zr_xO_3$) crystals are used primarily for displacement transducers and accelerometers. Barium titanate ($BaTiO_3$) is commonly used to transduce signals from microaccelerometers. Quartz crystals, on the other hand, are used for oscillators in ultrasonic transducers.

   Most piezoelectric materials are brittle. Special consideration should be given to packaging this material to avoid brittle fracture. Size and machinability are two problems in using piezoelectric materials. Piezoelectrics are suitable for use in accelerometers for measuring dynamic or impact forces that exist over short periods of time, as sustained piezoelectric action will cause overheating of the crystals and thus deteriorate their conversion capability. PZT crystals are the most popular piezoelectric used in industry because of their high piezoelectric coefficient, as shown in Table 7.14.

3. **Capacitance:** We learned the working principles of capacitance signal transduction and electrostatic actuation in Chapter 2. While this method is particularly attractive for high-temperature applications, the nonlinear input/output relationships between the change in the gap of the electrodes and the output voltage, as demonstrated in Equation (2.3), require special compensation for errors in interpretation of the output signals. This method also requires a larger space in the microsystem, as the output capacitance of a parallel-plate capacitor is directly proportional to the overlapped area of the plate electrodes, as shown in Equation (2.2).

4. **Resonant Vibrator:** The working principle of signal transduction using resonant vibration was presented in Section 2.2.5. While this technique can offer higher resolution and accuracy for signal transduction in micropressure sensors, its application to other microdevices is limited by the complexity of fabrication, such as fusion bonding of silicon beam to the substrate, as in the case of micropressure sensors. The required space for the vibrating members is another drawback of this transduction method.

5. **Electroresistant Heating:** This technique is widely used in microactuation, such as in microvalves and pumps in fluidics and micro–heat pipes (see Section 2.6). The technique is simple and straightforward. However, the requirements for precise control of the heating of the actuating element with thermal inertia may affect the timely response of the intended actuation. Also, contact of the heated element with the working medium can cause local heat transfer, which may in turn alter the flow pattern of the working medium in microvalves. The technique thus has serious drawbacks in liquid-based microfluidic systems. Sealing of heat-transmitting fluids often causes problems in the packaging and operations of these microsystems.

6. **Shape Memory Alloy:** A shape memory alloy (SMA) is a good actuating material when it is used in conjunction with electroresistant heating. A major drawback is the limited availability of SMA, and often the deformation of SMA cannot be accurately predicted because of its sensitivity to temperature.

10.2.5 Electromechanical System

No microsystem can function without electrical power. Electrical circuitry that provides the flow of electric current or maintains voltage and/or current supply in the case of actuators is an integrated part of the system. In microsensors, the electronic signals produced by the transducers need to be led outside of the device and be conditioned and processed by a suitable electrical system. Whatever the electrical system for the intended product, preliminary assessment of the interface between the mechanical action and the electrical system is needed in order to configure the product. For example, in a micropressure sensor design, the layout of the piezoresistors and the leads for connecting circuitry are to be deposited on the surface of the silicon die, as illustrated in Figure 2.9. These layouts may affect the overall configuration of the die.

10.2.6 Packaging

As mentioned earlier in this book, the cost of packaging a micropressure sensor can be as low as 20% of the overall cost of the product with simple plastic encapsulation or as high as 95% of the overall cost with complex passivation and stainless steel or tungsten casings for special-purpose units. The design parameters that affect packaging thus need to be considered at this early stage in the design process. Design engineers need to consider the following major factors that will affect the packaging of the product:

- Die passivation
- Media protection
- System protection
- Electrical interconnect
- Electrical interface
- Electromechanical isolation
- Signal conditioning and processing
- Mechanical joints (anodic bonding, eutectic bonding, TIG welding, adhesion, etc.)
- Processes for tunneling and thin-film lifting
- Strategy and procedures for system assembly
- Product reliability and performance testing

A more detailed description of each of the above items and other assembly, packaging, and testing design considerations will be provided in Chapter 11.

10.3 PROCESS DESIGN

Once the design engineer has selected a specific manufacturing technique, whether it is bulk micromanufacturing or surface micromachining or the LIGA process, a search for appropriate microfabrication processes begins. We will classify the various microfabrication processes involved in micromanufacturing in three categories: (1) photolithography,

(2) thin-film fabrications, and (3) geometry shaping. The LIGA process includes two additional stages: electroplating and injection molding.

10.3.1 Photolithography

Photolithography is the only viable way to produce micropatterns that depict the three-dimensional structural geometry of microsystems by the current state of the art. It is also used in the process for producing masks (or *mask sets*) for all micromanufacturing techniques, including masks for etching in bulk manufacturing, for thin-film deposition and for etching in surface micromachining (see, e.g., Figure 9.12), and for micromolds in the LIGA process. The detailed working principle of photolithography was presented in Section 8.2. There are three tasks that need to be carried out in the systems design process: (1) design of patterns for the substrates, (2) design of masks for lithography, and (3) fabrication processes for the mask sets.

Let us look at an example of the silicon die used in a micropressure sensor as illustrated in Figure 10.3. In this case, the pressure of the medium is applied at the backside, or the cavity side, of the silicon die. The front side of the die has four piezoresistors diffused beneath its surface. The location and orientations of these resistors are shown in the top view (mask for SiO_2) of the die in Figure 10.4.

We can visualize that two masks are required for the silicon die, one to etch the cavity and the other for the diffusion of the piezoresistors and the deposition of thin films for electrical conductors connecting these four resistors. As we learned in Chapter 8, both SiO_2 and Si_3N_4 are suitable candidate materials for this purpose. However, because deep etching is required for the production of many cavities in silicon dies, Si_3N_4 is a more suitable masking material.

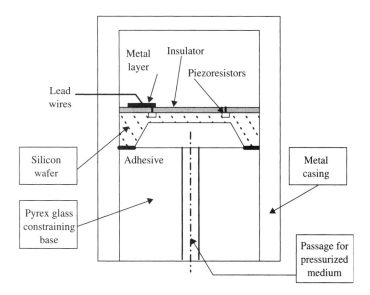

Figure 10.3. Cross section of a micropressure sensor.

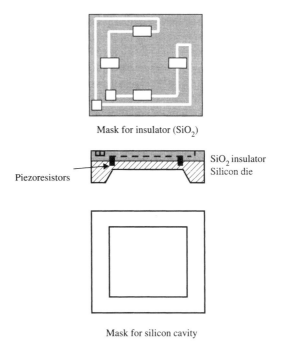

Figure 10.4. Patterns in masks for a micropressure sensor die.

As illustrated in Figure 10.4, the patterns for both masks are quite different. The pattern for cavity etching is simply a square opening, whereas the pattern for the doping of piezoresistors and the connecting leads is quite complicated. The white lines for the SiO_2 mask are for the electrical connection of the resistors in a Wheatstone bridge, as described in Chapter 2. The two square pads at the lower left corner of the top mask are the leads to signal conditioning and processing units outside the pressure sensor.

10.3.2 Thin-Film Fabrications

There are several ways to produce thin films over the substrate's surface, as described in Chapter 8. The design engineer may use Table 10.1 for the selection of particular processes for the production of thin films for microsystems. Many of these processes require high-temperature environments, which would likely result in residual stresses and residual strains as described in Section 9.3.3. Every effort should be made to minimize these residual effects.

The electroplating process was presented in Section 9.4.3. The rate of deposition for a common metal, that is, nickel over photoresist material, can be estimated from the following expression (Madou, 1997) for 100% yield:

$$m = \frac{1}{z} i A t \frac{1}{F} M \tag{10.1}$$

TABLE 10.1. Summary of Thin-Film Production for Microsystems

Processes	Principal Applications	Building Up or Building-In	High or Low Temperature	Approximate Rate of Production
Ion implantation (Section 8.3)	For doping p–n junctions or other impurities	In	Low	Equation (8.1)
Diffusion (Section 8.4)	For doping p–n junctions or other impurities	In	High	Equation (8.4)
Oxidation (Section 8.5)	For SiO_2 layers using O_2 or steam	In	High	Equations (8.9) and (8.10)
Deposition (Section 8.6)	Physical deposition for metals; chemical deposition (APCVD, LPCVD, PECVD) for SiO_2, Si_3N_4, and polysilicon	Up	Moderate to high	Equation (8.23) for APCVD
Sputtering (Section 8.7)	Thin metal films	Up	Moderate	Madou (1997)
Epitaxy deposition (Section 8.8)	Thin films of substrate material	Up	High	Table 8.9
Electroplating	Thin metal films over polymer photoresist materials in LIGA or SLIGA manufacturing process	Up	Low	Equation 10.1

where

A = electrode surface

t = electroplating time

F = Faraday constant (96,487 Å/s-mol)

i = current density

z = electrons involved in reaction $Ni^{2+}+2e^- \rightarrow Ni$

M = atomic weight of Ni

10.3.3 Geometry Shaping

The complex geometry of silicon-based microdevice components can be produced either by depositing thin films of various materials over the substrates, as described in Section 10.3.2, or by removing portions of material from the bulk substrates.

An effective process for removing material from substrates is etching, as described in Sections 8.9 and 9.2. Tables 9.1 and 9.2 and the description in Section 9.2.3 for chemical (wet) and Section 9.2.5 for plasma-assisted (dry) etching can be used to estimate the rates of etching.

10.4 MECHANICAL DESIGN

The principal objective of mechanical design is to ensure the structural integrity and reliability of microsystems when they are subjected to a specified loading at both normal operating and overload conditions. The latter condition relates to possible mishandling or unexpected surges of load due to system malfunction. Design methodologies developed for machines and structures at the macro- and mesoscales are used here with provisions to accommodate for the necessary modifications according to the scaling laws presented in Chapter 6 as well as the mechanical engineering design principles in Chapters 4 and 5. These design principles may require significant modifications for components of submicrometer size because of size-dependent material properties, as will be described in Chapter 12.

10.4.1 Geometry of MEMS Components

Despite the many different kinds of microdevices that have been developed and marketed, as described in Chapter 2, a closer look at these devices reveals a limited geometry for most components involved in the devices: (1) *beams* for microrelays (Figure 2.19), *gripping arms* in microtongs (Figure 1.4), and *beam springs* in microaccelerometers (Figures 1.6, 2.34, and 2.35); (2) *plates* for diaphragms in pressure sensors (Figures 1.22, 2.9, 2.12, and 2.14), *plate springs* in microaccelerometers (Figure 2.37), and *diaphragms* in microvalves and pumps (Figures 2.42 and 2.43); (3) *tubes* in electrokinetic pumping (Figure 2.4); and (4) *channels* of either closed or open cross section in microfluidics (Figure 2.40).

The complication in the geometry of MEMS and microsystems components, however, is that many of these components are made of multiple layers, for example, laminated beams and tubes and channels, coated with dissimilar materials, which increases the degree of difficulty in the design analyses.

10.4.2 Thermomechanical Loading

Most of the loads that a microsensor or actuator is subjected to are common to macrostructures. These can be categorized in the following way:

1. Concentrated forces, such as the contact forces between the actuating members and the fluid passages in microvalves as illustrated in Figures 2.41–2.43 and Example 4.7.

2. Distributed forces, such as the pressure loads on the diaphragms in micropressure sensors as illustrated in Figure 2.8 and Example 4.4.

3. Dynamic or inertia forces, as in the microaccelerometers in Figures 2.35–2.37 and Example 4.14.

4. Thermal stress induced by mismatch of coefficients of thermal expansion in layered structures (see Figure 2.18 and Examples 4.17–4.19).

5. Friction forces between moving components in a microsystem; examples are the bearing of a rotary micromotor (Figure 2.32), a linear motor (Figure 2.31), or a microturbine (Figure 1.11).

The following forces are unique to microsystems structures. While some of these forces can be quantified with reasonable accuracy, others could only be qualitatively assessed. Mathematical modeling of structures subjected to these forces thus requires special attention. Empirical formulas are often used for estimating these forces:

1. Electrostatic forces for actuation—Evaluation of these forces is available in Section 2.3.4 and Examples 2.3 and 2.4. One needs to realize that the formulas used to evaluate electrostatic forces in Equations (2.8), (2.10), and (2.11) only give the average forces generated by the matching plate electrodes. The distribution of these forces over the electrode's surface is by no means uniform.

2. Surface forces produced by piezoelectricity—These forces are generated by the mechanical deformation of piezoelectric crystals with the application of electric voltages. They are used to drive actuators such as those illustrated in Figure 2.21 and also in micropumping as described in Section 5.6.3 and Example 7.5.

    Piezoelectric forces are a common source for microactuation. These forces normally are transient in nature, as sustained piezoelectric force generation with prolonged application of electric voltage to the generating crystal causes undesirable heat in the material. The piezoelectric force determined by Equations (7.10)–(7.12) represents the peak value of the generated force. The time factor, that is, the rising time of the peak force can become a critical parameter in the design analysis involving high-precision microactuators.

3. Van der Waals forces that exist between closely spaced surfaces as described in Section 9.3.3—A van der Waals force is a form of electrostatic force but at the molecular level. Accurate estimation of the force is not straightforward as a result of the physical change of atomic cohesion involved in generating these forces. An empirical method may be used to estimate the magnitude of these forces, as will be presented in Chapter 11.

10.4.3 Thermomechanical Stress Analysis

Stress analysis is a major effort in design analysis. For microsystems that either involve fabrication at high temperature (see Table 10.1) or are expected to operate at elevated temperature, a thermal analysis should proceed the stress analysis.

Thermomechanical stress analysis of microsystems can be handled by the formulations provided in Section 4.4.3 for most microcomponents with geometries as described in Section 10.4.1 or by a finite element method (Hsu, 1986) for cases involving more complicated geometries and loading and boundary conditions. However, thermomechanical stress analysis of microsystems can significantly differ from that of macrosystems. In microsystems, there is a strong presence of intrinsic stresses resulting from the microfabrication processes, for example, residual stress and strain, as illustrated in Sections 4.6 and 9.3.3. These inherited residual stresses must be evaluated and included in the subsequent stress analysis (Hsu and Sun, 1998). Other possible sources for intrinsic stresses induced in thin films on thick substrates include the following (Madou, 1997):

- Doping of substrates with impurities that would cause intrinsic stresses because of lattice mismatch and variation of atomic sizes
- Atomic peening due to ion bombardment by sputtering atoms and working gas densification of the thin film
- Microvoids in the thin film as result of the escape of working gases
- Gas entrapment
- Shrinkage of polymers during cure
- Change of grain boundaries due to change of interatomic spacing during and after deposition or diffusion

An important consideration in using finite element analysis for thermomechanical analysis of MEMS and microsystems structures is that we must ensure that proper finite element formulations are chosen. Many constitutive laws, and thus constitutive equations derived for continua at the macroscale, require substantial modifications for structures at the submicometer scale (i.e., structures with dimensions less than 1 μm). For example, the heat conduction equation (5.31) for solids at the macroscale requires significant modification for the same solids at the submicrometer scale, as will be described in Chapter 12. Properties of most materials at the submicrometer scale become size dependent, which invalidates most commercial finite element codes for MEMS and microsystems components.

10.4.4 Dynamic Analysis

Dynamic analysis is conducted for microsystems that involve motion. The primary reason for the analysis is to find (1) the inertia forces that act on the device or the components of the device and (2) the natural frequencies of the moving structures at several modes of vibration.

Inertia forces due to acceleration or deceleration on the components must be accounted for in the stress analysis. The natural frequencies obtained from the modal analysis, on the other hand, will be used to avoid resonant vibration, as described in Chapter 4. However, in some microsystems design, resonant vibrations of beams or plates are used to enhance the output signals of sensors, such as demonstrated in Section 2.2.5 for a certain type of pressure sensor.

In general, excessive vibration can be a major cause of structure failure due to fatigue of the materials. The finite element method is widely used for dynamic analysis of microsystems.

Example 10.1. This example illustrates the intricacy of the mechanical and electrical coupling aspects in the design of fragile microstructures. The case is not meant to be realistic, as it is unlikely that a simple capacitor with parallel plates will be used to generate the necessary actuation forces in a microgripping device. This example is used for a comprehensive demonstration on how mechanical behavior of a structure can influence the electrical effects and vice versa.

A microgripper is actuated by a capacitor made of two square plates with an initial gap of 4 μm as illustrated in Figure 10.5. The gripping arms are made of single-crystal silicon. The gripper is designed to pick up a rigid object with a mass of 3 mg. The capacitor operates in an atmospheric environment, so a value of 1.0 for the relative permittivity is used in the capacitor. Determine the following:

1. The required electric voltage to pick up the weight. Assume a friction coefficient of 1.0 between the gripping points and the object surface.
2. The maximum stress in the gripping arms.
3. The maximum deflection in the gripping arms.
4. The effect of arm deflection on the performance of this microdevice.

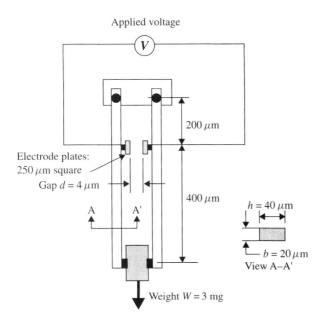

Figure 10.5. A microgripper.

Solution

1. To find the necessary voltage required to grip the weight, we need to first determine the necessary electrostatic force for gripping. This force can be determined by static equilibrium of the forces acting on the gripping arms.

We assume that each arm carries equal weight, that is, $W/2 = 1.5\,\text{mg}$. Forces acting on the arm are as shown in the free-body diagram in Figure 10.6

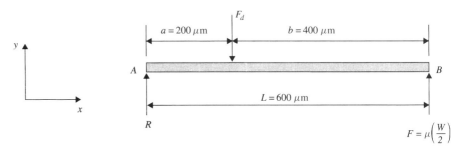

Figure 10.6. Free-body diagram of forces on a gripping arm.

In Figure 10.6, $R =$ reaction at hinged support, $F_d =$ required electrostatic force generated by capacitor, $F =$ required gripping force, $\mu =$ coefficient of friction (1.0), and $W =$ weight of mass to be picked up (3 mg).

From the first static equilibrium of forces, that is, $\sum F_y = 0$;

$$R + F - F_d = 0 \tag{a}$$

and from the other static equilibrium condition on moments, that is, $\sum M_A = 0$;

$$200F_d - 600F = 0 \tag{b}$$

We solve for $F_d = 3F = 3(W/2) = 1.5W$ from Equations (a) and (b)

Since the design weight to be picked up by the gripper is given as $W = 3\,\text{mg} = 29.43 \times 10^{-6}\,\text{N}$, we may compute the required electrostatic force $F_d = 44.15 \times 10^{-6}\,\text{N}$ from the above expression. We have also learned from Section 2.3.4 that the electrostatic force F_d produced by a capacitor with plate sizes as shown in Figure 10.7 can be related to the applied voltage by Equation (2.8):

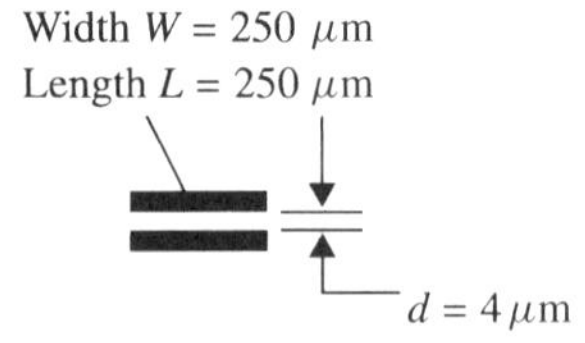

Width $W = 250\ \mu\text{m}$
Length $L = 250\ \mu\text{m}$

$d = 4\,\mu\text{m}$

Figure 10.7. Gap between two plate electrodes.

$$F_d = -\frac{1}{2}\frac{\varepsilon_0\varepsilon_r\, WL}{d^2}V^2$$

where W and L are respectively the width and length of the plate electrodes, the permittivity of free space $\varepsilon_0 = 8.85 \times 10^{-12}$ C/N-m^2, and the relative permittivity $\varepsilon_r = 1.0$ in air.

By substituting the dimensions of W and L, the given permittivity, and the required gripping force $F_d = 44.15 \times 10^{-6}$ N into the above expression, we obtain the required voltage supply: $V = 50.54$ V.

2. To determine the maximum stress in the gripping arm, the arm is subjected to combined bending and tensile forces as illustrated in Figure 10.8, where $F = F_w = 0.5W = 14.72 \times 10^{-6}$ N, and from Equation (a), we obtain the reaction $R = F_d - F = 29.43 \times 10^{-6}$N.

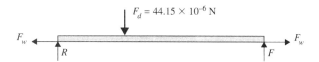

Figure 10.8. Forces in a loaded gripping arm.

We need to determine the maximum bending moment in the beam for the maximum stress. The moment distribution in a simply supported beam subjected to a concentrated force, such as in the present case, can be found in many engineering mechanics textbooks. Thus, by referring to the bending moment distribution diagram in Figure 10.9, we come up with the following expressions:

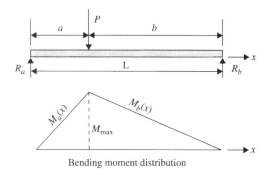

Figure 10.9. Bending moment diagram of a loaded gripping arm.

- Moment distributions:

$$M_a(x) = \frac{Pbx}{L} \quad \text{for} \quad 0 \leq x \leq a \tag{c}$$

$$M_b(x) = Pa\left(1 - \frac{x}{L}\right) \quad \text{for} \quad a \leq x \leq L \tag{d}$$

The maximum bending moment in the beam occurs at $x = a$ with $M_{max} = Pba/L$.

- Deflections:

$$\delta_a(x) = -\frac{Pbx}{6EIL}(L^2 - b^2 - x^2) \quad \text{for} \quad 0 \leq x \leq a \tag{e}$$

$$\delta_b(x) = -\frac{Pbx}{6EIL}(L^2 - b^2 - x^2) - \frac{P(x-a)^3}{6EI} \quad \text{for} \quad a \leq x \leq L \tag{f}$$

The maximum deflection of the gripping arm occurs at

$$x_m = \sqrt{\frac{L^2 - b^2}{3}} \tag{g}$$

The deflection under load P is determined as

$$\delta_p = -\frac{Pba}{6EIL}(L^2 - a^2 - b^2) \tag{h}$$

We are now ready to calculate the maximum stress in the beam using the following expression for beams subject to combined bending and tensile loads:

$$\sigma_{max} = \sigma_{m,b} + \sigma_t \tag{i}$$

where $\sigma_{m,b}$ is the maximum bending stress and σ_t is the tensile stress in the beam.

To determine both stresses, we need to find the cross-sectional area A and the area moment of inertia I of the beam for the cross section of the beam, as shown in Figure 10.10. From Figure 10.10, we can perform the following computations: The cross-sectional area $A = 8 \times 10^{-10}\,\text{m}^2$ and the area moment of inertia $I = 1.0666 \times 10^{-19}\,\text{m}^4$. The maximum bending stress is determined as

$$\sigma_{m,b} = \frac{M_{max}C}{I} = \frac{M_{max}h}{2I}$$

Figure 10.10. Cross section of the gripping arm.

where $M_{max} = Pba/L = F_d ba/L = 5.887 \times 10^{-9}\,\text{N-m}$ and $h = 20 \times 10^{-6}\,\text{m}$, and we have $\sigma_{m,b} = 1,103,800\,\text{N/m}^2$, or $1.1038\,\text{MPa}$. The tensile stress is given as $F_w/A = 0.0184 \times 10^6\,\text{N/m}^2 = 0.0184\,\text{MPa}$. Thus the maximum stress in the gripping arm is

$$\sigma_{max} = \sigma_{m,b} + \sigma_t = 1.1038 + 0.0184 = 1.1222\,\text{MPa}$$

We realize that σ_{max} is much less than the yielding strength of silicon (7000 MPa, as shown in Table 7.3).

3. The maximum deflection in the beam: The maximum deflection of the gripping arm occurs at x_m, given in Equation (g), or $x_m = 258.2 \times 10^{-6}$ m in the present case. The corresponding maximum deflection can be calculated from Equation (f) as $\delta_{max} = 0.00833$ μm using a Young's modulus of 1.9×10^{11} N/m^2.

4. As we may realize from Figure 10.5, the effect of the deflection of the gripping arm is to narrow the gap d in the capacitor, which produces the electrostatic force for the gripping. Significant narrowing of the gap will drastically increase the magnitude of F_d, which will in turn further deflect the arm. The situation is clearly a self-accelerating process that can lead to the collapsing of the gripper. It is thus critical to assess the deflection-induced gap change in the capacitor in this design case. Consequently, we need to calculate the deflection under the force F_d using Equation (h). In this case, the corresponding deflection is computed to be 0.007746 μm, which means a narrowing of the gap d by 0.015 μm, or a 0.375% change of the gap. Such a change is insignificant for this case.

A mechanical stopper which will prevent the continuous closing of the gap of the capacitor due to the coupled electromechanical effect as described in this design case may become necessary to avoid the instability of the gripper structure.

10.4.5 Interfacial Fracture Analysis

Microsystems components often involve inclusions of foreign substances. Many are also made of layers of thin films of a variety of materials. These foreign substance inclusions are produced in the substrates either by diffusion or by ion implantation, whereas thin films are deposited on the surface of the substrate by various deposition techniques. All three micromanufacturing techniques described in Chapter 9 involve multilayer thin films. Interfaces between dissimilar materials are thus common in microsystems structures.

Multilayer structures made of dissimilar materials present serious mechanical problems. In addition to excessive thermal stresses and strains due to mismatch of coefficients of thermal expansion, they are vulnerable to interfacial fracture. Delamination of interfaces is a principal concern in the mechanical design of microstructures.

In Section 4.5.3, we learned how to analyze the mechanical strength of an interface of two dissimilar materials by computing the coupled stress intensity factors relating to the opening and shearing modes of fracture of the interfaces. A major problem, however, is the lack of available values of the corresponding fracture toughness, which is required for determining whether the computed stress intensity factors are below the safe limits set by the fracture toughness, as illustrated in Figure 4.52.

10.5 MECHANICAL DESIGN USING FINITE ELEMENT METHOD

We have demonstrated that theories and principles of continuum mechanics and heat transfer as presented in Chapters 4 and 5 can be used in the mechanical design analysis of many MEMS and microsystems components. However, there are times when more sophisticated analytical tools are needed to handle microsystems involving components with complex geometry and loading/boundary conditions. Also, as we have learned from Chapters 8 and 9, many of these design analyses require inclusion of the thermomechanistic effects induced on the finished components by microfabrication and manufacturing processes. In such cases, finite element analysis becomes the only viable tool that is available to design engineers.

An overview of the working principle of the finite element analysis was presented in Section 4.7. What will be covered in the following are the basic formulation of this popular and effective method and how the method can be used to handle mechanical design of microsystems. Detailed derivation of the finite element method is available in several excellent reference books, for example, Zienkiewicz (1971), Segerlind (1976), Bathe and Wilson (1976), Hsu (1986), and Heinrich and Pepper (1999).

10.5.1 Finite Element Formulation

Discretization. This process begins with subdividing a continuum in Figure 10.11*a* into an assembly of a finite number of elements interconnected at the nodes, as illustrated in Figure 10.11*b*. We realize that, after such discretization, the original curved

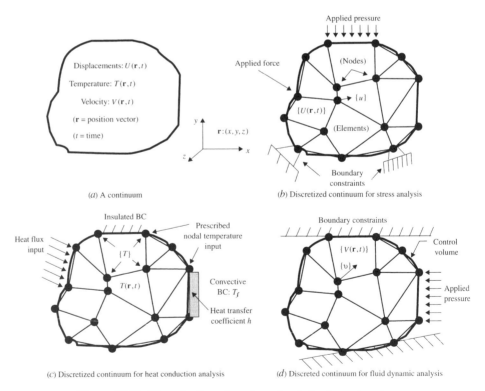

Figure 10.11. Discretization of continua for finite element analyses.

boundaries are approximated by straight edges of the elements along the boundary. Once satisfied with the discretized model, the user will relate the *primary unknown quantities* in the elements to those at the associated nodes. The primary unknown quantities are those used as the basic quantities in the finite element formulations. The computed values of these unknown quantities will lead to the values of other unknown quantities required in the analyses. For instance, the primary unknown quantities in a stress analysis are the displacement components in the elements. The element displacements can be used to compute the element strains and stresses using the constitutive equations. The primary unknown quantities vary from application to application. For instance, as shown in Figure 10.11*a*, $\{U(\mathbf{r}, t)\}$ designates the element displacement components in a stress analysis, element temperatures $T(\mathbf{r}, t)$ are for a heat conduction analysis, and the element velocity components $\{V(\mathbf{r}, t)\}$ are for a fluid dynamic analysis. In these mathematical designations, we represent the spatial coordinates with a position vector $\mathbf{r}$ and time t. However, for most finite element analyses, discretization is performed in the spatial domain with these quantities integrated with the time variable by using various integration schemes for transient analyses (Hsu, 1986). Thus, the mathematical expressions for discretization for the above three common mechanical engineering analyses can be shown as

$$\{U(\mathbf{r})\} = [N(\mathbf{r})]\{u\} \qquad (10.2a)$$

for stress analysis, where $[N(\mathbf{r})]$ is the *interpolation function*, with $\mathbf{r}$ denoting the position vector that defines the elements. This position vector represents the spatial coordinates, for example, (x,y,z) for cartesian coordinates, or the cylindrical polar coordinates (r,z) for an axisymmetric solid. The vector $\{u\}$ represents the corresponding nodal displacement components.

Likewise, as shown in Figure 10.11*c*, the discretization for the heat conduction and fluid dynamics analyses can be expressed as

$$T(\mathbf{r}) = [N(\mathbf{r})]\{T\} \qquad (10.2b)$$

for heat conduction analysis, with T the corresponding nodal temperature, and

$$\{V(\mathbf{r})\} = [N(\mathbf{r})]\{v\} \qquad (10.2c)$$

for fluid dynamic analysis with v the corresponding velocity components at the nodes, as illustrated in Figure 10.11*d*.

The values of the primary unknown quantities at the nodes of an element can be related to the applied nodal loads by the following *element equations*:

$$[K_e]\{q\} = \{Q_e\} \qquad (10.3)$$

where

$[K_e]$ = element coefficient matrix

$\{q\}$ = primary unknown quantities at nodes

$\{Q_e\}$ = applied loads at nodes

The element coefficient matrix contains constants that are related to the shape of the elements—the interpolation functions used in the discretization and the material properties of the elements. Material properties used in the finite element analyses include the elastic constants and Poisson's ratio in a stress analysis or thermal conductivity in a heat conduction analysis.

The applied nodal loads in Equation (10.3) are the applied forces at the nodes and/or the equivalent nodal forces of the applied pressure on the surface of the elements in a stress analysis (Figure 10.11b). They can also be the heat flux and/or prescribed nodal temperature in a heat conduction analysis (Figure 10.11c). The convective boundary condition in Figure 10.11c can be handled by the relationship in Equation (5.38), with a conversion to equivalent thermal forces at the adjacent nodes of the corresponding elements along the solid–fluid boundary (Hsu, 1986). In fluid dynamic analysis, the applied nodal load can be the pressure at the surface of a control volume (Figure 10.11d).

The primary unknown quantities at all nodes in a discretized continuum $\{q\}$ can be obtained from the overall equilibrium equations as shown below:

$$[K]\{q\} = \{Q\} \tag{10.4}$$

where the overall coefficient matrix $[K]$ is constructed by summing up all element coefficient matrices $[K_e]$ in the discretized model:

$$[K] = \sum_{1}^{M}[K_e] \tag{10.5a}$$

and the overall load matrix $\{Q\}$ is obtained by

$$\{Q\} = \sum_{1}^{n}\{Q_e\} \tag{10.5b}$$

where

M = total number of elements

n = total number of nodes in discretized model

The primary unknown quantities $\{q\}$ at all nodes in the discretized model can be obtained by solving the simultaneous algebraic equations given in Equation 10.4 by using a developed method such as Gaussian elimination with back substitution (Allaire, 1985).

Derivation of Element Equations. It is apparent that the element equation 10.3 is the key to a finite element analysis. This equation can be derived in several ways, depending on the nature of the analysis. For problems that can be described by differential equations, such as the heat conduction equations (5.31) or the Navier–Stokes equation (5.20) for fluid dynamics analysis, the *Galerkin method* is used for such derivations. On the other hand, for problems that cannot be described by distinct differential equations, such as the case of stress analysis, the *Rayleigh–Ritz method* is the viable way to derive the element coefficient matrix. Derivation of element equations using both these methods is available in Hsu (1986). We will outline these methods here with suggestions as to how they can be modified to account for submicrometer-scale structures.

Galerkin Method. This method is often referred to as the *weighted residuals* method. It is based on the fact that a continuum is no longer a continuum after the discretization illustrated in Figure 10.11. The primary quantities obtained from a discretized model in a finite element analysis, in fact, give the approximated values of those in the original continuum.

Consequently, let us consider "exact" solutions to physical problems to be those for problems that can be described mathematically in the form of differential equations with proper boundary conditions:

$$D(\phi) = 0 \quad \text{in domain } V \tag{10.6a}$$

$$B(\phi) = 0 \quad \text{on boundary } S \tag{10.6b}$$

where D and B are the respective differential operators for the differential equation and the boundary conditions. The function ϕ represents the primary unknown quantities as indicated in Figure 10.11a for the three common finite element analyses used in microsystems design.

Equation (10.6a) represents the differential equations for field problems such as Equation (5.20) for fluid dynamics and Equation (5.31) for the temperature field in a solid. The boundary conditions in Equation (10.6b) represent such expressions in Equations (5.36)–(5.38) for heat conduction analysis.

The above system, with ϕ the primary unknown quantity in Equation (10.6), can be replaced by an integral equation:

$$\int_v W D(\phi) \, dv + \int_S \overline{W} B(\phi) \, dS = 0 \tag{10.7}$$

where W and $\overline{W}$ are arbitrary weighting functions.

Equation (10.7) represents the "exact" solution to the continuum represented by Equations (10.6a) and (10.6b). A similar equation for the discretized continuum following Equations (10.2b) and (10.2c) with the generic expression $\phi = \sum N_i \phi_i$, where N_i and ϕ_i represent respectively the interpolation function and the primary unknown quantities at the nodes, will lead to the solution

$$\int_v W_j D\left(\sum N_i \phi_i\right) dv + \int_s \overline{W}_j B\left(\sum N_i \phi_i\right) dS = R \tag{10.8}$$

where W_j and $\overline{W}_j$ are the respective weighting functions for the discretized continuum and its boundary conditions.

We notice that the substitution of the unknown quantities ϕ in Equation (10.7) with the discretized values $\sum N_i \phi_i$, where i is the total number of nodal components in the element, leads to a *residual* value R as shown in Equation (10.8). The existence of the residual value is due to the approximate nature of the solution offered by the discretized continuum. In a physical sense, the residual represents the difference between the exact solution to the problem for a continuum and the approximate solution obtained from the same continuum but with discretization.

A good discretized system, of course, requires the condition that $R \to 0$. Choosing weighting functions that are identical to the interpolation function, that is, letting $W_j = \overline{W}_j = N_i$, one may arrive at the element equation (10.3) after performing a variational process of the residual with respect to the variable ϕ_i in Equation (10.8).

For structures that are in submicrometer sizes, differential equations derived for continua such as the heat conduction equation (5.31) need to be replaced by those for continua at submicrometer and nanometer scales as will be presented in Chapter 12. Likewise, the Navier–Stokes equation (5.20) needs to be substituted by the Boltzmann equation (Beskok amd Karniadakis, 1999).

Rayleigh–Ritz Method. This method involves the variation of a *functional* for deriving the element equations in a finite element analysis. It is a viable alternative to the Galerkin method and is specifically suitable for problems that cannot be described by distinct differential equations, such as the stress analysis of solid structures. A functional is defined as a function of functions.

For stress analysis of solid structures, the functional used in the variation is the *potential energy* $P(\{u\})$ of the deformed structure, where $\{u\}$ denotes the unknown nodal displacement components as shown in Figure 10.11b. Derivation of the potential energy for elastically deformed solids is available in many finite element books. The potential energy for elastic-plastically deformed solids is presented in Hsu (1986). The following relation is used to derive this energy function:

$$P(\{u\}) = U(\{u\}) - W_k(\{u\}) \tag{10.9}$$

where $U(\{u\})$ is the strain energy and $W_k(\{u\})$ is the work done to the solid by the applied body forces and surface tractions.

An element equation in the form of Equation (10.3) is derived by applying the variational principle to the potential energy in Equation (10.9) with respect to the nodal displacements $\{u\}$. Consequently, the following expression is obtained:

$$[K_e]\{u\} = \{F\} \tag{10.10}$$

where $[K_e]$ and $\{F\}$ are respectively the element stiffness and nodal force matrices.

The element stiffness matrix can be determined from the integral

$$[K_e] = \int_v [B(\mathbf{r})]^{\mathrm{T}} [C][B(\mathbf{r})]\, dv \tag{10.11}$$

and the nodal force matrix can be obtained by the integrals

$$\{F\} = \int_v [N(\mathbf{r})]^{\mathrm{T}} \{f\}\, dv + \int_s [N(\mathbf{r})]^{\mathrm{T}} \{t\}\, ds \tag{10.12}$$

where

v = volume of element

s = surface of element

$\{f\}$ = body force acting on element

$\{t\}$ = surface traction acting on element surface s

The matrix $[B(\mathbf{r})]$ in Equation (10.11) relates the element strain components $\{\varepsilon\}$ to the nodal displacements $\{u\}$. The matrix $[C]$ is the elasticity matrix involving Young's modulus and Poisson's ratio of the material. For elastic–plastic stress analysis, an elastic–plasticity matrix $[C_{ep}]$, derived from incremental plasticity theory, replaces the $[C]$ matrix in Equation (10.11) (Hsu, 1986).

Once the displacements $\{u\}$ are determined at the nodes, one may find the corresponding element displacements using Equation (10.2a) and the element strains and stresses through appropriate formulations in the theory of elasticity and the constitutive relations, such as the generalized Hooke's law for elastic stress analysis.

As in thermofluids analyses, the finite element analysis for deformable structures at the submicrometer scale require significant modification to accommodate the geometry and size-dependent anisotropic material properties and the distinct constitutive laws and equations. Much research effort is needed in these areas, as will be covered in Chapter 12.

10.5.2 Simulation of Microfabrication Processes

We have mentioned throughout this book the inherent effects of microfabrication on the mechanical behavior of MEMS and microsystems structures. Integration and simulation of microfabrication processes in finite element analyses have thus become a critical requirement. A finite element algorithm was proposed to simulate micromanufacturing of microsystems (Hsu and Sun, 1998). The algorithm that uses pseudoelements in the discretization of the continua is involved in the analysis. This simulation technique requires the finite element (FE) mesh in the microstructure to include two parts: (1) the part with real elements for the substrate material and (2) the part with *pseudoelements* for simulating the materials being added or removed in microfabrication processes. In these micromanufacturing processes, materials are added to substrates by *deposition*, *electroplating*, or *epitaxial growth* of substrate crystals in either surface micromachining or the LIGA process, whereas portions of the subtracted materials are removed by *etching* in bulk micromachining. By varying the assigned material properties to the pseudoelements before and after micromanufacturing, one can simulate the three micromanufacturing processes described in Chapter 9. The working principle of this simulation technique is described below.

Simulation of Surface Micromachining and LIGA Process. Surface micromachining is an additive process, and the pseudoelements used are referred to as *A-elements*, meaning the elements that are to be added to the substrate in a microfabrication process.

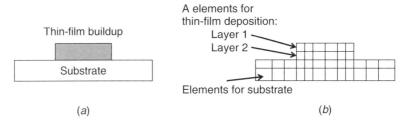

Figure 10.12. Finite element discretization of microstructure in additive microfabrication process: (*a*) microstructure; (*b*) discretized microstructure.

Following is the simulation procedure:

Step 1   The A-elements are included in the initial FE mesh as shown in Figure 10.12. The thickness of the layer made of A-elements is equal to the thickness of the layers of the additive materials to be built on the substrate.

Step 2   The initial material properties assigned to the A-elements are low Young's modulus $E = 1$ Pa; high yield strength $\sigma_y = 10^6$ MPa; the same coefficient of thermal expansion $\alpha_{\text{substrate}}$; and low mass density $\rho = 10^{-6}$ g/m^3.

Step 3   The real properties of the layer material are assigned to the A-elements after one pass of deposition or electroplating or fusion bonding is completed.

Step 4   Compute the stress distribution in the overall FE mesh with the current set of A-elements using the real layer material properties while the remaining A-elements still have pseudoproperties. In this case, the element stiffness matrix is calculated according to Equation (10.11).

Step 5   Update stresses, strains, and nodal coordinates in the FE mesh.

Step 6   Update the stiffness matrix $[K]$.

Step 7   Repeat the computation for the next surface "layer" in the fabrication process.

Simulation of Bulk Micromachining. Bulk microfabrication is a subtractive process with removal of substrate material at desired localities. We call the pseudoelements in this process the *S-elements*, the elements in the finite element discretization which are to be subtracted in the analytical process.

Step 1   The S-elements are included in the FE mesh for the parts that are to be etched, as illustrated in Figure 10.13.

Step 2   The initial properties assigned to the S-elements are identical to those of the substrate. The etch front in the S-elements are identified according to the etching rate used in the process.

Step 3   The $[B(\mathbf{r})]$ matrix in Equation (10.11) is adjusted according to the current nodal positions coinciding with the etching front with temporary assigned

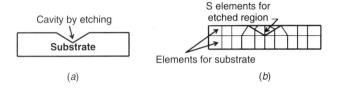

Figure 10.13. Finite element discretization of microstructure in subtractive microfabrication process: (*a*) microstructure; (*b*) discretized microstructure.

> nodal coordinates. The reduced element stiffness is computed according to Equation (10.11) as

$$[K_e] = \int_v [B(\mathbf{r}')]^T [C][B(\mathbf{r}')] \, dv$$

> where $\mathbf{r}'$ represents the position vector in the $[B]$ matrix involving temporary nodal coordinates.

Step 4 The entire S-element is switched to pseudoelement with pseudomaterial properties once the etching front passes from one end to the other in the S-element. Pseudomaterial properties in this case are low element stiffness matrix $[K_e] = 10^{-6}$; low Young's modulus $E = 1$ Pa; high yield strength $\sigma_y = 10^6$ MPa; the same coefficient of thermal expansion $\alpha_{\text{substrate}}$; and low mass density $\rho = 10^{-6}$ g/m^3.

Step 5 Repeat the same procedure for the subsequent etching process.

Both A- and S-elements can be used in conjunction to simulate micromanufacturing processes as described in Chapter 9, as illustrated in Figure 10.14.

Hsu (1986) developed an algorithm based on a similar principle to that presented above for the simulation of crack growth in solids (Kim and Hsu, 1982). A popular commercial FE code, ANSYS was used to simulate the etching process involving a silicon substrate with an SiO$_2$ layer (Hsu and Sun, 1998). The "birth" and "death" elements in that ANSYS code, which presumably were constructed using principles similar to the proposed A- and S-elements, were used to simulate the oxidation of silicon substrates. The residual stresses and strains obtained from the simulation were accounted for by the initial stresses and strains in the subsequent stress and strain analysis of a micropressure sensor die/diaphragm subjected to specified operating loading conditions.

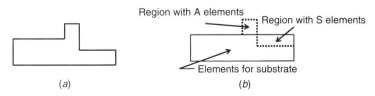

Figure 10.14. Finite element discretization for a microstructure produced by both additive and subtractive processes: (*a*) microstructure; (*b*) discretized microstructure

10.6 DESIGN OF SILICON DIE OF A MICROPRESSURE SENSOR

We will demonstrate the mechanical design of microdevices by a case study relating to the design of silicon dies used in pressure sensors. Silicon dies are key components in micropressure sensors. Proper mechanical design is necessary to ensure the proper functioning of these sensors.

General Description. The working principles of various types of micropressure sensors were described in Chapter 2. Here, we will take a closer look at how the mechanical design of one type of pressure sensor is carried out. The cross section of a typical square silicon die is illustrated in Figure 10.15. The die is made of a square silicon substrate with a cavity produced by etching from one side of the substrate. The thinned portion of the die is the diaphragm, which deflects when a medium applies pressure at the back face (the bottom face) of the die. Associated with the induced deflection are the bending and shear stresses in the diaphragm. Piezoresistors are planted in the diaphragm to sense the induced stresses. The die is attached to a constraint base, with die attachment by such means as epoxy adherents or special bonding techniques such as anodic bonding, as will be described in Chapter 11.

Design Considerations. There are a number of factors that a design engineer needs to be aware of before embarking on the actual design analysis of the die:

1. **Applications:** Different applications will mean different user groups. Sensors to be used in automobiles and machine tools will be required to survive in a tougher environment than those used in high-precision instruments in laboratories.
2. **Working Medium:** The pressure side of the die is normally in contact with the pressurizing medium. These contacts can be tolerated if the medium is inert and harmless to the die material. Unfortunately, many of these media are hostile and toxic, such as the exhaust gas of an automobile. In such cases, *die passivation* becomes necessary and the mechanical design of the die must provide protection.

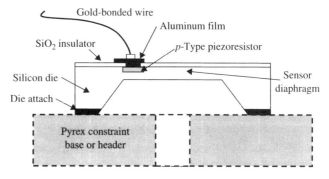

Figure 10.15. Silicon die in a typical micropressure sensor.

3. **Constraint Base:** Sensors are normally attached to a base or a header. A vibrating base can cause the die to fail in fatigue. The natural frequency of the die should be designed to be much higher than expected excitation frequencies during operation. The lowest natural frequency for the die of a pressure sensor is 100 Hz–2 kHz for automobile applications.

4. **Other Considerations:** Considerations should be given to unexpected conditions, such as mechanical shock due to accidentally dropping the sensor by mishandling it. Thermal shock may be another unexpected condition that should be considered in the design.

Geometry and Size of Die. Most pressure sensor dies are square in geometry. Rectangular and circular dies were used in early applications, but they rarely appear in recent products.

Careful consideration should be given to the size of the die, as it will affect not only the cost of the sensor but also the sensitivity in the performance. Figure 10.16 indicates the key dimensions of a square silicon die.

In this figure, the overall size of the die is $A \times A \times H$. The thickness of the die, H, is limited by the standard thickness of silicon wafers, for example, the thickness of 100-mm-diameter wafers is 500 μm, (the thicknesses of other size wafers are available in Section 7.4.2). The height of the die needs to be properly established, as it will provide the necessary isolation for the die from mechanical effects induced by the constraint base. The diaphragm has the size of $a \times a \times h$, where the length a is determined by the dimension b and the etch angle, 54.74°. This angle exists if wet etching is applied to the (100) plane of the silicon substrate. The dimension b, on the other hand, depends on the size of the die "foot print," c. The latter must be large enough so that the stresses in the die attach (see Figure 10.15) can be kept below plastic yielding strength. It is thus clear that all dimensions of the die are interdependent.

Another factor that affects the size of the die is the size of the signal transducer. In general, two types of transducers are used for pressure sensors: piezoresistors and capacitors. In either case, adequate space on the die surface must be provided for these transducers. For best results, these transducers must transmit signals from localized points, which means that either the die is made large enough that the signals can be treated as localized or the transducers are made as small as possible. The die size is thus also constrained by the transducer size.

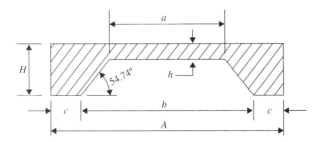

Figure 10.16. Key dimensions of a square die for a pressure sensor.

One factor that determines the cost of the die, of course, is its overall size. It is desirable to keep the die as small as physically possible for low material cost and conservation of space.

Strength of Die. We have shown that the maximum bending stress and deflection of the diaphragm can be estimated from Equations (4.10) and (4.11). These equations are expressed in slightly different forms below:

$$\sigma_{\max} = C_1 P \left(\frac{a}{h}\right)^2 \tag{10.13}$$

$$W_{\max} = C_2 P \left(\frac{a^4}{h^3}\right) = C_2 P a \left(\frac{a}{h}\right)^3 \tag{10.14}$$

where C_1 and C_2 are constant coefficients and P is the applied pressure.

From the point of view of sensor performance, it is desirable to generate sufficient $\sigma_{\max}$ in the diaphragm for significant signals from the piezoresistors. This would mean thin diaphragms, that is, low h as in Equation (10.13). However, a low h will increase the maximum deflection $W_{\max}$ much faster than $\sigma_{\max}$, as indicated in Equation (10.14). Large deflections of the diaphragm will result in a nonlinear relationship between the applied pressure and the bending stress and hence output signals as illustrated in Figure 10.17. Such nonlinear relationships are obviously not desirable in sensor design. Engineers are thus faced with a dilemma in the diaphragm design; on the one hand, they need large enough bending stress but, on the other hand, the deflection of the diaphragm needs to be kept as small as possible.

A possible way out of this dilemma is to introduce _boss stiffeners_ in the diaphragm. These stiffeners can keep the diaphragm from excessive deflection but allow maximum stresses in the diaphragm in localities away from the stiffeners. A diaphragm with a square boss stiffener is illustrated in Figure 10.18.

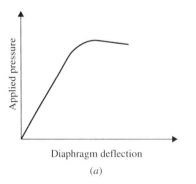
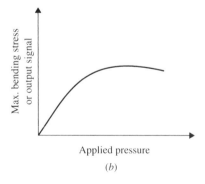

Figure 10.17. Linearity of output signal in a pressure sensor: (_a_) Pressure–deflection; (_b_) Pressure–output signal.

Design for Operating Pressures. Depending on the application, micropressure sensors are designed to operate at very low pressures to very high pressures. In the automotive industry, micropressure sensors are typically operated at 10 MPa for gasoline engine cylinders, 100 MPa for diesel engine cylinders, and 20 kPa for brake fluids and hydraulic suspensions.

Flat diaphragms are best for pressure sensors operating in the medium-pressure range, that is, between 35 kPa and 3.5 MPa. A linear relationship between the piezoresistor output and the applied pressure (Figure 10.17*b*) exists. However, this desirable situation does not always exist for applied pressures beyond this range.

Low operating pressure ($P < 30$ kPa) implies the need for a thin diaphragm [i.e., low h in Equations (10.13) and (10.14)]. A thin diaphragm also means small diaphragm size (the dimension a in Figure 10.16), which leads to low bending stress and hence low signal output. Thus, in order to generate significant output signal, it is necessary to enlarge the diaphragm. A large, thin diaphragm will result in nonlinear signal output, as indicated above in the discussion of 'strength of the die.' One solution to this problem is to use boss stiffeners as in Figure 10.18. It has been reported that such an arrangement was made in a micropressure sensor used to measure low pressure at 7 kPa with a full-scale output of 100 mV and a linearity better than 0.1%.

For *high operating pressures* ($P > 35$ MPa) thicker diaphragms are used. This will mean low *a/h* ratio for the diaphragm. This low ratio can alter the dominant stress in the diaphragm from bending to shearing stress. Consequently the resulting piezoresistive signal versus applied pressure exhibits a nonlinear relationship. One solution to this problem is to carefully locate the piezoresistors in the region where shearing stress is minimal. Using local boss stiffeners can control the distribution of bending and shearing stresses in the diaphragm. Finite element analyses such as demonstrated in Figure 4.53 can aid such design.

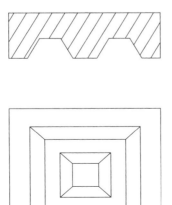

Figure 10.18. Silicon die with square boss stiffener.

Design for Overpressure. For safe operations, micropressure sensors are often designed for overpressure at as high as 10–30 times the design operating pressure. Such high pressure will obviously cause excessive deflection of the diaphragm. To avoid such occurrence, a deflection limiter, or a *stopper*, must be provided in the die to prevent the diaphragm from excessive deflection under the overpressure conditions.

Design with Operating Temperature. Temperature will affect the material strength and induce thermal stresses in the microstructure, as described in Section 4.4. Another critical factor that needs to be considered in the design is the temperature effect on the piezoresistors that are part of the silicon die. We will leave the design of a signal transduction system to Chapter 11. The fact that temperature can have a serious effect on transduction, however, should not be overlooked.

10.7   DESIGN OF MICROFLUIDIC NETWORK SYSTEMS

We mentioned elsewhere in this book that bioMEMS is a rapidly growing field of research, with new applications reported frequently by the industry. Microfluidic devices have become principal components in many of these new bioMEMS. In particular, the *capillary electrophoresis (CE) on-a-chip* has gained increasing popularity in the pharmaceutical, genetic, and human genome industries in recent years. Capillary electrophoresis on-a-chip is a microfluidic system that offers a means of miniaturizing chemical and biomedical analysis systems and achieves the sensing of the constituents in minute samples with almost instant results. As the complete analysis systems are integrated with signal processing in a single chip, they can be produced in batches in large quantities, thereby drastically reducing the production cost. Furthermore, these systems are disposable after use, which eliminates all maintenance costs. Most of these systems are so minute that assemblies with 1000 channels for parallel CE analyses on a single chip are made possible by micromanufacturing technologies. These arrangements have made CE on-a-chip the most effective and efficient analytical tool for the aforementioned purposes.

Capillary electrophoresis involves the combined applications of electro-osmostic and electrophoretic processes as described in Chapter 3. Both these processes involve fluid movements under the influence of applied electric fields. The electro-osmotic process moves an aqueous solution past a stationary solid surface. It requires the creation of a charged double layer at the solid–liquid interface. The resultant velocity profile across the passage cross section is nearly flat. The electrophoretic process, on the other hand, does not move bulk fluid. Instead, it moves charged species in the buffer solution under the applied electric field.

Typical CE chips are made of a network of capillary tubes or channels. Common cross sections of these capillaries are illustrated in Figure 10.19. Typical dimensions of these capillaries are less than 100 μm in diameter for the tubes and 30–50 μm wide by 10–30 μm deep for the channels.

The length of the tubes or channels may vary from a few millimeters to a few centimeters. Tubes and channels are typically fabricated by etching, and epoxy seals are used for closing the conduits for the passage of fluids, as shown in Figure 10.19. The closed

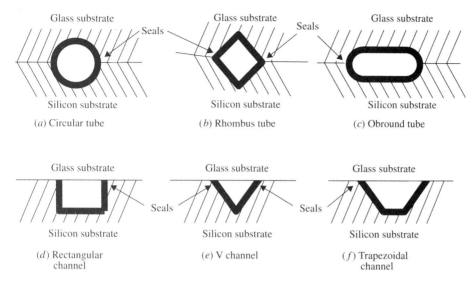

Glass substrate

Seals

Silicon substrate

(a) Circular tube

Glass substrate

Seals

Silicon substrate

(b) Rhombus tube

Glass substrate

Silicon substrate

(c) Obround tube

Glass substrate

Seals

Silicon substrate

(d) Rectangular channel

Glass substrate

Seals

Silicon substrate

(e) V channel

Glass substrate

Silicon substrate

(f) Trapezoidal channel

Figure 10.19. Typical cross sections of capillaries.

capillaries shown in Figures 10.19*a*, *b*, and *c* are produced in halves by microfabrication techniques, and the two halves are then bonded together by various bonding techniques, as will be described in Chapter 11. The open channels in Figures 10.19*d*, *e*, and *f* are bonded to glass substrates and sealed at the joints by epoxy (Harrison and Glavina, 1993). These tubes and channels must be electrically insulated from the substrates in order to avoid voltage breakdown in the electrophoretic and electro-osmotic processes during the analysis.

10.7.1   Fluid Resistance in Microchannels

We realize that cross sections of capillary tubes and channels may vary, as shown in Figure 10.19. Channels with noncircular cross sections are used because they can easily be fabricated by microfabrication techniques such as etching. These noncircular cross sections require special formulas for estimating the flow resistance in these channels.

Fluid resistance R is defined as the ratio of pressure drop ΔP to the volumetric flow rate of the medium, Q, along a specific length of the channel. Mathematically, it can be expressed as (Kovacs, 1998)

$$R = \frac{\Delta P}{Q} \tag{10.15}$$

The fluid resistance for a circular tube with a radius a can be readily obtained from Equation (5.17) as

$$R = \frac{8\mu L}{\pi a^4} \tag{10.16a}$$

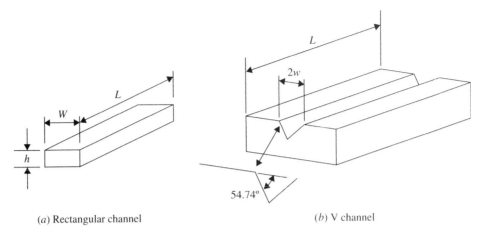

(a) Rectangular channel   (b) V channel

Figure 10.20. Noncircular microchannels.

The same resistance for the rectangular channel in Figure 10.20a is given as

$$R = \frac{12\mu L}{wh^3} \tag{10.16b}$$

for a cross section with a high aspect ratio and

$$R = \frac{12\mu L}{wh^3} \left\{ 1 - \frac{h}{w} \left[\frac{192}{\pi^5} \sum_{n=1}^{\infty} \frac{1}{n^5} \tanh\left(\frac{n\pi w}{h}\right) \right] \right\}^{-1} \tag{10.16c}$$

for a cross section with a low aspect ratio.

The fluid resistance for the V channel in Figure 10.20b is

$$R = \frac{17.4 L\mu}{w^4} \tag{10.16d}$$

If we use the units newtons per square meter for the pressure drop in Equation (10.15) and cubic meters per second for the volumetric flow rate, then the fluid resistance R has units of newton seconds per meters to the fifth power. More revealing units for the fluid resistance would be pascals per cubic meter per second, which reflects the physical meaning of R as the pressure drop in pascals per unit volumetric fluid flow in cubic meters per second.

The fluid resistance R is an important indicator that assists engineers in selecting the geometry of flow channels in a microfluidic system. The following example will illustrate this selection.

Example 10.2. Evaluate the resistance to water flow in microchannels of three distinct cross sections of the same hydraulic diameter d_h as illustrated in Figure 10.21. The circular capillary tube has an inside radius $a = 15$ μm. The dynamic viscosity of water μ at room temperature is given in Table 4.3 as 1001.65×10^{-6} N-s/m^2.

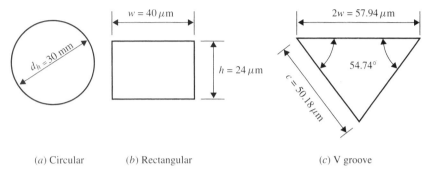

(*a*) Circular   (*b*) Rectangular   (*c*) V groove

Figure 10.21. Microflow channels.

Solution: We will first determine the dimensions of the rectangular channel in Figure 10.21*b* and the V-groove channel in Figure 10.21*c* so that both these channels will have the same hydraulic diameter d_h as the circular tube in Figure 10.21*a*.

The hydraulic diameter of noncircular cross sections can be found using Equation (5.19). We can thus compute this diameter for both rectangular and V-groove channel cross sections as

$$d_h = \frac{4(wh)}{2(w + h)} = \frac{2wh}{w + h}$$

for the rectangular cross section in Figure 10.21*b* and

$$d_h = \frac{4wh}{2c + 2w}$$

for the V-groove channel in Figure 10.21*c*. It is readily proven that $d_h = 2a$ for the circular tube by Equation (5.19).

Letting $d_h = 2a = 30$ μm and w for the rectangular channel be 40 μm in the above expressions, we will find h to be 24 μm and w for the V-groove channel to be 28.9 μm, as shown in Figure 10.21.

With the dimensions of the capillary tube and two microchannels determined, we may proceed to calculate the fluid resistance R for these conduits as follows:

1. For the capillary tube (Figure 5.12): We may use Equation (10.16a) to evaluate the fluid resistance per unit length of the tube:

$$\frac{R}{L} = \frac{8\mu}{\pi a^4} = \frac{8 \times 1001.65 \times 10^{-6}}{3.14 \times (15 \times 10^{-6})^4} = 5.0416 \times 10^{16} \text{ Pa/m}^3/\text{s/m}$$

2. For the rectangular channel (Figure 10.21*b*): We realize that the aspect ratio of the cross section is $W/h = 1.6667$, which is regarded to be low. Consequently, Equation (10.16c) is used for the evaluation of the fluid resistance per unit length of the channel:

$$\frac{R}{L} = \frac{12\mu}{wh^3} \left\{ 1 - \frac{h}{w} \left[\frac{192}{\pi^5} \sum_{n=1}^{\infty} \frac{1}{n^5} \tanh\left(\frac{n\pi w}{h}\right) \right] \right\}^{-1}$$

$$= \frac{12 \times 1001.6 \times 10^{-6}}{(40 \times 10^{-6})(24 \times 10^{-6})^3} \left\{ 1 - \frac{24}{40} \left[0.6274 \sum_{n=1}^{\infty} \frac{1}{n^5} \tanh(5.233n) \right] \right\}^{-1}$$

$$= 3.5183 \times 10^{16} \text{ Pa/m}^3/\text{s/m}$$

3. Fluid resistance in the V-groove channel (Figure 10.21*c*): From Equation (10.16d), we have

$$\frac{R}{L} = \frac{17.4\mu}{W^4} = \frac{17.4 \times 1001.65 \times 10^{-6}}{(28.97 \times 10^{-6})^4} = 2.4744 \times 10^{16} \text{ Pa/m}^3/\text{s/m}$$

We summarize these results in Table 10.2. It is interesting to note that the circular capillary tube appears to be the least desirable choice among the three selected geometries for microfluidic systems because of its high fluid resistance. It appears that the V-grooved channel generates the least resistance to microflows.

TABLE 10.2. Fluid Resistance in Various Microchannel Cross Sections with the Same Hydraulic Diameters

Channel Cross Sections	Fluid Resistance per Unit Length, 10^{16} Pa/m^3/s/m
Circular	5.0416
Rectangular	3.5183
V-groove	2.4744

10.7.2 Capillary Electrophoresis Network Systems

As mentioned at the beginning of this section, the CE system is an effective (and powerful) tool for fast and accurate analysis of chemical and biological samples. The simplest CE network consists of two channels intersecting at an angle. Figure 10.22 is a plan view of a CE network. One channel, called the *injection channel*, is for the passage of analyte that contains the species to be separated and measured. The other channel, called the *separation channel*, is where separation of the species occurs. The injection channel is connected to the analyte reservoir at the left and the analyte waste reservoir at the

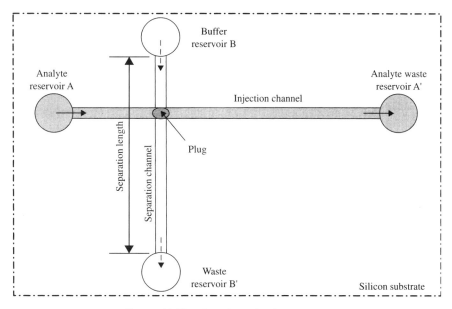

Figure 10.22. Plan view of a CE network.

right. The separation channel is connected to the buffer reservoir at the top and a waste reservoir at the bottom.

The general procedure of CE analysis is as follows:

1. An analyte solution with measurand species is injected at the injection channel from the analyte reservoir A by the application of an electric field ranging from 150 to 1500 V/cm between the analyte reservoir A and the analyte waste reservoir A'. The applied voltage forces the analyte solution to flow from reservoir A to reservoir A' by an electro-osmotic process. A *sample plug* is developed at the intersection of the two capillary channels, as shown in Figure 10.22. Sample separation does not occur during the injection process.

2. An electric field is then applied between the buffer reservoir B and the waste reservoir B'. Consequently, the sample plug starts moving in the separation channel. The flow becomes electrophoretic, in which the various species in the fluid will flow at different velocities according to their respective *ion* (or electro-osmotic) *mobility*. Separation of species in the measurand fluid thus takes place in this portion of the separation channel. The ion mobility ω_i for the ions of species i can be evaluated from the following formula (Manz et al., 1994):

$$\omega_i = \frac{z_i q}{6\pi r_i \mu} \tag{10.17}$$

where z_i is the charge of ion i, r_i is the radius of the ion, and μ is the dynamic viscosity of the ion in the buffer solvent. The charge q of an electron is 1.6022×10^{-19} C, as indicated in Equation (3.12).

3. The separated species in the buffer solution can be detected by either an ampero-metric electrochemical detector or a fluorescence detector, as described by Harrison and Glavina (1993) and Woolley and Mathies (1994).

10.7.3 Mathematical Modeling of Capillary Electrophoresis Network Systems

Mathematical modeling and analysis of the CE process, as outline above, is extremely complex. It involves the coupling of *advection*, *diffusion*, and *electromigration*. Advection is a phenomenon that involves the movement of a substance that causes changes in temperature or in other physical properties of the substance. One may well imagine the advection in both electrophoretic movement of the analyte during the injection and the subsequent separation of the species in the buffer solution. Diffusion takes place between the analyte substances in the plug and the buffer solution. This process, however, may not be significant if little time is allowed for such mixing (Patankar and Hu, 1998). Electromigration is required to model the movement of ions in the solutions induced by the applied electrical fields.

Researchers have made numerous efforts in the mathematical modeling of the CE process. Following is a partial list of references for such modeling and analyses: Saville and Palusinski (1986), Manz et al. (1994), Jiang et al. (1995), Williams and Vigh (1996), Patankar and Hu (1998), and Krishnamoorthy and Giridharan (2000).

Following are the governing equations for modeling the electrophoretic flow of the analyte involving various measurand species mixing with the buffer solvent at the intersection of the capillary network (where the plug is developed) (Krishnamoorthy and Giridharan, 2000).

The advection equation for the rate of change of concentration of species i is

$$\frac{\partial C_i}{\partial t} = -(\nabla \cdot \mathbf{J}_i) + \mathbf{r} \tag{10.18}$$

where C_i is the concentration of species i in the solution, t is the time into the process, $\mathbf{r}$ is the rate of production of the species. The flux vector $\mathbf{J}_i$ has the form

$$\mathbf{J}_i = \mathbf{V} C_i - z_i \omega_i C_i \nabla\phi - D \nabla C_i \tag{10.19}$$

where

$\mathbf{V} =$ velocity vector of species i in solution, e.g., with components $V_x(x, y)$ and $V_y(x,y)$ in respective x and y directions in flow defined by x–y plane

$z_i =$ valence of ion i

$\omega_i =$ electro-osmotic mobility of ith species as shown in Equation (10.17)

$\phi =$ applied electrical potential

$D_i =$ diffusion coefficient of ith species in solution

The rate of production of species i, $\mathbf{r}$ in Equation (10.18), is usually neglected in modeling a stable process. The electric field in Equation (10.19) can be obtained by solving the following differential equation:

$$\nabla \cdot (\sigma \nabla \phi) = 0 \tag{10.20}$$

where the electrical conductivity σ is defined as

$$\sigma = F \sum_i z_i^2 \omega_i C_i \tag{10.21}$$

where F is the Faraday constant with a numerical value of 9.648×10^4 C/mol.

The bulk fluid velocity due to electro-osmotic mobility is given as

$$V_0 = \omega_0 \nabla \phi \tag{10.22}$$

with V_0 the imposed slip velocity at the channel wall and ω_0 the electro-osmotic mobility of the species.

10.7.4 Design Case: Capillary Electrophoresis Network System

Following is a hypothetical design case used to illustrate the CE process developed by Krishnamoorthy of CFD Research Corporation, Huntsville, Alabama (private communication, 2000). The geometry and dimension of the CE network are shown in Figure 10.23. Both channels have rectangular cross section of 20 μm wide $\times$ 15 μm deep. The analyte contains three species with distinct electro-osmotic mobility of $\omega_1 = 2 \times 10^{-8}$ m^2/V-s for species A, $\omega_2 = 4 \times 10^{-8}$ m^2/V-s for species B, and $\omega_3 = 6 \times 10^{-8}$ m^2/V-s for species C. All species are assumed negatively charged. For the sake of simplicity, the flow is assumed to take place in a two-dimensional pattern in the x–y plane as shown in Figure 10.23. The process takes place after an equilibrium condition is reached in the analyte, so that the production rate in Equation (10.18) can be omitted. The CFD-ACE$^+$ code marketed by the CFD Research was used in this design case.

Since the flow is assumed to be confined in the x–y plane, the velocity vector $\mathbf{V}$ in Equation (10.18) consists of two components, $V_x(x,y)$ and $V_y(x, y)$, along the respective x and y coordinates. Consequently the two-dimensional form of Equation (10.18) becomes

$$\frac{\partial C_i}{\partial t} + (V_x + V_{ex})\frac{\partial C_i}{\partial x} + (V_y + V_{ey})\frac{\partial C_i}{\partial y} = \frac{\partial}{\partial x}\left(D_i \frac{\partial C_i}{\partial x}\right) - \frac{\partial}{\partial y}\left(D_i \frac{\partial C_i}{\partial y}\right) + \dot{r}_i \tag{10.23}$$

where

C_i = concentration of species i in solution

t = time into process

V_{ex} = x component of electromigration (drift velocity) = $-\omega_i z_i (\partial \phi / \partial x)$

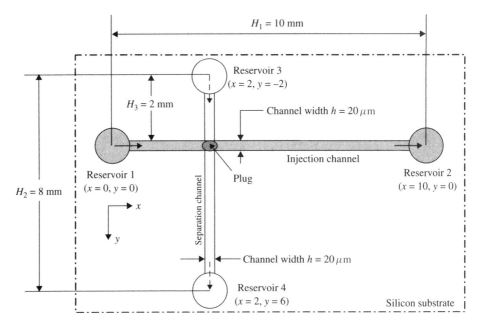

Figure 10.23. Capillary electrophoresis for a design case study.

$V_{ey} = y$ component of electromigration (drift velocity) $= -\omega_i\, z_i\, (\partial\phi/\partial y)$

$z_i =$ valence of ion of species i

$\omega_i =$ electro-osmotic mobility of ith species in Equation (10.17)

$\phi =$ externally applied electrical potential

$D_i =$ diffusion coefficient of ith species in solution

$\dot{r}_i =$ rate of production of species i

The rate of production $\dot{r}_i$ in Equation (10.23) is omitted in the computation for the reason mentioned above.

The electric field ϕ, as presented in V_{ex} and V_{ey} in Equation (10.23), can be obtained by solving the differential equation

$$\frac{\partial}{\partial x}\left(\sigma\frac{\partial\phi}{\partial x}\right) + \frac{\partial}{\partial y}\left(\sigma\frac{\partial\phi}{\partial y}\right) = 0 \qquad (10.24)$$

where the electrical conductivity σ is defined as

$$\sigma = F\sum_i z_i^2\omega_i C_i \qquad (10.25)$$

where F is the Faraday constant.

A close look at the above formulation will reveal the fact that Equations (10.23), (10.24), and (10.25) are all coupled. The solution of these coupled equations by a classical technique is extremely tedious and time consuming. Consequently, the CFD-ACE$^+$ code built on the principles of CFD was used for the numerical solutions for the design case.

In using the CFD-ACE$^+$ code, the separation is accomplished with the help of the electrokinetic switching technique discussed under the general procedure for the CE process in Section 10.7.2. This technique consists of two cycles:

1. **Injection Cycle:** An electric field is maintained in the injection channel until the sample flows past the intersection.
2. **Separation Cycle:** An electric field is maintained in the separation channel for the required length of time, until separation is completed.

The magnitude of the applied electric fields depends upon the analyte property, volume of the sample plug desired at the intersection, and geometry of the system.

Figure 10.24 shows the contour map of the sample, predicted by the CFD-ACE$^+$ code, at the end of the injection cycle. During this cycle, analyte reservoir 1 in Figure 10.23 is grounded and the analyte waste reservoir 2 is maintained at 250 V. The buffer reservoir 3 and the buffer waste reservoir 4 are maintained at 30 and 0 V, respectively. The strong electric field in the injection channel causes the sample to flow

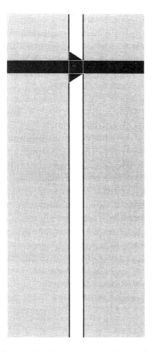

Figure 10.24. Contour profiles of sample that consists of species A, B, and C at end of injection process.

from reservoir 1 to 2. As it reaches the junction, it is squeezed because of the weak electric field in the separation channel. Consequently, the sample at the injection cross has a trapezoidal shape (see Figure 10.24). This method of generating a sample of specified volume at the intersection by using a "pinch" in the flow field is known as *pinched loading technique* (Culbertson et al., 1998) or *electrokinetic focusing*. The pinch in the flow field is caused by the nonuniform electric field at the intersection created by the channel geometry and orientation.

The next cycle in the CE process is the separation of species in the sample. After the sample reaches the intersection, the voltages are switched from the *sample-loading* mode to *sample separation* mode. A potential of 250 V is applied to buffer waste reservoir 4 while buffer reservoir 3 is grounded. The analyte and analyte waste reservoirs (1 and 2) are maintained at 70 and 100 V, respectively. The reason for maintaining a weak electric field in the injection channel during the separation process is to avoid leakage of sample from the injection channel into the separation channel so that a clean separation may be achieved. The results from the CFD simulation are shown in Figures 10.25a–c for three different time levels, 0.1, 0.3, and 0.5 s, respectively. These figures show the contour map of species A, B, and C and a line plot of concentration of these species along the center of the separation channel.

During the separation cycle, the sample at the junction starts flowing into the separation channel. As the sample continues its journey, species A, B, and C migrate with different speeds since their electro-osmotic mobilities are different. As expected, Figure 10.25 clearly shows that, in a given length of time, the distance traveled by a species is directly proportional to its mobility, and species C, which has the highest electro-osmotic mobility, will be detected by the amperometric electrochemical or fluorescence detector first.

10.7.5   Capillary Electrophoresis in Curved Channels

We notice from the line plots in Figure 10.25 that the peak values of the concentration of the species decrease with time. This phenomenon is called *band-broadening* or *dispersion*, which refers to the process of dilution of analyte concentration as it moves across the microchannel. Dispersion of species is more phenomenal in the CE processes involving curved channels, as the concentrations of the same species near the inner walls are higher than those near the outer walls. The latter geometry is a practical solution to the need for long separation lengths in many CE processes. For many applications involving a large number of species, typical separation times are on the order of 200 s and separation lengths on the order of 1–2 m. Consequently there are systems that involve numerous tight turns in the flow passage in order to conserve the footprint of the required space. Figure 10.26 illustrates a CE system involving a curved passage.

Dispersion/diffusion is the main cause of broadening in the CE process. A nonuniform electric field may also cause the different regions of the sample to migrate at different velocity (Culbertson et al., 1998) and thus cause band broadening. The nonuniform electric field occurs primarily because of the curvature in the channel geometry. For example, consider a rectangular-shaped sample plug traveling in a curved channel by

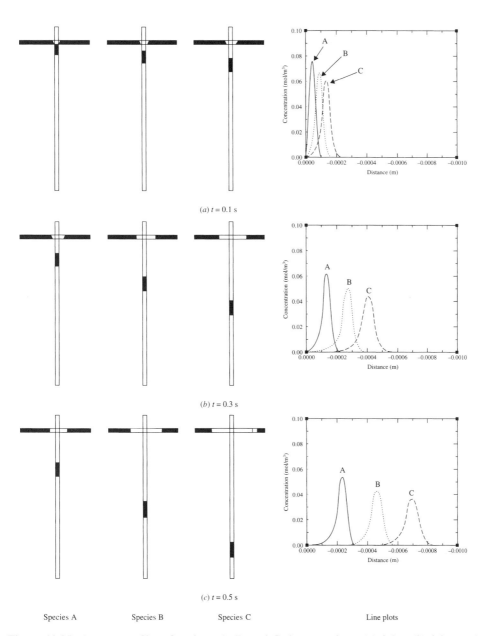

Figure 10.25. Contour profiles of analytes A, B, and C shown at times (*a*) 0.1 s, (*b*) 0.3 s, and (*c*) 0.5 s.

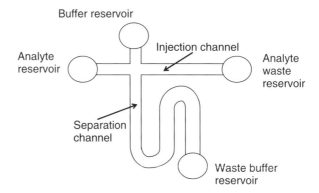

Figure 10.26. Curved separation channel for CE process.

electrophoresis. If the two ends of the curved channel are maintained at constant voltage, then higher current density along the inner wall of the bend will cause the inner region of the sample plug to migrate faster than the outer region. Consequently, the sample may attain a shape of a highly distorted parallelogram and eventually cause band broadening. Krishnamoorthy and Giridharan (2000) have studied this phenomenon in S-shaped microchannels with rectangular cross section.

10.7.6   Issues in Design of CE Processes

A time accurate CFD calculation like the one presented in Section 10.7.4 will help engineers to design and optimize CE devices. The following design issues are noted:

1. Parameters defining an optimal electrokinetic injection system are the shape and width of the sample in the separation column. The broader the sample shape at the injection cross, the longer will be the separation channel to achieve desired resolution. Longer channels may cause significant dispersion due to diffusion. Hence a thin and narrow sample is desired at the cross.
2. The sample plug shape is greatly influenced by the geometry of the intersection and the electric field. The CFD simulation will enable the designer to select appropriate voltage settings and channel geometry, such as length, cross-sectional area, and orientation of the cross, to achieve a narrow and thin band of the sample at the intersection.
3. The CFD analysis will also help the designers select appropriate column length and time duration of each cycle in the CE process to achieve the desired resolution during separation.

Thus, understanding the interaction among various physical phenomena by using computer-based simulation will help to optimize the microchip design on the basis of the trade-off between compactness and separation performance of the CE on-the-chip microfluidic system.

10.8 COMPUTER-AIDED DESIGN

10.8.1 Why CAD?

By now we have learned the many facets and the complexity of engineering design of MEMS and microsystems. It used to take an average of 5 years to develop a new microsystem product and another 5 years to have the product reach the marketplace. These lengthy development and production cycles have been drastically reduced in recent years with commercially available computer-aided design (CAD) software packages. The principal advantage of CAD is to expedite the design process of microsystem products to meet the critical time-to-market (TTM) condition. A good CAD package can also help design engineers in swiftly assessing the effect of design changes and evaluating manufacturing and the yield of the products. The solid modeling and animation capabilities of many CAD packages can provide design engineers with virtual prototypes that can simulate the performance of the desired functions of a real product.

One of the original efforts toward producing a CAD tool to simulate MEMS devices began with the development of the MEMCAD package at the Massachusetts Institute of Technology (MIT) in the late 1980s/early 1990s (Gilbert et al., 1996). This effort consisted of linking a few existing non-MEMS commercial CAD packages with features that are related to microstructure design. Much effort has since been made by code developers to develop commercial CAD packages specifically for MEMS and microsystems design. IntelliSense Corporation released the first commercial CAD tool developed specifically for MEMS in 1995 with the name IntelliCad (now IntelliSuite). In 1996, Microcosm Technologies licensed the code from MIT and sold it commercially under the name MEMCAD. Among other commercially available CAD tools, which do not include device analysis, are MEMSPro from Tanner and MEMSCaP.

10.8.2 What Is in a CAD Package for Microsystems?

While CAD for microsystems is still an emerging effort, a viable CAD package should include at least three major interactive databases: (1) *electromechanical design database*, (2) *material database*, and (3) *fabrication database*. The content of a generic CAD package including the above databases is shown in Figure 10.27.

As we can see from Figure 10.27, the design database provides the necessary information and tools, such as the inference machine for design synthesis, codes for FEA (finite element analysis) and BEA (boundary element analysis), and tables and charts for other design considerations. The need for a material database in a CAD for microsystems is obvious, as the properties of many materials used in microsystems are not available from traditional materials handbooks. This database should contain complete information on material properties, such as those presented in Table 7.3. Many of these properties should be presented as two- or three-dimensional graphs, as illustrated in Figures 10.28 and 10.29. It should also include properites for transduction components such as piezoelectric and piezoresistive materials.

Figure 10.29 shows the Young's modulus of silicon nitride at various temperatures and pressures in a three-dimensional representation.

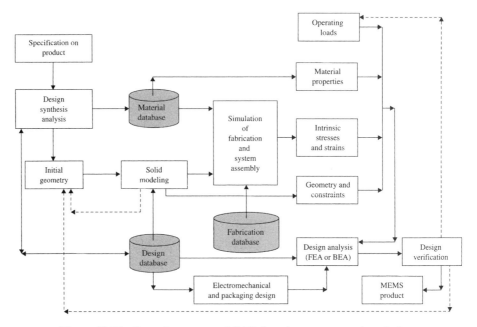

Figure 10.27. General structure of CAD for microsystem product design.

The fabrication database, which is unique for microsystems design, involves all fabrication process simulations required for specifically selected fabrication and manufacturing processes as described in Chapters 8 and 9. This database should also include wafer treatment such as the required cleaning processes for photolithography and thin-film depositions. The results of these fabrication process simulations often include the inherent residual stresses and strains and other intrinsic stresses, which are used as input to the subsequent design analysis under normal operating and overload conditions. Engineers can visualize the designed product in three-dimensional perspectives by using the solid model option provided by the CAD package. Most CAD packages have provisions for

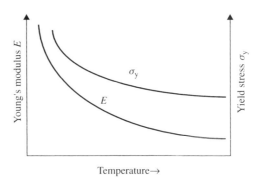

Figure 10.28. Two-dimensional material property representation.

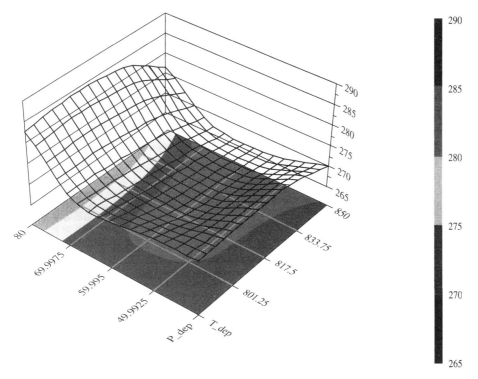

Figure 10.29. Three-dimensional material property representation. (Courtesy of IntelliSense Corporation, Wilmington, MA.)

animations, which allow engineers to visualize the functions "performed" by the virtual prototype, such as comb-driven actuators.

The flowchart in Figure 10.27 is self-explanatory. Design engineers will first establish a "process flow table" by selecting a substrate material once the product is configured from the design synthesis analysis. The CAD package will offer a possible prephotolithographic substrate treatment process from the fabrication process database. A mask is then either imported from external sources or created by the built-in design database for subsequent photolithography on the substrate. The same database is then used to determine the appropriate fabrication process flow or steps that may include oxidation, diffusion, ion implantation, etching, deposition, and other processes such as bonding, as selected by the designer. The CAD package offers detailed information on the selected processes, for instance, the etchants for the etching process with an estimated required time for each process. The CAD package also provides automatic flow of information between the material database and the fabrication database. Once the fabrication processes have been established, electromechanical design begins. Here, the design engineer uses the solid model constructed by the package for automatic mesh generation for the electromechanical analysis using the finite element method. Depending on the nature of the product, the CAD package can perform the finite element analysis for thermal conditions and mechanical strength of the structure as well as electrostatic and electromagnetic analyses in cases that involve actuation by the products. The latter analyses require the input of electrical potential and current to the finite element analysis. In addition to graphical displays for the analytical results, many CAD packages also offer animation of the designed product for kinematic and dynamic effects. Engineers may either terminate the design at this stage if the outcome of the design process is satisfactory or make any necessary revisions to the configuration or loading or boundary constraints until all design objectives and criteria are met.

10.8.3 How to Choose a CAD Package

Commercial CAD packages available in the MEMS community differ from one package to another. Following are general guidelines of what engineers should look for in selecting a suitable package for their specific MEMS needs:

1. User friendliness
2. Adaptability of the package to various computers and peripherals
3. Easy interfacing of CAD package with other software, for example, nonlinear thermomechanical analyses and integration of electric circuit design
4. Completeness of material database in the package
5. Versatility of the built-in finite element or boundary element codes
6. Pre- and postprocessing of design analyses by the package
7. Capability of producing masks from solid models
8. Provision for design optimization
9. Simulation and animation capability
10. Cost in purchasing or licensing and maintenance

10.8.4 Design Case Using CAD

The commercial CAD package IntelliSuite was used in the design of a microcell gripper in a student project (Griego et al., 2000). A general description of the package is available in the references (He et al., 1997). The package consists of three major databases similar to those shown in Figure 10.27: (1) a material database, (2) an electromechanical database, and (3) a fabrication process database. This CAD package offers material selection from the following categories:

1. Substrates: silicon (Si), polysilicon, gallium arsenate (GaAs), quartz, sapphire, alumina, doped semiconductors
2. Interconnects and mask materials: aluminum (Al), gold (Au), silver (Ag), chrome (Cr), silicon dioxide (SiO_2)
3. Other substrate and insulating materials: silicon dioxide (SiO_2), silicon nitride (Si_3N_4)
4. Photoresist materials

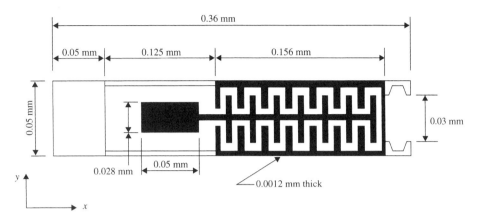

Figure 10.30. A MEMMS cell gripper.

The overall dimension of the comb-drive-actuated gripper in a plan view is presented in Figure 10.30, and the detailed arrangement between the electrodes is shown in Figure 10.31.

The following design case illustrates the steps involved in the design process. The design was a student project, so no effort was made by the designers to optimize the cost and time required for the fabrication of the cell gripper. Only a single gripper was designed for production from a single wafer. In reality, of course, one would seek a maximum yield of MEMS components produced from a single wafer.

The strategy adopted in this design case was to construct all required layers on the substrate wafer by oxidation, deposition, and sputtering. These layers are necessary for the electrical insulation, gripper structures, and electrical terminals. A bulk micromanufacturing technique with etching was then used to remove materials and thereby shape the gripper as needed. The IntelliSuite package provided the solid modeling of the substrate after each step with the estimated required time for each process. The solid model was also used for an electromechanical analysis assessing the reliability of the microstructure.

The steps involved in this design case are outlined below:

Step 1   **Substrate Selection:** A silicon wafer is chosen to be the substrate material for the gripper. The wafer is a standard 100 mm diameter sliced 500 μm thick from a single-silicon crystal boule produced by the Czochralski method

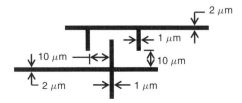

Figure 10.31. Gaps between electrodes in a MEMS cell gripper.

Silicon wafer substrate:

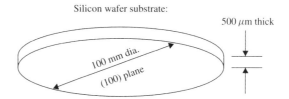

Figure 10.32. Silicon wafer for a microgripper.

as described in Chapter 7. The surface of the wafer is normal to the $<100>$ orientation, as illustrated in Figure 10.32.

Step 2 **Substrate Cleaning:** The designers chose Pirahna solvent to clean the wafer surface. This is one of several options offered by the CAD package. This solvent contains 75% H_2SO_4 and 25% H_2O_2. The substrate is submerged in the solvent for 10 min. The cleaned wafer is ready for oxidation on one of its surfaces.

Step 3 **Creating a SiO$_2$ Layer by CVD:** A 1-μm-thick SiO_2 layer is deposited on the surface of the wafer to serve as an electrical insulator between the anode and the cathode in the electrostatic actuation of the cell gripper. The deposition takes place in a furnace at a temperature of 1100° C and a pressure of 101 kPa as indicated by the CAD package.

Step 4 **LPCVD Deposition of Polysilicon Structure Layer:** Polysilicon is chosen to be the material for the cell gripper structure. A 1.2-μm-thick polysilicon layer is deposited over the oxide layer with a medium-temperature LPCVD process, with detail parameters provided by the IntelliSuite package. The deposition temperature is in the range of 500–900°C, with an annealing temperature of 1050°C. The CAD package also specifies 60 min as the required time for this process.

Step 5 **Aluminum Sputtering:** An aluminum film is deposited for the lead wire for conducting electrical current through the electrodes. A 3-μm-thick film is sputtered onto the polysilicon layer. The estimated time for this process is 10 min.

Step 6 **Application of Photoresist:** Positive photoresist is applied to the aluminum layer. A chuck spinning at 4000 rpm as illustrated in Figure 8.3 is used to spread the photoresist. The photoresist-covered substrate assembly is baked at 115°C, yielding a 3-μm-thick layer. All films, including the photoresist, deposited on the silicon wafer are shown in Figure 10.33.

Step 7 **Photolithography by UV Exposure:** A photolithographic process using a UV light source at 250 W with a wavelength $\lambda = 436$ nm is used in the process over a mask created for the anode and cathode. The exposure time in this case is 10 s. This mask is provided by the CAD package.

Step 8 **Wet Etching to Remove Photoresist:** The solvent KOH, described in Chapter 8, is used as the etchant to remove the exposed photoresist. The unexposed resist stays attached to the aluminum layer.

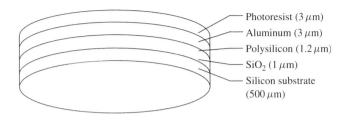

Figure 10.33. Thin-film depositions on silicon substrate for a microgripper.

Step 9   **Wet Etching on Aluminum:** A special etchant is selected to remove the unprotected aluminum from the surface. This etchant contains 75% H_2SO_4, 20% $C_2H_4O_2$, and 5% HNO_3. The depth of the aluminum layer to be removed is 3 μm. The estimated time for this process is 15 min.

Step 10   **Wet Etching to Remove Photoresist from Aluminum:** Once again, KOH is used to remove the photoresist left on the surface of the aluminum anode and cathode.

Step 11   **Photoresist Deposition and Photolithography of Gripper Structure:** Positive photoresist is applied to the entire surface of the wafer by the procedure in step 6. Another mask that outlines the gripper structure to comply with what is shown in Figure 10.30 is used for photolithography by the procedure in step 7.

Step 12   **Removing Photoresist by Wet Etch:** The procedure described in step 10 is used.

Step 13   **Etching Polysilicon by RIE:** Because of the relatively high aspect ratio of the gripper structure, RIE is chosen to remove the unprotected region of the polysilicon layer for the net shape of the gripper structure. The reactive chemical species with chlorine or fluorine in plasma are involved in this process.

Step 14   **Removing SiO$_2$ Sacrificial Layer:** This process involves the use of wet etching in conjunction with a laser photochemical etching process. The latter etching process uses a SiH_4 etchant and a KrF laser at an intensity of 0.3 J/cm^2. The combined etching provides an etching rate of 40 Å/s. The process in this step releases the gripper arms and tips from the SiO$_2$ layer.

Step 15   **Separation of Gripper and Substrate:** The net shape of the structure after step 14 is the gripper structure shown in Figure 10.30 attached to the silicon substrate of the same structural outline bonded by a thin SiO$_2$ film. Separation of the gripper structure from the substrate requires the removal of the in-between SiO$_2$ layer (a sacrifice layer). The removal of this thin layer can be accomplished either by using a thin diamond saw or by using the "etch pit" technique described by Kim et al. (1991).

Step 16   **Electromechanical Analysis:** The purpose of this analysis is to assess whether the gripper fabricated by the above processes would perform the desired functions.

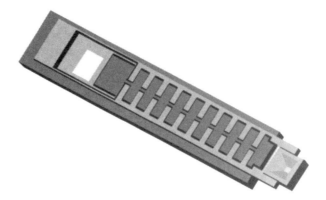

Figure 10.34. Solid model of a cell gripper.

A *charge density analysis* is performed first with the applied voltage input to the aluminum terminals to assess if the highest charge buildup in the comb drive will produce sufficient electrostatic forces for the gripping function. The CAD package offers a graph of the distribution of such charges with color-coded charge density at various parts of the gripper structure. Engineers are assured that the electrostatic charges at the comb fingers have sufficient magnitudes to generate the desired gripping forces.

Finite element strength analysis is followed after the charge density analysis to assure the following: (1) The maximum stress in the gripper structure is not excessive in the x, y, or z direction. The coordinate system used in this case is shown in Figure 10.30, with the z coordinate being normal to the x–y plane. These stresses and the von Mises stress expressed in Equation (4.71) must be kept below the yield strength of the materials. (2) The deformation-induced distortion of the structure is not large enough to affect the function of the comb-drive actuation. The coupled mechanical and electrostatic effect of the actuation is illustrated in Example 10.1. The CAD package offers automatic finite element mesh generation for the required analyses with results indicated in coded colors over the solid models of the gripper structure.

The solid model of the cell gripper as produced by the IntelliSuite package is shown in Figure 10.34.

PROBLEMS

Part 1 Multiple Choice

1. A major difference between the design of microsystems and the design of traditional products at the macroscale is that the design of microsystems requires (a) the integration of design and fabrication, (b) the integration of design and marketing, (c) the integration of chemical and mechanical forces.

2. Microsystem components are fabricated by (a) precision machine tools, (b) physical–chemical processes, (c) micromachine tools.

3. Microsystems design involves (a) the single task of design analysis, (b) the coupling of two tasks, (c) the coupling of three tasks.

4. *Process flow* is part of microsystem (a) design, (b) manufacturing, (c) production.

5. An effective way to shorten the microsystem design cycle is to (a) involve more engineers, (b) use a better design method, (c) use computer-aided design.

6. One critical design consideration for microsystems is TTM, which stands for (a) total time management, (b) time to market, (c) targeted time management.

7. Three critical design considerations related to environmental conditions are (a) thermal, mechanical, and chemical; (b) electrical, mechanical, and chemical; (c) electrical, mechanical, and material.

8. There are (a) one, (b) two, (c) three types of substrate materials used in microsystems.

9. The most commonly used substrate material is (a) silicon, (b) GaAs, (c) quartz.

10. The most suitable material for micro-optical components is (a) silicon, (b) GaAs, (c) quartz.

11. The most dimensionally stable material for microcomponents is (a) silicon, (b) GaAs, (c) quartz.

12. The cheapest thermal and electrical insulation material is (a) silicon dioxide, (b) silicon nitride, (c) silicon carbide.

13. The most chemically stable material for microsystems is (a) silicon dioxide, (b) silicon nitride, (c) silicon carbide.

14. The most suitable material for the masks used in deep etching is (a) silicon dioxide, (b) silicon nitride, (c) silicon carbide.

15. In general, the cheapest way to produce MEMS is (a) bulk micromanufacturing, (b) surface micromachining, (c) the LIGA process.

16. In general, the most expensive way to produce MEMS is (a) bulk micromanufacturing, (b) surface micromachining, (c) the LIGA process.

17. In general, the technique that offers the most flexibility in micromanufacturing is (a) bulk micromanufacturing, (b) surface micromachining, (c) the LIGA process.

18. In general, the micromanufacturing technique that offers high aspect ratio in geometry is (a) bulk micromanufacturing, (b) surface micromachining, (c) the LIGA process.

19. The principal advantage of using piezoresistors is (a) small size, (b) high sensitivity, (c) low cost in production.

20. The most serious drawback of piezoresistor transducer is (a) its high cost, (b) its slow responses, (c) its strong temperature dependence on sensitivity.

21. The principal advantage of a piezoelectric transducer is (a) its fast response, (b) its low cost in production, (c) its high sensitivity.

22. A serious drawback of a piezoelectric transducer is (a) its high cost, (b) its vulnerability to brittle fracture, (c) its strong temperature dependence on sensitivity.

23. The principal advantage of a capacitance transducer is (a) its suitability for high-temperature applications, (b) its simplicity, (c) its low cost in production.

24. A serious drawback of a capacitance transducer is (a) its bulky sizes, (b) its nonlinear input/output relationship, (c) its low sensitivity.

25. The resonant vibration transducer has the advantage of (a) simplicity, (b) reliability, (c) high sensitivity and precision.

26. A serious drawback of a resonant vibration transducer is (a) complexity in fabrication, (b) the strong temperature dependence on sensitivity, (c) unreliable input/output.

27. The problem with actuation using electro-resistant heating is (a) temperature, (b) heat source, (c) the control of heating and cooling rates.

28. SMA stands for (a) smart material actuator, (b) shape memory alloy, (c) shape memory actuator.

29. Microsystem packaging needs to be considered in the early stage of design to ensure (a) low packaging costs, (b) high customer demand, (c) acceptable product appearance.

30. Photolithography means to (a) make a photograph of a microsystem, (b) create the pattern of the geometry of a microsystem, (c) produce the label on a microsystem.

31. Selection of thin-film deposition processes in the early stage of microsystem design is important because these processes (a) are expensive, (b) are delicate, (c) may result in adverse consequences in the performance of the microsystem.

32. For closely spaced microstructural components, the force that causes most problems is (a) thermal force, (b) electrostatic force, (c) the van der Waals force.

33. Intrinsic stresses in microstructures are induced by (a) microfabrication processes, (b) applied loads, (c) thermomechanical effects.

34. Resonant vibration in microsystems (a) should always be avoided, (b) is necessary in some microdevices, (c) is irrelevant, as it never happens in microsystems.

35. Interfacial fracture is an important design consideration in microsystems that involves (a) layers of dissimilar materials made by thin-film deposition, (b) high-temperature environment, (c) microstructures with cracks.

36. The finite element method is (a) a universal, (b) a viable, (c) an irrelevant analytical tool for microstructure analysis.

37. As a rule of thumb, commercial finite element codes developed for macroscale structures can be used for microscale structures of size greater than (a) 0.1 micrometer, (b) 1 micrometer, (c) 100 micrometers.

38. The interpolation function in a finite element analysis relates (a) element and the corresponding nodal quantities, (b) element quantities in the entire structure, (c) nodal quantities and those in the entire structure.

39. The Galerkin method used to derive the element equation in a finite element analysis requires (a) the geometry of the element, (b) the governing differential equation, (c) the potential energy in the discretized continuum.

40. The Rayleigh–Ritz method used to derive the element equation in a finite element analysis requires (a) the geometry of the element, (b) the governing differential equation, (c) the potential energy in the discretized continuum.

41. The "birth," or "A," elements are used to simulate the microfabrication process of (a) etching, (b) deposition, (c) molding in a finite element analysis.

42. The "death," or "S," elements are used to simulate the microfabrication process of (a) etching, (b) deposition, (c) molding in a finite element analysis.

43. In designing a die for micropressure sensors, one of the critical design considerations is (a) die isolation, (b) mechanical strength, (c) contamination.

44. Better die isolation in a micropressure sensor can be achieved by (a) a thinner structure, (b) passivation by coating protective materials, (c) flexible die attach.

45. Passivation of the die in a pressure sensor is achieved by (a) plastic encapsulation, (b) coating of protective materials, (c) keeping the pressurized medium away form the die.

46. To achieve maximal sensitivity of a pressure sensor, one would maximize the (a) stress, (b) strain, (c) deformation in the diaphragm.

47. Bosses are introduced in the diaphragm of a pressure sensor die to provide (a) extra stiffness, (b) better appearance, (c) uniform stress distribution in the diaphragm.

48. In sensor design, including that for pressure sensors, the (a) magnitude, (b) linearity, (c) fast response of output signals is of primary concern to the design engineer.

49. Capillary electrophoresis (CE) on-a-chip is primarily used in (a) microfluidic actuator, (b) micropressure measurements, (c) biomedical analysis.

50. The principal advantages of using CE on-a-chip is that it (a) provides a cheap method of biomedical analysis, (b) involves minute sample size and fast responses, (c) is an easy-to-use analytical method.

51. Typical capillary electrophoresis involves a network of (a) capillary tubes and microchannels, (b) capillary tubes and microvalves, (c) capillary tubes and micropumps.

52. Fluid flow in capillary electrophoresis is prompted by (a) application of surface forces from conduit wall, (b) application of volumetric pumping forces, (c) application of electrical fields.

53. Two principal parts in a capillary electrophoresis network are (a) injection and separation channels, (b) injection and sample flow channels, (c) sample separation channels.

54. *Advection* is a physical phenomenon that involves a moving substance that changes its (a) pressure, (b) temperature, (c) phase during the movement.

55. Design analysis for capillary electrophoresis requires the coupling of (a) advection and diffusion, (b) advection and electromigration, (c) advection, diffusion, and electromigration.

56. Separation of various species in a biosample during capillary electrophoresis is due to the difference of (a) density, (b) buoyancy, (c) electro-osmotic mobility of the species.

57. The principal purpose of using computer-aided design (CAD) in microsystems design is to (a) make the design beautiful, (b) make it more accurate, (c) shorten the time required in the design process.

58. Modern CAD for microsystems requires the integration of (a) design and analysis databases and prototyping, (b) design, material, and analysis databases, (c) design, material, and fabrication databases.

59. The design database in a CAD for microsystems design involves (a) mechanical and electrical design analyses, (b) mechanical and chemical analyses, (c) electrical and chemical analyses.

60. One major advantage of using CAD in microsystem design is the capability in (a) graphics representation of the results, (b) animation of the device, (c) obtaining fast results.

Part 2 Descriptive Problems

1. What are the principal sources of intrinsic stresses induced in the microstructures?

2. Why is modal analysis important in the design of microsystems involving motion?

3. Assess the improvement of the flexibility of a micropressure sensor die with its cross section illustrated in Figure 10.16. Assuming the same overall dimension of the die, that is, dimension A and footprint size c, what will be the induced maximum bending stress and deflection in the diaphragm if the height H of the die is doubled and is subjected to the same applied pressure loading?

4. Evaluate the resistance to water flow in microchannels of the cross sections in (a) a rhombus, (b) an obround shape, and (c) a trapezoid with the same hydraulic diameter of 30 μm as indicated in Example 10.2. Tabulate the computed resistance to water flow for all these three cross sections and the other three cases as presented in Example 10.2.

5. Reformulate Equations 10.18–10.22 for a capillary electrophoresis network similar to that shown in Figure 10.23 but with a 45° angle between the two channels.

CHAPTER 11

ASSEMBLY, PACKAGING, AND TESTING OF MICROSYSTEMS

11.1 INTRODUCTION

Most MEMS and microsystems involve delicate components with size on the order of micrometers. These components are vulnerable to malfunctions and/or structural damage if they are not properly packaged and tested for reliability after being assembled into products. Reliable packaging of these devices and systems is a major challenge to the industry because microsystems packaging technology is far from being as mature as that for ICs and microcircuitry. Microsystems packaging, in a broader sense, include the three major tasks of *assembly*, *packaging*, and *testing*, abbreviated *AP&T* in a special report (National Research Council, 1997). Indeed, as we will notice in Figure 11.1, for the production of silicon die for micropressure sensors shown in Figures 2.9 and 4.6, a large portion of the process involves assembly (steps 6, 7, and 12), packaging (steps 3–5, 9, 10, 13, and 16), and testing (steps 2, 8, 11, 14, and 15).

It is thus not surprising that the AP&T of MEMS make up a large portion of the overall cost of production. The cost for the AP&T of micropressure sensors, for example, can be 20% of the overall production cost with plastic passivation or as high as 95% of the total cost for special pressure sensors for high-temperature applications with toxic pressurized media. The AP&T costs for MEMS and microsystems vary from product to product. Currently, this cost represents on average 70% of the total production cost. A more serious note is that AP&T are usually the source of failure in most microsystems. Thus AP&T technology is a key factor in MEMS and microsystems product design and development.

The high costs of AP&T in microsystems production is attributed to the lack of standards developed and established for the industry. In most microsystems the AP&T process is based on customer needs. The production industry does not share its knowledge and experiences with others because of proprietary interests. Standardization of AP&T such as what has happened in the microelectronics industry can drastically reduce the cost

407

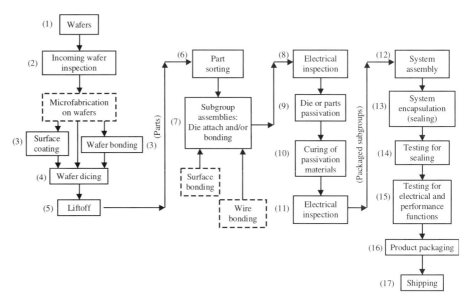

Figure 11.1. Production process flow for micropresssure sensor Die. (After Hsu and Custer, 2004.)

in microsystems production, but such standardization appears a long way from reality, as shown in Figure 11.2.

Another reason for the high cost of AP&T relates to the awkward technologies associated with this critical part in microsystems manufacturing. Lack of R&D effort in this critical area of microtechnology not only has resulted in high cost of production but also has reduced the value of miniaturization of MEMS and microsystems as stipulated in Section 1.6 because most of these products are often at the mesoscale after packaging.

A well-known fact is that the purpose of IC packaging is merely to protect the silicon chip and the associated wire bonds from environmental effects. Microsystem packaging, on the other hand, is expected to not only protect the delicate components such as silicon

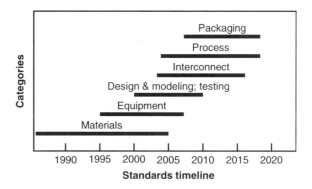

Figure 11.2. Timeline for standardization in microtechnology. (Walsh et al., 2002.)

dies from hostile environment but also allow these dies to probe that environment at the same time. For example, microsensors are required to sense the pressure of exhaust gas from an internal combustion engine as well as the composition of the gas. The sensors therefore need to be exposed to these highly corrosive pressurized gases at high temperature. Proper protection of the silicon die and the delicate transducers and associated wire bonds from excessive heat and corrosive chemicals attack is essential. Microsystems packaging is thus much more challenging to engineers than microelectronics packaging.

Unlike microelectronics packaging, there are few comprehensive publications on the AP&T of microsystems. Microsystems packaging was a subject in two early books (Madou, 1997; Tummala, 2001). A more recent book specifically devoted to the AP&T of microsystems is available (Hsu, 2004). We will use the latter as the principal reference for this chapter.

11.2 OVERVIEW OF MICROASSEMBLY

In general, microassembly involves the assembly of components of sizes varying from 1μm to a few millimeters with microscale tolerances for bonding and joining to make microdevices (Yang and Nelson, 2004). The assembly of components at the nanometer scale will be described in Chapter 12.

Assembly of microsystems is involved mainly with high-precision pick-n-place components that are required to produce primarily three-dimensional microstructures. Table 11.1 shows the number of units of microsystems produced and the revenues earned by the MEMS and microsystems industry from 1996 to 2002 by NEXUS. The established products include hard-disk read/write heads and inkjet printer heads, whereas the emerging products include such items such as drug delivery systems, optical switches, and lab-on-chip. The numbers in the last two columns are derived from linear extrapolation of the numbers from the 1996–2002 records. We thus project the number of microsystems to be assembled in 2006 to be 12 billion units. Of these units, approximately 2 billion units will be the read/write heads and the inkjet printer heads, which have achieved a high degree of automatic assembly. This would lead to close to 10 billion units being assembled by either a semiautomatic process or human hands. The latter form of assembly is clearly prohibitively expensive, tiresome, and time consuming. Often the products will not meet the extremely stringent requirements in precision and thus the necessary quality and reliability of the finished products.

TABLE 11.1. Production Volume of Microsystems

MST Products	1996 Units (millions)	1996 Revenue ($millions)	2002 Units (millions)	2002 Revenue ($millions)	2006 Units (millions)	2006 Revenue ($millions)
Established products	1,595	13,033	6,807	34,290	10,282	48,461
Emerging products	33	107	1,045	4,205	1,720	6,937
Total	1,628	13,140	7,852	38,495	**12,002**	55,398

Source: NEXUS, hhtp://www.wtec.org/loyala/mcc/mes.eu/pages/chapter-6.html.

11.3 HIGH COSTS OF MICROASSEMBLY

The assembly and packaging of microsystems in mass quantities as shown in Table 11.1 present major challenges to engineers in terms of reliability and cost. There are many reasons for the high cost associated with microassembly. Following are a few obvious ones:

1. There is a lack of standard procedures and established rules for such assemblies. Products are assembled according to procedures developed on the basis of either individual customer requirements or the personal experience of the design engineer.

2. There is a lack of effective tools for microassemblies. Tools such as microgrippers, manipulators, and robots are still being developed. Microassemblies also require reliable visual and alignment equipment such as stereo electron microscopes and electron beam, UV-stimulated beam, or ion beam imaging systems specially designed for microsystems assembly.

3. There is a lack of established methodology in setting proper tolerances of parts in insertion and assembly. There are many sets of tolerances involved in the assembly, as will be described in Section 11.4. The strategies for setting tolerances in parts feeders, grasping surface to mating surfaces, and fixture surface to mating surfaces are not well understood and established for microassemblies.

4. Assembly of microcomponents is mostly by physical–chemical processes that have strong material dependence. Traditional assembly techniques are not suitable for microdevices because of the minute sizes of the components to be assembled and close tolerances on the order of submicrometers. Moreover, chemical and electrostatic forces dominate in microassembly, whereas gravity and physics are primary considerations in macroassembly. There is little theory or methodology developed to deal with these problems in microassemblies.

The lack of effective tools and assembly strategies has resulted in the lengthy time required for all microsystem assemblies. Figure 11.3 shows a hand-assembled gear train

Press-fit pin transmission:
  Tool steel gauge pins>0.0059 in. dia.
  Aluminum substrate
  Ni gears

Figure 11.3. A hand-assembled gear train. (Courtesy of Sandia National Laboratories.)

at the Sandia National Laboratory (Feddema et al., 1999). One may well imagine the amount of time required for even highly skilled personnel to manually assemble these minute machine components.

11.4   MICROASSEMBLY PROCESSES

Despite the many differences between microassembly and traditional assembly in terms of methodologies and tools, the major steps involved are similar:

1. **Part Feeding:** Some part feeding techniques, such as for common tape-bonded feeders, used in microelectronics can be adopted in assembling some microcomponents.

2. **Part Grasping:** Microgrippers, manipulators, and robots are desirable tools for this task. However, these tools cannot properly handle minute parts without intelligent end effectors. An intelligent end effector requires the integration of gripping positioning, sensing, and orientation. The latter function is required for accurate alignment of mating parts. There are major problems associated with releasing minute parts from microgrippers, as will be described in detail in Section 11.5.3.

3. **Part Mating:** As mentioned earlier, at the micrometer scale, electrostatic and chemical forces dominate the interaction between the grippers and parts and between the mating parts. Special gripper design that can discharge these forces for possible stiction is necessary for easy releasing of parts from the gripper and for mating of small components. Possible solutions to these problems will be presented in Sections 11.5.2 and 11.5.3.

4. **Part Bonding and Fastening:** Various bonding techniques are available, as will be described in Section 11.14. Most of these bonding techniques involve micro-fabrication processes. Automated assembly is possible with batch fabrication of the bonding parts. Other methods of joining parts, including pulse laser deposition, welding, soldering, and snap-fitting based on surface chemistry and thin-film chemistry, can be used for fastening microparts.

5. **Encapsulation and Passivation:** Once the parts and components are placed and joined, encapsulation of delicate components is necessary. An example is the protection of delicate silicon dies in micropressure sensors as described in Sections 11.16 and 11.20 (silicon die with gel). Once all components are assembled, joined, and encapsulated, passivation of the assembled device using either plastic or metal cases is carried out. Typical encapsulated microdevices are shown in Figure 1.2 for micropressure sensors and Figure 1.8 for microaccelerometers.

6. **Sensing and Verification:** Three-dimensional machine vision systems such as stereomicroscopy are effective for visual identification of parts and parts alignment. Other types of microsensors, for example, tactile and thermal sensors, are also required for assembly and inspection. Near- and far-field infrared (IR) sensors can be used for microthermal feedback for monitoring welds or solder joints. Ideally these sensors should be integrated with grippers and/or other process tooling. One common problem with most of these sensors is the short depth of field near the wavelength of the light source in optical sensing systems. Another problem is the requirement for the sensing head to access the parts that need to be sensed.

As in the traditional automated assembly, both sequential and parallel assembly processes have been used in microassemblies. Sequential microassembly requires the use of micromanipulators and sensory feedback. At any time, only one or a few parts are being assembled. Parallel microassembly can either be deterministic or stochastic (Madou, 2002). In parallel microassembly, many parts are assembled simultaneously using distributed physical means such as electrostatic force, capillary force, centrifugal force, mechanical vibration, and fluid dynamic force (Yang and Nelson, 2004). Parallel assembly is attractive in the mass production of microsystems. One such microassembly is the self-assembly of certain component groups of a microsystem. An example is the work reported on the self-orienting fluidic transport (SOFT) technique (Smith, 1999). It allowed the placement of millions of microscopic objects with $\pm 1\,\mu\mathrm{m}$ accuracy each minute. The working principle of this assembly technique involves objects that are micromachined with a trapezoidal cross section and held in a slurry. Matching recesses are micromachined in a target substrate, and assembly of the objects and the recesses takes place randomly but with high yield accuracy. The objects can be as small as $30\,\mu\mathrm{m}$ in size. This assembly technique provides self-alignment of mating parts at high yield.

A parallel assembly workcell was developed at Sandia National Laboratory (Feddema et al., 1999). The workcell was originally intended to assemble microdevices made of components with high aspect ratio in geometry. The cell consists of a cartesian robot equipped with a microgripper, a visual servoing, and a microassembly planning unit.

The cartesian robot was used to press pins of 385 and $485\,\mu\mathrm{m}$ in diameter into a substrate (see Figure 11.4) and then place a 3-in-diameter wafer with gears made by the LIGA process onto the pins. Upward- and downward-looking microscopes were used to locate holes in the substrate, the pins to be pressed into the holes, and the gears to be

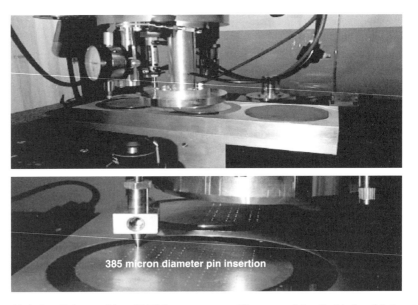

Figure 11.4. Parallel assembly of LIGA components. (Courtesy of Sandia National Laboratory.)

TABLE 11.2. Comparison of Sequential and Stochastic Parallel Microassembly

	Sequential Microassembly	Stochastic Parallel Microassembly
Allowable object complexity	High	Low
Assembly tool	Micromanipulator and microgripper	Distributed manipulation using, e.g., electrostatic forces
Sensory feedback requirement	Microscopic vision and microforce	Minimal
Interaction force control for damage minimization	Direct, precise control feasible	Indirect, precise control difficult
Process flexibility in terms of 3D manipulation and assembly	High	Low
Requirement of relative positioning accuracy between parts	Can be compensated by sensing	Cannot be compensated
Process efficiency	Low to medium	High
Error recovery	Possible	Difficult
Assembly system complexity	High	Low
Assembly system cost	High	Low

Source: Yang and Nelson, 2004.

placed on the pins. This vision system can locate parts within $3\,\mu$m, whereas the robot can place the parts within $0.4\,\mu$m.

Table 11.2 presents a comparison of sequential and stochastic parallel microassemblies which may help engineers to make viable decisions in developing automated microassembly systems.

11.5 MAJOR TECHNICAL PROBLEMS IN MICROASSEMBLY

Several major technical problems are associated with microassembly. Many of these problems relate to high precision engineering, and others are related more to a human factor–the operators (Van Brussel et al., 2000). Microassembly, whether it is fully automatic or machine-assisted manual operations, the operator is required to use special tools such as tweezers equipped with special end effectors under high-power microscopes to perform even a simple pick-n-place operation. Because the tools used in the operation are orders of magnitude larger than the parts being handled, the much needed hand–eye coordination in delicate assembly operations is often lost. The need to use a high-power microscope for the assembly operation inevitably results in a short working distance between the objective lens of the microscope and the substrate on which the assembly takes place. Moreover, the use of a microscope for high-power magnification of the assembled objects can seriously reduce the viewing area. Consequently, the operator loses global perspective of the area in which various parts are assembled.

Presented below are technical problems in three categories: (1) setting proper tolerances for microassembly, (2) developing proper tools and fixtures, and (3) precision control of tool and parts movements and gripping forces.

11.5.1 Tolerances in Microassembly

Tolerance is necessary for joining any pair of parts or components in all devices, whether they are at the macro- or microscale. Proper tolerancing is also critical to the quality of the finished products. Insufficient tolerances lead to either no-fit or forced-fit components. The latter situation may lead to overstress in the fitting and thus to premature failure of the joints. On the other hand, excessive tolerances can result in a loose or sluggish fit, which in turn results in poor fittings and thus poor product quality. The issue of setting proper tolerances is of critical importance in the assembly of MEMS and microsystems because of their minute sizes and lack of control in the surface topography of the components after microfabrication processes, as described in Chapter 8.

There are generally three types of tolerances involved in microassembly (Hsu and Custer, 2004): (1) dimensional tolerances, (2) geometric tolerances, and (3) alignment tolerances.

Dimensional Tolerances. Dimensional tolerances are associated with the dimensions of the parts that are designed for manufacturing. The design engineer sets such tolerances and applies them to the dimensions of the parts for the assembly. It is common to set the dimensional tolerance at 1% of the overall dimensions of the work piece in macroscale manufacturing. This rule, however, is not practical for MEMS and microsystems, as most parts of these systems are in micrometers. A tolerance at 1% of overall dimensions would make it on the order of nanometers for most microcomponents. Such small tolerances would be meaningless, as the surface roughness of most of these parts from microfabrication processes would have exceeded this amount. The issue of setting proper dimensional tolerances for microparts and components thus becomes a challenging task for the industry. Much research in this area is needed, as these tolerances will significantly affect the quality and performance of many MEMS and microsystems as well as the reliability of these products. Information on dimensional tolerances for microsystems is rarely available in the literature. Table 11.3 presents limited but typical tolerances for materials from three microfabrication processes.

Geometric Tolerances. Setting this tolerance relates to discrepancy of the geometry of microcomponents produced by microfabrication processes and the intended application of the microsystem. The consequence of this geometric discrepancy can significantly affect the performance of microsystems as demonstrated by a case study by Hsu and Custer (2004). The case involves a resonator actuated by a comb drive as illustrated in Figure 11.5a.

TABLE 11.3. Dimensional Tolerances for Typical Microfabrication Processes

Fabrication Processes	Materials	Minimum/Maximum Sizes	Dimensional Tolerance (μm)
Wet anisotropic etching	Si, GaAs, quartz, SiC, InP	Few micrometers/maximum wafer size	1.0
Dry etching	Si, GaAs, quartz, SiC, InP	Submicrometers/maximum wafer size	0.1
Polysilicon surface micromachining	Poly-Si, Al, Ti	Submicrometers/maximum wafer size	0.5
Silicon-on-insulators	Si crystal	Submicrometers/maximum wafer size	0.1
LIGA process	Ni, PMMA, Au, ceramics	0.2 μm/10 × 10 cm or larger	0.3

Source: Madou, 1997.

Tang et al. (1990) pointed out that the gap d between the moving and fixed electrodes in the comb drive in Figure 11.5a has a more pronounced effect on the performance of the comb drive than the dimensions and geometry in the width w and length L. We will thus focus our attention on the geometry and dimensions of the electrode, or "fingers."

As we learned from Section 9.2.5, deep reactive ion etching (DRIE) is a common process used to fabricate microcomponents with deep trenches such as the fingers in a comb drive. This process can produce trenches of high aspect ratios with close to vertical walls at a small tapered angle $\theta = \pm 2°$, as illustrated, with exaggeration, in Figure 11.5b.

Now, let us consider a comb drive with nominal dimensions of a finger with length $L = 40\,\mu$m, width $w = 2\,\mu$m, and gap $d = 3\,\mu$m, as described in Tang et al. (1990) and illustrated in Figure 11.5a. If we consider the worst case with the width of the finger $2\,\mu$m at the tip, we may expect the width of the base of the same finger to be $2\,\mu$m $+ 2 \times 40\,\mu$m $\times \tan(2°) = 4.8\,\mu$m, as shown in Figure 11.5b. When these fingers overlap by $36\,\mu$m, we may expect the gap between the matching fingers to be reduced to $1.74\,\mu$m at the root of the finger from the nominal 3-μm gap, as shown in Figure 11.5c. Since the output capacitance of a parallel-plate capacitor is inversely proportional to the gap between these electrodes, the variation of the output capacitance, and thus the pulling forces, between the two ends of the finger is $d_b/d_t = 3/1.74 = 1.72$, or approximately a factor of 2—a significant amount of deterioration in performance.

Alignment Tolerances. Proper alignment tolerance is necessary in the insertion and placement of parts in the microassembly. No established rule is available for setting the alignment tolerance in the microassembly, and this tolerance varies according to the specific application of the microdevice. For instance, the alignment tolerances for micro-optical systems may vary substantially from that of bioMEMS. We realize that the alignment tolerance for the fingers in a comb-drive resonator illustrated in Figure 11.5 cannot be less than $1.74\,\mu$m in order to avoid electrically short circuiting the device. For most parts, the alignment tolerance ranges from less than $1\,\mu$m to a few micrometers.

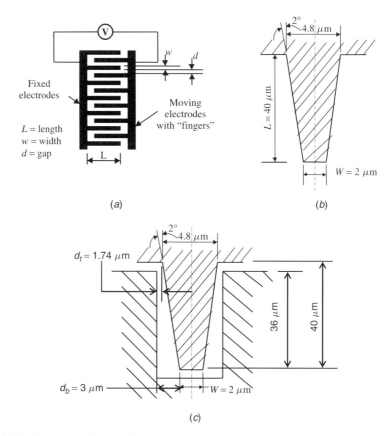

Figure 11.5. Geometric tolerance in a resonator actuated by comb drive: (*a*) resonator actuated by comb drive; (*b*) electrode finger with $2°$ tapered edges; (*c*) variation in gaps.

Alignment tolerance is most critical in the assembly and packaging of optoMEMS switches involving fiber optics, in which light beams from input fibers must be properly aligned to switching units (mirrors or lenses) that can redirect these light beams to the intended fibers. It is not unusual to align a laser beam with a spot size of $2–3\,\mu$m into a fiber with a core size of $9\,\mu$m. A special technique called *in-packaged microaligner* developed by the Boeing Company could achieve a tolerance of 200 nm from an initial placement accuracy of approximately $5\,\mu$m based on a total motion range of $10\,\mu$m.

Machines and fixtures with capability of alignment tolerance within $0.5\,\mu$m are available commercially. Alignment tolerances in typical three-dimensional microassemblies of less than $10\,\mu$m in the vertical direction and $20\,\mu$m in the horizontal direction were reported by Yang and Nelson (2004).

Other Tolerances in Microassembly. There are other tolerances involved in quasi-automated microassembly, as we will learn in Section 11.6. These include tolerances for parts feeders, grasping surface to mating parts, and fixture surface to mating surfaces.

11.5.2 Tools and Fixtures

Microscopes with high magnification are essential parts of microassembly. Magnification at 300X or 400X is a common requirement for clear visual definitions of microcomponents, such as those in the microaccelerometer in Figure 1.6. Both the working distance, which is the distance from the bottom of the microscope to the part that is in focus, and the diameter of the field of view reduce to less than 0.5 mm by most compound or zoom microscopes. This short working distance presents a major problem in the development of proper tools for microassembly operations.

Microtweezers are the most commonly used tools in microassembly. Because of the short working distance between the microscope and the work piece, these tweezers are typically long with arms on the order of a few hundred micrometers to centimeters. In contrast, the thickness of these tweezers rarely exceeds 100 μm. The large aspect ratio of the tweezer arms presents a mechanical design problem in terms of the structural stiffness required for precise positioning of the end point.

Microgrippers are used for manipulating larger parts in a microassembly. The working principle of microgrippers actuated by electrostatic forces is described in Section 2.4.1, with a typical structure illustrated in Figure 2.26. Other actuation methods have been used for microgrippers: hydraulic or pneumatic (Peirs et al., 1998), piezoelectric (Breguet et al., 1997), thermal (Greitmann and Buser, 1996), and shape memory alloys (SMAs) with local laser heating (Ikuta, 1990). Working principles of these actuation methods are described in Section 2.3.

A critical requirement for both microtweezers and microgrippers is accurate control of gripping force during the assembly operation. Too much force can damage the delicate work piece and insufficient force can result in the part slipping or damage to the mating parts due to close tolerances. Consequently almost all tweezers and grippers are equipped with tactile sensors. Closed-loop feedback control systems using the accurate sensed information are developed to control the gripping forces.

11.5.3 Contact Problems in Microassembly Tools

A major task involved in microassembly is the pick-n-place operation. We have learned that microtweezers and microgrippers are used to pick the work pieces in this operation. A serious problem, however, is release of the part. Figure 11.6a shows the grasping of a cylindrical object with forces that need to overcome the gravitation, that is, the weight of the object. In a microassembly, the gravitation of the part is no longer dominant because of its minute size. Moreover, the minute gaps that may exist in the contact surfaces between the gripping and the object can induce adhesive forces that are not present in traditional pick-n-place operations. Consequently, the work pieces that are picked up by the tool may not be released with the presence of these forces, as illustrated in Figure 11.6b. These unique adhesive forces may include (1) electrostatic force, (2) van der Waals forces, and (3) surface tension.

Electrostatic Force. We learned from Chapter 3 that atoms consist of three types of particles: the negatively charged electrons, positively charged protons, and neutrons. Atoms are ionized by external influences to carry either positive charge with deficit in

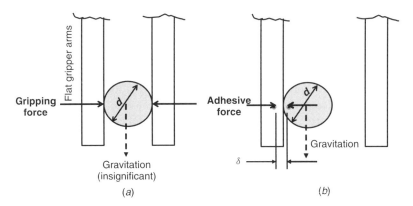

Figure 11.6. Gripping force in microassembly: (*a*) grasping; (*b*) releasing.

electrons or negative charge with gained electrons. When opposite charges are developed in two dielectric or non-electric-conducting materials in contact, electrostatic force is produced by the Coulomb effect, as described in Section 2.3.4. External influences that produce ionized atoms can be the doping processes described in Chapters 3 and 8 or mechanical means such as rubbing the two dielectric substances in contact. In many cases, two nonconductive surfaces can become charged by just being placed one on top of the other. The charge that is generated during contact electrification is stored on the surface of each object. The storage of these charges on the contacting surfaces is the cause for generating electrostatic attractive force in microgrippers.

Quantification of the electrostatic attractive forces induced in microgrouping depends on the geometry of the gripping device and the part to be gripped. In the case of sphere/flat plate combination such as illustrated in Figure 11.6, this force can be approximated by the equation

$$F_e = \frac{q^2}{4\pi \varepsilon d^2} \tag{11.1}$$

where

q = electrostatic charge, $\approx 1.6 \times 10^{-6}\, C/m^2$ for microgripping

ε = permittivity of medium, $= 8.85 \times 10^{-12}\, C^2/N\text{-}m^2$ in air

d = diameter of sphere, in the range $10\,\mu m$–$1\, mm$

Van der Waals Force. We mentioned that the van der Waals force is a form of the molecular bonding force in Section 3.4. This force is generally attractive in nature, existing between two intimately contacting surfaces. It is a short-range force that decays rapidly away from the contacting surfaces. However, this force contributes to the adhesive force used in grasping microparts, as illustrated in Figure i1.6. The van der Waals force between spheres with diameter d in contact with a flat gripping arm can be estimated as

$$F_v = \eta \frac{Ad}{12\delta^2} \tag{11.2}$$

where

A = Hamaker constant, $= 10^{-20}$–10^{-19} J

δ = atomic separation between contacting surfaces, typically 4–10 Å

η = correction factor for roughness of contacting surfaces, ≈ 0.01

Surface Tension. We learned that surface tension becomes the dominant force in the capillary flow of liquid. As illustrated in Figure 11.6, this form of adhesive force in microgripping appears when the operation takes place in a humid environment. The microvoids at the contacting surfaces can produce the surface tension that forms the adhesive forces. Following the description of surface tension in Section 5.6.1, the adhesive force between the sphere and the flat gripping arm, as illustrated in Figure 11.6, can be approximated by (White, 1994)

$$F_s = s\gamma \tag{11.3}$$

where

s = perimeter of microvoid in contacting area

γ = coefficient of surface tension

The total adhesive force that prevents the micropart from being released from a microgripper is thus equal to

$$F = F_e + F_v + F_s$$

where F_e, F_v, and F_s are given in Equations (11.1), (11.2), and 11.3, respectively.

The adhesive forces that prevent the micropart from being released from the gripper usually are not as large as that of stiction, as described in Chapter 9. There are several ways by which one may alleviate these forces and thereby drop the part on the right spot on the substrate (Van Brussel et al., 2000): (1) Glue the part to the substrate on the right place. In this case, the adhesion of the part to the substrate overcomes the adhesive forces. (2) Inject a puff of gas to blow loose the stuck part in the open gripping arm. (3) Set free the part by inserting a small needle to the contact surfaces. (4) Shake loose the stuck part with mild vibration of the gripper.

11.6 MICROASSEMBLY WORK CELLS

Other than a few MEMS and microsystems, for example, inertia sensors for automobile airbag deployment and inkjet printer heads with established markets for mass production, most products are assembled either manually using microassembly tools such as tweezers or microgrippers or quasi-automatically using *microassembly work cells*. These cells have been developed to assist human operators for quality assembly and packaging of microproducts. Most assembly work cells require a clean environment in a portable clean room of class 100. A work cell for pin-hole assembly by Sandia Laboratories was described in Section 11.4

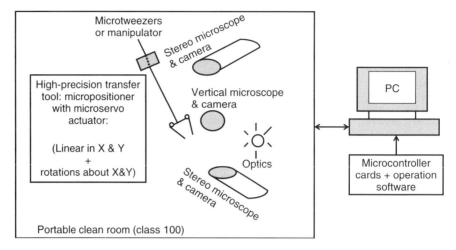

Figure 11.7. Typical microassembly work cell.

A typical microassembly work cell is illustrated in Figure 11.7.
Following are the essential components in a typical microassembly work cell:

1. **Integrated Microposition:** An integrated *micropositioner* carries the parts to be assembled and transports them to the assembly site. Micropositioners are designed with a resolution in linear movements at 40 nm. An acceptable positioner can produce linear movement with step size of 300 nm in the x and y directions and 70 nm in the z direction. The positioner can also rotate about the x and y axes at 0.0003° per step.

2. **Microscope Optics and Imaging Unit:** Stereomicroscopes with a camera for imaging are commonly used to provide real-time three-dimensional visual monitoring of the assembly operations. It is desirable to select microscopes that offer a long working distance up to 30 mm, a wide field of view, and high resolution at the 1-μm range. Proper illumination and high-resolution computer vision are necessary for avoidance of collision and occlusion during the operation.

3. **Micromanipulation Unit:** This unit may include microgrippers with specially designed end effectors or microtweezers with long gripping arms. The end effectors of the microgrippers or microtweezers are equipped with tactile sensors and real-time feedback control systems. These end effectors are capable of overcoming the adhesive forces mentioned in Section 11.5.3 for releasing the micropart at the right position on the substrate.

   Some work cells use microrobots for the pick-n-place operations in microassembly. The robots developed by Fatikow (1998) have a size of 50–80 mm. They are attached to a glass plate which can be set to motion by piezoelectric actuators in three degree of freedom (two in-plane and one rotational). The robot can move at a linear speed up to 30 mm/s. The attached gripper has three degree of freedom too, all rotational.

A major challenge in developing micromanipulators is the lack of small, lightweight actuators that can generate high-precision motions and sufficient force/ torque to withstand the static loads of the subsequent joints and the end effector (Yang and Nelson, 2004).

4. **High-Precision Transfer Tool:** The function of a transfer tool is to transport discrete parts or parts in trays from parts feeders to the assembly site. As shown in Figure 11.7, this unit offers four-degree-of-freedom motion with linear movement in the x and y directions and two rotational movements about the x and y axes. The distance of traveling on the $x-y$ plane can be from 30 to 80 cm, but with stringent repeatability at $1\,\mu$m in both directions. Position feedback with a resolution of 100 nm is provided by two linear encoders, and the rotary movement of the transfer tool is provided by microstepping motors with a maximum resolution of $0.0028°$ per step (Yang and Nelson, 2004).

5. **Real-Time Computer with Vision:** It is required to control the movements of the transfer tool, micropositioner, and micromanipulators in the assembly operation. Proper operation software is needed to perform the functions of implementation of the assembly strategies, process monitoring, diagnosis of system malfunctions, and error recovery during the assembly operation.

11.7 CHALLENGING ISSUES IN MICROASSEMBLY

We have shown throughout this book that cost-effectiveness is a primary concern in the microsystems industry, and microassembly is no exception. However, a number of technical issues need to be dealt with by scientists and engineers:

1. **Sensing:** There is need for reliable three-dimensional machine vision and visual servoing systems for objects at the microscale. These systems function with submicrometer resolutions and precise repeatability. Tactile and touch sensors that can be integrated in continuous closed-loop control systems are essential in any automatic microassembly process.

2. **Intelligent End Effector:** Microgrippers with intelligent end effectors that can locate and mate other microcomponents with automatic adjustment of gripping forces and ejection of parts are very much in need. The concept of having a microrobot carried by a microrobot is a viable way to accomplish this task.

3. **Material and Parts Delivery:** Reliable hardware and software with high precision motion control are required in transportation and delivery of materials or parts to the assembly site swiftly and accurately. All functions performed by these systems must be repeatable with submicrometer resolution.

4. **Part Insertion:** Special tools are required for insertion of parts at the microscale and free of adhesive force upon releasing.

5. **Material Science:** Issues related to this development include the following:
   - Material database for parts design, tools and fixtures, and gripper design and fastenings
   - Surface science related to the adhesive forces associate with microgripping

- Properties of materials at the microscale subjected to temperature, pressure, and humidity

6. **Design for Microassembly:** Develop standards and strategies for alignments and tolerancing of mating parts at the microscale and the assembly and packaging processes.

7. **Process Modeling and Simulation:** Develop mathematical models for microassembly processes using scaling laws involving adhesive forces discussed in Section 11.5.3. Apply computer simulation and animation for automatic assembly processes.

8. **Systems Engineering:** Develop strategies and techniques for the transition of microassembly processes and equipment from the laboratory to the factory floor, for example, from experimental microassembly work cells to the mass production environment.

9. **Test Methods:** Develop reliable testing procedures for thermal shock, vibration, humidity, and electromagnetic susceptibility in microassembly. Refer to Section 11.22 for the development of testing for reliability during and after the assembly.

10. Tools and techniques for design verification of microassembly processes.

11.8 OVERVIEW OF MICROSYSTEMS PACKAGING

We have learned from Chapter 1 and elsewhere in this book that MEMS and microsystems technologies are the result of evolution from microelectronics technology. As such, most microfabrication techniques presented in Chapter 8 for microelectronics can be used in microsystems manufacturing as described in Chapter 9. We will find that certain techniques that were developed for microelectronics packaging can also be used for microsystems packaging.

The purpose of microelectronic packaging is to provide mechanical support, electrical connection, and protection of the delicate ICs from all possible attacks by mechanical and environmental sources and removal of heat generated by the ICs. Other major roles that microelectronics packaging play are power delivery to constituents and signal mapping from outside a package to within and between the constituents (Lyke and Forman, 1999). The packaging of MEMS and microsystems, however, is further complicated by additional requirements such as ensuring light transport and mapping for many optical MEMS, fluidic transport and mapping, and environmental access for chemical and biological sensing in microfluidics.

Packaging is a key factor of the reliability of ICs in microelectronics and microsystems. Major reliability issues in a typical IC illustrated in Figure 11.8 include:

1. Silicon die and passivation cracking
2. Delamination between the die, die attach, die pad, and plastic passivation
3. Fatigue failure of interconnects
4. Fatigue fracture of solder joints
5. Warping of printed circuit board

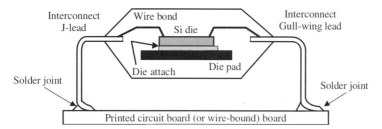

Figure 11.8. Cross section of typical integrated circuit.

Most of the failures listed above are due to excessive thermomechanical forces induced from the following principal sources (Hsu, 1992):

1. Mismatch of coefficients of thermal expansion between the attached materials
2. Fatigue fracture of materials due to thermal cycling and mechanical vibration
3. Deterioration of material strength due to environmental effects such as moisture
4. Intrinsic stresses and strains from fabrication processes as described in Chapter 8

All the above sources of problems that appeared in microelectronics packaging are present in microsystems packaging.

A major challenge to MEMS and microsystems packaging is that the core elements of these products, such as microsensors and actuators, usually involve delicate, complex three-dimensional geometry of layers of dissimilar materials. They are often required to interface with environmentally unfriendly working media such as hot pressurized fluids or toxic chemicals. Yet they are expected to generate a variety of signals, for example, mechanical, optical, biological, and chemical, as shown in Table 11.4 (National Research Council, 1997). These incoming signals need to be converted into electronic signals for swift and accurate interpretation of the results. In the case of actuators, the interface of the actuating elements with electrostatic or thermal power sources presents another major challenge to design engineers.

TABLE 11.4. Input and Output Signals in Microsystems

Signals	Input	Output
Chemical	Yes	Yes
Electrical	Yes	Yes
Fluid/hydraulic	Yes	Yes
Magnetic	—	Yes
Mechanical	Yes	Yes
Optical	Yes	Yes

11.9 GENERAL CONSIDERATIONS IN PACKAGING DESIGN

Proper packaging of MEMS and microsystems products is a critical factor in the overall product development cycle. There are many aspects to be considered in such an endeavor. Following are the principal design requirements that engineers should consider before embarking on the design analyses:

1. The required costs in manufacturing, assembly, and packaging of the components
2. The expected environmental effects, such as temperature, humidity, and chemical toxicity, that the product is designed for
3. Adequate overcapacity in the packaging design for mishandling and accidents
4. Proper choice of materials for the reliability of the package
5. Achieving minimum electrical feed-through and bonds in order to minimize the probability of wire breakage and malfunctioning

Because of the great variety of market demands for MEMS and microsystems, the structural geometry, materials, and configuration for MEMS and microsystems are far from being standardized (see Figure 11.2). We will thus focus our attention on the packaging of some more common devices and systems, such as micropressure sensors and actuators, as presented in Chapter 2. We will further assume that substrate materials are made of silicon unless otherwise specified. The packaging of most components produced by the LIGA process is thus excluded.

11.10 THREE LEVELS OF MICROSYSTEMS PACKAGING

Unlike electronics packaging with a hierarchy of four levels, microsystems packaging can be categorized in three levels:

Level 1: die level
Level 2: device level
Level 3: system level

The relationships among these three levels of packaging are illustrated in Figure 11.9. The major tasks involved in each of these levels of packaging are described below.

11.10.1 Die-Level Packaging

Die-level packaging involves the assembly and protection of many delicate components in microdevices. Examples are:

- The die/diaphragm/constraint base assembly of a pressure sensor in Figures 2.8 and 2.9
- The cantilever of an actuator in Figures 2.19 and 2.21
- The diaphragm-proof mass assembly in an accelerometer in Figures 2.34 to 2.37

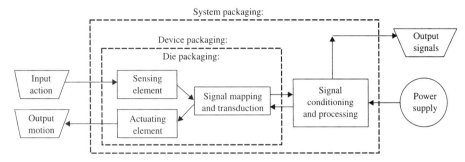

Figure 11.9. Three levels of microsystem packaging.

- The electrodes in micromotors in Figures 2.31 and 2.32
- The microfluidics components that include microvalves in Figures 2.41 and 2.42 and micropumps in Figure 2.43 and the microchannels in Figures 10.19 and 10.22

The primary objectives of this level packaging are:

1. To protect the die or other core elements from plastic deformation and cracking
2. To protect the active circuitry for signal transduction of the system
3. To provide necessary electrical and mechanical isolation of these elements
4. To ensure that the system functions at both normal and overload conditions

Die-level packaging for many MEMS and microsystems also involves wire bonds for electronic signal transmission and transduction, such as the embedded piezoresistors in a pressure sensor die and the circuits that connect them (see Figures 2.9 and 10.4). In some cases, that includes bonding the lead wires to the interconnects as illustrated in Figure 11.10 for micropressure sensors and Figure 11.11 for microaccelerometers.

By a comparison of Figure 11.8 and Figures 11.10 and 11.11, one will realize that there are indeed common features between the ICs and MEMS packaging at this level. The similarities of the two packaging technologies include (1) both use silicon dies, (2) die attaches are involved, and (3) there are wire bonds between the die and interconnect. Consequently, the technologies developed and used in IC packaging in these three areas can be used for the MEMS packaging. Encapsulation of dies and wire bonds in microsystems packaging can vary from plastics to stainless steel casings in MEMS packaging (Figure 11.10). The latter encapsulation material is not used in IC packaging.

There are several other issues relating to die isolation and protection in this level of packaging. We will deal with these issues using specific packaging design cases in Section 11.20.

11.10.2 Device-Level Packaging

Device-level packaging as categorized in Figure 11.9 requires the inclusion of proper signal conditioning and processing, which in most cases involves electric bridges and

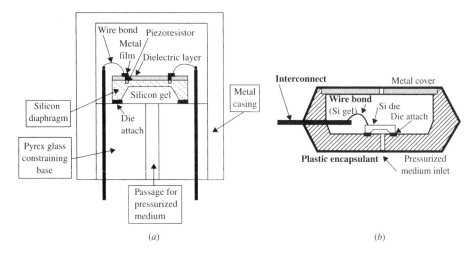

Figure 11.10. Die-level packaging of micropressure sensors: (*a*) with metal casing; (*b*) with plastic encapsulation.

signal conditioning circuitry for sensors and actuators. Proper regulation of input electric power is always necessary. Typical device-level packaging is shown in Figure 1.6 for a microaccelerometer.

A major challenge to design engineers at this level of packaging is the problems associated with interfaces. There are two aspects to interface problems:

1. The interfaces of delicate die and core elements with other parts of the packaged product at radically different sizes
2. The interfaces of these delicate elements with the environment, particularly in regard to such factors as temperature, pressure, and toxicity of the working and contacting media

The issues of interfaces will be described in more detail in Section 11.11 as well as in the packaging design cases in Section 11.20.

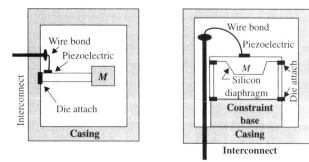

Figure 11.11. Die-level packaging of microaccelerometers.

11.10.3 System-Level Packaging

System-level packaging involves the packaging of primary signal circuitry with the die or core element. System packaging requires proper mechanical and thermal isolation as well as electromagnetic shielding of the circuitry. Metal housings usually give excellent protection for mechanical and electromagnetic influences.

The interface issue in this level of packaging is primarily in fitting components of radically different sizes. Assembly tolerance, such as described in Section 11.5.1, is a more serious problem in this level of packaging. Figures 1.2 and 1.22*b* show the system-level package of micropressure sensors, whereas Figure 1.8 shows the system-level packaging for a microaccelerometer.

11.11 INTERFACES IN MICROSYSTEMS PACKAGING

As we pointed out in Section 11.10, interfaces between MEMS components and their operating environment often present major challenges to design engineers. These interfaces often make the selection of proper materials for the components, the signal transduction, and, in some cases, the sealing of working media and other electromagnetic fields critical factors in the design of a successful microsystem. We will take an in-depth view of the interface problems associated with various kinds of microsystems in the following paragraphs (National Research Council, 1997).

Biomedical Interfaces. Interfaces in packaging of biomedical sensors and biosensors as described in Section 2.2.2 are much more critical than all other types of microsystems. Microbiosystems are subject to vigorous government regulation, being treated as medical products. The packaged systems need to be biologically compatible with human systems and they are expected to function for a specific lifetime. Every microbiosystem must be built to satisfy the following requirements related to interfaces:

1. Be inert to chemical attack during the useful lifetime of the unit
2. Allow mixing with biological materials as described in Section 2.2.2 in a well-controlled manner when used as a biosensor
3. Cause no damage or harm to the surrounding biological cells in instrumented catheters such as pacemakers
4. Cause no undesirable chemical reactions such as corrosion between the packaged device and the contacting cells

Optical Interfaces. There are two principal types of optical MEMS used at the present time. One type relates to the direction of light in such devices as microswitches involving mirrors and reflectors, as illustrated in Figure 1.12, and the other type is optical sensors, as described in Section 2.2.4. Both these optical MEMS require:

1. Proper passages for the light beams to be received and reflected
2. Proper coating of the surface on which light beams are received and reflected
3. Enduring quality of the coated surface during the lifetime of the device

4. Freedom from contamination of foreign substances on the exposing surfaces

5. Freedom from moisture in the enclosure; moisture in the environment may cause stiction of minute delicate optomechanical components.

All these requirements must be adequately met to ensure a quality optical MEMS.

Mechanical Interfaces. Mechanical interfacing is a design issue with moving parts in MEMS. These parts need to be interfaced with their driving mechanisms, which may be thermal, fluidic, or magnetic, such as microvalves and pumps in Figures 2.41–2.43 and other microactuators and robots. Improper handling of the interfaces may cause serious malfunctions or damage to these device components. Obvious examples for this are thermally actuated microvalves or pumps, in which the diaphragms are expected to be in contact with the working media. Thermal contact conditions at the interfaces can result in a negative effect on the performance or overstress of the diaphragm structure. Mechanical sealing at the interfaces is another major problem to be overcome in this type of MEMS.

Electromechanical Interfaces. Electrical insulation, grounding, and shielding are typical problems associated with MEMS and microsystems. These problems are more obvious in the systems that operate at low voltage levels. The intimate electromechanical interfaces in design were illustrated in Example 10.1. Selection of materials for electrical terminals and the shielding of electrical conductors for microdevices are other major considerations in the packaging.

Interfaces in Microfluidics. The prominence of microfluidics in chemical and biomedical applications was described elsewhere in this book. The value of CE in these applications was presented in Section 10.7. These systems require precise fluid delivery, thermal and environmental isolation, and mixing. Material compatibility for applications, such as the compatibility of materials of the microchannel and contained solvent, is another critical requirement. The sealing of the fluid flow and the interfaces between the contacting channel walls and the fluid are two major packaging issues relating to the interfaces.

11.12 ESSENTIAL PACKAGING TECHNOLOGIES

With respect to Figure 11.1, the packaging of microsystems components follows the pick-n-place process of microparts in microassembly operations as described in Section 11.4. Microsystems packaging mainly involves bonding of the microcomponents and sealing the devices as required. As mentioned at the beginning of this chapter, there is lack of industry standard in the assembly and packaging of MEMS and microsystems despite new techniques, and procedures as well as materials are being developed continuously by researchers and engineers. It is not possible to cover all these newly developed packaging technologies in this chapter. What we will cover in subsequent sections are the essential technologies that are necessary for packaging most of these products. These technologies include (1) wafer dicing for individual substrates, (2) bonding of parts and components to substrates, (3) wire bonding of signal transducers to interconnects and processing circuitry, and (4) sealing and encapsulation of microdevices.

11.13   DIE PREPARATION

It is a rare practice to use an entire silicon wafer to produce just one die for a MEMS or microsystem or use one wafer to produce one device. In reality, hundreds of these tiny dies required for various parts of microdevices are produced from a single wafer, as illustrated in Figure 11.12. These dies may all be of the same type and size, as in batch production, or may have different sizes and shapes (shown in solid black rectangles) to be cut along the dotted lines shown in the figure.

Sawing the wafer (or in the industrial terminology, *wafer dicing*) along the dotted lines as shown in Figure 11.12 is required for the production of individual dies. A common practice for wafer dicing is to mount the thin wafer on a sticky tape. The saw blade is made of a diamond/resin or diamond/nickel composite materials. The standard cutting procedure adopted by the microelectronics industry is cutting between the dies imprinted on the silicon wafer. The dies are spaced according to the thickness of the saw blades. For example, the space may be set to be on the order of $25\,\mu$m with a typical saw blade thickness of $20\,\mu$m. The cutting wheel is 75–100 mm in diameter. Wafer dicing is normally carried out at a cutting speed of 30,000–40,000 rpm. Laser dicing using CO_2 and YAG laser has become popular for dicing wafer of up to 300 mm diameter in recent years.

11.14   SURFACE BONDING

Bonding of microsystem components is a great deal more challenging than that of microelectronics. It is one of the most challenging issues in microsystems packaging because MEMS and microsystems are three-dimensional in geometry and made up of layers of dissimilar materials. Many of these microstructures are constructed by bonding thin layers of dissimilar materials. Another factor is that many microsystems contain fluids or

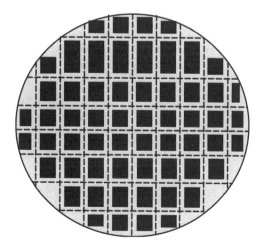

Figure 11.12. Silicon wafer with dies.

environmentally hostile substances. Hermetic sealing of these media is required at many of these bonding surfaces. In many cases, bonding of dissimilar materials is expected to simultaneously achieve hermetic sealing and provide flexibility in the sealed surface for die isolation, as will be described in more detail in the case of micropressure sensor packaging in Section 11.20.

Much effort is being made by the microsystems industry to develop new and more effective bonding techniques and procedures. Presented in the following sections is some of the commonly available surface bonding techniques. In-depth description of these techniques is available in an article by Hsu and Clatterbaugh (2004).

11.14.1 Adhesives

Adhesives are primarily used for attaching dies onto the supporting constraint bases. Typically, the silicon die in a pressure sensor is attached to the glass constraint base by adhesion (see Figure 4.6). There are two common adherents used for this type of bonding: epoxy resins and silicone rubbers.

Epoxy resin bonds provide flexibility for the bonded dies as well as good sealing. Good bonding depends on surface treatment and control of the curing process. Unfortunately, epoxy resins are also vulnerable to thermal environments. The bond should be kept below the glass transition temperature, which is normally around 150–175°C.

When flexibility of the bonding surfaces is a primary requirement, soft adhesives are used. The softest commercially available die bonding material is silicone rubber. One such adhesive material is room temperature vulcanizing silicone rubber (RTV). This material cures at room temperature. The soft nature of this type of adherent makes it the most flexible for the die bonding and thus provides the best die isolation. Unfortunately, the chemical resistance of this type of material is not as good, and peeling and flaking develop when they get in contact with air. It is not suitable for high-pressure applications.

A generic arrangement of this bonding technique is illustrated in Figure 11.13. Adhesive bonding takes place inside a chamber and heats the substrate to the desired bonding temperature. An automated dispenser is used to drop the right amount of thin adhesive on top of the substrate. The part to be bonded is then placed on top of the adhesive film. Mechanical pressure is normally applied to ensure the quality of bonding. The chamber

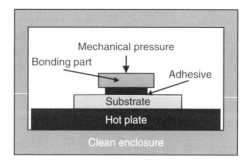

Figure 11.13. Typical setup for bonding by adhesion (After Hsu and Clatterbaugh, 2004.)

must be kept clean and free from dust or other solid contaminants. Often the bonding operation is carried out in a vacuum.

There are a number of adhesives commercially available for surface bonding of microsystems components. Following are a few of the desirable features for microsystems applications:

1. High resistance to mechanical shock and vibrations; thermal cycling; elevated temperatures; chemical attacks; working media such as water and oil
2. Low viscosity for producing thin film
3. Good physical strength in shear and tension
4. Good bonding to a variety of materials of plastics, metals, and glass
5. Outstanding electrical properties in conductivity for some applications or insulation for other applications
6. Good dimensional stability with low shrinkage during and after curing
7. Fast tacking and curing
8. Flexible if it also serves for hermetic sealing
9. Easily removed by chemical solvents
10. Durable

11.14.2 Eutectic Bonding

Bonding of silicon die to the base, or other similar component bonding, can be achieved using *eutectic solders*. This bonding technique has the advantage of being chemically inert. It also provides stable and hermetic seals.

Eutectic bonding involves the diffusion of atoms of eutectic alloys into the atomic structures of the materials to be bonded together and thus forms solid bonding of these materials. For this bonding to take place, one must first select a candidate material that will form a eutectic alloy with the materials to be bonded. A commonly used material to form a *eutectic alloy* with silicon is a thin film made of gold (Au) or alloys that involve gold.

Eutectic bonding takes place when the assembly of the two bonding surfaces (e.g., silicon substrates with a eutectic alloy component such as gold) is heated to above the *eutectic temperature*. A eutectic temperature is the lowest fusion temperature of an alloy with the melting point lower than that of any other combination of the same components. At such time, the atoms of the interface material (e.g., Au) start to diffuse rapidly into the contacting substrates of silicon. Sufficient migration of these atoms into the bonding substrate surface will result in the formation of a eutectic alloy (e.g., Au–Si). As the temperature continues to rise above the eutectic temperature, more eutectic alloy is formed. This process will continue until the atoms (e.g., gold) in the eutectic alloy at the interface material reach near-complete depletion. The newly created eutectic alloy at the interface serves as a solid bond as well as a hermetic seal for many MEMS and microsystems.

Like adhesive bonding, eutectic bonding of microcomponents to silicon substrate also takes place in a chamber that provides the required eutectic temperature for bonding, as shown in Figure 11.14. The bonded surface is separated by a thin film of eutectic alloy such as Au–Sn. The sandwiched structure is subjected to a pressure of approximately

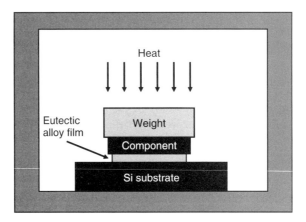

Figure 11.14. Schematics of a eutectic bonding chamber.

1 MPa provided by the weight. The process can take place from one to a few hours, followed by cooling in the same chamber for a few hours.

Bonding by eutectic means depends on the choice of interface component material with which a eutectic alloy with the bonding materials is formed. Gold is a popular choice for bonding silicon substrates. A common eutectic alloy is 97% Au–2.83% Si with a eutectic temperature of 363°C. Another common eutectic alloy is 62% Sn–38% Pb with a eutectic temperature of 183°C. The latter alloy is less costly than the Au–Si alloy, and it can bond most metallic materials. The nonmetallic bonding materials must be sputtered with thin films of metal on the order of a few micrometers on their bonding surfaces to facilitate such bonding.

11.14.3 Anodic Bonding

This process is reliable and effective for attaching silicon wafers to several other materials. It provides a hermetic seal and is an inexpensive way for die bonding. It is also called *electrostatic bonding* or *field-assisted thermal bonding* in the literature. Anodic bonding is popular in microsystems packaging because of the relatively simple setup with inexpensive equipment. It provides reliable hermetic sealing that is important for applications such as in microvalves and channels in a microfluidics network and for micropressure sensor dies. Another major advantage of this bonding technique is that bonding can take place at low to moderately high temperatures in the range of 180–500°C, which results in low risk of residual stress and strain in the bonded materials after the bonding.

Anodic bonding has been used to bond wafers made of the following materials: (1) glass to glass, (2) glass to silicon, (3) glass to silicon compounds, for example, GaAs, (4) glass to metals, and (5) silicon to silicon.

The most common application of anodic bonding is to bond glass wafers to silicon wafers. Figure 11.15 illustrates the setup for anodic bonding of a sodium-rich wafer (e.g., Pyrex 7740 glass) to a silicon wafer. A modest weight is placed on top of the Pyrex glass wafer to ensure good contacting pressure. An electric field of 200–1000 V DC is applied

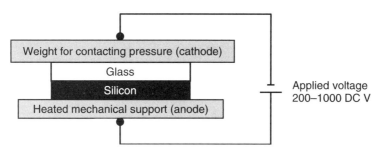

Figure 11.15. Anodic bonding of silicon to glass.

to the system with the silicon wafer in contact with the positive electrode. The sandwiched dielectric Pyrex and semiconductive silicon wafers between the two electrodes form an effective parallel-plate capacitor. Consequently, the voltage applied to the electrodes causes the two substrates to come into contact with the induced electrostatic force.

The bonding of silicon and glass is accomplished by the formation of an extremely thin layer of SiO_2 interface as the result of the applied electric field. Under the influence of the applied electric field, the sodium ions (Na^+) in the glass are attracted to the negatively charged cathode, which leaves behind Na^+ ions, and the negatively charged O_2^- ions in the depletion zone, as illustrated in Figure 11.16. These O_2^- ions can be chemically bonded to the contacting Si^+ ions with the supplied heating of the system and form a very thin layer of SiO_2 at approximately 20 nm at the interface. This thin film of SiO_2 serves as the bond of the silicon and glass wafers. Pyrex 7740 glass is a popular choice as the glass wafer for bonding to silicon wafers because this type of glass is rich in sodium.

Successful anodic bonding requires the maintaining of constant temperature, applied voltage, and current density conditions during the bonding process. The time required

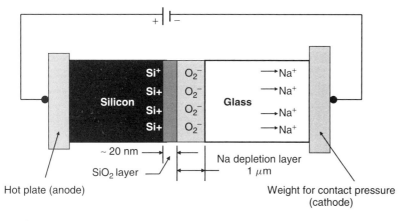

Figure 11.16. Formation of SiO_2 layer in anodic bonding. (After Hsu and Clatterbaugh, 2004.)

for bonding depends on (1) the surface condition of the bonding materials, (2) the temperature, and (3) the applied voltage.

Anodic bonding normally takes 10–20 min to have the 100-mm-diameter silicon and glass wafers bonded in an evacuated chamber at 450°C with applied electric field at 1000 V DC. The completion of bonding is judged by observing the change of color of the silicon wafer through the transparent Pyrex wafer into the dark gray area spreading from the center to the entire observed surface.

11.14.4   Silicon Fusion Bonding

Silicon fusion bonding (SFB) is an effective and reliable technique for bonding two silicon wafers or substrates without the use of intermediate adhesives. The concept of joining two silicon wafers was first developed in early 1960s for bonding discrete transistor chips (Barth, 1990). It was not until 1988 that the application of this technique in MEMS was first reported (Petersen et al., 1988). Silicon fusion bonding has been used to bond the following combinations of wafers (Madou, 1997): (1) silicon to silicon, (2) silicon with oxide to silicon, (3) silicon with oxide to silicon with oxide, (4) GaAs to silicon, (5) quartz to silicon, and (6) silicon to glass. However, SFB is most commonly used for bonding silicon to silicon.

There are two popular SFB techniques available for bonding silicon wafers: (1) SFB by hydrophilic and (2) SFB by hydration (Hsu and Clatterbaugh, 2004).

SFB by Hydrophilic Process. This involves having the wafer surfaces polished followed by soaking both wafers in boiling nitric acid for thorough cleaning and also to create hydrophobic surfaces for bonding. These two surfaces are naturally bonded even at room temperature. However, stronger bonding occurs with an annealing process at high temperature in the neighborhood of 1100–1400°C. Hydrogen-induced van der Waals forces are believed to be the principal bonding forces.

SFB by Hydration Process. This process involves first having the wafers cleaned in a liquid bath containing sulfuric acid and hydrogen peroxide, as illustrated in Figure 11.17. An oxygen (O)–hydrogen (H) bond, called a *silanol bond*, at the interface of the two silicon wafers is formed after the wafers are removed from the acid solution and rinsed with deionized water. The two wafers are then placed one on top of the other to form a bond. Heating is applied to the stacked wafers to drive the oxygen and hydrogen molecules in the contacted O–H bond to form the Si–O–Si bond between the two wafers, as illustrated in Figure 11.17.

11.14.5   Overview of Surface Bonding Techniques

The quality of surface bonding of microcomponents is judged by its hermeticity and reliability. Low or moderate temperature is also desirable. A recent survey offers an excellent overview of available bonding mechanisms, as shown in Table 11.5.

The integrated process bonding method in Table 11.5 involves selected microfabrication processes such as CVD for both bonding and sealing, as described in the literature (Lin, 2000).

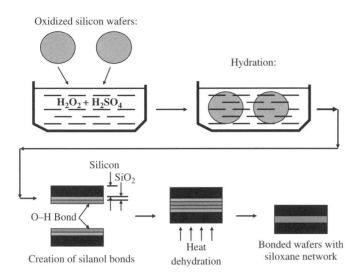

Figure 11.17. Silicon fusion bonding by hydration. (After Hsu and Clatterbaugh, 2004.)

TABLE 11.5. Overview of Bonding Methods

Bonding Methods	Temperature	Hermeticity	Reliability
Adhesive bonding	Low	No	Not certain
Eutectic bonding	Medium	Yes	Not certain
Anodic bonding	Medium	Yes	Good
Fusion bonding	Very high	Yes	Good
Integrated process	High	Yes	Good
Low-temperature bonding	Low	Not certain	Not certain
Brazing	Very high	Yes	Good

Source: Lin, 2000.

11.14.6   Silicon-on-Insulator: Special Surface Bonding Techniques

We learned from Table 7.1 that silicon is a semiconductor. As such, it has the ability to conduct electricity when it is subjected to high electric potentials or at an elevated temperature environment. For example, silicon becomes increasingly electrically conductive at a temperature above 125°C (Maluf, 2000). This transformation of silicon from a semiconductor to a conductor limits silicon sensor elements to being effective in elevated temperature applications. The *silicon-on-insulator* (SOI) process offers a viable solution to resolve this problem.

The SOI process is used in microelectronics to avoid leakage of charges in p–n junctions (Sze, 1985). It involves bonding the silicon with an amorphous material such

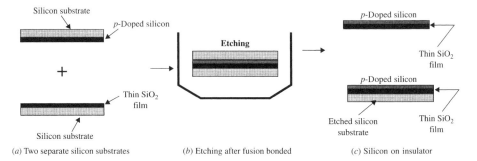

(a) Two separate silicon substrates   (b) Etching after fusion bonded   (c) Silicon on insulator

Figure 11.18. Silicon-on-insulator process.

as SiO_2 by way of epitaxy crystal growth as described in Section 8.8. An amorphous material does not have long-range order of crystal structure (Askeland, 1994).

Figure 11.18 illustrates a SOI process. The process uses two silicon substrates; one has its surface heavily doped with boron atoms to produce a layer of p-silicon, and the other has a thin silicon oxide film on one of its faces, as shown in Figure 11.18a. The two substrates are then mounted one on the top of the other, as shown in Figure 11.18b. The process of silicon fusion bonding joins the two substrates together. The bonded substrates are then exposed to etch the exposed surfaces of the bonded substrates. The heavily p-doped region can act as an etch stop, as described in Section 9.2.4. Consequently, one may obtain either a p-silicon layer on the SiO_2 insulator or sandwiched silicon substrates with an SiO_2 insulator in between, as shown in Figure 11.18c. This technique, which involves bonding of substrates followed by etching of the bonded substrates, is termed the *bonding-and-etchback* technique.

Instead of doping boron to form a p-silicon layer in the so-called donor wafer, one may implant H_2 ions to the contacting surface. The more active hydrogen ions allow the donor wafer to be bonded to the receptor wafer with SiO_2 grown on its surface at a lower temperature. There is no loss of materials in etching in this case compared with the bonding-and-etchback process, as described above.

This technique was used to introduce a SiO_2 insulation layer over a microchannel (Yun and Cheung, 1998), as illustrated in Figure 11.19. It illustrates the process of introducing

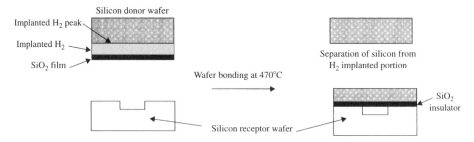

Figure 11.19. SOI for an embedded channel.

a SiO_2 insulator over the top of a silicon substrate. The substrate, called the receptor wafer, contains an embedded microchannel of a depth of 1 μm. A 60-nm-thick oxide film is thermally grown over the surface of the donor wafer. It is then implanted with H^+ ions with a dose of 8×10^{16} cm^{-2} with 40 keV of energy. The two wafers are then bonded face to face at room temperature after thorough cleaning of the surfaces. The bonded pair was then briefly heated to a temperature of $470°C$. This temperature is sufficient to cause crack at the peak of the implanted hydrogen region, as shown in the figure. The cracking at this face resulted in the bonded insulator adhering to the receptor wafer after being separated from the donor wafer, as illustrated. Consequently, a SiO_2 film insulator is produced to cover the silicon substrate, creating an embedded channel.

A similar process was used to bond wafers made of positively doped silicon and quartz (Lee et al., 1997). We have noticed from the previous case that the SiO_2 insulator thickness is usually in the range of single-digit micrometers. Thin SiO_2 layers were necessary to avoid the buildup of thermal stresses at the interface due to significantly different coefficients of thermal expansion of the mating materials. In this case, however, a quartz wafer 525 μm thick is bonded to a silicon wafer with a thickness greater than 300 μm. From Table 7.3 quartz has a coefficient of thermal expansion coefficient that is 3 times higher than that of silicon. Annealing the bonded pair at a temperature above $200°C$ was necessary in order to reduce the excessive residual thermal stresses. Peeling stress, which is primarily responsible for the debonding of the pair, could also be avoided if the thickness ratio expressed by Equation 11.4 is followed. This equation is useful in determining the thickness of the bonding wafers:

$$\frac{t_1}{t_2} = \sqrt{\frac{E_2(1 - v_1^2)}{E_1(1 - v_2^2)}} \tag{11.4}$$

where t_1 and t_2 are respective thicknesses of the mating substrates, E_1 and E_2 are the Young's moduli, and v_1 and v_2 are the Poisson's ratios of the mating materials.

11.15 WIRE BONDING

Wire bonding provides electrical connection to or from the core elements, such as the silicon diaphragm in the pressure sensor illustrated in Figure 11.10 or the actuating members in the microaccelerometer in Figure 11.11. Common wire materials are gold and aluminum. Other wire materials include copper, silver, and palladium. Wire sizes are on the order of 20–80 μm in diameter. In the IC industry wire bonding techniques are used for microsystems. Three wire bonding techniques are commonly used by the industry: (1) thermocompression, (2) wedge-wedge ultrasonic wire bonding, and (3) thermosonic.

The ultrasonic bonding method is used for very fragile structures. The working principles of these three techniques are outlined in the subsequent paragraphs. Detailed description of the equipment and the procedures used for wire bonding can be found elsewhere (Pecht et al., 1995).

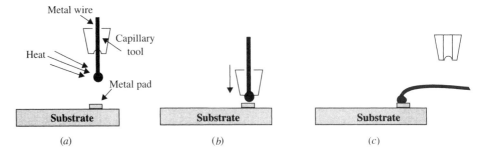

Figure 11.20. Wire bonding by thermocompression: (*a*) heating; (*b*) pressing; (*c*) after bonding.

Thermocompression Wire Bonding. The principle of thermocompression wire bonding is to press heated metal balls onto metal pads, as illustrated in Figure 11.20. In Figure 11.20*a*, the metal wire is fed through a capillary bonding tool with the tip of the wire above the metal pad. A torch heats the wire tip to about 400°C, at which temperature the wire tip turns into the shape of a ball. At this point, the tool, which has a semispherical mold shape at the end, is lowered to press the ball onto the pad (Figure 11.20*b*). After typically 40 ms of pressing, the tool is retracted and the wire is bonded to the pad, as illustrated in Figure 11.20*c*. The solid bonding of the wire ball to the flat metal pad is accomplished by two physical actions: the plastic deformation of the ball onto the pad by the tool and the atomic interdiffusion of the two bonded materials.

Wedge–Wedge Ultrasonic Bonding. Unlike thermocompression bonding, ultrasonic wire bonding is a low-temperature process. The bonding tool is in the shape of a wedge, through which the wire is fed. The energy supplied for the bonding is ultrasonic wave that is generated by a transducer that vibrates the tool at a frequency of 20–60 kHz. The bonding tool travels in parallel to the bonding pad, as illustrated in Figure 11.21*a*. After moving over the top of the desired bonding pad, the wedge tool is lowered to the pad's surface. A compressive force is applied and a burst of ultrasonic energy is

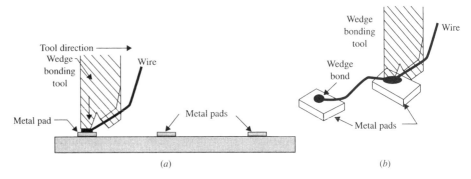

Figure 11.21. Wedge–Wedge wire bonding: (*a*) during bonding; (*b*) after bonding.

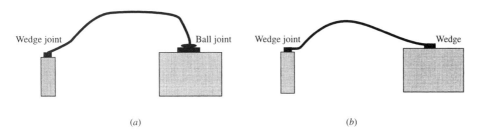

Figure 11.22. Wire bonding for microsystems: (*a*) wedge–ball wire bond; (*b*) wedge–wedge wire bond.

released to break down the contacting surfaces and achieve the desired surface bonding (Figure 11.21*b*). The typical time required for bonding is about 20 ms.

Thermosonic Bonding. This technique combines the previous two wire bonding techniques. Ultrasonic energy is used with thermocompression at a moderate temperature of 100–150°C. A capillary tool is used to provide ball joints.

For MEMS application, wire bonding is often combined with ball–wedge joints or wedge–wedge joints, as illustrated in Figure 11.22.

11.16 SEALING AND ENCAPSULATION

Hermetic sealing of assembled microcomponents is both necessary and desirable in microdevices. It is also essential that the packaged microdevices are protected by proper encapsulation from electromechanical and environmental influences from external sources.

Sealing for MEMS and microsystems is much more complicated than that of the ICs because of their complex three-dimensional structure and the many different requirements in sealing. For example, low process temperature is required for microsystems that involve chemical or biological substances, good optical paths must be provided for optoMEMS, and vacuum sealing is necessary for some mechanical devices such as comb-driven resonators. Another factor that makes MEMS microsystems sealing difficult is the interfacing of the key components with working media as described in Section 11.11. Devices, whether they are microsensors or accelerometers with damping, need to be interfaced with the measurand media. Many other types of microdevices, such as microchannels, valves, and pumps, in microfluidic systems need to be in contact with working fluid. All these devices and systems require that the contacting fluid be hermetically sealed off from the package, for the purpose of either protecting delicate core components such as sensing elements, check valves, or actuating beams, or ensuring the proper functioning of the device. Sealing is thus a major challenge to design engineers.

Figure 11.23 illustrates situations in which sealing of fluid is necessary. Hermetic seals for microchannels are indicated in Figure 10.19.

Mechanical sealing of the interfaces of mating components by epoxy resins is common in microchannels in microfluidic systems, as illustrated in Figure 10.19. This method of sealing is usually adequate for systems that do not operate at elevated temperature

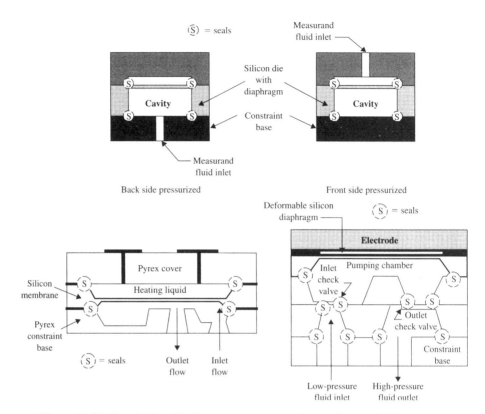

Figure 11.23. Required sealing in pressure sensors, micro valves, and micro pumps.

and/or with toxic working media, and they are not expected to last for a long period of time. In microvalves and pumps, micro-O-rings in conjunction with tongue-and-groove contacting surfaces are an effective sealing method. Reliable sealing of many other types of microsystem components requires the application of complicated physical–chemical processes.

Researchers are making great efforts to develop new processes for sealing micropackages. Many of the sealing techniques reported tend to be too complicated and specific with regard to materials and applications. A comprehensive description of various sealing and encapsulation techniques can be found in Chiao and Lin (2004). Here we will only present techniques that appear to be more generic in nature.

11.16.1   Integrated Encapsulation Processes

It is customary to seal and encapsulate MEMS and microsystems using microfabrication processes as described in Chapter 8. This practice has the advantage of sealing micro-components in situ at the wafer level prior to wafer dicing. Following are two examples of such sealing methods.

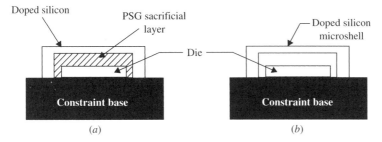

Figure 11.24. Sealing by microshell: (*a*) with sacrificial layer; (*b*) with sacrificial layer etched away.

Sealing by Microshells. Microshells are produced to protect the delicate sensing or actuating elements in microdevices. Figure 11.24 illustrates how these shells are produced.

A surface micromachining technique is used to produce microshells. The procedure involves depositing a sacrificial layer over the die to be protected, as shown in Figure 11.24*a*. A shell material is then deposited over the sacrificial layer. An etching process then follows to remove the sacrificial layer. Consequently, a gap space between the die and the microshell is created, as shown in Figure 11.24*b*. These gaps can be as small as 100 nm.

Reactive Sealing Technique. This technique relies on specific chemical reactions to produce the necessary sealing of the mating components. As illustrated in Figure 11.25, the encapsulant cover for the die is initially placed on the top of the die/constraint base with small gaps (Figure 11.25*a*). The unit is subject to a chemical reaction such as the thermal oxidation process described in Section 8.5. The growth of SiO_2 from both ends of the silicon encapsulant and the constraint base can provide a reliable and effective seal for the encapsulated die.

11.16.2 Sealing by Wafer Bonding

This technique allows the sealing of core elements by an encapsulating cover without an intermediate layer as in the reactive sealing illustrated in Figure 11.25. Surface bonding

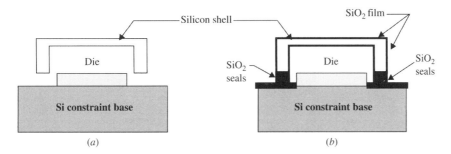

Figure 11.25. Reactive sealing: (*a*) unsealed encapsulant; (*b*) encapsulation by oxide seals.

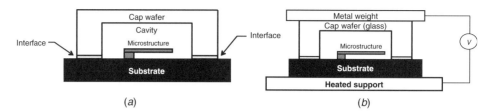

Figure 11.26. Sealing by wafer bonding: (*a*) by adhesive and SFB; (*b*) by anodic bonding.

techniques, described in Section 11.14, are used to seal the interfaces. Figure 11.26 illustrates the application of this technique to sealing of microdevices.

Two wafers, one with a microstructure attached on its surface (the substrate) and the other (the cap wafer) with a cavity, are bonded at the interfaces using surface bonding techniques such as adhesive or silicon fusion bonding in Figure 11.26*a*, or by anodic bonding in Figure 11.26*b*.

11.16.3 Vacuum Sealing and Encapsulation

There are ample cases in which the sealed space in MEMS and microsystems is a vacuum. The performance of such devices as the microaccelerometers in Section 2.5.1 and gyroscopes in Section 2.5.2, involving resonant vibration of comb drives, relates to the damping property of the air between the vibrating electrodes. Operating these devices in a vacuum can significantly enhance their performance. Vacuum sealing is also desired for some micro-optical switches in which the presence of air can result in unwanted resistance to the moving components.

Vacuum sealing is a challenge to many process design engineers. There are two techniques reported by Chiao and Lin (2004) for vacuum sealing of microsystems: (1) sealing by rapid thermal processing (RTP) and (2) sealing by localized CVD.

Sealing by RTP Bonding. This bonding technique was developed for the vacuum sealing of a microdevice involving aluminum-to-nitride bonding using (RTP) (Chiao and Lin, 2004). The RTP is a popular method in microelectronics bonding. With respect to the schematic in Figure 11.27, both the device and cap wafers are prebaked in vacuum at 300°C for 4 h in a vacuum quartz tube. The prebaking is necessary in order to drive out

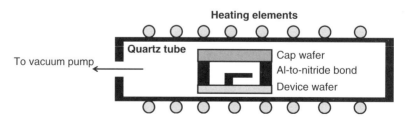

Figure 11.27. Vacuum sealing with RTP bonding.

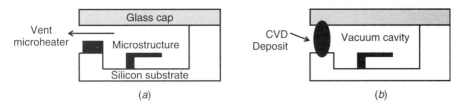

Figure 11.28. Vacuum sealing by localized CVD: (*a*) prevacuum sealing; (*b*) after vacuum sealing.

the gases inherent in the materials from microfabrication. The two wafers are assembled and loaded into a sample holder and placed in the vacuum heating tube again. The set is then placed inside the RTP equipment and the base pressure is pumped down to about 1 mTorr. The vacuum is held steady for about 4 h to drive out entrapped gas inside the cavity. The sealing is completed by RTP heating in 10 s at 750°C.

Sealing by Localized CVD Process. This method was first reported by Chang-Chien and Wise (2001). As illustrated in Figure 11.28, a cavity is produced in a silicon substrate by an etching process. The microstructure is then assembled and bonded to the substrate. The silicon substrate is bonded to a glass cap wafer by anodic bonding. A vent hole is left between the glass and silicon substrate, as shown in Figure 11.28*a*. At the vent hole is also a microheater made of polysilicon. The set is put into a vacuum chamber at about 250 mTorr with the flow of silane gas. The intense heat released by the microheater decomposes silane for localized polysilicon deposition to seal the venting hole, as shown in Figure 11.28*b*. The CVD process that seals the vent hole is described in Equation (8.16).

11.17 THREE-DIMENSIONAL PACKAGING

Three-dimensional packaging is a relatively mature technology in the microelectronics industry. It involves the stacking of ICs and multichip modules (MCMs) in compact configurations. Following are a few desired features in the three-dimensional packaging approach (Lyke and Forman, 1999). Some of these features can also be applied to the three-dimensional packaging of MEMS and microsystems:

1. High volumetric efficiency
2. High-capacity layer-to-layer signal transport
3. Ability to accommodate wide range of layer types
4. Ability to isolate and access a fundamental stackable element for repair, maintenance, or upgrading
5. Ability to accommodate multiple modalities, for example, analog, digital, RF power
6. Adequate heat removal among the package layers
7. High pin-count delivery to the next level of packaging with high electrical efficiency

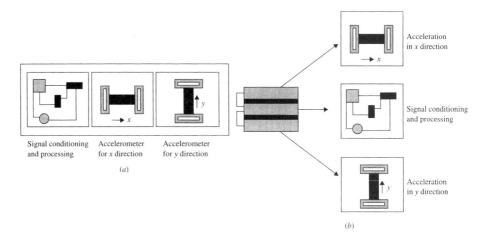

Figure 11.29. Conceptual 3 dimensional packaging: (*a*) planar (two-dimensional) packaging; (*b*) three-dimensional packaging.

The term *three-dimensional microsystems packaging* may appear ambiguous to many engineers, as we have long recognized the fact that MEMS and most microsystems are three-dimensional in geometry, as indicated in Table 1.1. *Three-dimensional packaging for these products, in a real sense, means packaging of microdevices and systems with distinct functions stacked up with signal processing units in compact configurations*.

The packaging of MEMS or microsystems together with signal processing has been practiced by industry under the term *lab-on-a-chip*. An example is the packaged microaccelerometers shown in Figure 1.6. A three-dimensional packaged microsystem would package the system in more compact forms, for example, by placing the two force-balanced accelerometers in *x*- and *y*-axis orientations, one on top of the other. The stacked accelerometers can also be placed on the top of the signal processing chip. This concept is illustrated in Figure 11.29. Most of the microfluidic devices using capillary electrophoresis for biological analyses are built on lab-on-a-chip structures.

Stacking layers of MEMS and microsystems of various functions and the signal processing unit requires the application of many of the essential packaging technologies described in Sections 11.12–11.16. However, the shielding of electromagnetic and thermal effects and the hermeticity of moving fluids in many of these systems become even more critical issues in such packaging. Electromechanical leads must be properly designed, and the interfaces must be properly sealed.

11.18 SELECTION OF PACKAGING MATERIALS

Section 7.11 presented an overview of packaging materials. We will take a closer look at these materials in this section. Properties of many of these materials may be used for design purposes. Table 11.6 presents a summary of materials that can be used for packaging various parts of a microsystem. Table 11.7 lists properties of some of the frequently used materials for die packaging.

TABLE 11.6. Common Materials for Microsystems Packaging

Microsystem Components	Available Materials	Remarks
Die	Silicon, polycrystalline silicon, GaAs, ceramics, quartz, polymers	Refer to Chapter 7 and Section 10.2.2 for selection
Insulators	SiO_2, Si_3N_4, quartz, polymers	Quality and cost are in ascending order of listed materials
Constraint base	Glass (Pyrex), quartz, alumina, silicon carbide	Pyrex and alumina are more commonly used materials
Die bonding	Solder alloys, epoxy resins, silicone rubber	Solder for better seal, silicone rubber for better die isolation
Wire bonds	Gold, silver, copper, aluminum, tungsten	Gold and aluminum are popular choices
Interconnect pins	Copper, aluminum	
Headers and casings	Plastic, aluminum, stainless steel	

Design engineers should be aware of the strong temperature-dependent properties of die bonding materials when performing design analyses. Table 11.8 and 11.9 provide temperature-dependent material properties of the solder and epoxy resins listed in Table 11.6. Interpolations and extrapolations can be used to determine properties within and beyond given temperature and strain ranges.

Measured stress-versus-strain relations for silicone rubber (RTV) show significant scattering. The mean value of Young's modulus for this material is about 1 Pa at all temperatures.

TABLE 11.7. Summary of Die Packaging Material Properties

Materials	Young's Modulus (MPa)	Poisson's Ratio	Thermal Expansion Coefficient (ppm/K)
Silicon	190,000	0.29	2.33
Alumina	344,830–408,990 (20°C) 344,830–395,010 (500°C)	0.27	6.0–7.0 (25–300°C)
Solder (60Sn40Pb)	31,000	0.44	26
Epoxy (Ablebond 789-3)	4,100		63 below 126°C, 140 above 126°C
Silicone rubber (Dow Corning 730)	1.2	0.49	370

Source: Schulze, 1998.

TABLE 11.8. Temperature-Dependent Properties of 60Sn40Pb Solder Alloy

Strain range (10^{-6})	Young's Modulus (MPa)	Poisson's Ratio	Yield Strength (MPa)
0–500	At −40°C: 46,100	0.32	60
	At 25°C: 27,700	0.43	38
	At 125°C: 17,000	0.43	14
500–1500	At −40°C: 27,800	Same as above	Same as above
	At 25°C : 16,200		
	At 125°C: 4670		
1500–3000	At −40°C: 5600	Same as above	Same as above
	At 25°C: 5290		
	At 125°C: 1140		
3000–10,000	At −40°C: 1490	Same as above	Same as above
	At 25°C: 700		
	At 125°C: 210		

TABLE 11.9. Temperature-Dependent Properties of Epoxy Resin (Ablebond 789-3)

Strain Range (10^{-6})	Young's Modulus (MPa)	Poisson's Ratio	Fracture Strength (MPa)
0–500	At −40°C: 7990	0.42	55
	At 25°C: 5930	0.42	60
	At 125°C: 200	0.49	1.5
500–2000	At −40°C: 4680	Same as above	Same as above
	At 25°C: 4360		
	At 125°C: 110		
2000–10,000	At −40°C: 3830	Same as above	Same as above
	At 25°C: 3620		
	At 125°C: 60		
10,000–20,000	At −40°C: 3610	Same as above	Same as above
	At 25°C: 2650		
	At 125°C: 40		
20,000–30,000	At 25°C: 1790		
	At 125°C: 30		

Source: Schulze, 1998.

11.19 SIGNAL MAPPING AND TRANSDUCTION

11.19.1 Typical Electrical Signals in Microsystems

Signal mapping relates to developing and establishing strategies in selecting both the types and positions of transducers for the microsystem. It is one of the most critical issues in packaging. As we have seen from Section 11.10.2, signal transduction is a major part in the device-level packaging of a microsystem. Electrical or mechanical signals generated from the die of a sensor or an actuator must be converted into a desired form that can be readily measured and processed. Various transducers have been used for that purpose, as illustrated in Figure 10.2. The electrical signals take the forms shown in Table 11.10.

Voltmeters and ammeters can be used to measure electrical voltage and current, as listed in Table 11.10. We will present ways for measuring resistance and capacitance with electric bridges.

11.19.2 Measurement of Resistance

A popular way to measure the change of electrical resistance from a strain gage or piezoresistor is to use a Wheatstone bridge. A typical bridge circuit is shown in Figure 11.30. The bridge consists of four resistors connected as shown in the figure. A constant voltage supply V_{in} is applied to the circuit. In this bridge circuit, a strain gage or piezoresistor whose resistance is to be measured after being deformed by a strain or stress replaces the resistor R_1. Let R_g be the resistance of that strain gage or piezoresistor. There are two measurement modes for the variable resistor R_g: the static balance mode and the dynamic deflection operation mode (Histand and Alciatore, 1999):

- **Static Balanced Mode:** The voltage V_o between terminal a and b is adjusted to zero to achieve a balanced bridge circuit. The following relation can be obtained in such a balanced condition:

$$\frac{R_g}{R_4} = \frac{R_3}{R_2} \tag{11.5}$$

TABLE 11.10. Common Transducers for Microsystems

Transducers	Electric Signals	Input or Output	Typical Applications
Piezoresistors	Resistance R	Output	pressure sensors
Piezoelectric	Voltage V	Input or output	Actuators, accelerometers
Capacitors	Capacitance C	Input or output	Actuators by electrostatic forces, pressure sensors
Electro-resistant heating/shape memory alloys	Current i	Input	Actuators

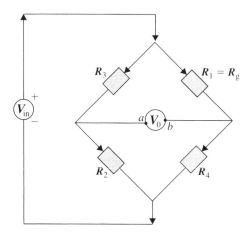

Figure 11.30. Typical Wheatstone bridge.

from which we can evaluate the resistance R_g as

$$R_g = \frac{R_3 R_4}{R_2} \tag{11.6}$$

- **Dynamic Deflection Operation Mode:** When the resistance in the strain gage or piezoresistor undergoes continuous variation due to changing loads, the dynamic deflection mode is used. In such a case, the output voltage V_o measured in a continuous mode can be expressed as

$$V_o = V_{in} \left(\frac{R_g}{R_g + R_4} - \frac{R_3}{R_2 + R_3} \right) \tag{11.7}$$

Thus, the change of the resistance in the resistor R_g can be determined by the formula

$$\frac{\Delta R_g}{R_1} = \frac{\frac{R_4}{R_1} \left(\frac{\Delta V_o}{V_{in}} + \frac{R_3}{R_2 + R_3} \right)}{1 - \frac{\Delta V_o}{V_{in}} - \frac{R_3}{R_2 + R_3}} - 1 \tag{11.8}$$

where R_1 is the original resistance of R_g.

11.19.3 Signal Mapping and Transduction in Pressure Sensors

Piezoresistors are used to convert the mechanical stresses induced in the diaphragm of a micropressure sensor to the corresponding change of electric resistance. There are a number of ways that these piezoresistors can be installed beneath the surface of the diaphragm.

Figure 11.31a shows four piezoresistors with the same resistivity placed in the midspan of the four edges of a square diaphragm. This arrangement is the most popular design. It has one resistor each on the right and left sides of the diaphragm, and they are oriented in such a way that they are subjected to tensile stress. The other two piezoresistors are placed at the top and bottom edges of the diaphragm. They are subjected to compressive stress due to Poisson's effect. This arrangement provides the most optimum gain in resistance measurements. Figure 11.31b shows a rectangular diaphragm with four resistors. All resistors are oriented in the same direction. Two resistors are placed parallel to and near the edge of the diaphragm, whereas the other two are placed near the center. This arrangement can increase the sensitivity of the resistors and allows the use of a single center boss or double bosses to concentrate the stresses and minimize the nonlinearity of the output. The arrangement in Figure 11.31c is for the measurement of shearing stress. However, due to the nonsymmetry of the piezoresistor, this arrangement is more vulnerable to thermal compensation than the other arrangements.

In pressure sensors with a square diaphragm containing four piezoresistors placed as shown in Figure 11.31a, proper wiring to the Wheatstone bridge is as shown in Figure 11.32b, with the piezoresistor designations shown in Figure 11.32a.

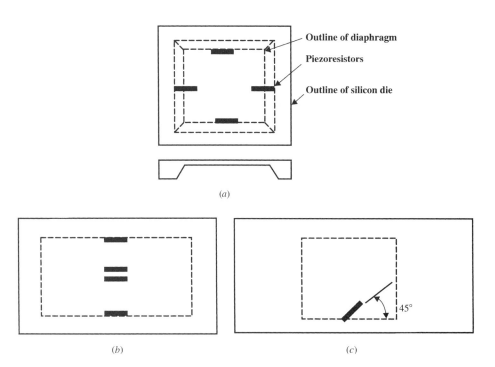

Figure 11.31. Arrangements of piezoresistors in pressure sensors: (a) square die/square diaphragm; (b) rectangular die/diaphragm; (c) for shear deformation measurements.

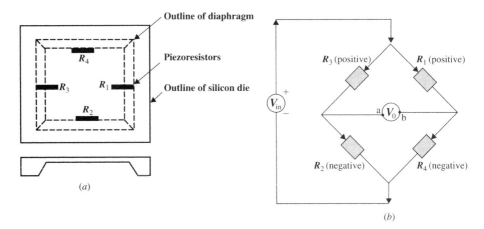

Figure 11.32. Electric bridge for pressure sensors with square diaphragms: (*a*) piezoresistors on square diaphragm; (*b*) corresponding Wheatstone bridge.

Correct measurements of the resistance change in piezoresistors (due to the application of pressure on the diaphragm) by the Wheatstone bridge circuitry require proper conditioning. Two situations warrant special attention:

1. Offset balancing and temperature compensation for the resistivity changes in piezoresistors due to a change in the temperature coefficient of resistance (TCR) and temperature coefficient of the piezoresistor (TCP) as described in Section 7.6
2. Sensitivity normalization, linearization, and temperature compensation due to thermal expansion of piezoresistors at elevated temperature

The offset voltage results from a mismatch of the Wheatstone bridge circuit or by possible residual stresses in either the diaphragm or the piezoresistors from a previous pressure loading cycle. The bridge circuit is unbalanced and the output is nonzero without any pressure applied to the diaphragm. Typically, a 1% full-scale output makes additional circuitry, either on or off the diaphragm, necessary, as shown in Figure 11.33 (Pourahmadi and Twerdok, 1990).

In Figure 11.33*a*, by proper arrangement of resistors $R_1, R_2, \ldots, R_5$ in the circuitry, one can minimize the offset error within the required range. Alternately, one can build the compensation in the piezoresistors embedded in the silicon diaphragm. In this case, the circuitry is much simpler, as can be seen in Figure 11.32*b*.

11.19.4   Capacitance Measurements

Capacitors are common transducers as well as actuators used in microsystems. The use of capacitance transducers in micropressure sensors is illustrated in Figures 2.10 and 2.12. The capacitance variation in a capacitor can be measured by simple circuits such as illustrated in Figure 11.34 (Bradley et al., 1991).

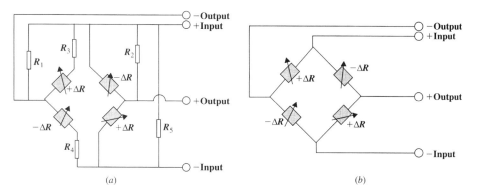

Figure 11.33. Typical circuitry for piezoresistors in pressure sensors: (*a*) standard circuitry; (*b*) circuitry with compensation. (From Pourahmadi and Twerdok, 1990.)

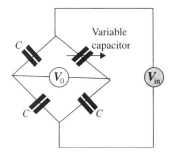

Figure 11.34. Typical bridge for capacitance measurements (From Bradley et al., 1991.)

The electrical bridge in Figure 11.34 is similar to that of Wheatstone bridge for resistance measurements. The variable capacitance can be determined by measuring the output voltage V_o. The following relationship can be used to determine the variable capacitance in the circuit:

$$\Delta C = \frac{4C v_o}{V_{\text{in}} - 2V_o} \tag{11.9}$$

where ΔC is the capacitance change in the microsystem and C is the capacitance in the bridge. The bridge is subjected to a constant voltage supply designated by V_{in}.

11.20   DESIGN CASE ON PRESSURE SENSOR PACKAGING

The following packaging design case for a micropressure sensor is presented to illustrate the major steps involved.

Die and Substrate Preparations. This level of packaging begins with the configuration of silicon dies and the production of these dies from wafers. The design engineer must bear in mind the constraints imposed by using the standard wafer sizes and dimensions as presented in Section 7.4.2. The thickness of silicon substrates or dies is limited by the wafer size chosen. A careful assessment of the maximum number of dies or substrates that one can cut from a standard wafer is thus necessary. Wafer dicing is described in Section 11.13. Many companies have their own established procedures for cutting thin wafers into small dies and substrates.

Primary Packaging Consideration. Dies are the most critical component for proper functioning of a MEMS device or microsystem. As illustrated in Chapter 2, dies are the components that generate or receive signals. It is essential that these dies are packaged in such a way that the surroundings and the environmental conditions will not affect their intended mechanical or electrical performance. *Die isolation* is thus the primary consideration in packaging a micropressure sensor and other similar MEMS devices or microsystems.

Die isolation often becomes a critical part in packaging design. Let us look at a typical micropressure sensor die, as illustrated in Figure 11.35. The silicon die of a pressure sensor needs to be attached to the Pyrex glass constraint base, as shown in Figure 11.35*a*. We realize that the CTE of the die, the die attach, and the base, as shown in Figure 11.35*b*, can be significantly different in magnitudes. The differences of the CTEs of the individual materials will result in substantial thermal stresses in the die, as indicated in Section 4.4.3. These undesirable stresses are termed *parasite stresses* because the thermal environment induces them in the diaphragm but they do not represent the applied pressure. These parasite stresses can make the output signals from the transducers inaccurate and introduce significant error in the measured results.

Die isolation may be accomplished by proper design of die configuration and also the die attach. The height *H* in Figure 11.35*a* is an important parameter in isolating the die from the mechanical and thermal effects from the base. A larger value of *H* will provide more flexibility for the diaphragm. However, the dimension *H* in a pressure sensor die is constrained by the standard thickness of silicon wafers, from which the cavity is created. A spacer, such as shown in Figure 11.36, can provide additional flexibility of the die support in order to improve die isolation. Unfortunately, this extra component increases the cost in packaging as well as the overall dimension of the die.

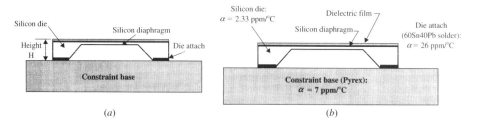

Figure 11.35. Schemes for die isolation: (*a*) typical geometry; (*b*) CTE of die attachment.

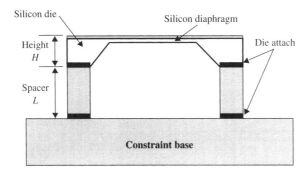

Figure 11.36. Spacer for die isolation.

Die Down. This procedure relates to the bonding of the die to the constraint base. The latter can be glasses or ceramics or metals. Die bonding requires the use of proper die attach, as illustrated on Figure 11.10. There are generally three ways to attach the die to the constraint bases: (1) anodic bonding, (2) eutectic soldering, and (3) adhesion, as described in Section 11.14. Each of these die bonding methods has its advantages and disadvantages, as described in Section 11.14.

Die Protection. The need for protecting the pressure sensor die from the contacting pressurized medium was described in Section 11.11. Since the dies in micropressure sensors are most likely to be in the situation described above, we may consider three common ways of protecting the dies in this type of sensor (Bryzek et al., 1991; Schulze, 1998):

1. **Vapor-Deposited Organic:** A material called Parylene is used for this purpose. This material adds a uniform layer of passivation over the die surface by a LPCVD process but at near room temperature. Annealing takes place after the deposition. The coating is usually thinner than $3.5\,\mu m$ for sensors operating up to $105°C$. The coating has a negative effect on the performance of the device, as it stiffens the diaphragm. Thus, the effect of the thickness of the coating on the diaphragm flexibility is a limiting factor in this practice. However, the technique is useful in protecting the die and the wire bonds of the sensor.

2. **Coating with Silicone Gel:** A common practice of protecting the die is to apply a coating silicone gel over the die surface, such as shown in Figures 2.12 and 11.37. The gel consists of one or two parts of siloxanes. After the gel is dispensed as a thick film in the millimeter range over the die surface, it is cured. This die protection technique works well for many pressure sensors. The very low modulus of elasticity of the gel minimizes the effect of the coating on the flexibility of the diaphragm, yet it protects the die from direct contact with the pressurized medium.

3. **Indirect Pressure Transmission:** There are times when micropressure sensors are designed to function in severe environments. The pressurizing media can be highly corrosive or aggressive with strong chemicals. Complete isolation of the die from the pressure medium becomes necessary. For example, micropressure

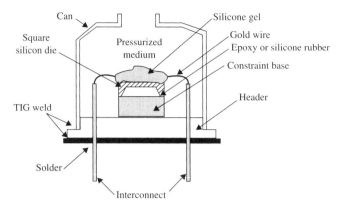

Figure 11.37. Die protection with silicone gel.

sensors have been produced to measure the air blast pressure in simulated nuclear bomb explosions. The measured pressure is in the range of 70 kPa–350 MPa with an impact force equivalent to $10g$ to $20,000g$ (g is gravitational acceleration). In addition to these extreme mechanical conditions, the measurements needed to be made in harsh environments with high-velocity dust particles, high intensity of light, and, above all, a temperature rise to $5000°F$ in a few milliseconds. In such cases, robust packaging is the only way to ensure the survival and proper functioning of these sensors (Bryzek et al., 1991). Lucas NovaSensor in California produced an unusual pressure sensor packaging for such an extreme environment. It involved the use of an oil-filled stainless steel diaphragm, as shown in Figure 11.38.

As observed from the arrangement in Figure 11.38, the die and the wire bonds are submerged in silicone oil. The medium pressure is applied to the thin stainless steel diaphragm instead, which in turn transmits the applied pressure to the silicon diaphragm

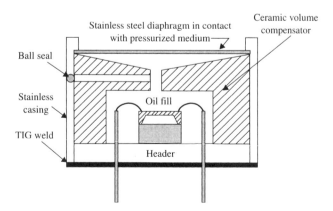

Figure 11.38. Oil-filled stainless diaphragm for die protection.

through the incompressible silicone oil. The stainless steel diaphragm is designed in such a way that its compliance is at least 100 times less than that of the silicon diaphragm, so that the thermal expansion of the oil inside the package has a minor effect on the mechanical response of the die.

The thermal expansion of the oil inside the package obviously can generate undesirable thermal stress on the silicon diaphragm and cause error in the measurements. For this reason, the volume of oil needs to be kept at a minimum. Special volume compensators made of low-thermal-expansion materials (e.g., ceramics) are inserted into the package to fill the space between the die and the stainless steel diaphragm.

The oil used to transmit the pressure has to be baked for outgassing before it is injected into the package. This type of pressure sensor is used to measure pressures up to 350 MPa. However, the cost of the packaging is very high because of the stringent requirements and the low demand for these special sensors.

Device-Level Packaging. Once the critical die and wire bonds are properly secured in the constraint bases, they are assembled to a passivation with plastics (see the package at the right of Figure 1.2), ceramics (the package at the middle of Figure 1.2), or metallic casings (the package at the left of Figure 1.2). Major tasks involved in this level of packaging are the design and packaging of signal transduction and the interconnection of the die and the required user connectors. Plastic encapsulation of the micropressure sensor is common for those operating in a moderate temperature environment. On the other hand, metal encapsulation is used for high-temperature environments. Injection molding is used for plastic-encapsulated packages as in IC manufacturing. The fabrication techniques used for metal packaging involve soldering and welding. There is no standard practice established for microdevice packaging at this time. Virtually every microsystem requires special design in the final product packaging.

System-Level Packaging. In many cases, the devices are shipped as shown in Figure 1.2 with internal arrangements such as illustrated in Figure 11.10. The customer normally includes the sensor and the conditioning electronics in a new package. Although this is a common practice, it is not always economical, as the microsystems package is usually built around existing engineering systems. A better approach is to include the primary signal conditioning circuitry into the package of the sensor. In such cases, the package has to provide the necessary mechanical and thermal insulation as well as electromagnetic shielding of the circuitry, such as shown in Figures 1.22a and 1.22b. The accuracy of the measurements will be compromised if this protection is not sufficient. A metal housing usually gives adequate protection for mechanical and electrical influences. However, plastic encapsulation of the sensor and the conditioning circuitry is better suited for mass production purposes. In this case, thin metal layers for the necessary electromagnetic shielding are usually required.

11.21 RELIABILITY IN MEMS PACKAGING

Cost-effective packaging and robust reliability are two key factors for successful commercialization of any MEMS and microsystem. The critical issue of reliability is that,

TABLE 11.11. Failure Mechanisms in MEMS and Microsystems

Failure mode	Causes	Probability
Mechanical	Local stress concentration due to surface roughness	Low
	Improper assembly tolerances	Moderate
	Vibration-induced high cycle fatigue failure	Low in silicon, moderate in plastic
	Delamination of thin layers	Moderate to high
	Thermal stresses by mismatch of CTE	High
Electromechanical breakdown	Collapse of electrodes due to excessive deformation	High
Deterioration of materials	Aging and degassing of plastic and polymers; corrosion and erosion of materials	Moderate
Excessive intrinsic stresses	Residual stresses and molecular forces inherent in microfabrication	High
Packaging	Improper bonding and sealing, poor die protection and isolation	High
Environmental effects	Temperature, humidity, dusts, and toxic gas	High

Source: Hsu, 2006.

no matter how sophisticated a product is designed and manufactured, it becomes useless if it fails to deliver the designed performance during the expected lifetime. Reliability of MEMS and microsystems is particularly critical as failure of many of these products can be catastrophic and devastating. Engineers cannot design reliable MEMS and microsystems without first understanding the many possible mechanisms that can cause the structure to fail and lead to malfunctioning of these products.

MEMS products are designed to perform a variety of electromechanical, chemical, optical, biological, and thermohydraulic functions. Mechanisms that cause failure of microdevices thus vary significantly from one type to another. Table 11.11 presents a few of these failure mechanisms for MEMS and microsystems.

11.22 TESTING FOR RELIABILITY

Reliability testing is developed to detect the failure mechanisms and the usable life of microdevices. Many of the failure mechanisms for microdevices listed in Table 11.11 cannot be accurately predicted in the design stage. Credible testing during fabrication, assembly, and packaging is needed to ensure the reliability of the devices. Unlike reliability testing for ICs with well-established procedures and criteria, reliability testing for microsystems is far from standardized. This is primarily because microsystems involve combined electromechanical, optical, chemical, and biological functions and many of these devices involve complex input/output in the forms of forces, inertia, displacements, pressure, fluid flows, and sources of chemical/optical/biological natures. Virtually all reliability testing procedures and criteria need to be developed and established for every new product by the industry. What we will present in this section are the essential elements that need to be considered in such development (Oliver and Custer, 2004). Some of these items are not technical. They are included in this section because they are related to the quality and reliability of the product.

Range of Acceptable Device Performance. Engineers must first set the pass/fail limits for the test results. A wide limit may reduce the needless rejection of microcomponents during or after fabrication but with the possibility of inferior quality. A narrow limit, on the other hand, would result in unnecessary waste. However, the limit should be just narrow enough so that the user can predict and expect the proper performance of the device.

Parametric Testing. Instead of waiting until all components of a product are fabricated, assembled, and packaged and then performing the reliability testing, a better and more economic way is to inspect key components during and after the fabrication. Consequently, the engineer needs to define the parameter(s) that properly describe the function of these components, from which one can design and implement the testing accordingly. Specific *parametric test structures* are developed for this purpose. These structures are attached to the device components during the assembly and packaging processes. Description of some of these special parametric test structures is available in Oliver and Custer (2004). Satisfactory parametric testing relates closely to the proper selection of *test points* on the work piece, which is another important part of the task of design for test.

Testing during Assembly. The purpose of assembly testing is to ensure only components free of defect can proceed for the assembly and packaging. Both visual and machine inspections on assembled components are performed. In many cases, performance of the partially assembled components is also tested with carefully designed instruments. Figure 11.1 illustrates the testing and inspection involved in the process of assembling and packaging a pressure sensor die.

Burn-in and Final Testing. Burn-in testing is a common practice in IC manufacturing for ensuring the reliability of the product. In MEMS and microsystems these tests usually involve loading to the device in the form of mechanical vibration, electric, and flow, such as listed in Table 11.11. In some cases, accelerated loading in thermal cycling, thermal shock, and other expected environmental conditions such as high humidity is applied to the finished devices before they are shipped to the customers.

Self-Testing. Many microdevices are designed to perform self-testing on a regular basis after they are shipped to the customers. A device equipped with self-testing capability can also ensure the reliability of the device in service. These tests are primarily to ensure the proper functionality of the device. Self-testing involves the use of electrical stimuli that mimic the real input loads.

Testing during Use. These tests are necessary to calibrate the functions of the devices, such as sensors.

It is apparent that much effort is needed in the design and development of strategies for conducting many of these tests on microsystems at various stages of production. The techniques of instrumentation and the tools and fixtures required for conducting these delicate tests, with many in situ tests, in no doubt need to be developed. One cannot overlook the importance of testing for reliability and the cost associated with this major undertaking in the micromanufacturing process.

PROBLEMS

Part 1 Multiple Choice

1. AP&T is the abbreviation of (a) applied physics and technology, (b) assembly planning technology, (c) assembly, packaging, and testing of MEMS and microsystems.

2. The single most costly stage of microsystems production is (a) design, (b) microfabrication, (c) assembly, packaging, and testing.

3. One major problem in microsystems assembly is lack of (a) standards, (b) practices, (c) funds.

4. One major problem in microassembly is lack of (a) a strategy on parts and gripper tolerances, (b) interest, (c) demand.

5. One serious nontechnical problem in microassembly is loss of (a) global perspectives of assembly area, (b) human–machine interaction, (c) human–work piece interaction.

6. The higher the magnification power of the microscope, (a) the shorter, (b) the longer, (c) no change in the working distance under the objective lens.

7. The higher the magnification power of the microscope, (a) the wider, (b) the narrower, (c) no change in the field of view.

8. One problem in the design of microtweezers or microgripper arms is the (a) high aspect ratio, (b) low aspect ratio, (c) odd shape of the structure geometry.

9. Tolerances involved in microassembly are often a few (a) micrometers, (b) millimeters, (c) centimeters.

10. Geometry tolerances are inherent from (a) microfabrication, (b) design setting, (c) customer specifications.

11. Dimensional tolerances of microcomponents are established by (a) microfabrication, (b) design setting, (c) customer specifications.

12. One principal requirement in gripping microcomponents is (a) adequate gripping forces, (b) accurate gripping position, (c) adequate releasing force.

13. Too high a gripping force would cause the (a) releasing of, (b) damaging of, (c) misplacing of the work piece.

14. Too light a gripping force would cause the (a) slipping of, (b) damaging of, (c) misplacing of the work piece.

15. The basic requirement for an intelligent end effector in microassembly is the ability to sense (a) the gripping force, (b) the temperature, (c) the humidity.

16. A serious problem in the pick-n-place operation in microassembly is (a) picking the right work piece, (b) accurately placing the work piece, (c) releasing the work piece.

17. The difficulty in releasing minute work pieces from a gripper is because of (a) the insignificant weight, (b) the insignificant mass, (c) the insignificant size of the work piece.

18. The adhesive force that causes the work piece from being released from a gripper is (a) electrostatic force, (b) van der Waals force, (c) the combination of the two forces.

19. Another component of adhesive forces induced by a humid environment is (a) hydrogen force, (b) surface tension, (c) viscous force of water.

20. Electrostatic forces are generated by (a) electric charges built up between two closely separated surfaces, (b) charges induced by an electric field, (c) charges introduced by a doping process.

21. Microassembly work cells are developed to perform (a) automatic assembly, (b) quasi-automated assembly, (c) fast assembly of microsystems.

22. Stereomicroscopes and cameras are used in microassembly work cells because of (a) better pictures of the microsystem, (b) better view of three-dimensional microstructures of the assembled work pieces, (c) better records of the assembly.

23. The primary function of a micropositioner is (a) proper positioning the work piece, (b) accurate positioning the work piece, (c) transporting the work pieces to assembly site.

24. Mirorobots are used in microassembly work cells to serve (a) micropositioner, (b) micromanipulator, (c) combined micropositioner and micromanipulator.

25. The top three priority research and development areas in microassembly are (a) sensing, intelligent end effector, and material and parts delivery, (b) sensing, part insertion, and process modeling, (c) testing method, material delivery, and intelligent tool design.

26. There are (a) two, (b) three, (c) four levels packaging in microsystems.

27. The softest die-attach (bonding) material is (a) solder alloy, (b) epoxy resin, (c) silicone rubber.

28. Silicone gel is used to (a) protect the die, (b) strengthen the die, (c) isolate the die.

29. A spacer is often used in the silicon die of pressure sensors for (a) extra strength, (b) extra size, (c) better die isolation.

30. Anodic bonding of silicon/glass substrate takes place under (a) high temperature, (b) high temperature and pressure, (c) high temperature and high electric voltage.

31. Anodic bonding is most commonly used in bonding (a) silicon to glass, (b) silicon to silicon, (c) silicon to aluminum.

32. The bonding of silicon to glass relies on the creation of (a) eutectic alloy layer, (b) silicon oxide layer, (c) oxide–hydrogen layer.

33. Eutectic bonding of silicon to silicon relies on the creation of (a) eutectic alloy layer, (b) silicon oxide layer, (c) oxide–hydrogen layer.

34. Silicon fusion bonding relies on the creation of (a) eutectic alloy layer, (b) silicon oxide layer, (c) oxide–hydrogen layer.

35. One of the reasons for applying a high-voltage electric field to the bonding pieces is to keep the bonding pieces (a) flat, (b) tight in contact, (c) ionized.

36. Silicon fusion bonding takes place under (a) high temperature, (b) high temperature and pressure, (c) high temperature and high electric voltage.

37. Silicon fusion bonding is used to bond (a) silicon to silicon, (b) silicon to glass, (c) silicon to quartz.

38. SOI stands for (a) splitting of ions, (b) silicon on insulator, (c) substrate on insulator.

39. The purpose of SOI is to prevent (a) leak of electric charge, (b) leak of thermal effect, (c) spreading of corrosion across the silicon substrates.

40. The SOI process normally takes place at (a) high temperature around $1000°C$, (b) medium temperature around $500°C$, (c) low temperature below $200°C$.

41. The "lift-off" process takes place at (a) high temperature around $1000°C$, (b) medium temperature around $500°C$, (c) low temperature below $200°C$.

42. The silicon fusion bonding process takes place at (a) high temperature around $1000°C$, (b) medium temperature around $500°C$, (c) low temperature below $200°C$.

43. The wire bonding technique that operates at room temperature is (a) thermocompression, (b) wedge–wedge ultrasonic, (c) thermosonic bonding.

44. Reliable hermetic sealing of microsystems can be accomplished by (a) mechanical means only, (b) electromechanical means, (c) physical–chemical processes.

45. Microshells can be created by (a) bulk micromanufacturing, (b) surface micromachining, (c) the LIGA process.

46. Packaging technologies for microelectronics and microsystems are (a) the same, (b) different, (c) interchangeable.

47. Packaging cost in microsystems is (a) trivial, (b) very significant, (c) somewhat significant in the cost of production.

48. A major challenge in microsystem packaging is to (a) protect the core elements from the contacting environment, (b) prevent these core elements from disintegration, (c) wire these elements correctly.

49. Package of bioMEMS must be (a) inert to biological attack of human systems, (b) inert to the body temperature, (c) inert to mishandling by the user.

50. Packaging optical MEMS requires (a) sensitivity to light, (b) adequate access to light beams, (c) reflection of light beams.

51. One serious packaging problem in microvalves is (a) speed, (b) sensitivity, (c) sealing of fluid.

52. Epoxy resins are (a) suitable, (b) not suitable, (c) not a matter of concern for high-temperature applications.

53. A reactive sealing process involves (a) chemical reactions, (b) physical treatment, (c) mechanical adherence.

54. Vacuum sealing is necessary for MEMS involving delicate moving components because it will offer the package (a) a clean environment, (b) freedom from resistance from air, (c) better thermal insulation.

55. Both vacuum sealing techniques offered in this book involve sealing the packages by (a) a vacuum pump, (b) heat treatment, (c) in situ sealing in evacuated microfabrication.

56. Three-dimensional microsystems packaging involves stacking up layers of (a) microcircuits, (b) devices for different functions, (c) different materials.

57. The primary objective of three-dimensional packaging is to achieve (a) volume efficiency, (b) material efficiency, (c) functional efficiency.

58. Signal mapping is related to (a) mapping out the transducers used in the system, (b) setting the strategy for selecting and positioning the transducers, (c) tracing transducers in the system.

59. A Wheatstone bridge is use to measure (a) electric current, (b) electric voltage, (c) electric resistance.

60. A Wheatstone bridge can be used to measure electric resistance in (a) static, (b) dynamic, (c) both static and dynamic situations.

61. Sensitivity normalization in electric resistance measurements is to compensate for (a) thermal expansion of the device, (b) change in resistivity of the gage due to temperature, (c) thermal expansion of the gage.

62. Offset balance in electric resistance measurements is required to compensate for (a) thermal expansion of the device, (b) change in resistivity of the gage due to temperature, (c) thermal expansion of the gage.

63. Silicone gel is a die protective material for (a) high temperature, (b) moderate temperature, (c) all temperatures of operation.

64. A piezoresistor is implanted in a rectangular silicon diaphragm with a $45°$ angle with the longer edge to measure (a) the normal bending strain, (b) the shearing stress, (c) both normal and shear strains of the diaphragm.

65. Microsystem packaging takes up on average (a) 60%, (b) 70%, (c) 80% of total production cost with current technology.

66. Two key factors for success in commercializing microsystems are (a) cost and appearance, (b) cost and reliability, (c) reliability and functionality.

67. Parametric test structures are (a) a structure to support the testing, (b) a microstructure for testing the design parameter of a component, (c) a strain gage structure.

68. The purpose of burn-in testing for microsystems is (a) to test the product's resistance to high-temperature, (b) to ensure the product can survive in high temperature environment, (c) to ensure the reliability of the product.

69. Self-testing of microsystems is desirable for (a) ensuring the reliability of, (b) updating the performance of, (c) ensuring the credibility of the product.

70. The provision for testing during use by a microsystem is desirable for (a) ensuring the reliability of, (b) updating the performance of, (c) calibration of the microsystem in its life cycle.

Part 2 Descriptive Problems

1. Plot the relationship between the magnification and the working distance of a typical microscope.

2. Design a microgripper similar to the one illustrated in Figure 2.45. Determine the number of pairs of electrodes (fingers) required to pick a microcylinder $200\,\mu$m long with a mass of 2 mg. The gripper is actuated by an applied voltage of 25 V. Also compute the deflection at the end point if the weight of the long gripping arms is accounted for. The total weight of the electrodes is assumed to be uniformly distributed on the gripping arms.

3. Estimate the adhesive forces induced in the gripping action described in problem 2 in a dry ambient environment. Would the microcylinder drop from the gripper upon releasing?

4. Assess the residual curvature of the anodically bonded quartz and silicon beam with dimensions given in Figure 11.39. Bonding of the beams takes place at 500°C. What is the consequence of excessive residual curvature of the bonded beams in terms of

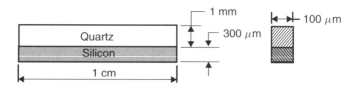

Figure 11.39. Anodically bonded beams.

alignment of possible dies in the silicon beam and the supporting bases in the quartz beam?

5. Design the spacer of a micropressure sensor die as illustrated in Figure 11.36 for the situation described in Example 4.4 using the finite element method. Plot the relationship between the maximum bending stress versus the height of the spacer, L.

CHAPTER 12

INTRODUCTION TO NANOSCALE ENGINEERING

12.1 INTRODUCTION

All matter, whether of hard metals or soft biological cells or tissues, that exists on Earth is made up of atoms. The periodic table presented in Chapter 3 shows that the atomic structure of individual elements (e.g., the total number of electrons,) characterizes their physical–chemical behaviors. We also showed in the same chapter that rearranging the atomic structure of matter, for example, by including atoms of another material with different number of electrons in its atoms as in doping a semiconductor, can alter the electric resistivity of that material. One may thus realize an important fact that properties of matter can be controlled by manipulating their atomic structures, or the arrangement of the atoms. If we approximate the size of atoms by their outer orbits on which electrons move, then the size of most atoms is in the range of 2×10^{-10}–3×10^{-10} m, or 0.2–0.3 nm, as described in Section 3.2 (1 nm $= 10^{-9}$ m, or 10^{-3} μm). This range defines the linear scale of the emerging *nanotechnology*. We may thus view nanotechnology as *the technology that involves manipulation and controlling of the composition and arrangement of atoms to satisfy human needs, leading to the design and fabrication of objects at the nanometer scale*.

Figure 12.1 shows the linear scale of selected objects at the micro- and nanoscale. We envisage that 1 nm is equivalent to about 5 shoulder-to-shoulder atoms of average size (or 10 shoulder-to-shoulder span for the smallest hydrogen atoms); a strand of DNA is about 2.5 nm wide; and biological cells are about thousands of nanometers in size.

Scientists and engineers attribute the concept of arranging the atoms and thereby controlling the material properties to a frequently quoted speech by Richard P. Feynman (1918–1988), a visionary Nobel Laureate in Physics in 1965 (Feynman, 1959): Dr. Feynman stated in his speech: "What would the properties of materials be if we could really arrange the atoms the way we want them? "…" I can hardly doubt that when

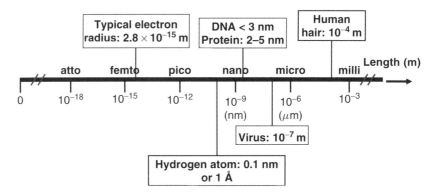

Figure 12.1. Length scales of micro- and nanoscale objects.

we have some control of the arrangement of things on a small scale we will get an enormously greater range of possible properties that substance can have, and of different things that we can do." It was his conviction that if humans can master the arrangement of atoms at will, then many marvelous things can happen as demonstrated by *the marvelous biological system*. Feynman said: "A biological system can be exceedingly small. Many of the cells are very tiny, but they are very active; they manufacture various substances; they walk around; they wiggle; and they do all kinds of marvelous things—all on a very small scale. Also, they store information. Consider the possibility that we too can make a thing very small which does what we want—that we can manufacture an object that maneuvers at that level!" Feynman's speech has since inspired many scientists and engineers in their efforts to further miniaturize devices and engineering systems. The enormous potential value of nanotechnology to humankind was concurred by another Nobel Laureate in Physics, Horst Stormer in his statement: "Nanotechnology has given us the tools . . . to play with the ultimate toy box of nature—atoms and molecules. Everything is made from it. . . . The possibilities to create new things appear limitless." [Amato, 1999] Ronald Hoffmann, a Nobel Laureate in chemistry viewed the nanotechnology as "the way of ingeniously controlling the building of small and large structures, with intricate properties; it is the way of the future, a way of precise, controlled building, with incidentally, environmental benignness built in by design." [Amato, 1999]

The coining of the term *nanotechnology*, however, did not happen until mid-1990. It has since caught the attention of many scientists and engineers as well as business and industry all over the world with an unprecedented enthusiasm and expectations. Applications of nanotechnology have been found in electronics, computing and data storage, materials and manufacturing, health care and medicine, energy and environment, transportation, and space exploration, as will be elaborated in Section 12.5. Revenue generated by this emerging technology has grown from a modest level of $50 million in 2001 to a staggering $26.5 billion in 2003. The National Science Foundation of the U.S. government projected the annual revenue generated by nanotechnology to reach $1 trillion by year 2015 ("Nano Technology," 2002). This enormous market potential has prompted governments of affluent nations to pour billions of dollars in the past decade in support of research and development in nanotechnology.

This chapter will cover the engineering aspect of nanotechnology. We will first learn the fundamental differences between micro- and nanoscale technologies, the basic principles in the nanoscale fabrication of device components atom by atom. It will be followed by a brief description of techniques for the production of basic nanoscale products of nanoparticles, nanowires, and nanotubes and their applications. We will then review the principles of existing tools for engineering design analyses in solid mechanics and thermal fluids at the nanoscale, with recommendations on modifications for their applications in nanoscale engineering design. The chapter will conclude with challenging issues to scientists and engineers in the design and manufacture of nanoscale device components and systems as well as major impacts of nanotechnology to the social and economic well-being of humanity.

12.2 MICRO- AND NANOSCALE TECHNOLOGIES

There appears to be a common conviction by some members of the technical community that nanotechnology (NT) is a natural evolution of the microsystem technology (MST) that produces MEMS and microsystems. Some others view nanoscale engineering as nothing more than simple scaling down of microsystem technology. These convictions and views are obviously incorrect; although both MST and NT share a common goal of miniaturization as illustrated in Figure 12.2, MST is viewed as a *top-down approach* to the design and manufacture of miniature devices of size 1 nm–1 mm, whereas NT is an alternative approach toward the same goal but from the *bottom up*.

There are other differences between MST and NT, as presented in Table 12.1. Despite the substantial effort and colossal amount of funding from many governments of affluent countries that has been spent in research and development, there remains a long way to capitalize the enormous potential benefits of NT with significant commercialization.

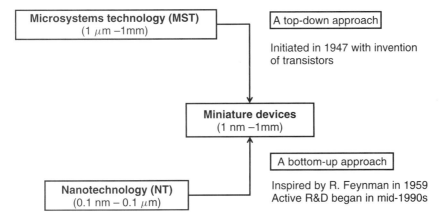

Figure 12.2. Microsystem technologies and nanotechnology.

TABLE 12.1. Comparison of MST and NT

Microsystem Technology	Nanotechnology
Top-down approach to miniaturization	Bottom-up approach to miniaturization
Miniaturization with micrometer and submicrometer tolerances	Miniaturization with atomic (subnanometer) and nanometer accuracy
Builds on solid-state physics	Builds on quantum physics and quantum mechanics
Evolved from IC fabrication technology	Undefined
Established techniques for producing electromechanical functions	Far from being developed
Established fabrication techniques	Fabrication techniques in rudimentary stage
Established manufacturing techniques	Available for limited types of products
Proven devices and engineering systems in marketplace	Limited to dots, wires, and tubes at the nanoscale made from limited number of materials
Success in commercialization of a few products	Success in some special-purpose materials and limited number of biomedical products

Major breakthrough in nanoscale actuation materials, for instance, such as piezoelectric at the nanoscale, is needed before practical and reliable *nanoelectromechanical systems* (NEMSs) can be built.

12.3 GENERAL PRINCIPLE OF NANOFABRICATION

It is reasonable to view the atoms or molecules of matter as the building blocks of nanoscale devices and components. Three major steps are involved in fabricating these components: (1) isolation of atoms or molecules, (2) assembly of loose atoms or molecules, and (3) rebonding of free atoms or molecules.

Step 1: Isolation of Atoms or Molecules. Several techniques can be used to isolate atoms from a base material—mechanical, electromechanical, and electrochemical—as presented in the following paragraphs.

Isolation of Atoms by Mechanical Means. The *atomic force microscope* (AFM) was first developed by Gerd Binnig and Christoph Gerber of the International Business Machines (IBM) Corporation and Calvin Quate of Stanford University in 1985. It has since been widely used to measure the surface topography of solids with atomic resolution. An AFM is a specially designed structure that consists of a cantilever beam with a hard tip, often made by diamonds, attached to the free end. The beam structure is attached

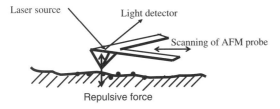

Figure 12.3. Scanning atomic force microscope.

to a cartridge which controls its scanning motion over the surface of the measurand solid, as illustrated in Figure 12.3. The tip of the beam can be in direct contact with the solid surface or with a minute gap over the surface it scans. Any irregularity of the solid surface can result in deflection of the cantilever beam structure and thus tilt of the minute optical reflector attached to the top surface of the scanning beam structure. The amount of tilt induced by the deflection of the tip can be measured by a reflected light detector, as illustrated in Figure 12.3. The detected signals can be related to the surface irregularities and hence mapping the surface topology of the solid with nanometer-scale accuracy. An AFM can be used to mechanically "scrap" and therefore free atoms from the surface of soft materials.

Isolation of Atoms by Scanning Tunneling Microscopy. The scanning tunneling microscope (STM) was invented by Gerd Binnig and Heinrich Rohrer of IBM Corporation in 1981. This important invention won both inventors the Nobel Prize in Physics for 1986. As illustrated in Figure 12.4, an electric conducting probe is placed over a rough solid surface with a narrow gap. Upon passing an electric field across the probe and the solid, electrons jump (movement) across the narrow gap, resulting in the creation of a *tunnel*. Consequently, electric current can flow in the tunnel with its intensity proportional to the gap and the applied voltage. The landscape of the substance surface can be mapped by adjusting the gap d to maintain constant current flow in the tunnel while the probe scans across the surface. The STM can be used to destroy the atomic bonds at the surface of the matter by localized bursting of electric current at the needle tip and thereby free the atoms near the solid surface.

Isolation of Atoms by Electrochemical Means. The idea of miniaturization by evaporation of materials was mentioned in Feynman's famous speech in 1959. Indeed, as we learned from Chapter 8, there are ways that we can ionize substances and set free

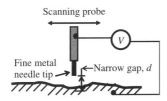

Figure 12.4. Scanning tunneling microscope.

these charge-carrying atoms to move around. The plasma etching technique described in Section 8.9.2 is another way to free atoms from substance surfaces. These techniques are widely used in the production of many nanoscale products, as will be described in Section 12.4.

Step 2: Assembly of Loose Atoms or Molecules. There are several ways available for assembling loose atoms or molecules freed from substances as described in step 1. The loose atoms can be "pushed" to the assembly site by mechanical means, such as the AFM-guided nanomaching system (AGN), or by a special controlled environment with controlled temperature or electromagnetic fields. Figure 12.5 shows how loose gold atoms are reassembled by IBM using its STM. The top left frame in the figure indicates the loose atoms. A STM guided these loose atoms one at a time to the final assembled order showing the company's name, "IBM," at the lower right frame.

Biochemical processes have been developed for assembling loose atoms and molecules for industrial-scale production. Natural self-assembly and replication of biological cells, for instance, provide a promising base for developing nanoscale assembly for volume production of nanoproducts, as will be described in Section 12.11.2.

Step 3: Rebonding of Free Atoms and Molecules. Synthetic processes have been developed to rebond loose atoms and molecules. Mechanical synthesis (e.g., vaporizing free atoms at high temperature in high vacuum) followed by condensation has been used in the production of nanoscale products. Chemical synthesis is a promising technique for consolidating loose atoms and molecules. Chemical vapor deposition involving diffusion and chemical reactions can provide high-quality rebonding of atoms and molecules. There has been active research in developing biochemical synthesis that can artificially duplicate and reassemble natural molecular machines for growing proteins, enzymes,

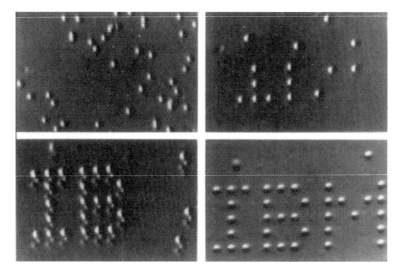

Figure 12.5. Assembly of loose atoms by STM. (Courtesy of IBM Corp.)

and antibodies. Some microfabrication processes (e.g., epitaxial growth technique such as described in Section 8.8) have been used for this purpose.

12.4 NANOPRODUCTS

As mentioned in Section 12.1, nanotechnology has offered enormous potential values as well as opportunities for the manufacturing industry. There have been rigorous research and development efforts in producing new products using nanotechnology. In recent years new applications of nanotechnology and nanoproducts have been reported by the scientific community and industries. Many of these products have been successful in the marketplace. It is not practical to report all these products in this chapter. What we will describe are the basic nanostructures that can be viewed as the building blocks of many current and future nanoproducts.

Buckyball. The first man-made nanostructure is the *buckyball* discovered by Harry Kroto, Richard Smalley, and Robert Curl, the three Nobel laureates in chemistry for 1985 ("Nano Technology," 2002). These balls are extracted by vaporizing graphite in extremely high temperature using a powerful laser in an inert gas environment. The name buckyball is after an American visionary and architect, Richard Buckminster Fuller (1895–1983), because the hexagonal bonding of carbon atoms to form the surface of the buckyballs illustrated in Figure 12.6 simulates the geodesic dome designed by this architect for the Montreal Expo in 1967. The buckyball contains 60 pure carbon (C_{60}) atoms. It has a shape of a soccer ball with a hollow cage with a diameter of 0.7 nm. It is extremely stable and can withstand very high temperatures and pressures. Much research effort has been made to trap smaller molecules inside the ball or attach other molecules to its outside surface.

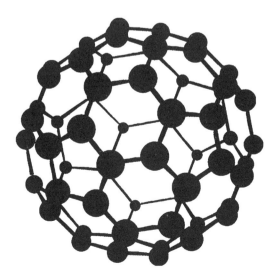

Figure 12.6. The buckyball.

Nanoparticles. *Nanoparticles*, or *quantum dots*, are microscopic particles with sizes in the nanometers. It is defined as a particle with at least one dimension to be less than 100 nm (Wikipedia encyclopedia). Two principal methods of fabricating nanoparticles are by (1) attrition and (2) pyrolysis:

1. **Attrition:** This method involves grinding down small particles in ball mills. Often, size reduction of the particles is accomplished by grinding the particles in decreasing orders in stages.
2. **Pyrolysis:** This method is effectively a vaporization process in which the substance is vaporized and then condensed into minute particles. Thermal plasma that generates temperature on the order of 10,000 K by either a high-voltage DC electric field or a RF source as described in Chapter 3 can be used to evaporate substances into fine powder in nanometers. Nanoparticles of these sizes are produced after cooling outside the plasma chamber.

Nanoparticles can offer distinct properties relative to their bulk materials depending on their sizes. Nanoparticles may be significantly harder, stronger, or more electrically and thermally conductive than their bulk materials. Another unique characteristic of nanoparticles is that the color of these particles is dictated by their sizes. For example, the color of silver nanoparticles begins with a light pink color at the size of 120 nm. It turns to light blue at 80 nm and dark blue at 40 nm. Gold nanoparticles change color from orange at 100 nm to olive green at 50 nm.

Nanowires. *Nanowires* typically have dimension on the order of 1 nm in the lateral direction. Their dimension in the length direction can be as high as a few millimeters. Nanowires can be made of metals (e.g., Ni and Au), semiconductors (e.g., Si and GaN), and compounds (e.g., SiO_2).

There are several ways nanowires can be produced. A synthesis process called *vapor-liquid-solid* (VLS) is commonly used by industry. It involves the use of minute particles made from a *source material* in the form of a gas. The source material is exposed to a catalyst made of liquid metal nanocluster. The catalyst facilitates the aggregation and growth of the source material on its surface. The source material solidifies once reaching the supersaturation state in the cluster. It then grows outward from the cluster to form nanowires. Figure 12.7 shows pure crystal zinc oxide nanowires grown vertically in aligned arrays from a sapphire wafer (Yarris, 2003). These wires have average diameter at 2.5 nm, with length ranging from 2 to 10 μm.

Nanowires can be made from conductive or semiconductive or dielectric bulk materials. They are the primary elements in molecular or nanoelectronics acting as conducting circuitry wires and as the gates for nanoscale transistors. The electrical conductivity of nanowires, in general, is less than that of the corresponding bulk material. The conductivity of these wires can be controlled by the surface condition, and more importantly, many of these nanowires can have their conductivities controlled by doping processes as described in Chapter 3.

Nanotubes. The lessons learned from producing the buckyball led to extending the use of the newly discovered pure carbon C_{60} to other nanoscale structures—the tubes, called

Figure 12.7. Nanowire of zinc oxide. (Courtesy of Peidong Yang.)

nanotubes. These tubes can have diameters in nanometers with wall thickness of just 1 nm. The length of these tubes, however, can be a hundred micrometers or millimeters. Nanotubes in theory can be made from a variety of bulk materials as in the case of nanowires. However, the success in producing of buckyballs has made *carbon nanotubes* (*CNTs*) a common form of nanotubes at the present time. Consequently, many call carbon nanotubes *buckytubes*.

The first synthetic carbon nanotube (CNT) was reported in 1991 (Iijima, 1991). Seamless CNTs can be produced by rolling graphite sheet approximately one atom thick into a tubular shape, as illustrated in Figure 12.8.

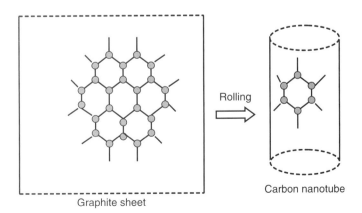

Figure 12.8. Production of carbon nanotube.

Figure 12.9. Nanotubes made of GaN. (Courtesy of Peidong Yang.)

Figure 12.8 shows a sheet of graphite (one of three forms of pure carbon) containing carbon atoms bonded in a hexagonal structure and rolled into a tubular shape with diameters in single-digit nanometers. The process requires extremely high temperature in high vacuum, as in the production of buckyballs. Several techniques have been used to mass produce single-wall CNTs. Three techniques are used by industry: arc discharge, laser ablation, and catalyzed chemical vapor deposition. The arc discharge method was used to generate the high temperature required to vaporize the bulk carbon and synthesize the free atoms into the desired shapes. It was used in the early stage of development. Laser ablation uses a pulse laser to vaporize a graphite target in a high-temperature chamber. The nanotubes are formed on the cooled surfaces of a nanowire mold. Catalyzed chemical vapor deposition shows the most promise for mass producing CNTs. It involves growing CNTs on substrates with two carrying gases that include a process gas (NH_3, N_2, H_2, etc.) and a carbon-containing gas (acetylene, ethylene, ethanol, methane, etc.) (Wikipedia encyclopedia). Carbon nanotubes on average are 20 times as strong as steel and can conduct electricity 1000 times better than copper. They can be bent without breaking, making this material a desirable component in nanoscale devices.

Figure 12.9 shows nanotubes made of gallium nitride (GaN) produced by the *epitaxial casting* technique (Yarris, 2003). The zinc oxide nanowires shown in Figure 12.7 are used as templates over which the vaporized gallium nitride atoms grow over the wire surface by chemical vapor deposition. Heat is then applied to evaporate away the zinc oxide cores to form the gallium nitride nanotubes. These tubes have diameters ranging from 30 to 200 nm with wall thickness between 5 and 50 nm. They are made of single crystal and are transparent and electrically and optically active.

12.5   APPLICATION OF NANOPRODUCTS

In Section 12.4, we learned that nanoparticles, nanowires, and nanotubes can be produced in various sizes from a number of materials. One striking phenomenon with these nanoscale products is that fundamental properties of materials change with the size, shape, and composition that are not observed in any of their corresponding bulk materials. These size-dependent properties include:

- Mechanical strength (e.g., maximum tensile strength and Young's modulus)
- Electrical properties (e.g., electrical resistivity)
- Chemical properties (e.g., reactivity to chemical reactions)
- Optical properties (e.g., reactivity to light)
- Thermophysical properties (e.g., thermal conductivity, thermal diffusivity, and coefficient of thermal expansion)
- Melting points and solubility
- Magnetic properties
- Catalytic properties (e.g., ability to enhance chemical reactions)
- Viscosity for liquid and surface tension

The unique size-dependent properties of nanoproducts offer many opportunities for new applications. Many view the size-dependent properties as the principal reason for the major impact of nanotechnology in science and industry. Following are some applications of nanoproducts in biomedicine, molecular electronics, energy, environment, and industrial products.

Biomedicine. Nanoparticles are extremely sensitive to biological materials. They can be used as sensitive disease detectors for intricate detection of damaged or dead cells as well as early diagnosis of complex cases such as Alzheimer's disease. Nanoparticles can also be used as drug carriers for target drug delivery. In such cases, the carriers are coated with nanosensors which can recognize diseased tissues and attach to them, releasing a drug where needed. Nanoparticles can also be used to enter damaged cells and release enzymes that tell the cells to autodestruct, or they can release enzymes to try to repair the cell and return it to normal functioning. Target drug delivery offers great value in treating cancerous disease because the treatment will be targeted on cancer cells and not the surrounding healthy ones. Nanoparticles change color with their sizes. This unique characteristic makes them ideal candidates as *biological markers*, or *bio-barcode*, for detecting early stages of major diseases. They are capable of detecting various biomarkers comprising nucleic acids or proteins at exceptionally low numbers. It is vital to the practice of diagnostic medicine. These biological markers can indicate the onset of a neurodegenerative disease, an infection by a virus, or the environmental presence of a potentially toxic or lethal pathogen. High-sensitivity detection is essential for early disease diagnosis, tracking therapeutic efficacy, DNA sequencing, blood and food supply screening applications, and tracking disease recurrence. Due to their minute size and controllability of material properties, nanoparticles are used as building blocks for the repair of skin, cartilage, and bone.

The extremely high surface area–volume ratio makes nanoparticles important in the diffusion process, a common and important process in drug discovery and production by the pharmaceutical industry.

Molecular Electronics. Nanowires are the basic building blocks for nanoscale electric circuits. They are also used in simple switches. New nanoscale memory devices using carbon nanotubes are proposed; these minute tubes can be connected in parallel or in junctions. Current passing these tubes can result in change of electrical properties that can serve as permanent memory in a nanoscale data storage system. A new concept of nanoscale transistors uses electric conductive molecules to connect the gate and the source and drain. The applied voltage changes the electric properties of these molecules and thereby regulates the current flow in the transistor.

The high thermal conductivity of carbon nanotubes has made this material an ideal candidate for thermal interface materials in nanoscale electronics. The thermal conductivity of individual multiwall carbon nanotubes has been measured to be 3000 W/m-K at room temperature, which is about eight times higher than that of copper and 20 times higher than that of silicon (Kim et al., 2001; Yu et al., 2005). Carbon nanotubes can thus effectively dissipate heat generated by ohmic heating of nanoscale circuits in molecular electronics.

The concept of *spintronics* has caught much attention of scientists and engineers in the past decade. It involves the development of a data storage system based on the "spin" properties of electrons. Spin can be used as an up—down binary code and offers a bright horizon for quantum computing, where elemental building blocks are manipulated based on the laws of quantum mechanics.

Energy. Energy is essential in sustaining quality living for humans in modern societies. Nanotechnology and nanoproducts have already had major impacts on the production of all forms of energy. The reader is referred to a good reference for detailed documentation on this subject (Gillett, 2002). Examples on the application of nanotechnology in selected forms of energies will be presented below:

1. **Fuel Cell:** Fuel cells provide direct conversion of chemical energy to electric energy at the present time. One principal application of nanotechnology is to produce catalysts that can enhance the ionization of hydrogen and oxygen in the respective anode and cathode in the cell. Nanoparticles with controlled electrochemical properties can be used to produce such catalysts.

2. **Batteries:** Batteries are the only available equipment for storing electrical energy. Nanotechnology allows the design and fabrication of new anodes and cathode using materials such as nanocarbon tubes. These anodes and cathods are able to substantially increase the amount and rate of energy that can be transferred to a battery and reduce the recharge times significantly as a result. Another application of nanotechnology is to control the crystal structure of lithium in lithiumion batteries with improved storage efficiency.

3. **Solar Photovoltaic:** Solar photovoltaic (PV) cells convert energy in solar rays directly into electrical energy. The working principle of solar PV cells is described in Section 2.2.4. The significant improvement of conversion efficiencies of silicon solar PV cells in recent years has prompted strong demand in the marketplace for these cells. The conversion efficiency of silicon solar PV cells can be significantly improved with the high density of electrodes embedded in the silicon wafers in order to capture all electrons that flow across the *p*- and n-wafers. The number

of electrodes are limited because they occupy the surface area that is required for solar ray exposure. Highly conductive nanowires can be used to maximize the collection of the free electrons generated by the photoelectric conversion with minimum physical space. Nanofabrication techniques can also be used to produce ultrathin wafers to minimize the use of expensive silicon wafers, as the photoelectric conversion of energy in solar PV cells is predominately a surface effect.

4. **In Situ Bitumen Extraction:** There is vast reserve of oil sands deposits in the world. One technique to extract the bitumen from these deposits is to inject steam into the deposit and thereby drive the heat-loosened bitumen into the collecting well. This in situ bitumen extraction method is more costly than conventional open-pit mining but is environmentally sustainable. Control of the flow of the bitumen is a major hurdle in this method of extraction. Based on the similar working principle of biological markers by nanoparticles, these particles can be injected into the oil sands deposit with steam. They can trace the in situ heat-loosened bitumen flow and thus enables engineers to determine where and when to inject additional steam for more effective production of bitumen.

5. **Integration of Solar/Fuel Cell Energies:** Research has begun on integrating solar panels and fuel cells for a new renewable energy production system that can complement the functions that each system performs. For example, when the solar cells are producing energy, the fuel cell is running in reverse to collect excess energy, convert it to hydrogen, and store it. When the sun goes down and the solar panel is no longer producing energy, the fuel cell will run forward and produce energy from the hydrogen it has stored.

Environment. Nanoparticles can absorb, trap, or break down pollutants and make fossil fuels less polluting and coolant more efficient. Nanotubes are being incorporated into sensors to detect the presence of dangerous gases. For examples, zinc oxide film emits visible green light when exposed to UV light. It grows dims when toxic chemicals such as chlorinated phenols and polychlorinated biophenols pass nearby. The pollutants react with the UV irradiation film and decompose into harmless chemicals.

Nanoparticles also have the ability to detoxifiy a wide variety of common contaminants. Nanocomposite material can absorb mercury and release the absorbed mercury when exposed to heat or vacuum treatment. It is thus an effective tool for recycling mercury.

Other Applications. We have learned from Section 12.4 that the properties of nanoparticles, nanowires, and nanotubes can be controlled by controlling their size. This characteristic of nanoparticles has led to the successful commercialization of some special paints and coating materials made by carefully controlled nanoparticle sizes of certain materials for smart and enduring structural surfaces. These surfaces can respond to a changing environment or act as a supertough protective layer. The paint and coating materials made of nanoparticles can also be made superstrong and tough for buildings and the exterior of a vehicle chassis.

In addition to molecular electronics, nanowires have been used in optoelectronics and nanoelectromechanical devices. They can also be used to produce nanocomposites for special applications and leads for nanoscale sensors and actuators.

Carbon nanotubes are extremely useful in nanostructures for the following reasons:

1. Superior strength. It has a tensile strength over 60 GPa, in comparison to just 1 GPa yield strength of silicon as given in Table 7.3. A CNT composite can further increase this strength to 138 GPa, or 20×10^6 psi. Materials with this superhigh strength can drastically reduce the weight of machine structures. It will make the dreamed super-lightweight spacecraft with the weight of a family car closer to reality.

2. High electric conductivity. The electric resistivity of CNTs can be controlled to give superhigh electric conductivity. With its small size, it will make such material ideal for nanoelectric circuitry and nanoscale relays and switches in logic gates.

3. Nanotubes made with other materials such as the gallium nitride tubes shown in Figure 12.9 have applications in chemical sensors and nanofluidics, including nanoscale electrophoresis for biomedical analyses.

12.6 QUANTUM PHYSICS

Many scientists and engineers view the continuous effort in developing nanotechnology will lead to the ultimate miniaturization of devices and engineering systems, as the visionary Feynman forecast over half a century ago (Feynman, 1959). The substances that we deal with in nanotechnology are in nanometers and fall in the range of the atomic scale. We can thus expect more pronounced influence of atomic effects on the design and fabrication of device components at the nanoscale. For example, the effects of van der Waals forces can no longer be neglected in the mechanical design of these systems. The laws and theories of physics that are derived for substances at the macroscale are no longer valid without significant modifications to accommodate these atomic effects.

In addition to the van der Waals forces, another critical atomic effect that can affect the engineering design of nanosystems is the dominant mechanism for the transportation of energy, that is, heat, in solids at the submicrometer and nanometer scales. Such an effect can be better illustrated by quantum physics.

A *quantum* represents the smallest amount of energy that any system can gain or lose. It occurs at the atomic scale. Maxwell Planck (1858–1947) postulated the quantization of energy around 1900 in his formulation of thermal radiation. This concept led to a fundamentally important topic in physics—quantum physics. Planck postulated that atoms in a substance oscillate back and forth from their neutral positions in a lattice (see Figure 3.5) when they are energized. These oscillations radiate energy that can be determined by the simple formula

$$E = nh\nu \tag{12.1}$$

where

n = integer number, $= 1, 2, 3, \ldots$

h = Planck's constant, $= 6.626076 \times 10^{-34}$ J-s

ν = frequency of oscillation, s^{-1}

The energy associated with the oscillation of atoms in Equation (12.1) is called a *phonon*, which is transmitted through a substance in the form of acoustic waves at an extremely high frequency, on the order of 10^{13} Hz.

Albert Einstein proposed his photon theory in 1905, in which he postulated that light behaves as if energy is concentrated into localized bundles consisting of a stream of particles, called *photons*. The energy associate with a single photon is

$$E = h\nu \tag{12.2}$$

where ν is the frequency of the light.

With Einstein's photon theory, we are able to explain many phenomena in optics that the classical wave theory fails to support. One such example is the photoelectric effect described in Chapter 2.

Both phonons and photons are *energy carriers*. Neither of them has mass, despite the fact that they are treated as "particles" in quantum physics. A familiar phenomenon, the light emitted from heating of a metal rod by electricity, demonstrates this. The higher the electric energy that is supplied to the metal, the brighter the emitting light becomes. If light is carried by photons, such as postulated by Einstein's photon theory, a brighter light should carry more photon energy. Thus, one should not be surprised that one can feel the emitted photons as the heat from a glowing metal heater, but not in the form of forces, as photons do not have mass.

Because both phonons and photons are energy carriers, we will expect these carriers to play major roles when dealing with heat transfer in substances that have the dimensions near the atomic scale. The mathematical modeling of heat transmission in solids at the submicrometer- and nanoscale with the influence of quanta energy carriers will be presented in Section 12.9.

12.7 MOLECULAR DYNAMICS

As we learned in Section 3.4, substances can deform into different shapes by externally applied forces or energy. Associated with the shape change of the substance is the change of the relative distances between the atoms or molecules inside the substance. Atoms and molecules in any substance (Figure 12.10*a*) in a natural state are aggregated together by elastic bonds (Figures 12.10*b,c*) that can be compressed or stretched by the application of forces or energy. To deform the substance, the applied forces or energy need to overcome the elastic bonding force, or the intermolecular forces, that exist between pairs of atoms or molecules (Figure 12.10*c*).

Intermolecular forces, such as illustrated in Figure 3.6, change their sense from attraction to repulsion, or vice versa, depending on the relative distances between pairs of atoms or molecules. For substances at the macroscale, there are zillions of atomic pairs and the effect of the sense of intermolecular forces is insignificant with the law of averaging. In the case of substances at the nanoscale, however, such effect can no longer be neglected.

As indicated in Section 12.5, Maxwell Planck postulated that atoms in a substance oscillate back and forth from their neutral positions in a lattice when they are energized

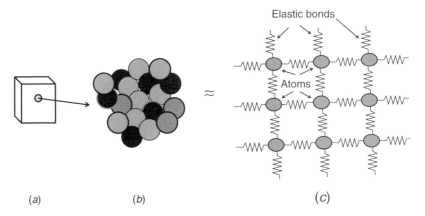

Figure 12.10. Elastic bonding of atoms: (*a*) solid plane; (*b*) aggregated atoms; (*c*) elastic bonds of atoms.

by externally applied forces or energy. We can simulate this postulation by replacing the chemical bonding between atoms in nature by elastic springs, as illustrated in Figure 12.10*c*. Any disturbance to any atom in the substance can be viewed as an initiation of a mechanical vibration involving mass and springs, as described in Section 4.3. It underlines an important fact in nanoscale engineering analyses that *the physical behavior of atoms should always be treated in a dynamic mode*, whether the induced cause is stationary or timevarying. Molecular dynamics attempts to postulate the physical movements of atoms of substances at the nanoscale.

We illustrate the dynamic motion of an individual atom in a matrix of atoms in a plane, as illustrated in Figure 12.11*a*. The atoms are bonded by elastic bonds represented by the dotted lines. Upon imposing an instantaneous displacement vector to the atom at the center of the matrix shown in grey, we expect this atom to respond by a movement shown in Figure 12.11*b* at instant t_1. Assuming the movement of the surrounding atoms to be negligible, we expect some of the elastic bonds associated with the moving atom to be stretched and some others to be compressed. The restoring of these deformed elastic

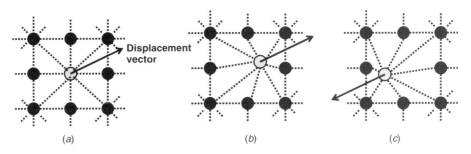

Figure 12.11. Oscillatory motion of an atom. (*a*) Initial equilibrium position. Position at (*b*) time t_1 and (*c*) time t_2.

bonds can cause the atom to reverse its motion in the opposite direction, passing its original equilibrium position, as illustrated in Figure 12.11c. We can well imagine such back-and-forth atomic movements, or oscillation, will take place for a while until the associated energy is completely dissipated.

The governing equation that could be used to describe the dynamics of atoms such as described above is the Schrödinger equation derived from nonrelativistic quantum mechanics in 1926. We will only show the form of the equation and the physical significance of the variables and parameters involved in the equation. Detailed description of this equation and its solution for nanoscale engineering design are available in an excellent reference (Drexler, 1992).

The position of a charged particle (e.g., an ionized atom), $\phi(\mathbf{r}, t)$ (e.g., the atom in grey in Figure 12.11a), in a force field of n charged particles (the other atoms in Figure 12.11) with masses $m_0, m_1, m_2, \ldots, m_{n-1}$ can be determined by solving the equation

$$-\frac{h^2}{2}\sum_j \frac{1}{m_j}\frac{\partial^2 \phi(\mathbf{r}, t)}{\partial \mathbf{r}_j^2} + U(\mathbf{r}, t)\phi(\mathbf{r}, t) = ih\frac{\partial \phi(\mathbf{r}, t)}{\partial t} \tag{12.3}$$

where

$\mathbf{r}$ = position vector, (x,y,z) in cartesian coordinates or (r, θ, z) in cylindrical polar coordinates

$U(\mathbf{r},t)$ = potential energy function

h = Planck's constant, $= 6.626076 \times 10^{-34}$ J-s

i = imaginary number, $= \sqrt{-1}$

The potential energy function in the Schrödinger equation (12.3) can be evaluated by the equation.

$$U(\mathbf{r}) = \sum_{i<j} \frac{q_i q_j}{4\pi \varepsilon_0 d_{ij}} \tag{12.4}$$

where q_i and q_j are charge intensities of particles i and j, respectively, d_{ij} is the distance between particles i and j, and ε_0 is the permittivity of free space.

The Schrödinger equation offers a solution of the instantaneous position of a particular atom in response to an external influence by a deterministic model. The corresponding instantaneous velocity and acceleration vectors can be obtained by its respective first- and second-order derivatives with respect to the time variable. The solution of this equation is by no means straightforward, and it requires horrendous effort to account for the potential function in Equation (12.4) because of the difficulty in determining the instantaneous interatomic distance d_{ij}. The potential energy in Equation (12.1) for charge-free atoms can be related to the potential energy of a deformed solid in the form of Equation (10.9). This formulation requires prior knowledge of the intermolecular forces of particular materials, which vary with the intermolecular distances according to what is illustrated in Figure 3.6. Such information is not readily available. The Schrödinger equation thus is not practical in assessing the dynamic behavior of solids at the nanoscale.

12.8   FLUID FLOW IN SUBMICROMETER- AND NANOSCALES

In Chapter 2, we predicted that bioMEMS that include miniature biomedical instruments and equipment and biosensors would be the next widely used MEMS products after pressure sensors. Indeed, the development of new bioMEMS products for biotechnology and pharmaceutical industry has been reported frequently. Many of these applications involve microfluidics for genome analyses or for separation or mixing microingredients in biomedical and pharmaceutical processes. Microfluidics in these applications typically involve fluid flow at low Reynolds numbers with low volumetric flow rates from a few microliters down to a few hundred nanoliters per minutes. The size of conduits for the flow varies from a few micrometers to a few hundred micrometers. Many of the involved fluids are gases, which behave quite differently from liquids because of their compressibility, and the more unique phenomenon of *rarefaction* in micro- and nanoscales. In this section, our attention will be focused on the behavior of gas flow in an extremely small space at low pressure and with low velocity.

12.8.1   Rarefied Gas

The meaning of the word *rarefaction* as found in *Merriam Webster's Dictionary* (Mish, 1995) is "a state or region of minimum pressure in a medium traversed by compression waves." The medium in the above statement is a gas that exists in a state of minimum pressure.

As described at the beginning of Chapter 5, fluids are aggregations of molecules. The molecules are much more widely spaced in gases than in liquids. Even in the natural state, gas molecules are randomly translating in the space and collide with each other or with the surface of an immersed body or container. The phenomenon is much like particles in random motion (Patterson, 1971). For a given gas at a given state, there is the *mean free path (MFP) length* λ, in which the gas molecules move without colliding with any obstacle. A value $\lambda = 65$ nm has been mentioned as a good approximation for most gases at atmospheric condition (Beskok and Karniadakis, 1999). The MFP for liquids is about twice the size of the molecules.

12.8.2   Knudsen and Mach Numbers

Apparently the validity of using the continuum fluid mechanics theories for fluid flow as stipulated in Sections 5.2–5.6 depends on the magnitude of the *Knudsen number*, which is defined as

$$\text{Kn} = \frac{\lambda}{L} \tag{12.5}$$

where L is the characteristic length scale, which is chosen to include the gradients of density, velocity, and temperature within the flow domain.

With MFP, $\lambda \approx 65$ nm, one may readily estimate the magnitude of the Knudsen number to be very small at 1.3×10^{-4} for the liquid flow in Example 5.5. On the other hand, $\text{Kn} = 1.3$ in the air entrapped in a space between a read/write head flying at a height of 50 nm and the surface of a fast-spinning disk in a computer data storage system.

The *Mach number* (Ma) of a moving gas is the measure of its velocity as well as the compressibility of the gas. It is defined as

$$\text{Ma} = \frac{V}{\alpha} \tag{12.6}$$

where V is the velocity of the moving gas and α is the speed of sound in that gas. The corresponding speed of the sound of an ideal gas can be obtained by the expression

$$\alpha^2 = \frac{c_p R T}{c_v} \tag{12.7}$$

where c_p and c_v are the respective specific heats of the gas at constant pressure and constant volume, R is the gas constant, and T is the absolute temperature of the gas.

For air at room temperature, the corresponding speed of sound is about 343 m/s. A gas is considered as incompressible if its Ma < 0.3.

We thus consider a rarefied gas flow at extremely low pressure to be incompressible because of its high Knudsen number (Kn > 0.1) and low Mach number (Ma < 0.3). Collision of gas molecules, which is characterized by the mean free path λ, is expected to be dominant in the behavior of gas flow.

12.8.3 Modeling of Micro- and Nanoscale Gas Flow

A critical question that engineers involved in the design of microfluidics would ask is how far can they "stretch" the use of the classical governing Navier–Stokes equations (5.20) in the micro- or nanoregime. The spectrum of gas flow in Figure 12.12 may provide some useful clue for the answer. From this spectrum, we envisage that the continuum theories and thus the Navier–Stokes equations can be applied to model gas flows with Kn < 0.01. This application can be further extended to the range $0.01 < \text{Kn} < 0.1^+$ with proper modifications to accommodate the slip boundaries. Beyond the range, that is, $\text{Kn} > 0.1^+$, other equations such as the Burnett and modified Boltzmann equations will be required (Patterson, 1971; Beskok and Karniadakis, 1999). These equations are beyond the scope of this book and will not be included in this chapter. The reader is referred to special books on gas dynamics such as Patterson (1971) for detailed description of these equations.

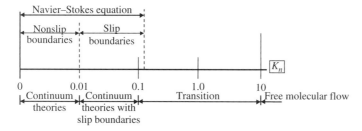

Figure 12.12. Spectrum of gas flow regimes.

As indicated in Figure 12.12, gas flow is in the form of free molecule movement for Kn > 10. The mass flow rate for free gas molecules in a pipe can be obtained using the formula (Beskok and Karniadakis, 1999)

$$\dot{m}_{fm} = \frac{4d^3 \Delta P}{3L} \sqrt{\frac{2\pi}{RT}} \tag{12.8}$$

where d and L are the respective diameter and length of the conduit, with L in the range of $d \leq L \leq \lambda$, the mean free path of the gas molecule. Both d and L have the unit of meter (m). The unit for the pressure drop ΔP is pascal (Pa), or kg_f/m^2. The units for the specific gas constant R and the temperature T are J/kg-K and kelvin (K), respectively. The units for $\sqrt{RT}$ in Equation (12.8), however, are meters per second. The units for the mass flow rate in the above equation are thus kilograms per second.

> **Example 12.1.** Estimate the flow rate of nitrogen gas in a section of minute tube 30 nm in diameter by 50 nm long. A pressure difference of 0.5 nm is applied to drive the flow. The flow is conducted at room temperature, 20°C.
>
> *Solution:* We will need the following parameters for using Equation (12.8) for the mass flow rate of the gas. The assumption of the specific gas constant $R = 286$ J/kg-K for air is close enough for the nitrogen gas in question. By using the average mean free path length $\lambda = 65$ nm for gas in the present case, we have the Knudsen number Kn = $\lambda/d = 65/30 = 2.17$. Other parameters are $L = 50$ nm, $d = 30$ nm, $\Delta P = 0.5$ Pa, and $T = 293$ K.
>
> Thus, by substituting the values of the above parameters into Equation (12.8), we have the rate of mass flow of the gas as
>
> $$\dot{m}_{fm} = \frac{4(30 \times 10^{-9})^3(0.5)}{3(50 \times 10^{-9})} \sqrt{\frac{2\pi}{286 \times 293}} = 3.116 \times 10^{-18} \text{ kg/s}$$
>
> This extremely low mass flow rate practically means a stagnant gas at the nanoscale.
>
> The same authors (Beskok and Karniadakis, 1999) proposed a unified model for two-dimensional gas flow that covers the entire regime of $0 \leq$ Kn $\leq \infty$ for both tubes and channels.
>
> For tubular flow, the normalized rate of gas flow can be obtained from the expression
>
> $$\frac{\dot{m}}{\dot{m}_{fm}} = \frac{3\pi}{64\text{Kn}}(1 + \alpha\overline{\text{Kn}})\left(1 + \frac{4\overline{\text{Kn}}}{1 - b\overline{\text{Kn}}}\right) \tag{12.9}$$
>
> where $\overline{\text{Kn}}$ is the Knudsen number evaluated at the average pressure, that is, at $\overline{P} = \frac{1}{2}(P_{in} + P_{out})$, where P_{in} and P_{out} are the respective gas pressures at the inlet

and outlet of the tube. The parameter α is related to the *rarefaction coefficient* $C_r(\text{Kn}) = 1 + \alpha\text{Kn}$ and is determined by experiments.

For tubular flow, the approximate value of α can be obtained from the expression

$$\alpha \approx \left[\frac{64}{3\pi(1-4/b)}\right] \tag{12.10}$$

The coefficient b is related to the slip boundary condition, where $b = 0$ for nonslip boundaries and $b < 0$ for slip boundaries. A value of $b = -1$ is used for gas flow in tubes, which leads to $\alpha = 1.358$ from Equation (12.10). Consequently, Equation (12.10) takes the form

$$\frac{\dot{m}}{\dot{m}_{fm}} = \frac{3\pi}{64\text{Kn}}(1 + 1.358\overline{\text{Kn}})\left(1 + \frac{4\overline{\text{Kn}}}{1+\overline{\text{Kn}}}\right) \tag{12.11}$$

We will find that the above equation will lead to the ratio, $\dot{m}/\dot{m}_{fm} = 1.001$ with an assumption of $\text{Kn} \approx \overline{\text{Kn}} = 2.17$ in Example 12.1.

Example 12.2. Estimate the rate of airflow in the section of a small tube 10 μm in diameter with 1 cm in length. Assume that a pressure difference of 5 Pa was maintained between the inlet and outlet of the tube section. The airflow takes place at room temperature.

Solution: We will first evaluate the Knudsen number to be

$$Kn = \frac{\lambda}{d} = \frac{65 \times 10^{-9}}{10 \times 10^{-6}} = 0.0065$$

The equivalent free molecular mass flow rate in this case can be obtained from Equation (12.8) as

$$\dot{m}_{fm} = \frac{4 \times (10 \times 10^{-6})^3 \times 5}{3 \times (10000 \times 10^{-6})}\sqrt{\frac{2 \times 3.14}{286 \times 293}} = 5.77 \times 10^{-15} \text{ kg/s}$$

Assume that $\overline{\text{Kn}} \approx \text{Kn} = 0.0065$. From Equation (12.11), we then have

$$\frac{\dot{m}}{\dot{m}_{fm}} = \frac{3 \times 3.14}{64 \times 0.0065}(1 + 1.385 \times 0.0065)\left(1 + \frac{4 \times 0.0065}{1 + 0.0065}\right) = 23.44$$

from which we have the mass flow rate of the air as

$$\dot{m} = 23.44 \times 5.77 \times 10^{-15} = 0.13524 \times 10^{-12} \text{ kg/s or } 0.487\mu\text{g/hr}$$

12.9   HEAT CONDUCTION AT NANOSCALE

In Section 12.6, we introduced phonons to be the energy carrier in a substance at the nanoscale, and this energy carrier is the result of the oscillation of atoms about the lattices connecting them in the substance. This phenomenon is illustrated in Figure 12.13, in which the atoms are bonded by lattices represented by dotted lines. The atom at the center of a matrix is energized by an input thermal energy. This input energy causes the lattices associated with this atom to vibrate back and forth from their initial natural position, as shown in Figures 12.13*b* and *c*. According to Planc's postulation, energy in the form of a phonon is created and it carries the energy in the direction of the vibrating atom, shown in Figures 12.13*b* and *c*.

12.9.1   Heat Transmission at Submicrometer- and Nanoscale

Since heat is a form of energy, the flow of heat requires a carrier that is similar to other energy carriers. Molecular heat transmission theory postulates that the carriers for heat vary from substance to substance. For example, the carriers for heat in metals are electrons and phonons, whereas the carrier in dielectric and semiconducting materials is predominantly phonons. We have learned from Section 12.6 that phonons may be viewed as a group of "virtual" mass particles that characterize the state of energy of the lattice in which atoms are situated. The vibration of the lattices that bond the atoms supplies the energy to phonons and thus the heat in the substance. Consequently, the transmission of heat in a substance relies on the movement of phonons in semiconductors, or the movement of both phonons and electrons in the case of metals.

The study of the movements of electrons and phonons associated with heat transmission in a substance is an extremely complicated subject and is beyond the scope of this book. What we hope to learn from this section is the basic science of molecular heat conduction in solids only.

The reader is referred to the literature (Tzou, 1997; Tien et al., 1998; Flik et al., 1992; Tien and Chen, 1994) for detailed descriptions of this subject.

One can well imagine that there are millions of phonon "particles" in a solid of any size. The movement of these heat-carrying phonon particles in the solid inevitably results in numerous collisions with other phonons. Figure 12.14 illustrates how heat is

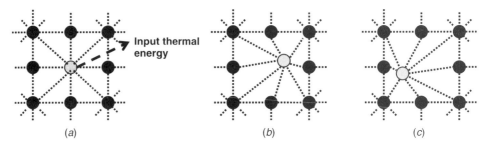

       (*a*)   (*b*)   (*c*)

Figure 12.13. Vibrations of lattices in a substance. (*a*) Initial state. (*b*) Lattice movement at time t_1 and (*c*) time t_2.

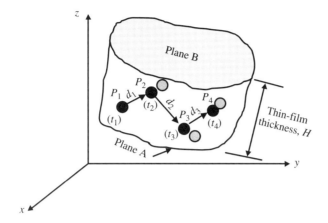

Figure 12.14. Collision of a traveling phonon.

transmitted from plane A to plane B. As heat transportation in a solid is predominantly done by the flow of phonon particles, we will hypothesize the traveling of phonons for this process. Because there are great many phonons present, collisions of free phonon particles traveling inside the solid are unavoidable.

With reference to Figure 12.14, a phonon (in black) begins its journey of transmitting heat at position P_1 on plane A at time t_1. It collides with another phonon (in gray), which happens to be on its course of travel, after the energy carrier phonon traveled by a distance d_1 to position P_2 at time t_2. The collision with a neighboring phonon results in altering the direction of its course (we call this *phonon scattering*), as can be seen in the figure. Similar forms of collisions and scattering of the phonon with other nearby phonon particles take place in its remaining journey to the top plane B. The following phenomena for individual phonons are worth noting:

1. The traveling phonon changes its course after each collision—the scattering effect.
2. The distance of free traveling, called the *free path*, varies between collisions.
3. Because of the change of free path, the time that the phonon travels between collisions also changes.

We will thus define the following key terms used in molecular heat transfer as illustrated in Figure 12.14:

1. The average *mean free path* (MFP):

$$\lambda = \tfrac{1}{3}(d_1 + d_2 + d_3)$$

2. The average *mean free time* (MFT):

$$\tau = \tfrac{1}{3}[(t_2 - t_1) + (t_3 - t_2) + (t_4 - t_3)] = \tfrac{1}{3}(t_4 - t_1)$$

The average MFP in the above expression for a heat carrier in solids means the average distance that results in the carrier's losing its excessive energy while traveling in a bulk material. Exact values of both average MFP and MFT are scarce. They depend strongly on the molecular structure of the materials, such as the congregation of molecules, grain geometry, and grain boundaries as well as the temperature environment. Following are the approximate values that engineers may use for qualitative assessments on heat transportation in solids at submicrometer- and nanoscales.

The average MFP for electrons is about 10^{-8} m at room temperature. The average MFP for phonon collision/scattering can be much longer. For example, the MFP in thin diamond films is measured at approximately 10^{-7} m. As mentioned earlier in Section 12.8.2, the MFP for gas is approximately 65 nm, whereas the MFP for liquids is about twice the size of their molecules. The MFP for both electrons and phonons is strongly dependent on temperature. It becomes shorter as temperature increases.

The typical value for the average MFT for metals is on the order of 10^{-12} s. It is in the range of nanoseconds to picoseconds for dielectric materials. The magnitude of average MFT obviously varies significantly with abrupt change of geometry, for example, cracks and interfaces of thin films.

In macroscale heat transmission, the film thickness H in Figure 12.14, or the space is large enough to allow hundreds and thousands of phonon collisions to take place. The MFP and MFT associated with the collisions of individual energy carriers such as phonons are not critical in the process, as these effects are average out over the size of the solid, as well as the time frame that is involved in macroscale heat transfer. However, the phonon collision/scattering effect becomes critical when the size of the solid shrinks to submicrometers and nanometers. Consequently, heat transfer in submicrometer- and nanosized solids needs to be considered in both *length* and *time* scales. All heat transmission activities in solids at the nanoscale are transient in nature.

The summary in Table 12.2 may be useful in understanding the principal heat carriers in substances (Tien and Chen, 1994). There are two significant changes in heat conduction in the submicrometer- and nanoscale domains: (1) the thermal conductivity and

TABLE 12.2. General Features of Heat Carriers

Feature	Free Electrons	Phonons	Photons
Substance of dominance	Heat conduction in metals	Heat conduction in dielectric and semiconductive materials	Thermal radiation
Sources of generation	Valence or excited electrons	Lattice vibration	Atomic or molecular transition
Propagation media	In vacuum or media	In media	In vacuum or media
Approximate velocity (m/s)	10^6	10^3	10^8

TABLE 12.3. Parameters for Thermal Conductivity of Thin Films

Parameter	Materials	Dielectric and Semiconductors
Specific heats, C	Specific heat of electrons, C_e	Specific heat of phonons, C_s
Molecular velocity, V	Electron Fermi velocity, $V_e \cong 1.4 \times 10^6$ m/s	Velocity of phonons (sound velocity), $V_s \cong 10^3$ m/s
Average mean free path, λ	Electron mean free path, $\lambda_e \cong 10^{-8}$ m	Phonon mean free path, $\lambda_s \cong$ from 10^{-7} m and up

Source: Flik et al. (1992) and Tien and Chen (1994).

(2) the heat conduction equation similar to Equation (5.31). The approximate demarcation between submicrometer- and micro/macroscale heat conduction is at $H < 7\lambda$ (Flik et al., 1992; Tien and Chen, 1994), where H is the thickness of thin films and λ is the mean free path of phonons. For semiconductors such as silicon with $\lambda \approx 10^{-7}$ m given in Table 12.3, a thin film of that material at a thickness $H < 0.7$ μm requires the use of nanoscale heat transfer principles in the design analysis.

12.9.2 Thermal Conductivity of Thin Films

There are several models proposed for estimating the thermal conductivity of materials at submicrometer scales. The following simple model for thermal conductivity k for thin films was derived from kinetic theory (Rohsenow and Choi, 1961):

$$k = \tfrac{1}{3}CV\lambda \qquad (12.12)$$

The definitions of C, V, and λ in Equation (12.12) for metals and dielectrics and semiconductors are presented in Table 12.3. Also indicated in the table are the approximate values of these parameters for qualitative assessment purposes.

A more straightforward model for estimating the thermal conductivity of thin films was proposed as follows (Flik and Tien, 1990):

$$\frac{k_{\text{eff}}}{k} = 1 - \frac{\lambda}{3H} \qquad (12.13a)$$

for thermal conductivity normal to the thin films and

$$\frac{k_{\text{eff}}}{k} = 1 - \frac{2\lambda}{3\pi H} \qquad (12.13b)$$

for thermal conductivity along the surface of the thin film. Here k is the thermal conductivity of the same bulk material at the macroscale.

Equation (12.13) offers a more straightforward way for engineers to evaluate thermal conductivity in heat conduction analysis of solids at submicrometer- and nanoscales. As expected, there are some discrepancies between the predicted values with the above formula and those measured by experiments. A 5% error is observed for thin films with $H < 7\lambda$ for thermal conductivity normal to the film and the same error for thin film with $H < 4.5\lambda$ with thermal conductivity along the films.

> **Example 12.3.** Estimate the thermal conductivity of thin silicon films 0.2 μm thick.
>
> *Solution:* From Table 12.3, we have the average mean free path length of phonons, $\lambda_e = 10^{-7}$ m. Thus, using Equation (12.13a), we obtain the following thermal conductivity for the thin silicon film with 0.2 μm thickness:
>
> $$k_{\text{eff}} = \left(1 - \frac{\lambda_e}{3H}\right) k = \left(1 - \frac{10^{-7}}{3 \times (0.2 \times 10^{-6})}\right) k = 0.833k$$
>
> with $k = 1.57$ W/cm-$^\circ$C from Table 7.3. The thermal conductivity of the thin silicon film normal to the film surface is $k_{\text{eff}} = 0.833 \times 1.57$ W/cm-$^\circ$C $= 1.308$ W/cm-$^\circ$C.
>
> One may use Equation (12.13b) to estimate the thermal conductivity of the same silicon film along the film surface to be $k_{\text{eff}} = 0.894k = 1.404$ W/cm-$^\circ$C.

12.9.3 Heat Conduction Equation for Thin Films

A common perception of heat transportation by most engineers is that heat flux $q(\mathbf{r},t)$ crossing the boundary of a solid results in a temperature difference or temperature gradient $\nabla T(\mathbf{r},t)$ in the same solid. Likewise, a temperature gradient maintained in a solid can result in heat flow in the solid. A critical question is how long it takes to produce the resulting temperature gradient or heat flow by the respective causes mentioned above. We assume intuitively that these results take immediate effect with the input causes, so there is no time lag between the causes and the resulting effects. This perception serves as the basis for our derivation of the classical Fourier law of heat conduction in Equation (5.28). However, this hypothesis of the simultaneous occurrence of the causes and the corresponding results between the heat flux and the temperature gradients is a clear contradiction to what we have postulated on the heat transmission in solids as illustrated in Figure 12.14. As we have envisaged from the illustration in that figure, finite time is required for phonons to carry heat from one location in the solid to another. The average MFT is used to establish such required time for heat to be transmitted from one location to another in a solid. The MFT, in essence, results in the delay of heat transportation in the solid and thus the delay of the resulting temperature gradient in applying heat flux across the boundaries of the solid.

The delay, or the *lag time*, between the heat flux and the corresponding temperature gradients is not a significant factor in heat transportation in solids at the macroscale, as a relatively large time scale is used in those analyses. However, for solids at the submicrometer- and nanoscale, such delay can introduce significant error in the analysis if it is not properly compensated for. Consequently, the heat conduction equation (5.31) needs to be modified with an additional term to account for the lag time. The modified heat conduction equation is expressed as

$$\nabla^2 T(\mathbf{r}, t) + \frac{Q}{k} = \frac{1}{\alpha} \frac{\partial T(\mathbf{r}, t)}{\partial t} + \frac{\tau}{\alpha} \frac{\partial^2 T(\mathbf{r}, t)}{\partial t^2} \qquad (12.14)$$

The last term on the right-hand side of Equation (12.14) represents the velocity of heat transmission in solids as illustrated in Figure 12.14. It is in the form of wave propagation of the temperature, $T(\mathbf{r}, t)$, and it is called *thermal wave propagation* in the solid. This term becomes insignificant when $H \gg \lambda$. The variable τ in the above equation is referred to as the *relaxation time*, which is the average time that a carrier (e.g., a phonon) travels between collisions. The following relationship is used for evaluating τ:

$$\tau = \frac{\lambda}{V} \qquad (12.15)$$

where λ is the average mean free path and V is the average velocity of the heat carrier.

One may readily envisage that τ is on the order of 10^{-10} s from Table 12.3 for semiconductors.

According to Flik et al. (1992), the time to use Equation (12.15) is when the observation time t is shorter than H^2/α, where H is the thickness of the thin film and α is the thermal diffusivity of the material.

12.10 MEASUREMENT OF THERMAL CONDUCTIVITY

As described in Section 5.7.1, the thermal conductivity k of a material is a measure of how well it can conduct heat. In general, metals are better conductors than semiconductors and insulators. We also learned that material properties (e.g., thermal conductivity) of matter become size dependent at the submicrometer- and nanoscale. Section 12.9 and Equation (12.13) have indicated that the k value of matter decreases with the reduction of their size. Example 12.3 illustrates the significant reduction of thermal conductivity of a thin silicon film.

The thermal conductivity of materials at the submicrometer- and nanoscale is a critical parameter in micro- and nanoscale engineering design. Excessive heating and electricity leakage are the two most critical problems in molecular electronics as gaps between electrical circuits scale down drastically in those cases. Accurate assessment of the thermal conductivity of the substrate material at the submicrometer- and nanoscale is a critical requirement in the mechanical design of nanoscale circuitries.

Equations such as (12.13) are derived primarily on data collected from measurements of heat conduction in thin films. Techniques for thermal conductivity measurements at the macroscale are well established. Figure 12.15 shows a typical setup for measuring k vales for conducting materials such as metals. It involves two cylinders A and B with the same diameter but with respective lengths L_A and L_B. Cylinder A has a known thermal

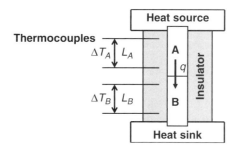

Figure 12.15. Measurements of thermal conductivity of conductive materials.

conductivity k_A, and the thermal conductivity in cylinder B, k_B, is to be measured. These two cylinders are placed in a test chamber as shown in the figure. The compound cylindrical sample is thermally insulated on its circumference to ensure one-dimensional heat flow in the longitudinal direction. One end of the cylinders (cylinder A) is in contact with a heat source and the other end of the sample (cylinder B) is in contact with a heat sink. Heat can thus flow from the heat source to the heat sink. The k value of cylinder B, k_B, can be obtained from the following expression derived from the Fourier law of heat conduction expressed in Equation (5.27):

$$k_B = \frac{\Delta T_A}{\Delta T_B} k_A \tag{12.16}$$

where the temperature differences in both rods, ΔT_A and ΔT_B, are measured by the thermocouples attached to cylinders A and B, respectively.

A different setup is used for measuring the thermal conductivity of semiconductor materials, mainly because of their low k values (Table 7.3 shows a 2.5 times higher k for copper than for silicon). With such low k values, the setup needs to provide more assurance of one-dimensional heat flow in the sample than that for conducting materials. Figure 12.16 shows such a setup.

With respect to Figure 12.16, samples made of wafers of semiconductor materials of thickness d are placed on both surfaces of a hot plate that serves as a heat source. The other surfaces of the samples are in contact with heat sinks. Temperatures at the interfaces

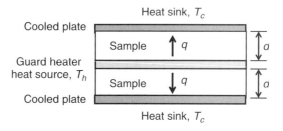

Figure 12.16. Setup for measuring thermal conductivity of semiconductors.

of sample/heat source/heat sinks are strictly controlled and accurately measured. The k value of the sample material can be obtained by the expression

$$k = \frac{d}{A(T_h - T_c)} q \qquad (12.17)$$

where d is the thickness of the samples, A is the contact surface area, q is the heat flux output of the hot plate, and T_h and T_c are the temperatures of the heat source and heat sink, respectively.

The same measurements for solids at the submicrometer- and nanoscale, however, are far more complicated. Neither of the setups in Figures 12.15 and 12.16 for measuring thermal conductivity for solids at the macroscale can be used to measure k values of thin-film materials because of the following reasons:

1. Thin films of semiconductors or dielectric materials have much lower k values than conducting materials at the macroscale. Consequently, the technique used for this type of material at the nanoscale must have high resolution and precision.

2. Samples are minute in size and extremely thin. It is not practically feasible to develop a proper fixture for proper positioning and stationing of these minute samples in the measurement system.

3. Being thin in the samples, it is not possible to ensure one-dimensional heat flow as in the case of conventional thermal conductivity measurements.

4. For samples of thin films, the temperature gradient along the sample thickness is too small to be measurable.

5. There is no place for thermocouples for temperature measurement in the minute samples.

Two techniques that have been developed for measuring the thermal conductivity of thin films at the submicrometer- and nanoscale are presented below.

The 3-Omega Method. This method was first proposed by Cahill (1990). The theoretical basis of this method is an early example presented in a classic heat conduction book (Carslaw and Jaeger, 1959).

Figure 12.17 illustrates a solid of half space, that is, the solid has infinite surface area and depth, containing a minute line heat source on the exposed surface. This line heat source produces heat periodically with an angular frequency ω in radians per second. If the line source is defined by an (x, y) coordinate system shown in Figure 12.17, the temperature change at a point $P(x,y)$ inside the solid volume at a distance r away from the center of the line heat source can be obtained by the expression

$$\Delta T(r) = \frac{1}{\pi k} \left(\frac{P}{L} \right) K_0(qr) \qquad (12.18)$$

where

r = linear distance between line heat source and point at which temperature change is measured, $= \sqrt{x^2 + y^2}$

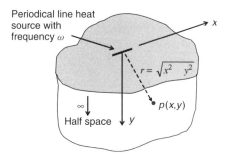

Figure 12.17. Semi-infinite solid with a line heat source.

P = amplitude of power generated by line heat source at angular frequency 2ω

L = length of line heat source

k = thermal conductivity of semi-infinite solid

$K_0(r)$ = modified Bessel function of second kind at zero order

The variable q in Equation (12.18) is given as

$$\frac{1}{q} = \sqrt{\frac{\alpha}{i2\omega}}$$

where α is the thermal diffusivity of the solid and $i = \sqrt{-1}$.

The relationship of temperature change ΔT in the semi-infinite solid and its thermal conductivity k in Equation (12.18) is used as the basis for the 3-omega method for the measurement of thermal conductivity of thin films.

Following Cahill (1990), we consider the situation illustrated in Figure 12.18.

A fine wire made of conductive metal such as gold or copper with length L and width $2b$ is deposited onto the surface of a thin film, of which the thermal conductivity is to be measured. The wire will function as a line heat source with passing alternate current $i(t) = P \sin(\omega t)$, where P is the maximum power, ω is the angular frequency of an AC power supply, and t is time. It can also be used as a fine thermocouple for the temperature

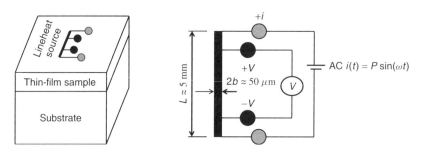

Figure 12.18. Power supply to a line heat source in semi-infinite solid surface.

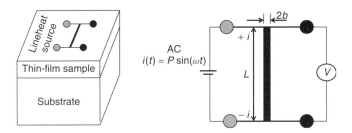

Figure 12.19. Alternative setup for measurement of thermal conductivity of thin film.

measurement of the line heat source. The thin film is deposited onto a substrate with low thermal conductivity. There are four terminals leading from the line heat source, two for the voltage measurements and the other two for power input, as shown in Figure 12.18.

The size of the line heat source is so small in comparison to the sample thin film in Figure 12.18 that the situation may be considered to be a good approximation to the case illustrated in Figure 12.17. Therefore, the temperature change in the thin film can be determined by Equation (12.18).

An alternative arrangement of the four terminals from the line heat source is presented in Figure 12.19, in which these terminals are located on either sides of the line heat source.

Cahill (1990) offered the following findings between the frequency of the heat input ω and those for the corresponding temperature change in the solid and the voltage change in the line heat source. He concluded that if the power supply to the line heat source in a semi-infinite solid at a frequency of 1ω, then the following correlations become valid following the solution in Equation (12.18):

1. The temperature change in the thin film, relating to the change of electrical resistance in the line heat source, oscillates with a frequency of 2ω.
2. The measured voltage across the two terminals from the line heat source contains a component that oscillates at a frequency of 3ω.

The above correlations lead to the measurement of the thermal conductivity of the sample material by measuring the resistance change in the heater with a frequency of 2ω for the temperature change of the sample and the voltage change that relates to a frequency of 3ω.

Consequently, the k value of a thin-film sample can be measured by attaching it to a substrate as shown in Figure 12.20.

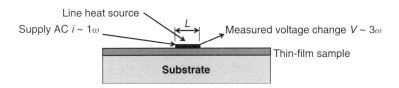

Figure 12.20. Thermal conductivity measurement by 3-omega method.

The thermal conductivity k of the thin film in the arrangement in Figure 12.20 can be obtained as

$$k = \frac{V^3 \ln (\omega_2/\omega_1)}{4\pi L R^2 (V_{3,1} - V_{3,2})} \frac{\Delta R}{\Delta T} \tag{12.19}$$

where

$\omega_1, \omega_2 =$ two input angular frequencies of power supply to line heat source

$R =$ electrical resistance of line heat source

$V =$ voltage across metal line heat source at 1ω

$V_{3,1}, V_{3,2} =$ Measured voltages across line heater at 3ω, with power supplies at ω_1 and ω_2, respectively (both measurements relate to 3ω)

$\Delta R, \Delta T =$ measured change of electrical resistance and temperature at given point in thin film during change of power supplies with angular frequencies ω_1 and ω_2 (related to 2ω)

The term $\Delta R/\Delta T$ in Equation (12.19) is the calibration of the temperature versus electrical resistance change of the line heater. This method was used to measure thermal conductivities of α-SiO$_2$ and two other materials, Pyroceram 9606 and Pyrex 7740 (Cahill, 1990), and thin SiO$_2$ films (Kim et al., 1999).

The 3-omega method was used to measure the thermal conductivity of vertically oriented carbon nanotubes (Hu et al., 2006). Carbon nanotube samples with diameters ranging from 10 to 80 nm with a length of 13 μm were used in the experiment. Measured thermal conductivities are 74 W/m-K at 295 K and 83 W/m-K at 323 K. The measured thermal conductivity appears to vary linearly with temperature. The scattering of measured data at each temperature is approximately ± 2 W/m-K.

Scanning Thermal Microscopy. This method involves the use of a special atomic force microscope (AFM) with a heated tip. As shown in Figure 12.21, the thin-film sample is fixed to an actuator with high precision that can move the sample in three perpendicular directions along the x, y, and z axis to provide the scanning of the surface topography of the sample under the AFM. The tip of the AFM heats the sample at the contacted points and senses the temperature of the sample at the same point, as illustrated in the diagrams of $T(x)$ and $T(y)$ at the lower left of the figure.

This arrangement can measure the thermal conductivity of thin-film samples in the z direction, k_z, by measuring the heat flow along that direction, whereas the thermal conductivities in the other two directions, k_x and k_y, may be measured by mapping the temperature variations along the surface, $T(x)$ versus x and $T(y)$ versus y, as illustrated in the figure.

There have been several other methods reported for the measurements of thermal conductivities of thin films, with spatial resolutions as presented in Table 12.4.

Fiege et al. (1999) developed a method for measuring the thermal conductivity of thin film by a setup that combines the 3-omega method and the scanning thermal microscopy. The latter is used to measure the in-plane thermal conductivities, k_x and k_y, whereas the

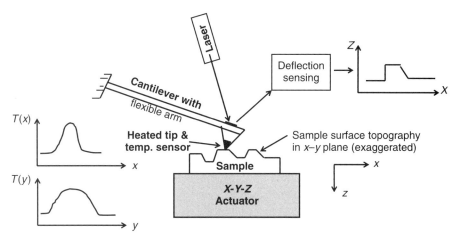

Figure 12.21. Setup for thermal conductivity measurement using scanning thermal microscope.

TABLE 12.4. Thermal Microscopy for Measurement
of Thermal Conductivity of Thin Films

Measurement Techniques	Spatial Resolution
Scanning thermal microscopy	<100 nm ($= 0.1$ μm)
Near-field optical thermometry	<1 μm
Laser surface reflectance	1 μm
Liquid crystals	1 μm
Raman spectroscopy	1 μm
Infrared thermometry	$1-10$ μm

3-omega method is used to measure the thermal conductivity k_z of the thin film in the normal direction.

12.11 CHALLENGES IN NANOSCALE ENGINEERING

The limitless opportunities for nanotechnology offered by Feynman 50 years ago has inspired enormous enthusiasm not only among the members of the science and engineering communities but also those in business and government funding agencies. Indeed, there have been significant accomplishments in new nanoproducts of nanoparticles, nanowires, and nanotubes and their applications in a great many areas as described in Section 12.5. However, there remain many challenges to scientists and engineers for further capitalizing on the potential benefits of this emerging technology. We view the following five areas to be major challenges to scientists and engineers.

12.11.1 Nanopatterning in Nanofabrication

We learned in Chapter 8 that the first task in microfabrication is to produce patterns of the shape of the microscale components and devices using a photolithographic method as described in Section 8.2. The working principle of the photolithographic process on substrate surfaces involves the creation of microscale master prints of microscale patterns using a photoreduction process. These microscale patterns are then projected onto a thin film of photoresist that is coated onto the substrate material by a light source. Etching the exposed photoresist with microscale patterns can create permanent microscale patterns on the substrate. This method works satisfactorily for pattern sizes greater than 100 nm.

For pattern sizes below 100 nm, the wavelength of the light sources used in photolithography, normally in the range of 300–500 nm (Section 8.2.3), is too long to produce sharp edge definitions of the minute patterns because of the diffraction of the light beams at these wavelengths. Consequently light sources with much shorter wavelengths, such as extreme ultraviolet light with 13.5 nm wavelength or X-rays with wavelengths ranging from 0.01 to 1 nm, are required for nanoscale photolithography. Optical lithographic techniques that use these light sources for nanoscale patterning are much too costly for mass production. Alternative techniques of nanoimprint lithography have been developed for nanopatterning.

The concept of *nanoimprint lithography* (NIL) was first proposed by Stephen Y. Chou (1998). It is often referred to as thermoplastic nanoimprint lithography (T-Nil), or *hot embossing lithography*, by the industry because it involves the use of thermoplastic material and heating of the thermoplastics in the process. The working principle of NIL is similar to the creation of a pattern on a piece of paper in a stamping process, with the desired pattern engraved on the surface of the stamp. With respect to Figure 12.22, a typical T-Nil process involves spin coating a thin layer of imprint thermoplastic polymer (e.g., PMMA) on the substrate surface similar to what is described in Section 8.2.2 (Figure 12.22*a*). A mold with the intended pattern created by such a method as electron beam engraving is bought into contact with the thermoplastic coating under certain

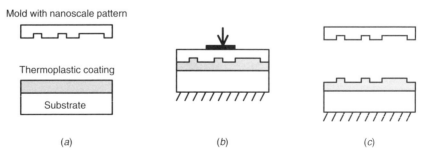

(a)   *(b)*   *(c)*

Figure 12.22. Thermoplastic nanoimprint process: (*a*) mold and sample; (*b*) hot pressing; (*c*) mold separation.

pressure (Figure 12.22*b*). The mold is pressed into the thermoplastic, which is now heated above its glass transition temperature in a hot embossing chamber. After being cooled, the mold is retracted and a permanent pattern is created in the thermoplastic resist. The transfer of patterns in the polymer resist in Figure 12.22*c* to the substrate is accomplished by using reactive ion etching of the part of the patterns on the resist until the substrate is exposed. The substrate is then patterned by etching using the patterned resist as mask. Line resolution of 250 nm is reported by this method.

UV–nanoimprint lithography, or UV-Nil, is used to create patterns with higher resolutions than those by the T-Nil method. The working principle of this method is similar to that of T-Nil. Instead of using heat to cure the imprinted polymer coating on the substrate, UV light is applied for the curing. Figure 12.23 illustrates the setup of UV-Nil. Transparent mold such as quartz is used in this case, as shown in Figure 12.23*a*. Polymer coating material of low viscosity, such as soft PDMS, is used in this type of nanoimprint process, so very light pressure is required to press the mold into the polymer coating, as shown in Figure 12.23*b*. Ultraviolet light is then shone on the pressed polymer coating through the transparent mold for curing (Figure 12.23*c*). The mold is separated from the imprint after curing, as shown in Figure 12.23*d*.

UV-Nil can be accomplished at room temperature. The problem associated with the dimensional stability of polymer coating materials around the transition temperature can be avoided. This method can result in feature sizes varying between 50 nm and 20 μm.

There are several challenging issues involved with nanopatterning using the two nanoimprint methods presented above:

1. **Overlay Accuracy:** Separation of mold from the polymer coating after the imprinting process often causes distortion, and therefore accurate edge definition of the minute imprinted patterns. Optical alignment and temperature control are other possible sources for producing quality nanopatterns.

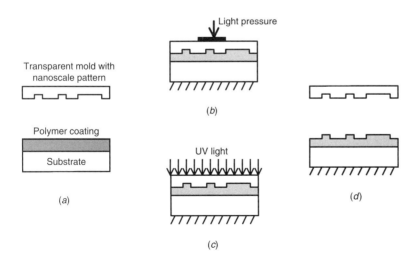

Figure 12.23. UV nanoimprint process: (*a*) mold and sample; (*b*) pressing; (*c*) UV curing; (*d*) mold separation.

2. **Selection of Polymer Coating or Resist:** It is desirable to use polymer resist suitable for low-temperature and low-pressure imprint with minimum shrinkage after curing. It must have sufficient mechanical strength and tearing resistance during mold separation.

3. **Templates:** These are molds with permanent nanopatterns on their contacting surface. They must be defect free from fabrication. For easy separation after imprint, suitable antistick coating on the template is necessary. Other issues involve the cleaning after each imprint and mitigation of thermal stress/strain induced by mismatch of the coefficient of thermal expansion of the template and the substrate, especially in the UV-Nil process.

12.11.2 Nanoassembly

The nanofabrication technology as described in Section 12.3 has prompted strong interest by engineers in the design and manufacture of nanoscale machine components such as rings, tubes, shafts, bearings, and gears. Indeed, we have succeeded in producing nanoscale wires and tubes as described in Section 12.4. However, the production of these nanoscale components of simple geometries remains in the rudimentary stage. The manufacturing of nanoscale components of more complex geometry such as stepped shafts or gears requires the assembly of many atoms, one at a time, into the desired shape, as illustrated in Section 12.3. This process is clearly an impossible task, as, for example, the production of a miniature device component with a volume of $1\,mm^3$. One will quickly reach an accounting of 10^{21} atoms that need to be assembled, based on an assumption of 10 atoms in the span of $1\,nm$. This obviously does not make any practical sense. Synthetic *self-assembly* and *replication* have thus become the only practical solutions to nanoscale assembly.

A major challenge to scientists and engineers is to develop techniques that will enable reproduction of atoms or molecules of specific materials into desired shapes in orderly and controllable ways. Biochemical processes similar to cellular replication and division and the assembly of nucleotides in DNA comprising particular genes that lead to the production of cellular proteins are potential models for the development of practical nanoassembly technology.

12.11.3 New Materials for Nanoelectromechanical Systems (NEMS)

A practical and useful electromechanical system requires materials that can perform electromechanical functions. Materials such as piezoelectric crystals and shape memory alloys described in Chapters 2 and 7 are essential elements in MEMS. A challenge to scientists and engineers is to develop materials with similar characteristics suitable for applications in NEMS.

Piezoelectric crystals as described in Section 7.9 have low energy conversion efficiencies. As indicated in a white paper by Gillett (2002), the conversion efficiency of electricity production by the popular PZT, a commercially available piezoelectric material, is only 10%. Miniaturizing this material to the nanoscale will further reduce its production of electricity with minute mechanical input. The challenge to engineers is thus to develop nanoscale piezoelectric materials that can produce electromechanical conversion with higher efficiency. An alternative solution for better piezoelectric performance by piezoelectric crystals is to use these materials in thin layers stacked up in a nanostructure (Gillett, 2002). This proposition would be a technical challenge to engineers in the fabrication, assembly, and packaging of these nanoscale thin piezoelectric layers.

Nanomagnets can be another useful material for nanoscale actuating materials. Ferromagnets, including iron, cobalt, and nickel, can be made magnetic when exposed to an external magnetic field. Nanoscale magnet disks made of supermalloy ($Ni_{80}Fe_{14}Mo_5$) of diameter ranging from 500 to 50 nm and thickness of 6–15 nm were fabricated using electron beam lithography (Cowburn et al., 1999). The challenge is to make these nanomagnets magnetized by external electric field instead. These electromagnetic nanodots or disks will be valuable actuating materials for NEMS. The success in developing $CoFe_2O_4$–$BaTiO_3$ ferroelectromagnetic nanocomposite (Zheng et al., 2004) has shown promise for making available materials for NEMS devices.

12.11.4 Analytical Modeling

We learned in Sections 12.4 and 12.5 that an important characteristic of nanoscale materials is the size-dependent properties, in particular the change of optical absorptivity and reflectivity of nanoparticles. Recent active research has also indicated that thermophysical properties of several materials at the nanoscale also vary with the size, such as the thermal conductivity of silicon as described in Section 12.9.2. Much effort has also been made in measuring the mechanical properties of nanoscale materials. Results have shown that mechanical properties such as Young's modulus are affected by the size of the samples depending on the materials. For ductile materials such as copper and gold, the Young's modulus increases by 5–20% from thin-film samples 1–8 nm in size (Liang et al., 2002). On the contrary, the Young's modulus of silicon decreases monotonically as the thickness of the specimen decreases (Park et al., 2005). Their measured values are compared with of that of bulk silicon presented in Table 7.2 in Table 12.5.

The size-dependent thermophysical and mechanical properties of materials at the nanoscale significantly increase the complexity of engineering analyses of thermophysical behavior of nanostructures. Solutions of all governing equations derived for such analyses require the coupling with mechanical deformation equations such as shown in Equation (12.3) for updating with the current size of the structure required to assign

TABLE 12.5. Size Effect on Young's Modulus of Silicon

Miller Index for Orientation	Young's Modulus[a] of Bulk Material (Table 7.2)	Young's Modulus[a] of Nanocantilever at 3 nm Thick	% Change
<100>	129.5	75	42.1
<110>	168.0	110	34.5
<111>	186.5	130	30.3

[a]Unit: GPa.

the proper material properties. The size-dependent material properties make governing equations such as those listed in Figure 12.12 for fluid dynamics analyses and Equation (12.14) for heat conduction analysis for nanostructure nonlinear. The size-dependent thermophysical and mechanical material properties also need to be included in the constitutive equations required in finite element formulations for engineering analyses (Hsu, 1986). Solution techniques for these nonlinear and coupled equations present a major challenge to scientists and engineers. Proper modifications of commercially available finite element codes to accommodate these complex modifications for nanostructures will be a significant challenge to developers.

12.11.5 Testing

The minute size of microscale device components has made testing a major challenge to engineers, as described in Hsu (2004). Further shrinking in size to the nanoscale of these device components, which do not exist at present but will emerge in the near future, will make such a task a major challenge to engineers.

12.12 SOCIAL IMPACTS OF NANOSCALE ENGINEERING

Nanotechnology and nanoscale engineering have the potential to offer humankind with limitless opportunities in improving their way of life. Following are a few products that could be realized in the next 10–20 years:

1. Health care

    New vaccines and medicines that cure many incurable diseases

    Synthetic antibody–like nanoscale drugs seeking out to destroy malignant cells in human or animal bodies

    Nanoscale devices that can deliver drugs to needy cells or organs in the human body and dispense drugs as needed

    Nanoscale devices that can perform surgeries inside the human body without incision

    Reliable and accurate in vivo medical diagnostic systems

2. Smart materials

    Smart surface coating materials with self-adjusting thermophysical properties for enduring and energy-efficient buildings and air conditioning

    Smart fabrics for self-cleaning cloths

    Superstrong and lightweight materials for ground and air transportation

3. Energy

    Clean energy supply, including new synthetic fuels

    High-efficiency energy conversion for fossil fuel or renewable energy such as solar energy and fuel cells

    Superlong life energy storage systems, such as new batteries

4. Agriculture

Environmentally friendly chemicals for effective pesticide control

Environmentally friendly fertilizers

New breed of crops and domestic animals that can feed the entire population in the world

We may thus expect many bright prospects that nanotechnology would bring to our society for better living conditions for all humankind. With the breakthroughs in health care listed above, one may expect human life span to be extended to 150 years or longer. The much improved technology in agriculture could virtually eliminate famine in world. The impact to our society by the advance of nanotechnology is thus immeasurable.

We may also come up with a list of some "dream products" by nanotechnology toward the end of the current century. These may include:

1. "Dust size" computers that are orders of magnitude more powerful than current available computers
2. "Needle tip" size robots that can perform biomedical and surgical functions inside the human body and for search and rescue in disastrous sites
3. Robots with artificial human intelligence becoming the mainstream workforce in our society
4. Unlimited supply of clean and affordable renewable energy for all humankind
5. Spacecraft weighing less than today's family cars, suitable for interplanet traveling

The above list, of course, is by no means complete. Many now-unthinkable products by nanotechnology will emerge as the technology advances further. We may well imagine the positive impacts of this technology to society and the global village in which we all live and enjoy.

There are possible negative impacts by nanotechnology too; this technology is more than capable of developing deadly vaccines or weapons for mass destruction that could wipe out the entire human civilization overnight. Another likely negative impact of nanotechnology is unethical human cloning and duplication. We single out these two developments because they can bring irreversible colossal grave consequences to the human race. It is therefore the responsibility of all scientists and engineers with good conscience to prevent such occurrences.

PROBLEMS

Part 1 Multiple Choice

1. Nanotechnology deals with objects at the scale of (a) subnanometers, (b) nanometers, (c) micrometers.

2. One nanometer is equal to (a) 10^{-3} m, (b) 10^{-6} m, (c) 10^{-9} m.

3. One nanometer has an approximate span length of (a) 5, (b) 10, (c) 15 shoulder-to-shoulder hydrogen atoms.

4. A DNA has a width less than (a) 2 nm, (b) 3 nm, (c) 4 nm.

5. The diameter of human hair is about (a) one-hundredth, (b) one-thousandth, (c) one-millionth of meter.

6. In general, a virus has the size of (a) 1 nm, (b) 10 nm, (c) 100 nm.

7. Modern-day nanotechnology was initiated by Richard Feynman, who was a (a) engineer, (b) chemist, (c) physicist.

8. Dr. Feynman delivered his famous speech on miniaturization in (a) 1939, (b) 1949, (c) 1959.

9. Dr. Feynman's vision on miniaturization that has led to the development of nanotechnology was inspired by (a) chemical systems, (b) physical systems, (c) biological systems.

10. The essence of nanotechnology is to manipulate and control the composition and structure of (a) atoms, (b) crystal structures, (c) grain boundaries of matter.

11. The principal benefit of nanotechnology is to control the (a) material properties, (b) geometry, (c) size of matter.

12. The projected annual revenue earned by nanotechnology by year 2015 is (a) $100 million, (b) $1000 million, (c) $1 trillion.

13. The scientific basis on which microsystems technology is developed is (a) solid-state physics, (b) quantum physics, (c) particle physics.

14. The scientific basis on which nanotechnology is developed is (a) solid-state physics, (b) quantum physics, (c) particle physics.

15. An atomic force microscope is used to (a) examine atomic structures, (b) measure the surface topography of solids, (c) measure the force required to move atoms.

16. A scanning tunneling microscope is used to (a) examine atomic structures, (b) measure the surface topography of solids, (c) measure the electric power required to move atoms.

17. A scanning tunneling microscope creates tunnels (a) for the flow of atoms, (b) for the passage for heat, (c) for the passage of electric charges.

18. The first human-made nanostructure is (a) the buckyball, (b) nanoparticles, (c) nanotubes.

19. Buckyballs are produced from (a) graphite, (b) carbon, (c) sapphire.

20. Buckyballs have a diameter of (a) 0.5 nm, (b) 0.7 nm, (c) 1 nm.

21. In general, nanoparticles have at least one dimension to be less than (a) 10 nm, (b) 100 nm, (c) 1000 nm.

22. A spectacular phenomenon of nanoparticles is the distinct changes in (a) strength, (b) shape, (c) color with the size.

23. Nanoparticles have extremely high surface area–volume ratio; make them easier in (a) diffusion, (b) evaporation, (c) advection processes.

24. Nanoparticles are used as biomarkers because of their minute sizes and (a) distinct shape, (b) distinct color, (c) distinct hardness.

25. Nanowires typically have diameter on the order of (a) 0.1 nm, (b) 1 nm, (c) 10 nm.

26. Nanowires are basic building blocks for (a) molecular electronics, (b) molecular biology, (c) molecular chemistry.

27. In general, nanowires have (a) less, (b) about the same, (c) higher electric conductivity than their bulk materials.

28. A common technique for producing carbon nanotubes is by rolling (a) carbon 60 sheets (b) graphite sheets, (c) diamond sheets.

29. In general, carbon nanotube walls are made by carbon atoms in (a) square, (b) diamond, (c) hexagonal bond structures.

30. In general, carbon nanotubes are (a) 10, (b) 20, (c) 30 times stronger than steel.

31. Carbon nanotubes are (a) 6, (b) 60, (c) 600 times stronger than silicon.

32. In general, carbon nanotubes are (a) 10, (b) 100, (c) 1000 times better electrical conductors than copper.

33. Gallium nitride nanotubes are good candidates for (a) biomedical, (b) optical, (c) molecular electronic applications.

34. Quantum physics is used to describe (a) physical movement of atoms, (b) energy transport, (c) collisions of atoms in MEMS and nanosystems.

35. A quantum represents the smallest amount of (a) mass, (b) volume, (c) energy that any system can gain or lose.

36. Photons have mass that equals that of (a) an electron, (b) a neutron, (c) zero mass.

37. A fundamental postulation in quantum mechanics is that atoms in matter are (a) packed shoulder to shoulder, (b) connected to each other by elastic bonds, (c) sporadically situated in space with no bonding.

38. A change of the arrangement of atoms in matter from the natural state is a change in (a) atomic size, (b) number, (c) overall size of the matter.

39. Molecular dynamics deals with movement of atoms in (a) free space, (b) elastically bonded matrix, (c) electrical field.

40. The Schrödinger equation solves for the instantaneous (a) position of atoms, (b) temperature, (c) electric field in a matter subject to external force or energy.

41. Solution of the Schrödinger equation offers the instantaneous (a) electric field, (b) temperature, (c) size of an object at the nanoscale.

42. A rarefied gas means the gas is at (a) extremely low pressure, (b) intermediate pressure, (c) vacuum.

43. A good estimate of the MFP for gases is (a) 65 nm, (b) 75 nm, (c) 130 nm.

44. The Knudsen number is defined as (a) the density of the gas over the physical size of the confinement, (b) the mean free path over the physical size of the confinement, (c) the velocity of the gas over the speed of sound.

45. Conventional fluid dynamics theories are valid for (a) very small, (b) large, (c) very large Knudsen numbers.

46. The Mach number is defined as (a) the density of the gas over the physical size of the confinement, (b) the mean free path over the physical size of the confinement, (c) the velocity of the gas over the speed of sound.

47. A gas is considered to be compressible if the Mach number is (a) less than, (b) equal to, (c) greater than 0.3.

48. The larger the Knudsen number, the (a) smaller, (b) larger, (c) no difference in the confinement of the gas.

49. A "rule of thumb" to classify a rarefied gas is (a) Kn < 0.1, Ma < 0.3, (b) Kn > 0.1, Ma < 0.3, (c) Kn > 0.1, Ma > 0.3.

50. The Navier–Stokes equations can be reasonably used for gas flow with Knudsen number that is less than (a) 0.01, (b) 0.1, (c) 1.0.

51. One could use the Navier–Stokes equation for gas flow between the values of 0.01 and 0.1 by using (a) any, (b) nonslip, (c) slip boundary condition.

52. MFP stands for (a) molecular free path, (b) minimum free path, (c) mean free path for energy carriers in a substance.

53. MFT stands for (a) molecular free time, (b) minimum free time, (c) mean free time for energy carriers in a substance.

54. The thermal conductivity of solids at the submicrometer- and nanoscale is (a) less than, (b) about equal to, (c) greater than the value of the same material at the macroscale.

55. The additional term in the heat conduction equation for solids of submicrometer- and nanoscales is related to (a) the extra time, (b) the extra velocity, (c) the extra heat in heat transmission.

56. The difficulty in measuring thermal conductivity of thin films is because of (a) the quantum effect, (b) the high cost, (c) the low k value and proper fixtures for the measurement.

57. The 3-omega method is used to measure (a) thermal diffusivity, (b) thermal conductivity, (c) thermal expansion coefficient of thin films.

58. The 3-omega method was derived from a classical heat conduction analysis based on a semi-infinite solid subject to a line heat source generating heat at a frequency of (a) ω, (b) 2ω, (c) 3ω.

59. The induced temperature in the solid in the 3-omega method is related to (a) ω, (b) 2ω, (c) 3ω.

60. The voltage across the line heat source in the 3-omega method is related to (a) ω, (b) 2ω, (c) 3ω.

61. Measurement of the thermal conductivity of thin films by scanning thermal microscopes involves the use of (a) atomic force microscopes, (b) scanning tunneling microscopes, (c) high-power zoom-in microscopes.

62. Scanning thermal microscopes in theory can measure the thermal conductivity of thin films in (a) one, (b) two, (c) three directions.

63. Conventional photolithography is not suitable to produce nanoscale patterns on substrates because (a) no suitable material, (b) cannot produce nanopatterns on template, (c) excessive diffraction by the long wavelength of light sources.

64. Nanoimprint lithography is primarily used to produce nanopatterns on substrates in (a) laboratory, (b) mass production, (c) new scientific discoveries.

65. The hot embossing lithographic process uses heat to (a) shorten the process time, (b) soften the resist and for curing, (c) produce better quality nanopatterns.

66. Ultraviolet-nanoimprint lithography uses extreme ultraviolet light to (a) produce better nanopatterns, (b) heat the resist faster, (c) cure the imprinted resist.

67. The most critical challenging issue to engineers in producing nanoscale machine components is (a) design, (b) fabrication, (c) assembly and packaging.

68. The most promising technique for nanoassembly is by (a) intelligent assembly tools, (b) biochemical process, (c) superfast evaporation technique.

69. Viable nanoelectromechanical systems (NEMS) require the availability of (a) solid theoretical design tools, (b) reliable material properties, (c) nanoscale electromechanical actuation materials.

70. Strong size-dependent material properties of nanoscale objects require major modifications in (a) manufacturing, (b) design analyses, (c) testing method.

Part 2 Descriptive Problems

1. What is the fundamental reason for developing nanotechnology?

2. Why can nanotechnology offer humankind limitless benefits and opportunities?

3. Offer three unique characteristics of nanoparticles.

4. Offer three unique characteristics of nanowires.

5. Offer three unique characteristics of nanotubes.

6. Offer three applications of nanoparticles.

7. Offer three applications of nanotubes.

8. List two advantages and two disadvantages of the size-dependent material properties of nanoscale substances.

9. What is the difference between solid-state physics and quantum physics?

10. Why was the first human-made nanostructure called a buckyball?

11. Conduct research to find how much mechanical force is required to free one atom from a gold substrate using an atomic force microscope.

12. Conduct research to find how high a voltage is required to free one atom from a gold substrate using a scanning tunneling microscope.

13. Conduct research to illustrate how nanowires are used as the gate of a nanoscale transistor.

14. Why does the high surface area–volume ratio of nanoparticles make them preferable to use in a diffusion process? Conduct research to find why this offers a unique advantage in the pharmaceutical industry.

15. Describe the material characteristics of carbon nanotubes in your own terms based on what has been described in this chapter.

16. Find the effective thermal conductivity of the thin films described in Example 12.3 for all the materials listed in Table 7.3.

17. Estimate the "lagging time" in heat conduction in the metal thin films listed in Table 7.3.

18. Formulate the element equation for a finite element analysis using the Galerkin method described in Section 10.5.1 in conjunction with the modified heat conduction equation (12.14) for solids at the submicrometer scale.

19. Of the listed positive social impacts by nanotechnology, what are the two most benevolent items in your view. Why?

20. What are the two potentially most devastating impacts by nanotechnology in your view. Why?

REFERENCES

Abramowitz, M., and Stegun, I., *Handbook of Mathematical Functions*, Dover Publications, New York, 1964.

Allaire, P. E., *Basics of the Finite Element Method*, William. C. Brown, Dubuque, IA, 1985.

Amato, I., Nanotechnology-Shaping the World Atom by Atom, NSTC report, http://itri .loyola.edu/nano/IWGN.Public.Brochure/, 1999.

American Society for Testing and Materials (ASTM), *Manual on the Use of Thermocouples in Temperature Measurement*, ASTM Special Technical Publication 470A ASTM, Philadelphia, PA, 1974.

Angell, J. B., Terry, S. C., and Barth, P. W., "Silicon Micromechanical Devices," *Scientific American*, vol. 248, no. 4, April 1983, pp. 44–55.

Arnold, D. P., Gururaj, S., Bhardwaj, S., Nishida, T., and Sheplak, M., "A Piezoresistive Microphone for Aeroacoustic Measurements," paper presented at the ASME International Mechanical Engineering Congress and Exposition, New York, November 11–16, 2001, pp. 1–8.

Askeland, D., *The Science and Engineering of Materials*, 3rd ed., PWS Publishing, Boston, 1994.

Arnold, D. P., Nishida, T., Cattafesta, L. N., and Sheplak, M., "A Directional Acoustic Arrary Using Silicon Micromachined Piezoresistive Microphones," *Journal of Acoustical Society of America*, Vol 113, No. 1, January 2003, pp. 289–298.

Avallone, E. A., and Baumeister, T., III, Eds., *Marks' Standard Handbook for Mechanical Engineers*, 13th ed., McGraw-Hill, New York, 1996.

Bart, S. F., Tavrow, L. S., Mehregany, M., and Lang, J., "Microfabricated Electrohydrodynamic Pumps," *Sensors and Actuators*, vol. A21–A23, 1990, pp. 193–197.

Barth, P. W., "Silicon Fusion Bonding for Fabrication of Sensors, Actuators and Microstructures," in *Transducers '89, Proceedings of the 5th International Conference on Solid-State Sensors and Actuators and Eurosensors III*, Elsevier Sequoia, Netherlands, Vol. 2, 1990, pp. 919–926.

Bathe, K-J., and Wilson, E. L., *Numerical Methods in Finite Element Analysis*, Prentice-Hall, Englewood Cliffs, NJ, 1976.

Bean, K., "Anisotropic Etching of Silicon," *IEEE Transactions on Electron Devices*, vol. ED-25, no. 10, October 1978, pp. 1185–1193.

Beckwith, T. G., Marangoni, R. D., and Lienhard V. J. H., *Mechanical Measurements*, 5th ed., Addison-Wesley, Reading, MA, 1993.

Beer, F. P., and Johnston, Jr., E. R., *Vector Mechanics for Engineers*, 5th ed. McGraw-Hill, New York, 1988.

Beskok, A., and Karniadakis, G. E., "A Model for Flows in Channels, Pipes, and Ducts at Micro and Nano Scales," *Microscale Thermophysical Engineering*, vol. 3, no. 1, 1999, pp. 43–77.

Bifano, T. G., Johnson, H. T., Bierden, P., and Mali, R. K., "Elimination of Stress-Induced Curvature in Thin-Film Structures," *Journal of Microelectromechanical Systems*, vol. 11, no. 5, October 2002, pp. 592–597.

Bley, P., "The LIGA Process for Fabrication of Three-Dimensional Microscale Structures," *Interdisciplinary Science Reviews*, vol. 18, no. 3, 1993, pp. 267–272.

Bley, P., "Polymers—An Excellent and Increasingly Used Material for Microsystems," paper presented at the SPIE 1999 Symposium on Micromachining and Microfabrication, Santa Clara, CA, September 20–22, 1999.

Boley, B. A., and Weiner, J. H., *Theory of Thermal Stresses*, Wiley, New York, 1960.

Bouwstra, S., and Geijselaers, B., "On the Resonance Frequencies of Micro Bridges," paper presented at the 1991 International Conference on Solid-State Sensors and Actuators (Transducers '91), San Francisco, CA, 1991, pp. 538–542.

Bowen, C. R., Stevens, R., Perry, A., and Mahon, S. W., "Optimization of Materials and Microstructure in 3-3 Piezocomposites", in *Proceedings of Engineering Design Conference 2000 on Design for Excellence*, Sivaloganathan, S., and Andrew, P. T. J., Eds., Professional Engineering Publishing, London, 2000, pp. 361–370.

Bradley, D. A., Dawson, D., Burd, N. C., and Loader, A. J., *Mechatronics*, Chapman & Hall, London, 1993.

Breguet, J. M., Henein, S., Mericio, R., and Clavel, R., "Monolithic Piezoceramic Flexible Structure for Micromanipulation," in *Proceedings of the 9th International Precision Engineering Seminar*, Brawnschweig, Germany, May 1997, pp. 397–400.

Brown, T. L., and LeMay, Jr., H. E., *Chemistry*, 2nd ed., Prentice-Hall, Englewood Cliffs, NJ, 1981.

Bryzek, J., Petersen, K., Mallon, Jr., J. R., Christel, L., and Pourahmadi, F. *Silicon Sensors and Microactuators,* Lucas Nova Sensors, Fremont, CA, 1991.

Bryzek, J., Petersen, K., Christel, L., and Pourahmadi, F. "New Technologies for Silicon Accelerometers Enable Automotive Applications," *Sensors and Actuators,* vol. sp-903, SAE International, 1992, pp. 25–32.

Bryzek, J., Petersen, K., and McCulley, W. "Micromachines on the March," *IEEE Spectrum*, May 1994, pp. 20–31.

Buerk, D. G., *Biosensors, Theory and Applications*, Technomic Publishing, Lancaster, PA, 1993.

Cahill, D. G., "Thermal Conductivity Measurement from 30 to 750K: The 3 Omega Method," *Review of Scientific Instruments*, vol. 61, no. 2, 1990, pp. 802–808.

Carslaw, H. S., and Jaeger, J. C., *Conduction of Heat in Solids*, Oxford University Press, Oxford, 1959, p. 193.

Chang-Chien, P. P. L., and Wise, K. D., "Wafer-Level Packaging Using Localized Mass Deposition," paper presented at Transducers '01. Eurosensors XV, 11th International Conference on Solid-State Sensors and Actuators, *Digest of Technical Papers*, vol 1, 2001, pp. 182–185.

Chau, K. H.-L., Lewis, S. R., Zhao, Y., Howe, R. T., Bart, S. F., and Marcheselli, R. G., "An Integrated Force-Balanced Capacitive Accelerometer for Low-G Applications," paper presented at Transducers '95, Proceedings, the 8th International Conference on Solid-State Sensors and Actuators, Stockholm, June 25–29, 1995, pp. 593–596.

Chiao, M., and Lin, L., "Sealing Technologies," in *MEMS Packaging*, Hsu, T. R., Ed., Institute of Electrical Engineers (IEE), Herts, United Kingdom, 2004, pp. 61–83.

Chilton, J. A., and Goosey, M. T., Eds., *Special Polymers for Electronics & Optoelectronics*, Chapman & Hall, London, 1995.

Chiou, J., "Pressure Sensors in Automotive Applications and Future Challenges," paper presented at IMECE'99, Nashville, TN, November 17, 1999.

Choi, I. H., and Wise, K. D., "A Silicon-Thermopile-Based Infrared Sensing Array for Use in Automated Manufacturing," *IEEE Transactions on Electron Devices*, vol. ED-33, no. 1, January 1986, pp. 72–79.

Chou, S.Y., Nanoimprint Lithography, U.S. Patent No. 5,772,905, June 30, 1998.

Chowdhury, S., Ahmadi, M., and Miller, W. C., "Design of a MEMS Acoustical Beamforming Sensor Microarray," *IEEE Sensors Journal*, vol. 2, no. 6, December 2002, pp. 617–627.

Conradie, E. H., and Moore, D. F., "SU-8 Thick Photoresist Processing as a Functional Material for MEMS Applications," *Journal of Micromechanics and Microengineering*, vol. 12, 2002, pp. 368–374.

Cotter, T. P., "Principles and Prospects for Micro Heat Pipes," in *Proceedings of 5th International Heat Pipe Conference*, Vol. 4, Tsukuba, Japan, May 14–18, 1984, pp. 328–334.

Cowburn, R. P., Koltsov, D. K., Adeyeye, A. O., and Welland, M. E., "Single-Domain Circular Nanomagnets," *Physical Review Letters*, vol. 83, no. 5, August 1999, pp. 1042–1045.

Culbertson, C. T., Jacobson, S. C., and Ramsey, J. M., "Dispersion Sources for Compact Geometries on Microships," *Analytical Chemistry*, vol. 70, no. 18, 1998, pp. 3781–3789.

Desai, C. S., *Elementary of Finite Element Method*, Prentice-Hall, Englewood Cliffs, NJ, 1979.

Doscher, J., "Accelerometer Design and Applications," Analog Devices, Inc., Norwood, MA, private communication, 1999.

Dove, R. C., and Adams, P. H., *Experimental Stress Analysis and Motion Measurement*, Charles E. Merrill, Columbus, OH, 1964.

Drafts, W., "Acoustic Wave Technology Sensors," *Sensors*, vol. 17, no. 10, 2000, pp. 68–77.

Drexler. K. E., *Nanosystems*, Wiley, New York, 1992.

Eargle, J., *The Microphone Book*, 2nd ed., Elsevier, Amsterdam, 2004.

Fan, L.-S., Tai, Y.-C., and Muller, R. S., "IC-Processes Electrostatic Micro-motors," paper presented at the IEEE International Electronic Devices Meeting, December 1988, pp. 666–669.

Fatikow, S., "Microrobot-Based Assembly of Microsystems," *NEXUS Research News*, no. 1, 1998, pp. 12–14.

Feddema, J., Christenson, T., and Polosky, M., "Parallel Assembly of High Aspect Ratio Microstructures," *SPIE Proceedings on Microrobotics and Microassembly*, vol. 3834, Boston, MA, September, 21–22, 1999, pp. 153–164.

Feynman, R. P., "There's Plenty Room at the Bottom," paper presented at the Annual Meeting of the American Physical Society, California Institute of Technology, Pasadena, CA, December 29, 1959.

Fiege, G. B. M., Altes, A., Heiderhoff, R. and Balk, L.J., "Quantitative Thermal Conductivity Measurements with Nanometer Resolution," *Journal of Physics D, Applied Physics*, vol. 32, no. 5, 1999, pp. L13–L17.

Flik, M. I., Choi, B. I., and Goodson, K. E., "Heat Transfer Regimes in Microstructures," *Journal of Heat Transfer, ASME Transactions*, vol. 114, August 1992, pp. 666–674.

Flik, M. I., and Tien, C. L., "Size Effect on Thermal Conductivity of High-Tc Thin-Film Superconductors," *Journal of Heat Transfer, ASME Transactions*, vol. 112, 1990, pp. 873–881.

French, P. J., and Evans, A. G. R., "Piezoresistance in Single Crystal and Polycrystalline Si," in *Properties of Silicon*, INSPEC, American Physical Society, Ridge, NY 1988, Section 3.4, pp. 94–103.

Gabriel, K. J., Trimmer, W. S. N., and Walker, J. A., "A Micro Rotary Actuator Using Shape Memory Alloys," *Sensors and Actuators*, vol. 15, no. 1, 1988, pp. 95–102.

Giachino, J. M., and Miree, T. J., "The Challenge of Automotive Sensors," *Proceedings of SPIE on Microlithography and Metrology in Micromachining*, vol. 2640, October 1995.

Gilbert, J. R., Ananthasuresh, G. K., and Senturia, S. D., "3D Modeling of Contact Problems and Hysteresis in Coupled Electro-Mechanics," paper presented at the 9th Annual International Workshop on Micro Electro Mechanical Systems, MEMS '96, San Diego, 1996, pp. 127–132.

Gillett, S. L., Nanotechnology: Clean Energy and Resources for the Future, White paper for the Foresight Institute, University of Nevada, Reno, NV, November 2002.

Gise, P., and Blanchard, R., *Modern Semiconductor Fabrication Technology*, Prentice-Hall, Englewood Cliffs, NJ, 1986.

Greitmann, G., and Buser, R. A., "Tactile Microgripper for Automated Handling of Microparts," *Sensors and Actuators*, vol. A53, 1996, pp. 410–415.

Griego, J., Smith, M., and Wood, J., "MEMS Cell Gripper," Senior project report, San Jose State University, San Jose, CA, May 2000.

Guerin, L., The SU8 homepage, www.geocities.com/guerinnlj/, 2005.

Halliday, D., Resnick, R., and Krane, K. S., *Physics*, Vol. 2, extended version, 4th ed., Wiley, New York, 1992.

Harrison, D. J., and Glavina, P. G., "Towards Miniaturized Electrophoresis and Chemical Analysis Systems on Silicon: An Alternative to Chemical Sensors," *Sensors and Actuators B*, vol. 10, 1993, pp. 107–116.

He, Y., Marchetti, J., and Maseeh, F., "MEMS Computer-Aided Design," paper presented at the European Design & Test Conference and Exhibition on Microfabrication, Paris, March 17–20, 1997.

Heinrich, J. C., and Pepper, D. W., *Intermediate Finite Element Method*, Taylor and Francis, Philadelphia, 1999.

Helvajian, H., and Janson, S.W., "Microengineering Space Systems," in *Microengineering Aerospace Systems*, H. Helvajian, Ed., American Institute of Aeronautics and Astronautics, Reston, VA, 1999, Chapter 2, pp. 29–72.

Henning, A. K., "Microfluidic MEMS," Paper 4.906, paper presented at the IEEE Aerospace Conference, Snowmass, CO, March 25, 1998.

Henning, A. K., Fitch, J., Hopkins, D., Lilly, L., Faeth, R., Falsken, E., and Zdeblick, M., "A Thermopneumatically Actuated Microvalve for Liquid Expansion and Proportional Control," in *Proceedings, Transducers '97, IEEE Conference on Sensors and Actuators*, IEEE, Chicago, June, 1997, pp. 825–828.

Higuchi, T., Yamagata, Y., Furutani, K., and Kudoh, K., "Precise Positioning Mechanism Utilizing Rapid Deformations of Piezoelectric Elements," in *Proceedings of IEEE, Micro Electro Mechanical Systems*, IEEE, Travemunde, Germany, February 1990, pp. 222–226.

Histand, M. B., and Alciatore, D. G., *Introduction to Mechatronics and Measurement Systems*, McGraw-Hill, Boston, 1999.

Howe, R., "Resonant Microsensors," paper presented at the 4th International Conference on Solid-State Sensors and Actuators, Tokyo, Japan, 1987, pp. 843–848.

Hsieh, W. H., Hsu, T. Y., and Tai, Y. C., "A Micromachined Thin-film Teflon Electret Microphone," paper presented at the 1997 International Conference on Solid-State Sensors & Actuators, Chicago, June 16–19, 1997, pp. 425–428.

Hsu, P. C., Mastrangelo, C. H., and Wise, K. D., "A High Sensitivity Polysilicon Diaphragm Condenser Microphone," paper presented at the IEEE MEMS'98 Conference, Heidelberg, Germany, February 1998, pp. 580–585.

Hsu, T. R., *The Finite Element Method in Thermomechanics*, Allen & Unwin, London, 1986.

Hsu, T. R. *On Nonlinear Thermomechanical Analysis of IC Packaging,* "Advances in Electronic Packaging," ASME, 1992, pp. 325–326.

Hsu, T. R., "Reliability in MEMS Packaging," paper presented at the 44th Annual IEEE International Reliability Physics Symposium, San Jose, CA, March 26–30, 2006, pp. 398–404.

Hsu, T. R., Ed., *MEMS Packaging*, The Institute of Electrical Engineers (IEE), United Kingdom, 2004.

Hsu, T. R., Chen, G. G., and Sun, B. K., "A Continuum Damage Mechanics Model Approach for Cyclic Creep Fracture Analysis of Solder Joints," in Advances in Electronic Packaging, Proceedings of the 1993 ASME International Electronics Packaging Conference, Binghamton, NY, September 29–October 2, 1993, pp. 127–138.

Hsu, T. R., and Clatterbaugh, J., "Joing and Bonding Technologies," in *MEMS Packaging*, Institute of Electrical Engineers (IEE), Herts, United Kingdom, 2004, Chapter 2, pp. 23–59.

Hsu, T. R. and Custer, J. S., "Fundamentals of MEMS Packaging," in *MEMS Packaging*, Hsu, T. R., Ed., Institute of Electrical Engineers, Herts, United Kingdom, 2004, pp. 1–21.

Hsu, T. R., and Sinha, D. K., *Computer-Aided Design—An Integrated Approach*, West Publication, St. Paul, MN, 1992.

Hsu, T. R., and Sun, N. S., "Residual Stresses/Strains Analysis of MEMS," in *Proceedings of MSM '98*, Santa Clara, CA April 6–8, 1998, pp. 82–87.

Hsu, T. R., and Zheng, X. M., "Tensile Creep Strength of Eutectic Solder Cores," Paper 93-WA/EEP-11, presented at the ASME Winter Annual Meeting, New Orleans, November 28–December 3, 1993.

Hsu, T. R., and Nguyen, L. "On the Use of Fracture Mechanics in Plastic-Encapsulated Microcircuits Packaging," Proceedings of ASME International Mechanics of Surface Mount Assemblies, November 1999, pp. 1–8.

Hu, X. J., Padilla, A. A., Fisher, T. S., and Goodson, K. E., "3-Omega Measurements of Vertically Oriented Carbon Nanotubes on Silicon," *Journal of Heat Transfer, ASME Transactions*, vol. 128, November 2006, pp. 1109–1113.

Iijima, S., "Helical Microtubes of Graphite Carbon," *Nature*, vol. 354, 1991, pp. 56–58.

Ikuta, K., "Micro/Miniature Shape Memory Alloy Actuator," in 1990 *IEEE Proceedings of ICRA*, Cincinnati, OH May 13–18, 1990, pp. 2156–2161.

Janna, W. S., *Introduction to Fluid Mechanics*, 3rd ed., PWS Publishing, Boston, 1993.

Janson, S., Helvajian, H., and Breuer, K., "Micropropulsion Systems for Aircraft and Spacecraft," in *Microengineering Aerospace Systems*, Helvajian, H., Ed., American Institute of Aeronautics and Astronautics, Reston, VA, 1999, Chapter 17, pp. 657–696.

Janusz, B., Petersen, K., Christel, L., and Pourahmadi, F., "New Technologies for Silicon Accelerometers Enable Automotive Applications," SAE Technical Paper Series 920474; SAE International, Detroit, MI, 1992, pp. 25–32.

Jerman, H., "Electrically-Activated, Micromachined Diaphragm Valves," *Technical Digest*, paper presented at the IEEE Solid-State Sensor and Actuator Workshop, Hilton Head Island, SC. June 1990, pp. 65–69.

Jerman, H., "Electrically-Activated, Normally-Closed Diaphragm Valves," *Technical Digest*, paper presented at the 6th International Conference on Solid-State Sensors and Actuators, San Francisco, June 1991, pp. 1045–1048.

Jiang, X. N., Zhou, Z. Y., Yao, J., Li, Y., and Ye, X.Y., "Micro-Fluid Flow in Microchannel," in *Proceedings of Transducers '95, the 8th International Conference on Solid-State Sensors & Actuators & Eurosensors IX*, Vol. 2, Stockholm, Sweden, June 25–29, 1995, pp. 317–320.

Johnson, D. H. and Dudgeon, D. E., *Array Signal Processing*, Prentice-Hall, Upper Saddle River, NJ, 1993.

Kasap, S. O., *Principles of Electrical Engineering Materials and Devices*, Irwin, Chicago, 1997.

Keneyasu, M., Kurihara, N., Katogi, K., and Tabuchi, K., "An Advanced Engine Knock Detection Module Performance Higher Accurate MBT Control and Fuel Consumption Improvement," in *Proceedings of Transducers '95, Eurosensors IX*, stockholm, Sweden, 1995, pp. 111–114.

Kim, Y. J. and Hsu, T. R. "A Numerical Analysis on Stable Crack Growth Under Increasing Load," *International Journal of Fracture*, vol. 20, 1982, pp. 17–32.

Kim, C. J., Pisano, A. P., and Muller, R. S., "Overhung Electrostatic Microgripper," *Technical Digest*, paper presented at the IEEE International Conference on Solid-State Sensors and Actuators, San Francisco, 1991, pp. 610–613.

Kim, J. H, Feldman, A., and Novotny, D., "Application of the three Omega Thermal Conductivity Measurement Method to a Film on a Substrate of Finite Thickness," *Journal of Applied Physics*, vol. 86, no. 7, October 1999, pp. 3959–3960.

Kim, P., Majumdar, A., and McEuen, P.L., "Thermal Transport Measurements of Individual Multiwall Nanotubes," *Physics Review Letters*, vol. 87, 2001, pp. 215502 (1–4)

Kovacs, G. T. A., *Micromachined Transducers Sourcebook*, McGraw-Hill, New York, 1998.

Kreith, F. (Ed.), *CRC Handbook of Mechanical Engineering*, CRC Press, Boca Raton, FL, 1997.

Kreith, F., and Bohn, M. S., *Principles of Heat Transfer*, PWS Publishing, Boston, 1997.

Krishnamoorthy, S., and Gridharan, M. G., "Analysis of Sample Injection and Band-Broadening in Capillary Electrophoresis Microchips," in *Proceedings of 2000 International Conference on Modeling and Simulation of Microsystems*, San Diego, March 27–29, 2000, pp. 528–531.

Kumar, S., and Cho, D., "Electric Levitation Bearings for Micromotors," *Technical Digest*, IEEE International Conference on Solid-State Sensors and Actuators, San Francisco, 1991, pp. 882–885.

Kwok, H. L., *Electronic Materials*, PWS Publishing, Boston, 1997.

LeBerre, M., Launay, S., Sartre, V., and Lallemand, M., "Fabrication and Experimental Investigation of Silicon Micro Heat Pipes for Cooling Electronics," *Journal of Micromechanics and Microengineering*, vol. 13, 2003, pp. 436–441.

Lee, T. H., Tong, Q. Y., Chao, Y. L., Huang, L. J., and Gosele, U., "Silicon on Quartz by a Smarter Cut Process," in *Proceedings of the 8th International Symposium on Silicon-on-Insulator Technology and Devices*, Crisstoloveanu, S., Ed., The Electrochemical Society, Pennington, NJ, 1997, pp. 27–32.

Liang, L. H., Li, J. C., and Jiang, Q., "Size-Dependent Elastic Modulus of Cu and Au Thin Films," *Solid State Communications*, vol. 121, 2002, pp. 453–455.

Lin, L., "MEMS Post-Packaging by Localized Heating and Bonding," *IEEE Transactions on Advanced Packaging*, vol. 23, no. 4, November 2000, pp. 608–616.

Lipman, J., "Microfluidics Puts Big Labs on Small Chips," *EDN Magazine*, December 1999, pp. 79–86.

Lober, T. A., and Howe, R. T., "Surface-Micromachining Processes for Electrostatic Microactuator Fabrication," *Technical Digest*, paper presented at the IEEE Solid-State Sensor and Actuator Workshop, Hilton Head Island, SC, June 1988, pp. 59–62.

Lyke, J., and Forman, G., "Space Electronics Packaging Research and Engineering," in *Microengineering Aerospace Systems*, Helvajian, H., Ed., The Aerospace Corporation, El Segundo, CA, 1999, Chapter 8, pp. 259–346.

MacDonald, G. A., "A Review of Low Cost Accelerometers for Vehicle Dynamics," *Sensors and Actuators*, vol. A21-A23, 1990, pp. 303–307.

Madou, M., *Fundamentals of Microfabrication*, CRC Press, Boca Raton, FL, 1997.

Madou, M. J., *Fundamentals of Microfabrication*, 2nd ed., Taylor & Francis CRC Press, Boca Raton, FL, 2002.

Maluf, N., *An Introduction to Microelectromechanical Systems Engineering*, Artech House, Boston, 2000.

Manz, A., Effenhauser, C. S., Burggraf, N., Harrison, D. J., Seller, K., and Fluri, K., "Electroosmotic Pumping and Electrophoretic Separations for Miniaturized Chemical Analysis Systems," *Journal of Micromechanics and Micro Engineering*, vol. 4, 1994, pp. 257–265.

Mehregany, M., Gabriel, K. J., and Trimmer, W. S. N., "Integrated Fabrication of Polysilicon Mechanisms," *IEEE Transactions on Electron Devices*, vol. ED-35, no. 6, 1988, pp. 719–723.

Mehregany, M., Senturia, S. D., and Lang, J. H., "Friction and Wear in Microfabricated Harmonic Side-Drive Motors," *Technical Digest*, paper presented at the IEEE Solid-State Sensor and Actuator Workshop, Hilton Head Island, SC, June 1990, pp. 17–22.

Micro Chem, "NANO™ SU-8," www.microchem/products/pdf/SU8_50-100/pdf, 2007.

Mish, F. C., Ed.-in-Chief, *Merriam Webster's Collegiate Dictionary*, 10th ed., Merriam-Webster, Springfield, MA, 1995.

Moroney, R. M., White, R. M., and Howe, R. T., "Fluid Motion Produced by Ultrasonic Lamb Waves," in *IEEE 1990 Ultrasonics Symposium Proceedings*, Honolulu, 1990, pp. 355–358.

Moroney, R. M., White, R. M., and Howe, R. T., "Microtransport Induced by Ultrasonic Waves," *Applied Physics Letters*, vol. 59, 1991, pp. 774–776.

"Nano Technology," *The BusinessWeek*, vol. 50, Spring 2002, pp. 181–182.

National Geographic, "Geographica on Global Economy," February 2005.

National Research Council, *Microelectromechanical Systems—Advanced Materials and Fabrication Methods*, Vol. 5, Assembly, Packaging, and Testing, No. NMAB-483, National Academy Press, Washington, DC, 1997, pp. 38–49.

Newell, W. E., "Miniaturization of Tuning Forks," *Science*, vol. 161, 1968, p. 1320.

Nguyen, L. T., Hsu, T. R., and Kuo, A.Y., "Interfacial Fracture Toughness in Plastic Packages," in *Application of Fracture Mechanics in Electronic Packaging*, AMD-vol. 222/EEP-vol. 20, American Society of Mechanical Engineers, New York, 1997, pp. 15–24.

Nyborg, W. L. M., "Acoustic Streaming," in *Physical Acoustics*, Mason, W. P., Ed., Academic, New York, 1965, pp. 265–331.

O'Connor, L., "MEMS: Microelectromechanical Systems," in *Mechanical Engineering*, American Society of Mechanical Engineers, New York, February 1992, pp. 40–47.

Ohnstein, T., Fukiura, T., Ridley, J., and Bonne, U., "Micromachined Silicon Microvalve," *IEEE Micro Electro Mechanical Systems*, February 1990, pp. 95–98.

Oliver, A. D. and Custer, J. S., "Testing and Design for Test," in *MEMS Packaging*, Hsu, T.-R., Ed., Institute of Electrical Engineers (IEE), Herts, United Kingdom 2004, pp. 141–157.

Osgood, W. R., *Residual Stresses in Metals and Metal Construction*, Reinhold Publishing, New York, 1954.

Ozisik, M. N., *Boundary Value Problems of Heat Conduction*, International Textbook Company, Scranton, PA, 1968.

Papila, M., Haftka, R. T., Nishida, T., and Sheplak, M., "Piezoresistive Microphone Design Pareto Optimization: Tradeoff between Sensitivity and Noise Floor," Paper No. AIAA-2003-1632, paper presented at the 44th AIAA/ASME/ASCE/AHS Structural Dynamics and Materials Conference, Norfolk, VA, April 7–10, 2003.

Park, S. H., Kim, J. S., Paark, J. H., Lee, J. S., Choi, Y. K., and Kwon, O. M., "Molecular Dynamics Study on Size-Dependent Elastic Properties of Silicon Nanocantilevers," *Thin Solid Films*, vol. 492, 2005, pp. 285–289.

Patankar, N. A., and Hu, H. H., "Numerical Simulation of Electroosmotic Flow," *Analytical Chemistry*, vol. 70, 1998, pp. 1870–1881.

Patterson, G. N., *Introduction to the Kinetic Theory of Gas Flow*, University of Toronto Press, Toronto, 1971.

Paulsen, J., and Giachino, J., "Powertrain Sensors and Actuators: Driving toward Optimized Vehicle Performance", paper presented at the 39th IEEE Vehicular Conference, vol. II, 1989, pp. 574–594.

Pecht, M. G., Nguyen, L. T. and Hakim, E. B., *Plastic Encapsulated Microelectronics*, John Wiley & Sons, Inc., New York, 1995.

Pedersen, M., Olthuis, W., and Bergveld, P., "High-Performance Condenser Microphone with Fully Integraged CMOS Amplifier and DC-DC Voltage Converter," *Journal of Microelectromecahnical Systems*, vol. 7, no. 4, December 1998, pp. 387–394.

Peirs, J., Reynaerts, D., and VanBrusssel, H., "A Microrobotic Arm for a Self Propelling Colonoscope," in *Proceedings of Actuator '98*, Bremen, Germany, 1998, pp. 576–579.

Petersen, K., Barth, P., Poydock, J., Brown, J., Mallon, Jr., J., and Bryzek, J., "Silicon Fusion Bonding for Pressure Sensors," *Technical Digest*, paper presented at the IEEE Solid-State Sensor and Actuator Workshop, Hilton Head Island, SC, June 1988.

Petersen, K., Pourahmadi, F., Brown, J., Parsons, P., Skinner, M., and Tudor, J., "Resonant Beam Pressure Sensor Fabricated with Silicon Fusion Bonding," *Proceedings of Transducers 91, the 1991 International Conference on Solid State Sensors and Actuators*, San Francisco, CA., June 24–27, 1991, pp. 177–180.

Petersen, K. E., "Silicon as a Mechanical Material," *Proceedings of IEEE*, vol. 70, no. 5, May 1982, pp. 420–457.

Pfahler, J., Harley, J., and Bau, H. H., Liquid and Gas Transport in Small Channels, *Microstructures, Sensors and Actuators*, vol. DSC-19, ASME, November 1990, pp. 149–157.

Pottenger, M., Eyre, B., Kruglick, E., and Lin, G., "MEMS: The Maturing of a New Technology," *Solid State Technology*, September 1997, pp. 89–96.

Pourahmadi, F., Christ, L., and Petersen, K., "Variable-Flow Micro-Valve Structure Fabricated with Silicon Fusion Bonding," *Technical Digest*, paper presented at the IEEE Solid-State Sensor and Actuator Workshop, Hilton Head Island, SC, June 1990, pp. 78–81.

Pourahmadi, F., and Twerdok, J. W., "Modeling Micromachined Sensors with Finite Elements," *Engineering & Technology Guide, Machine Design*, July 26, 1990, pp. 44–60.

Powers, W. F., and Nicastri, P. R., *Automotive Vehicle Control Challenges in the Twenty-First Century*, International Federation of Automatic Control, Beijing, 1999.

Putty, M. W., and Chang, S.-C., "Process Integration for Active Polysilicon Resonant Microstructures," *Sensors and Actuators*, vol. 20, 1989, pp. 143–151.

Rebeiz, G. M. *RF MEMS-Theory, Design, and Technology*, John Wiley & Sons, Hoboken, N. J., 2003.

Rice, J. R., "Elastic Fracture Mechanics Concept for Interfacial Cracks," *Journal of Applied Mechanics, ASME Transactions*, vol. 55, March 1988, pp. 98–103.

Riethmuller, W., Benecke, W., Schnakenberg, U., and Heuberger, A., "Micromechanical Silicon Actuators Based on Thermal Expansion Effects," paper presented at Transducer '87, 4th International Conference on Solid-State Sensors and Actuators, Tokyo, Japan, June 1987, pp. 834–837.

Roark, R. J., *Formulas for Stress and Strain*, 4th ed., McGraw-Hill, New York, 1965.

Rohsenow, W. M., and Choi, H. Y., *Heat, Mass and Momentum Transfer*, Prentice-Hall, Englewood Cliffs, NJ, 1961, Chapters 11 and 20.

Ruska, W.S., *Microelectronic Processing*, McGraw-Hill, New York, 1987.

Saville, D. A., and Palusinski, O. A., "Theory of Electrophoretic Separations," *AIChE Journal*, vol. 32, no. 2, February 1986, Part I, pp. 207–214; Part II, pp. 215–223.

Schulze, V., "Design and Analysis of Typical Micro Pressure Sensors for Automotive Application," M.S. Thesis, San Jose State University, San Jose, CA, 1998.

Segerlind, L. J., *Applied Finite Element Analysis*, Wiley, New York, 1976.

Seidemann, V., Butefisch, S., and Buttgenbach, S., "Fabrication and Investigation of In-Plane Compliant SU8 Structures for MEMS and Their Application to Micro Valves and Micro Grippers," *Sensors and Actuators*, vol. A97-98, 2002, pp. 457–461.

"Smart Cars," cover story, *Business Week*, June 13, 1988, pp. 67–77.

Smith, J. S., "Massively Parallel Assembly Using SOFT (Self Oriented, Fluidic Transport)," in *Proceedings of the Integration of Nano-to Millimeter Sized Technologies Workshop*, DARPA/NIST, Arlington, VA, March 11–12, 1999.

Starr, J. B., "Squeeze-Film Damping in Solid-State Accelerometers," *Technical Digest*, paper presented at the IEEE Solid-State Sensor and Actuator Workshop, Hilton Head Island, SC, June 1990, pp. 44–47.

Sulouff, Jr., R. E., "Silicon Sensors for Automotive Applications," in *Proceedings of Transducers '91, 1991 International Conference on Solid-State Sensors & Actuators*, San Francisco, CA, June 24–27, 1991, pp. 170–176.

Sze, S. M., *Semiconductor Devices—Physics and Technology*, Wiley, New York, 1985.

Tang, W. C., Nguyen, T. H., Judy, M. W., and Howe, R. T., "Electrostatic-Comb Drive of Lateral Polysilicon Resonators," in *Proceedings of the 5th International Conference on Solid-State Sensors and Actuators and Eurosensors III*, Elsevier Sequoia, Netherlands, Vol. 2, June 1990, pp. 328–331.

Teshigahara, A., Watnabe, M., Kawahara, N., Ohtsuka, Y., and Hattori, T., "Performance of a 7-mm Microfabricated Car," *Journal of Microelectromechanical Systems*, vol. 4, no. 2, June 1995, pp. 76–80.

Tien, C. L., and Chen, G., "Challenges in Microscale Conductive and Radiative Heat Transfer," *Journal of Heat Transfer, ASME Transactions*, vol. 116, November 1994, pp. 799–807.

Tien, C. L., Majumdar, A., and Gerner, F. M., Eds., *Microscale Energy Transport*, Taylor & Francis, Washington, DC, 1998.

Timoshenko, S., "Analysis of Bi-metal Thermostats," *Journal of Optical Society of America*, vol. 11, 1925, pp. 233–255.

Timoshenko, S., and Woinowsky-Krieger, S., *Theory of Plates and Shells*, 2nd ed., McGraw-Hill, New York, 1959.

Trim, D. W., *Applied Partial Differential Equations*, PWS-Kent Publishing, Boston, 1990.

Trimmer, W. S. N., "Microrobots and Micromechanical Systems," *Sensors and Actuators*, vol. 19, no. 3, 1989, pp. 267–287.

Trimmer, W. S. N., Ed., *Micromechanics and MEMS—Classic and Seminal Papers to 1990*, IEEE Press, New York, 1997.

Trimmer, W. S. N., and Gabriel, K. J., "Design Considerations for a Practical Electrostatic Micro-Motor," *Sensors and Actuators*, vol. 11, 1987, pp. 189–206.

Tummala, R. R., Ed. *Fundamentals of Microsystems Packaging*, McGraw-Hill, New York, 2001.

Tzou, D. Y., *Macro- to Microscale Heat Transfer*, Taylor & Francis, Washington, DC, 1997.

VanBrussel, H., Peirs, J., Reynaerts, D., Delchambre, A., Reinhart, G., Roth, N., Weck, M., and Zussman, E. "Assembly of Microsystems," Keynote paper, *Annals of the CIRP*, vol. 49, no. 2, 2000, pp. 451–472.

Van der Spiegel, J., Tau, J. F., Alaima, T. F., and Lin, P. A., "The ENIAC—History, Operation and Reconstruction in VLSI," paper presented at the International Conference on the History of Computing, HNF Paderborn, Germany, August 14–16, 1988.

Van Zant, P., *Microchip Fabrication*, 3rd ed., McGraw-Hill, New York, 1997.

Waanders, J. W., *Piezoelectric Ceramics, Properties and Applications*, Philips Components, N.V. Philips Gloeilampeenfabrieken, Eindhoven, The Netherlands, 1991

Walsh, S. T., Elder, J., and Roessger, W., "Commercial Importance of Microsystems Standards Pondered," *Micro-Nano Newsletter*, vol. 7, no. 3, March 2002, pp. 10–11.

White, F. M., *Fluid Mechanics*, 3rd ed., McGraw-Hill, New York, 1994.

White, R. M., "A Sensor Classification Scheme," *IEEE Transactions on Ultrasonic Ferroelectric Frequency Control*, vol. UFFC-34, 1987, pp. 124–126.

Williams, B. A., and Vigh, G., "Fast, Accurate Mobility Determination Method for Capillary Electrophoresis," *Analytical Chemistry*, vol. 68, no. 7, 1996, pp. 1174–1180.

Williams, K., "New Technology & Applications at Lucas NovaSensor," in *Tribology Issues and Opportunities in MEMS*, Bhusshan, B., Ed., Kluwer Academic, Norwell, MA, 1998, pp. 121–135.

Wise, K. D., "Integrated Microelectromechanical Systems: A Perspective on MEMS on the 90s," paper presented at the *IEEE MEMS '91*, Nara, Japan, 1991, pp. 33–38.

Wong, W. S., Wengrow, A. B., Cho, Y., Salleo, A., Quitoriano, N. J., Cheung, N. W., and Sands, T., "Integration of GaN Thin Films with Dissimilar Substrate Materials by Pd-In Metal Bonding and Laser Lift-off," *Journal of Electronics Materials*, vol. 28, no. 12, 1999, pp. 1409–1413.

Woolley, A. T., and Mathies, R. A., "Ultra-High-Speed DNA Fragment Separations Using Microfabricated Capillary Arrays Electrophoresis Chips," *Proceedings of National Academy of Science U.S.A*, vol. 91, November 1994, pp. 11348–11352.

Wright, J. I., Fillion, R., Meyer, L., and Shaddock, D., "Packaging of Devices for Topside Cooling by Replacing Air-Bridge with SU-8 Polymer Bridge," paper presented at the 9th International Symposium on Advanced Packaging Materials, IEEE, Atlanta, GA, 2004, pp. 63–68.

Yang, G., and Nelson, B. J., "Automatic Microassembly," in *MEMS Packaging*, Hsu, T. R., Ed., Institute of Electrical Engineers, Herts, United Kingdom, 2004, Chapter 5.

Yarris, L., Gallium Nitride Makes for a New Kind of Nanotube, *Sciencebeat*, Berkeley laboratoty, May 12, 2003.

Yu, C., Shi, L., Yao, Z., Li, D., and Majumdar, A., "Thermal Conductance and Thermopower of an Individual Single-Walled Carbon Nanotube," *Nano Letters*, vol. 5, no. 9, 2005, pp. 1842–1846.

Yun, C. H., and Cheung, N. W., "SOI on Buried Cavity Patterns Using Ion-Cut Layer Transfer," in *Proceedings of 1998 IEEE International SOI Conference*, Stuart, FL, October 1998, pp. 165–166.

Zengerle, R., Richter, M., and H. Sandmaier, "A Micro Membrane Pump With Electrostatic Actuation," in *Proceedings of IEEE Micro Electro Mechanical Systems Workshop*, Travemunde, Germany, February 1992, pp. 19–24.

Zhang, X., and Misra, A. "Residual Stresses in Sputter-Deposited Copper/330 Stainless Steel Multilayers," *Journal of Applied Physics*, vol. 96, no. 12, 2004, pp. 7173–7178.

Zheng, H., Wang, J., Lofland, S. E., Ma, Z., Mohaddes-Ardabili, L., Zhao, T., Salamanca, L., Shinde, S. R., Ogale, S. B., Bai, F., Viehland, D., Jia, Y., Schlom, D. G., Wuttig, M., Roytburd, A., and Ramesh, R., "Multiferroic $BaTiO_3$-$CoFe_2O_4$ Nanostructures," *Science*, vol. 303, January 2004, pp. 661–663

Zienkiewicz, O. C., *The Finite Element Method in Engineering Science*, McGraw-Hill, New York, 1971.

Zum Gahr, K.–H., "Microtribology," *Interdisciplinary Science Reviews*, vol. 18, no. 3, 1993, pp. 259–266.

RECOMMENDED UNITS FOR THERMOPHYSICAL QUANTITIES

Length	Meter (m)
Area (A)	Square meter (m^2)
Volume (v)	Cubic meter (m^3)
Time (t)	Second (s)
Temperature (T)	Degree Celsius ($^\circ$C) or Kelvin (K)
Force (F)	Newton (N)
Weight (W)	Kilogram (kg$_f$) $= 9.81$ N
Pressure (P)	Pascal (P) $=$ N/m^2
Mass (m)	Gram (g)
Mass density (ρ)	g/cm^3
Energy	Joule (J) $= 1$ N-m
Power	Watt (W) $= 1$ J/s $= 1$ N-m/s
Stress (σ)	Megapascal (MPa) $= 10^6$ Pa $= 10^6$ N/m^2
Velocity (V)	m/s
Acceleration (a)	m/s^2
Specific heat (c)	J/g-$^\circ$C
Thermal conductivity (k)	W/m-$^\circ$C
Heat transfer coefficient (h)	W/m^2-$^\circ$C
Thermal diffusivity (α)	m^2/s
Coefficient of linear thermal expansion (α)	$^\circ$C^{-1}
Dynamic viscosity (μ)	N-s/m^2

APPENDIX 2

CONVERSION OF UNITS

Length	1 meter (m) = 39.37 inches (in.) = 3.28 feet (ft)
	1 centimeter (cm) = 10^{-2} m
	1 millimeter (mm) = 10^{-3} m
	1 micrometer (μm) = 10^{-6} m
	1 nanometer (nm) = 10^{-9} m
	1 angstrom (Å) = 10^{-10} m
Area	1 square meter (m^2) = 10.76 ft^2
Volume	1 cubic meter (m^3) = 35.29 ft^3
	1 liter (l) = 1000 cm^3
Temperature	degree Celsius, $^\circ$C = $\left(\frac{5}{9}\right)$[degree Fahrenheit ($^\circ$F) − 32]
	K = $^\circ$C + 273
Force	1 N = 0.2252 pound force (lb$_f$) = 1 kg-m/s
Weight	1 kilogram force (kg$_f$) = 2.2 lb$_f$
Pressure and stress	1 MPa = 145.05 pound per square inch (psi)
	1 Pa = 1 N/m^2 = 10 dynes/cm^2 = 2.089 × 10^{-2} lb$_f$/ft^2
Mass	1 g = 68.5 × 10^{-6} slug
	1 kg = 1000 g = 2.2 pounds mass (lb$_m$)
Energy	1 J = 0.2389 calories (cal) = 10^7
	ergs = 0.7376 ft-lb$_f$ = 9.481 × 10^{-4} Btu
	1 kilowatt-hour (kW-h) = 3.6 × 10^6 J = 3413 Btu
Power	1000 watts (W) = 1.341 horsepower
	(hp) = 737.6 ft-lb$_f$/s = 0.9483 Btu/s
	1 W = 1 J/s
Velocity	1 m/s = 3.28 ft/s

Specific heat	$1 \text{ J/g-}°\text{C} = 0.2394 \text{ Btu/lb}_m\text{-}°\text{F}$
Thermal conductivity	$1 \text{ W/m-}°\text{C} = 13.3816 \times 10^{-6} \text{ Btu/in.-s-}°\text{F}$
Heat transfer coefficient	$1 \text{ W/m}^2\text{-}°\text{C} = 1.1151 \times 10^{-6} \text{ Btu/in.}^2\text{-s-}°\text{F}$
Thermal diffusivity	$1 \text{ m}^2/\text{s} = 1545 \text{ in.}^2/\text{s}$
Coefficient of linear thermal expansion	$°\text{C}^{-1} = 0.5555/°\text{F}$
Dynamic viscosity	$\text{N-s/m}^2 = 10 \text{ poise} = 0.02089 \text{ slug/ft-s}$

INDEX